Lecture Notes in Physics

Springer
Berlin
Heidelberg
New York
Barcelona
Hong Kong
London
Milan
Paris
Tokyo

Physics and Astronomy ONLINE LIBRARY

http://www.springer.de/phys/

Editorial Policy

The series *Lecture Notes in Physics* (LNP), founded in 1969, reports new developments in physics research and teaching -- quickly, informally but with a high quality. Manuscripts to be considered for publication are topical volumes consisting of a limited number of contributions, carefully edited and closely related to each other. Each contribution should contain at least partly original and previously unpublished material, be written in a clear, pedagogical style and aimed at a broader readership, especially graduate students and nonspecialist researchers wishing to familiarize themselves with the topic concerned. For this reason, traditional proceedings cannot be considered for this series though volumes to appear in this series are often based on material presented at conferences, workshops and schools (in exceptional cases the original papers and/or those not included in the printed book may be added on an accompanying CD ROM, together with the abstracts of posters and other material suitable for publication, e.g. large tables, colour pictures, program codes, etc.).

Acceptance

A project can only be accepted tentatively for publication, by both the editorial board and the publisher, following thorough examination of the material submitted. The book proposal sent to the publisher should consist at least of a preliminary table of contents outlining the structure of the book together with abstracts of all contributions to be included.
Final acceptance is issued by the series editor in charge, in consultation with the publisher, only after receiving the complete manuscript. Final acceptance, possibly requiring minor corrections, usually follows the tentative acceptance unless the final manuscript differs significantly from expectations (project outline). In particular, the series editors are entitled to reject individual contributions if they do not meet the high quality standards of this series. The final manuscript must be camera-ready, and should include both an informative introduction and a sufficiently detailed subject index.

Contractual Aspects

Publication in LNP is free of charge. There is no formal contract, no royalties are paid, and no bulk orders are required, although special discounts are offered in this case. The volume editors receive jointly 30 free copies for their personal use and are entitled, as are the contributing authors, to purchase Springer books at a reduced rate. The publisher secures the copyright for each volume. As a rule, no reprints of individual contributions can be supplied.

Manuscript Submission

The manuscript in its final and approved version must be submitted in camera-ready form. The corresponding electronic source files are also required for the production process, in particular the online version. Technical assistance in compiling the final manuscript can be provided by the publisher's production editor(s), especially with regard to the publisher's own Latex macro package which has been specially designed for this series.

Online Version/ LNP Homepage

LNP homepage (list of available titles, aims and scope, editorial contacts etc.):
http://www.springer.de/phys/books/lnpp/

LNP online (abstracts, full-texts, subscriptions etc.):
http://link.springer.de/series/lnpp/

P. Nielaba M. Mareschal G. Ciccotti (Eds.)

Bridging Time Scales: Molecular Simulations for the Next Decade

Springer

Editors

Peter Nielaba
Universität Konstanz
Lehrstuhl für Theoretische Physik
Fachbereich Physik
78457 Konstanz, Germany

Giovanni Ciccotti
Universitá La Sapienza
Dipartimento di Fisica
Piazzale Aldo Moro
00185 Roma, Italy

Michel Mareschal
CECAM
Ecole Normale Supérieure de Lyon
46 Allée d'Italie
69364 Lyon Cedex 07, France

Cover Picture: (see contribution by Jean-Pierre Hansen and Hartmut Löwen in this volume)

Cataloging-in-Publication Data applied for

A catalog record for this book is available from the Library of Congress.

Bibliographic information published by Die Deutsche Bibliothek

Die Deutsche Bibliothek lists this publication in the Deutsche Nationalbibliografie; detailed bibliographic data is available in the Internet at http://dnb.ddb.de

ISSN 0075-8450
ISBN 3-540-44317-7 Springer-Verlag Berlin Heidelberg New York

Springer-Verlag Berlin Heidelberg New York
a member of BertelsmannSpringer Science+Business Media GmbH

http://www.springer.de

Typesetting: Camera-ready by the authors/editor
Camera-data conversion by Steingraeber Satztechnik GmbH Heidelberg
Cover design: *design & production*, Heidelberg

Printed on acid-free paper
SPIN: 10893324 54/3141/du - 5 4 3 2 1 0

Preface

"Bridging the Gap!": We had been discussing the challenges to be met by the atomistic simulation community for a few hours when someone came up with this expression. As often happens in animated but exhausting discussions, there was at the same time both enthusiasm and relief. People were relieved by those three simple words which aptly described the main common trend in the approaches analyzed by the participants at the meeting. The meeting itself was held in a small town near Amsterdam, named Bussum, in order to get Daan Frenkel with us, and its aim was to lay the foundations of a large-scale European network in computational condensed matter statistical physics. This was not the end of the story which saw the building of a large collaboration in the form of the European Science Foundation (ESF) program known as SIMU. This program, in fact, required further discussion and effort, but it is probably correct to say that this simple formula helped initiate the network because it succinctly expressed the intellectual attitude shared by the participants in their effort to meet the actual challenges of the field.

First, which attitude? Molecular Dynamics and Monte Carlo simulation techniques are nowadays well accepted theoretical tools to predict, by heavy computing on realistic models, physical properties and dynamical processes in materials. Their scope has steadily increased in the years since the pioneering work of the fifties. Applications are common from the most simple liquid or solid materials to cover also, at least in principle, complex materials like colloids, polymers or poly-electrolytes, not to mention proteins or biological membranes. Most of those materials are studied experimentally, with atomic scale resolution techniques, and are used in many industrial processes. The theoretical understanding of their behavior is crucial in materials science also to analyze the experiments. However, those behaviors extend over length and time scales which are orders of magnitude longer and larger than the ones that can be achieved by brute force simulations in a fully atomistic description. Thus the challenge is to be able to reach scales which can be of the order of micrometers and seconds, starting from a fundamental level of description.

Second, which challenge? There is an agreement on the analysis that most of the progress made in recent years in the atomistic simulation of condensed matter originate in the development of new methods of simulation more than in the increase of available computing power, however impressive the latter has been. No foreseeable increase in computing power will ever be sufficient to give access to the large physical scales needed to describe material properties of realistic complex materials. There is a strong need for further development of methods able to address and, possibly, solve physical problems which are multi-scale in nature. Biased Monte Carlo and *ab initio* MD techniques are two beautiful examples of very successful progress.

There are, however, deeper connections in our community of "simulators" which make possible the transfer of techniques. For example, people working in polymer physics can easily understand the numerical issues which arise in Quantum Monte Carlo techniques and the progress in one field can be quickly transferred to the other. To give another example, the numerical approach behind the Car and Parrinello method can and has been used in classical statistical mechanics, via classical density functional theory, to get "exact" thermodynamic equilibrium averages over solvent configurations. In other words, the technical culture, often based on concepts of statistical physics, is shared, and thus allows an easy exchange of ideas and an efficient form of interdisciplinary collaboration.

This provides another reason (other than the more important one of maintaining the ambition and pride of the community) for the variety of subjects in this book which reproduces articles written after the conference *Bridging the time-scale gap* was held, at the University of Konstanz, in September 2001. The conference was organized within a series of activities supported by the 5-year ESF program SIMU (web site : http://simu.ulb.ac.be/). It focused on the subject of the time scale issue and got a large and enthusiastic participation: besides the 42 invited talks, there were more than one hundred posters and around 250 participants. There have been of course several large conferences dedicated to computational physics, but the peculiarity of this one was its focus on a well-defined theme, one however allowing interdisciplinary participation because of the variety of approaches and levels of description. It had similar spirit to some of the advanced schools organized previously, such as the summer school in Como in 1995 [1] preceded by similar but more restricted initiatives such as the collective book on Monte Carlo methods in statistical physics in 1986 [2] or the proceedings of the Varenna

[1] *Monte Carlo and Molecular Dynamics of Condensed Matter Systems*, edited by K. Binder and G. Ciccotti, SIF, Bologna, (1996).

[2] *Monte Carlo Methods in Statistical Physics*, edited by K. Binder, Topics in Current Physics 7, Springer-Verlag (1986).

school in 1985,[3] which have been important references in the community for many years. At the end of the conference, the scientific committee (the editors plus Daan Frenkel) discussed the possibility of offering selected speakers the opportunity to contribute to a book which would be representative of conference topics and discussions and could remain a good testimony of ideas and techniques on which to build progress in the forthcoming years.

The result goes well beyond our best expectations both for the number and the quality of the contributions that provide a fair picture of the state-of-the-art in the field! We have reproduced the book chapters in the order chosen for the conference, and it is amazing to see that the order follows a kind of logic, starting with the largest scale, where proteins fold and unfold, and ending with Quantum Monte Carlo simulations where, as it was once said, one is bridging the gap in the other direction!

The book starts with contributions dealing with biological and polymer physics. All-atoms and lattice models are used to investigate protein folding dynamics and some of its mechanisms (Eugene I. Shakhnovich et al.) while coarse-grained models are developed in order to describe lipid mono-layers and bi-layers (Steve O. Nielsen and Michael L. Klein) on the relevant time scales. The contribution by Doros N. Theodorou presents a more methodological approach, with various fast (bridging!) algorithms allowing to equilibrate polymers. In his chapter, Alexander Grosberg introduces the new concept of commitor in order to deal with dynamics in conformation space : this concept was elaborated from an analysis of Monte Carlo simulations of protein folding and it is hoped that it could initiate new ideas in the simulation community. Kurt Kremer et al., in turn, describe micro-meso mapping schemes for polymeric materials and present results of a combined approach of mesoscale model simulations and quantum mechanical density functional theory calculations for polycarbonates near surfaces.

The next chapters deal with the statistical mechanics of complex materials. First, the coarse-graining through effective interactions allows Jean-Pierre Hansen and Hartmut Löwen to describe equilibrium properties of polymer and colloid fluid mixtures. The slow dynamic of glasses require not only coarse-graining but also some specific techniques like parallel tempering (Kurt Binder et al.). This problem is examined in a more systematic way by Nigel Wilding and David P. Landau who review several methods allowing faster convergence in lattice and continuous models. The hydrodynamic evolution is then investigated by Christopher P. Lowe and Sauro Succi who apply lattice-Boltzmann and hybrid techniques to various flow problems.

---

[3] *Molecular-Dynamics Simulation of Statistical-Mechanical Systems*, edited by G. Ciccotti and W.G. Hoover, SIF, Amsterdam, North-Holland (1986).

Multi-scale methods are also described and applied to the problem of solid friction where a direct simulation inspection has permitted progress in the basic mechanisms involved (Martin H. Müser).

Three chapters of a more methodological nature follow: they are the contributions on the transition path sampling (Christoph Dellago and David Chandler), on the stochastic difference equation (Ron Elber et al.) and, finally, on the proper treatment of long range interactions. Transition path sampling was explained as *throwing ropes over a mountain path in the dark* and it deals with the computation of rate constants when the reaction mechanisms are not precisely known. Stochastic dynamics is being introduced in order to generate long-time trajectories. Problems with long-range Coulombic and dipolar systems are then treated by Dominique Levesque.

The last part of the book deals with simulation techniques involving a quantum aspect. It starts with a description of *ab initio* MD recent advances by Glenn J. Martyna and Mark E. Tuckerman. The use of this technique is heavily time-consuming to create a serious time-scale problem. Ways to overcome the time-scale barrier are described in the contribution by Ursula Röthlisberger, Michiel Sprik, and Jürg Hutter: bias potentials and electronic bias potentials are being introduced, together with the explanations on how to apply the method, and to compute rate constants. Often it is necessary to treat part of the system classically and, in the presentation by Raymond Kapral and Giovanni Ciccotti, the embedding of a quantum system interacting with classical degrees of freedom is studied in a systematic way. The book ends with a contribution by David Ceperley, Mark Dewing, and Carlo Pierleoni where a classical Monte Carlo simulation for the ions is coupled to a Quantum Monte Carlo simulation for the electrons in order to describe liquid and metallic behavior of quantum hydrogen. Time scales in this approach are an order of magnitude smaller than in the first chapter, however the numerical problems to overcome are very familiar, as one knows from the similarity with polymer physics.

It is our belief that multi-scale and hierarchical modeling will be used more and more in the future. Our ambition in assembling these contributions is not only to show the great vitality of the field with the many different approaches to the time-scale problem, but also to help readers to understand what are the real issues and difficulties in applying those techniques to the many problems arising in the microscopic description of the thermodynamical properties of matter. Let us hope that the ideas and methods presented in this book will have a lasting impact.

The conference could not have taken place without the support provided by the European Science Foundation. This support came through the program *Challenges in Molecular Simulations* (SIMU) which has been approved for 5 years (1999–2003). It is a pleasure to extend our thanks to Professor J. Rojo, the chairman of the ESF committee, PESC, who accepted our invitation to come and talk at the conference.

The members of the steering committee of the SIMU program have been decisive to the success of the conference, both for their enthusiastic support and financial generosity. Even more important to the success of the meeting has been their participation in chairing the sessions and enlivening the discussions which took place during them. Many thanks to all.

Financial and logistic support is also acknowledged to CECAM (Centre Europeen de Calcul Atomique et Moleculaire, Lyon), the University of Konstanz and the Landesbank of Baden-Württemberg.

The staff in Konstanz has been extraordinarily efficient: Sabine Becker-Weimann, Markus Dreher, Dominik Fischer, Yolanda Fischer, Kerstin Franzrahe, Martin Frick, Peter Henseler, Jochen Hoffmann, Guido Günther, Hans Knoth, Marc Lohrer, Ulrich Mack, Günther Schafranek, Ralf Stadelhofer, Wolfram Strepp. In particular we appreciated their participation and response to all the emotional and organizational strains of September 11th. This is a lasting memory of the qualities and generosity of Peter's group. Particular thanks go to Yolanda Fischer: all financial, hosting and booking problems were handled by her in a masterful way! Additional help by the usually overburdened Emmanuelle Crespeau will remain unforgettable.

Konstanz, Lyon and Rome,
August 2002

Peter Nielaba
Michel Mareschal
Giovanni Ciccotti

Fig. 1. Photo taken at the conference *Bridging the time-scale gap* in Konstanz, 10–13.9.2001

Table of Contents

Part III Polymer Structure and Dynamics

Part V Slow Dynamics and Reactivity

Part VI Lattice Models

List of Contributors

Cameron Abrams
Max-Planck-Institute
for Polymer Research
Postfach 3148
Ackermannweg 10
55021 Mainz, Germany

Jörg Baschnagel
Institut Charles Sadron
Université Louis Pasteur
6, rue Bossingault
67083 Strasbourg, France

Kurt Binder
Institut für Physik
Johannes-Gutenberg-Universität
55099 Mainz, Germany
kurt.binder@uni-mainz.de

Alfredo Cardenas
Department of Computer Science
Cornell University
Ithaca, NY 14850, USA

David Ceperley
Department of Physics
University of Illinois
at Urbana–Champaign
1110 West Green Street
Urbana, Illinois 61801, USA
ceperley@pop.ncsa.uiuc.edu

David Chandler
Department of Chemistry
University of California
Berkeley, California 94720, USA
chandler@gold.cchem.berkeley.edu

Giovanni Ciccotti
INFM and Dipartimento di Fisica
Università "La Sapienza"
Piazzale Aldo Moro, 2
00185 Roma, Italy
giovanni.ciccotti@roma1.infn.it

Cristoph Dellago
Department of Chemistry
University of Rochester
Rochester, New York 14627, USA
dellago@chem.rochester.edu

Luigi Delle Site
Max-Planck-Institute
for Polymer Research
Postfach 3148
Ackermannweg 10
55021 Mainz, Germany

Mark Dewing
INTEL
1906 Fox Dr.
Champaign, IL, 61820, USA

Ron Elber
Department of Computer Science
Cornell University
Ithaca, NY 14850, USA
ron@cs.cornell.edu

Avijit Ghosh
Department of Computer Science
Cornell University
Ithaca, NY 14850, USA

Alexander Grosberg
Department of Physics
University of Minnesota
116 Church Street SE
Minneapolis, MN 55455, USA
and
Institute of Biochemical Physics
Russian Academy of Sciences
Moscow 117119, Russia
grosberg@physics.umn.edu

Jean-Pierre Hansen
Department of Chemistry
Lensfield Road
Cambridge CB2 1EW, UK
jph32@cus.cam.ac.uk

Jürg Hutter
Physical Chemistry Institute
University of Zürich
8057 Zürich, Switzerland
hutter@pci.unizh.ch

Raymond Kapral
Chemical Physics Theory Group
Department of Chemistry
University of Toronto
Toronto, ON M5S 3H6, Canada
rkapral@gatto.chem.utoronto.ca

Michael L. Klein
Center for Molecular Modeling
and
Department of Chemistry
University of Pennsylvania
Philadelphia PA 19104-6323, USA
klein@cmm.upenn.edu

Walter Kob
Laboratoire des Verres
Université Montpellier II
34095 Montpellier, France
kob@ldv.univ-montp2.fr

Kurt Kremer
Max-Planck-Institute
for Polymer Research
Postfach 3148
Ackermannweg 10
55021 Mainz, Germany
k.kremer@mpip-mainz.mpg.de

Edo Kussell
Department of Biophysics
Harvard University
240 Longwood Ave
Boston, MA 02115, USA

David P. Landau
Center for Simulational Physics
The University of Georgia
Athens, Georgia 30602, USA
dlandau@hal.physast.uga.edu

Dominique Levesque
Laboratoire de Physique Théorique
Bâtiment 210, Université Paris-Sud
91405 Orsay Cedex, France
dominique.levesque@th.u-psud.fr

Hartmut Löwen
Institut für Theoretische Physik II
Heinrich-Heine-Universität
Universitätsstraße 1,
40225 Düsseldorf, Germany
hlowen@thphy.uni-duesseldorf.de

Christopher P. Lowe
Department of Chemical Engineering
University of Amsterdam
Nieuwe Achtergracht 166
1018 WV Amsterdam
The Netherlands
c.p.lowe@tn.tudelft.nl

Glenn J. Martyna
IBM TJ Watson Lab
PO Box 218
Yorktown Heights, NY 10598, USA

Martin H. Müser
Institut für Physik
Johannes Gutenberg-Universität
55099 Mainz, Germany
martin.mueser@uni-mainz.de

Steve O. Nielsen
Center for Molecular Modeling
and
Department of Chemistry
University of Pennsylvania
Philadelphia, PA 19104-6323, USA

Wolfgang Paul
Institut für Physik
Johannes-Gutenberg-Universität
55099 Mainz, Germany
wolfgang.paul@uni-mainz.de

Carlo Pierleoni
INFM
and
Department of Physics
University of L'Aquila
Via Vetoio
L'Aquila, Italy
carlo.pierleoni@aquila.infn.it

Ursula Röthlisberger
Institute of Molecular and Biological Chemistry
Swiss Federal Institute of Technology
EPFL
1015 Lausanne, Switzerland
ursula.roethlisberger
@inorg.chem.ethz.ch

Eugene I. Shakhnovich
Department of Chemistry and Chemical Biology
Harvard University
12 Oxford Street
Cambridge, MA 02138, USA
eugene@belok.harvard.edu

Jun Shimada
Department of Chemistry and Chemical Biology
Harvard University
12 Oxford Street
Cambridge, MA 02138, USA

Michiel Sprik
Chemistry Department
University of Cambridge
Cambridge CB2 1EW, UK
sprik@theor.ch.cam.ac.uk

Sauro Succi
Istituto di Applicazioni Calcolo
viale Policlinico 137
00161 - Roma, Italy
succi@iac.rm.cnr.it

Doros N. Theodorou
Department of Materials Science and Engineering
School of Chemical Engineering
National Technical University
Zografou Campus
157 80 Athens, Greece
and
Molecular Modelling of Materials Laboratory
Institute of Physical Chemistry
National Research Centre for Physical Sciences "Demokritos"
153 10 Ag. Paraskevi Attikis, Greece
doros@sequoia.chemeng.upatras.gr

Mark E. Tuckerman
Chemistry Department
and
Courant Institute
New York University
New York, NY 10003, USA
mark.tuckerman@nyu.edu

Nigel Wilding
Department of Physics
University of Bath
Bath BA2 7AY, UK
N.B.Wilding@bath.ac.uk

Part I

Protein Folding

1 Sidechain Dynamics and Protein Folding

Edo Kussell[1], Jun Shimada[2], and Eugene I. Shakhnovich[2,3]

[1] Department of Biophysics, Harvard University, 240 Longwood Ave, Boston, MA 02115
[2] Department of Chemistry and Chemical Biology, Harvard University, 12 Oxford Street, Cambridge MA 02138,
[3] corresponding author, tel: 617-495-4130, fax: 617-496-5948, email: eugene@belok.harvard.edu

Abstract. The processes by which protein sidechains reach equilibrium during a folding reaction are investigated using both lattice and all-atom simulations. We find that rates of sidechain relaxation exhibit a distribution over the protein structure, with the fastest relaxing sidechains being involved in kinetically important positions. Traversal of the major folding transition state corresponds to the freezing of a small number of residues, while the rest of the chain proceeds towards equilibrium via backbone fluctuations around the native fold. The post-nucleation processes by which sidechains relax are characterized by very slow dynamics, and many barrier crossings, and thus resemble the behavior of a glass. At optimal temperature, however, the nucleated ensemble is energetically very close to equilibrium; slow relaxation is still observed. At lower temperatures, sidechain relaxation becomes a significant and very noticeable part of the folding reaction.

1.1 Introduction

Protein folding is a complex, single molecule process in which a polypeptide backbone with diverse sidechain groups efficiently searches a vast conformational space and finds a unique native fold. Most theoretical attempts to understand the folding process have modeled the polymer, in one way or another, as a chain of beads which undergoes a backbone freezing transition. The internal degrees of freedom of each sidechain (the χ angles) add another layer of difficulty to understanding the folding process.

In unfolded conformations, the barriers between rotamer states of sidechains are low [1] and sidechains easily convert between them. Upon folding, buried sidechains are usually found in a single, well-defined rotamer state [2–5], and interconversion between rotamers, when energetically allowed, is slow due to high barriers [6,7]. Because protein folding is thought to be a sidechain-driven process, finding the native rotamers is an integral part of the folding reaction. Do sidechains reach their native conformations simultaneously with the backbone, or is sidechain ordering a separate process that occurs after the native fold has been reached? This question poses a challenge for experimentalists and theoreticians alike.

We study the dynamics of sidechains during the folding process via a simplified model, which captures the basic physical aspects of the problem, as well as an all-atom protein folding simulation, in which the full complexity of sidechain shapes and mobility is represented. We start by modifying the classic lattice model of proteins to account for internal sidechain states of each amino-acid. This new model, like its predecessors, is found to fold in a cooperative all-or-none manner. The presence of internal degrees of freedom at each site, however, allows the backbone to reach the native conformation before all of the sidechains have become properly ordered. Since energy comes from sidechain-sidechain interactions, completion of the folding reaction requires both the backbone and the sidechains to reach their native states. We find that, depending on simulation temperature, the backbone can become native-like long before the internal sidechain states reach equilibrium. By measuring the rate of sidechain ordering for each monomer, we find that there is a cluster of residues whose sidechains become ordered very fast, while the relaxation rates of other positions is up to an order of magnitude slower. The fast cluster turns out to be very close to the folding nucleus identified previously for the structure we use.

A small number of kinetically important residues thus freeze in their native "rotamer" state on a fast timescale, pulling the backbone strongly toward its native conformation, while the rest of the protein relaxes on a slower timescale toward its equilibrium energy. This slow phase could not be observed in previous lattice simulations because it arises entirely from the presence of the internal sidechain states of each monomer. We find that the slow-phase relaxation of energy to equilibrium follows stretched-exponential kinetics, suggesting that the dynamics are exhibiting some glass-like properties due to sidechains. We note that backbone-only lattice models have been shown to be free of a glass transition over a wide range of temperatures [8]. Various experimental and theoretical studies have suggested that some aspects of protein folding might be interpreted as glassy behavior [9–12], though whether these phenomena are attributable to backbone, sidechains, or solvent remains to be seen.

Because the lattice model is computationally very fast, one can observe relaxation to equilibrium even at lower temperatures. This becomes impossible once more realistic models are used. Our previous work using an all-atom simulation, with a simplified Gō potential [13,14], demonstrated that folding to the native backbone topology (rms deviation < 1 Å from crystal structure) happened within a reasonable amount of time, but that full relaxation to equilibrium could not be observed at temperatures below a certain threshold. Thus the slow phase observed in the lattice model is also present in the all-atom model, but cannot be fully characterized therein due to prohibitively long simulation times. This does not prevent us from identifying the residues that exhibit fast transition to the native state, because the fast phase in which the native backbone conformation is reached is fully accessible to our simulation.

For Protein G, the folding transition state was rigorously characterized using the all-atom simulation [14]. For this study, we have reanalyzed the folding trajectories obtained from the protein G study to identify the residues whose sidechains exhibit fast relaxation to their native state. We find that these same residues play a key role in the transition state ensemble. As in the simpler lattice model, we find a wide distribution of sidechain relaxation times. Two very different models are thus in marked agreement; together they provide a clear picture of sidechain dynamics during the folding process.

1.2 Results

In order to mimic the internal degrees of freedom of protein sidechains (the χ angles), we modified the standard lattice model by adding n sidechain states to each residue. This is consistent with the observation that protein sidechains usually populate discrete rotameric configurations [2,3]. The state of each residue at any given time is a number between 0 and $n-1$. Of these n states only one state (the 0 state) was designated as native for each residue. When two residues came into contact during simulation, they interacted only if both were in their native state - a contact formed with one or both residues in a non-native sidechain state did not contribute to the energy of the conformation (see Methods). While there are other ways to model a native vs. non-native rotamer interaction using a lattice model (for example, we could have assigned some fraction of the native energy when non-native monomers interact), we chose the present scheme for simplicity. Previous lattice models have added sidechain degrees of freedom by letting sidechains occupy a lattice site [15–17]. In our model, sidechain states are treated implicitly, resulting in a considerable computational advantage.

An important aspect of sidechain motion in real proteins is that upon compactification of the polypeptide chain, sidechain motion is restricted due to the excluded volume effect [1,18]. In order for sidechains to repack in the protein interior, the backbone must perform a "breathing motion" [6], allowing sidechains some extra room to move, and thus making certain sidechain configurations momentarily available. Any model of sidechain dynamics must incorporate this effect in some way. Our all-atom simulation contains this effect explicitly. In the lattice simulation we mimic this effect by our choice of moves. In addition to the usual lattice backbone moves, we allow the sidechain states of a given residue to interconvert when there are no other residues in contact with it (see Methods). Thus, when the chain is fully compact, the sidechain states are frozen until a backbone fluctuation frees some residues, and allows their states to change.

We tested the lattice model with 1, 2, and 4 internal states per monomer using a 48-mer sequence designed to fold into a 3 x 4 x 4 cube. The $n = 1$ model corresponds to the standard lattice model and is shown here only as a control. The thermodynamics of these models is shown in Fig. 1.1. All are

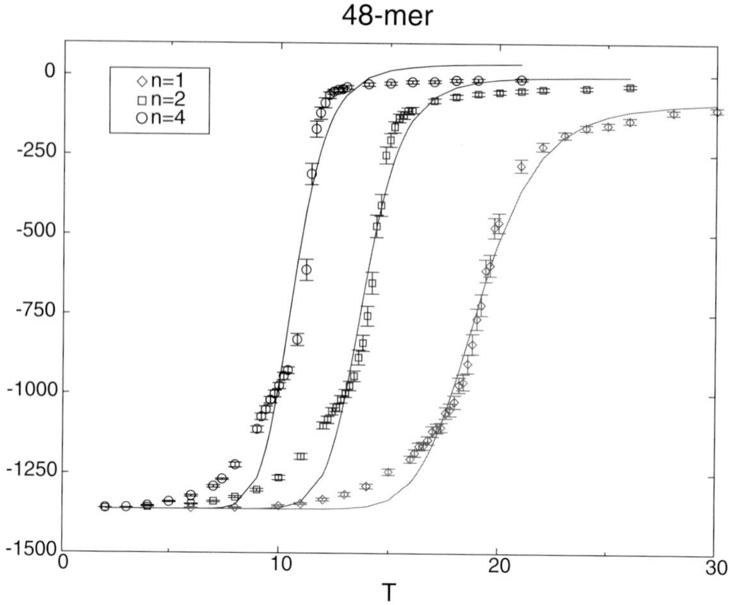

Fig. 1.1. Thermodynamics of 48-mer lattice models. Each point corresponds to an average of energy over a lattice simulation started at the native state. Each simulation was run for 3×10^8 steps and energy was sampled every 3×10^5 steps. The correlation time of energy was found to be much less than our sampling interval at all temperatures. The number of internal sidechain states for each model is indicated by the color and shape of the points, as in the legend. Fits to a two-state thermodynamic model are given in solid lines colored to match the corresponding lattice model that was used. Parameters for these fits are given in Table 1.1.

Table 1.1. Two-state Fits to Thermodynamic Data. Thermodynamics shown in Fig. 1.1 was fit using the form $f(x) = a_3 + (a_0 - a_3)\exp(a_2 - a_1/T)/(1+\exp(a_2 - a_1/T))$, where a_3 is the native state energy of -1361.

n	a_0	a_1	a_2
	48-mer		
1	-82	263	13.8
2	-3.6	222	16.0
4	38	151	14.2

seen to exhibit a cooperative temperature transition, with the transition temperature getting progressively lower as the number of internal states of each monomer increases. The lowered transition temperature is to be expected as the increased entropy of the model (due to more internal states) necessarily leads to some destabilization. The transition region becomes narrower as n

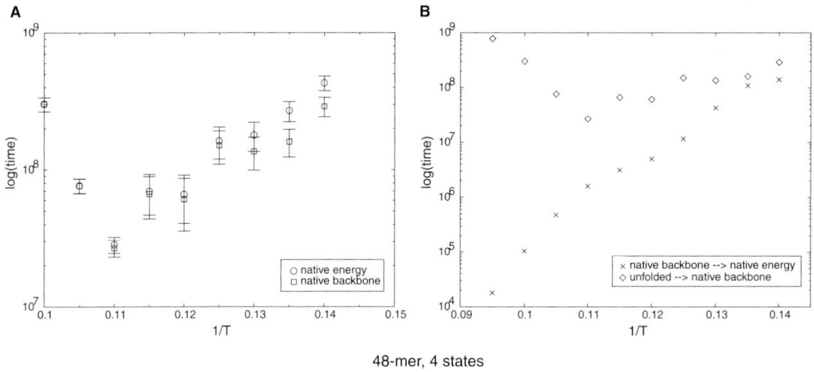

Fig. 1.2. Kinetics of the 48-mer model with 4 internal states per monomer. Panel **A** indicates the MFPT for reaching native energy by circles, and the average time to reach the stable native backbone by squares. Panel **B** shows the time to reach the stable backbone with diamonds, and the amount of time to go from the native backbone to the native energy with x-marks.

increases, due to the increase in entropy of the unfolded state relative to the folded state.

We studied the kinetics of these models by plotting the average folding time as a function of temperature (see Fig. 1.2). The models with $n > 1$ possess the property that the backbone can reach full nativity before all of the sidechains have become native. This leads to the interesting question of how the polymer chain reaches its native energy. That is, does the formation of the native backbone lead to immediate sidechain ordering, or do sidechains relax slowly to equilibrium after the chain has folded?

To answer this question, we plotted both the average time to reach the native energy (which corresponds to full sidechain ordering), and the average time to reach the native backbone in Fig. 1.2A. We find that at temperatures above the optimal folding temperature, the native energy is reached immediately after the native backbone is found, and thus sidechain ordering is fast. At low temperatures, on the other hand, the native backbone is reached long before native energy is achieved, and sidechain ordering is slow.

We consider a kinetic model with three steps:

$$\text{Unfolded Backbone} \longrightarrow \text{Folded Backbone} \longrightarrow \text{Sidechain Ordering}$$

Figure 1.2B shows the average time of the first and second steps plotted as diamonds and crosses, respectively. We immediately see that for high temperatures, the sidechain ordering step is several orders of magnitude faster than the backbone folding step. As temperature becomes lower, the sidechain ordering time becomes comparable to the backbone folding time. At $1/\mathrm{T} = 0.13 = 85\%\ T_{opt}$, the rate of sidechain ordering becomes significant as it comes within an order of magnitude of the rate of backbone folding (T_{opt} is the temperature of fastest folding).

The mechanism for reaching full nativity (backbone and sidechains) at temperatures lower than T_{opt} thus seems to be one in which the native backbone structure is formed, followed by sidechain ordering via backbone fluctuations around the native structure. It is entirely possible, however, that the native backbone structure is reached during the folding trajectory but unfolds immediately because too few sidechains are native. This, in fact, is the case even at low temperatures. At some point in time, however, the native backbone structure is reached with enough native sidechains to remain stable long enough to allow the rest of the sidechains to become ordered. It is the ordering of sidechains after this *stable* native backbone is reached that we identify as an important kinetic step at temperatures below T_{opt}. Accordingly, we plot the average time of the *last* pass to the native backbone conformation in all of our figures. The time of the last pass is defined as the first time the chain reached the native backbone without losing more than 50% of its native contacts before reaching the native energy. We found that our results did not change significantly when we varied the fraction of native contacts used in this definition.

Having observed that folding to the correct backbone structure occurs significantly before the native energy is reached, we asked the following questions: Do some residues reach their native sidechain state faster than others? If so which ones are fast, which ones are slow, and why?

In order to obtain individual sidechain ordering rates for each residue, we performed many long folding runs. For each residue we averaged its sidechain state over all folding runs: we assigned a value of 1 to the native internal state of a given residue, and a value of 0 to all other internal states, and at each timestep averaged these values over runs. Two traces obtained after averaging are shown in Fig. 1.3. We fit a single exponential (see Methods) to each trace, and obtained time constants for each of the 48 residues.

The distribution of rate constants for two temperatures is given in Fig. 1.4, and the fast residues are labelled by number. The first striking feature is that these distributions span two orders of magnitude. At the lower of the two temperatures (T = 7.4 = 81% T_{opt}), most residues exhibit slow relaxation rates, as seen by the sharp peak near zero. At the higher temperature of $T_{opt} = 9.1$, the height of the peak is reduced and more residues are seen with faster rates.

At both temperatures, a small number of residues have very fast rates. Many of these fast residues belong to the folding nucleus for this structure that was identified in another study [19,20] using the standard lattice model. In Fig. 1.5 we show the 48-mer structure colored by rate of sidechain freezing at $T = 7.4$, and we indicate the original nucleus by large spheres. Of the 10 fastest residues that become fully ordered at T = 7.4, 7 belong to the folding nucleus. While some of the nucleus positions are no longer kinetically

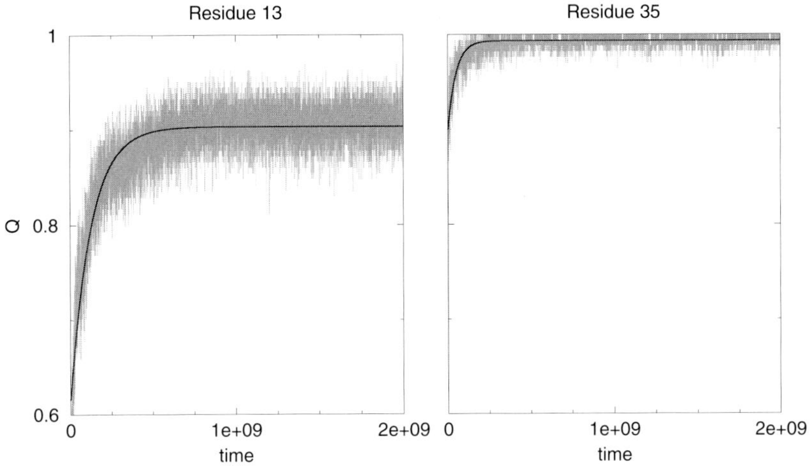

Fig. 1.3. Average time traces for two representative residues in the lattice 48-mer. Residue 35 is a nucleus residue exhibiting fast freezing, while residue 13 is a non-nucleus residue with an average freezing rate. The black line is the best single-exponential fit to the data.

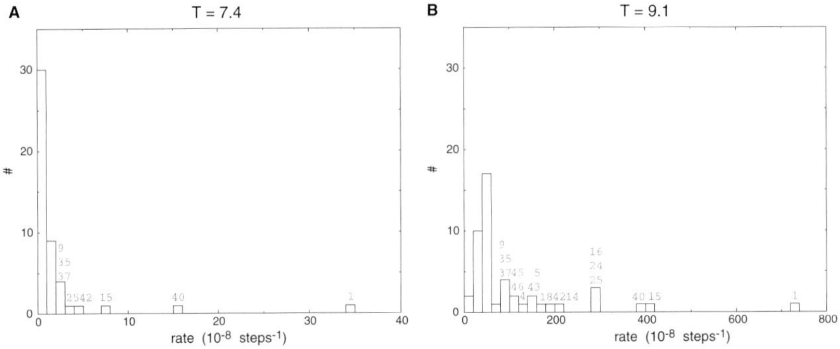

Fig. 1.4. Histogram of residue relaxation rates for 48-mer with 4 internal states. Histograms for low temperature (T = 7.4) and optimal folding temperature (T = 9.1) are shown. Each residue was assigned a value of 1 if it was in its native sidechain state, and 0 otherwise, and these numbers were averaged at each time step over 130 long runs. Rates were calculated by fiting a single exponential relaxation to the resulting native occupancy curves for each residue. At T = 7.4, runs of length 2×10^9 were used; at T = 9.1, run length was 2×10^8. The fast positions at each temperature are labelled by numbers on the histograms. Red numbers correspond to positions which are more than 90% ordered in the native state, while green numbers are less than 90% ordered.

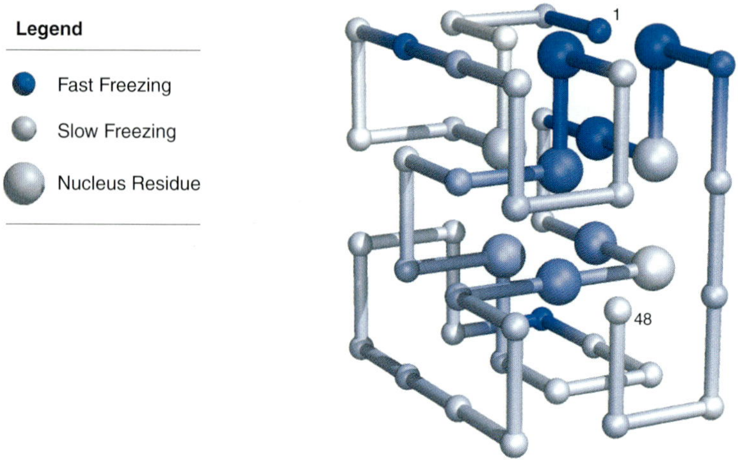

Fig. 1.5. Lattice 48-mer structure colored by rate of freezing at $T = 7.4$. Nucleus positions, determined in [19], are indicated by large spheres. Colors range from white (slow-freezing) to blue (fast-freezing).

important in the present model, a strong signature of the old nucleus has remained. Importantly, with the exception of residue 9, all of the fast positions that reach full nativity are located in or near the original nucleus was found.

It appears, then, that at temperatures at or below T_{opt}, a small group of residues reaches full nativity quickly, thus organizing a critical piece of structure which remains fully stable, allowing the rest of the chain to gradually order its sidechains. At $T = 7.4$, the formation of the stable piece traps many sidechains in non-native states which take a very long time to reorganize via backbone fluctuations. On the other hand, at T_{opt}, as seen in Fig. 1.4, more residues are found in the fast tail of the rate distribution, indicating that backbone fluctuations are sufficient to allow sidechain ordering to occur more quickly once enough native structure has formed. Additionally, the higher temperature requires a larger amount of native structure to be formed in order to remain stable. These two effects act to eliminate sidechain ordering as a relevant kinetic step at T_{opt}. This is seen clearly in Fig. 1.2 where at T_{opt} the sidechain ordering step is an order of magnitude faster than the backbone folding step.

Another way to see that sidechain dynamics becomes markedly different as temperature is lowered is given in Fig. 1.6. The red line indicates the equilibrium energy at each of the two temperatures, while the solid green line is a time trace of the average energy over all runs. The average time to form the stable native backbone is 2.7 x 10^7 at $T = T_{opt} = 9.1$, and 1.6 x 10^8 at $T = 7.4$, and is marked by an arrow in the figure. For T = T_{opt}, the arrow indicates that at the time of native backbone formation, the energy of the chain is already very close to its equilibrium value. That is not the case

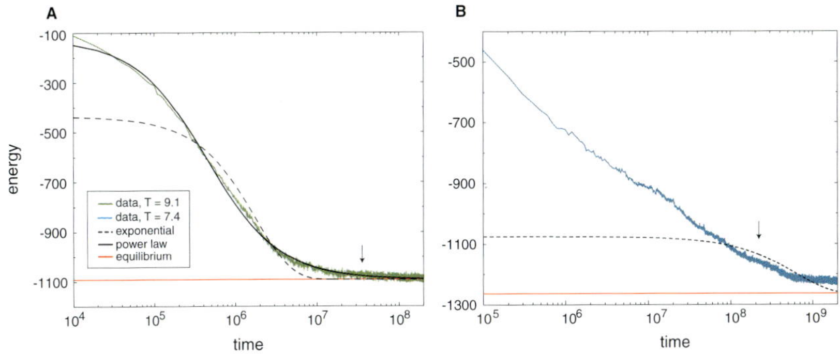

Fig. 1.6. Relaxation of energy to equilibrium. By averaging energy at each time step over 130 long runs, we obtained energy relaxation curves at temperatures 9.1 (panel A) and 7.4 (panel B). The same runs were used in Fig. 1.4. The green curves in each figure are the average energy obtained from simulations. The red line corresponds to the average energy at equilibrium, and was obtained for each temperature by averaging over a long run started at the native lowest energy state, as in Fig. 1.1. The dashed line is a fit to a three-state exponential model (see Methods). A fit using a two-state exponential model yielded a nearly identical curve. The arrows indicate the average time to reach the stable native backbone at each temperature. The solid curve in panel **A** is a fit for T = 9.1 using a power-law $(-1092 + a_0(x - a_1)^{a_2})$ with $a_0 = 1.73 \times 10^8$, $a_1 = -4.1 \times 10^5$, and $a_2 = -0.94$. The value of -1092 for the equilibrium energy corresponds to the red line.

at low temperature, at which there is a significant gap between the energy of the folded chain and the equilibrium energy.

We tried to fit the relaxation of energy by a standard, single-barrier process (single exponential) as well as a double exponential fit - both fits converged to the same curve which is shown as a dashed line in Fig. 1.6. The fit is not appropriate at any timescale. In particular, we note that at short times, the trajectory may resemble a single-exponential process, but it develops a very long tail at long times. A power-law fits the entire curve very well. The power-law fit was done at $T = 9.1$ and is shown as the solid black line in Fig. 1.6A. At $T = 7.4$ we did not have enough data at long times to see relaxation to equilibrium and therefore a fit would not be meaningful. The short-time behavior can be modeled equally well by a single-exponential, and thus corresponds to the classic nucleation mechanism in which the backbone topology becomes organized [21]. Long-time non-exponential behavior is a signature of glassy dynamics associated with the sidechain degrees of freedom and will be discussed below.

Because lattice models can give only a schematic view of the folding process, we proceeded to investigate sidechain dynamics in an all-atom simulation of protein G, an alpha/beta protein that has featured in numerous experiments [22–24]. The details of the simulation and a full characterization of the folding kinetics of this protein are given elsewhere [14]. Our goal in the present study is to see how the results obtained from our simplified lattice model compare with a much more realistic representation of a protein, and whether the same kind of analysis can shed light on the kinetics of a real protein. In the lattice model we had to postulate a set of microscopic dynamics for the internal states of each residue. In the all-atom simulation, we model all sidechain atoms and torsions explicitly. We use the simulation methodology described previously [25]. Because rotations around sidechain χ angles are continuous, interconversion between sidechain rotamers can become restricted if a residue is buried. Slowing down of sidechain dynamics upon compactification emerges from the excluded volume interaction in this model, and does not have to be included phenomenologically as in the lattice model.

In a previous study [14], 50 folding trajectories each for the wild type and D46A mutant of protein G, starting from random backbone and sidechain conformations, were obtained. All runs were terminated after 2×10^9 steps, by which time $> 90\%$ had reached the native backbone fold. We then applied a time series analysis to the mutant data similar to the one we used for the lattice model. Specifically, at each time step and for each residue we recorded a value of 1 if the sidechain was in its native rotameric state, and a 0 otherwise. We averaged these values for each residue over all trajectories, and then fit a kinetic model to the resulting traces. The parameters for the fits for each residue are given in Tables 1.2 and 1.3.

Both the wild type and mutant protein G folds in simulations via partially folded kinetic intermediates. Two of the most populated intermediates feature nativity for only hairpin 1 and the helix or only hairpin 2 and the helix. A two-state fit was therefore not appropriate for some of the residues. We used a three-state fit for all residues, and found that for some of the residues one rate constant was at least an order of magnitude larger than the other. Such residues were classified as two-state, while the others were classified as three-state. Figure 1.7 shows representative fits for two-state and three-state residues. The relaxation rates given in Tables 1.2 and 1.3 span an order of magnitude. The equilibrium level of ordering of each residue (parameter d in the Tables) was obtained directly as an average over a long simulation started in the native state, and was not obtained by fitting. Some residues are seen to be highly ordered in the native state, while others are not. We looked at the fastest residues whose equilibrium level was at least 70% ordered (bold residues in the Tables).

The four fastest two-state residues are shown in Fig. 1.8. These four make key contacts between the first hairpin and the helix. Phenylalanine 30 and leucine 5 have a strong hydrophobic interaction that secures the first strand of the hairpin against the helix, while threonines 18 and 25 lock in the second beta strand. The fastest two-state residues are thus seen to be important in forming the kinetic intermediate. All residues involved in inter-

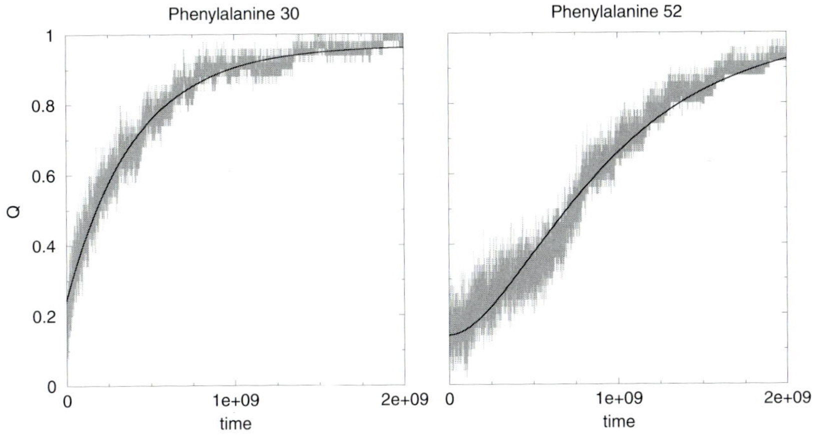

Fig. 1.7. Average time traces for two representative residues in the D46A mutant of protein G. F30 and F52 are typical two-state and three-state residues, respectively. The black line is the best fit as described in Tables 1.2 and 1.3.

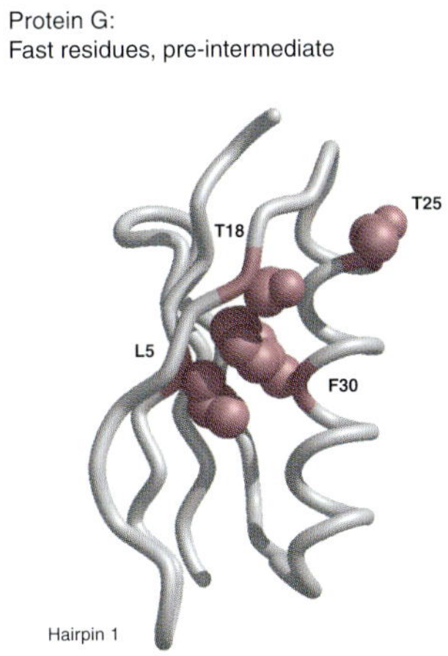

Fig. 1.8. D46A mutant of protein G residues exhibiting fastest two-state relaxation to a highly ordered state. The four fastest residues whose relaxation curves fit well to a two-state kinetic model, and whose equilibrium conformation is at least 70% ordered are shown in pink. These residues occupy key positions in the intermediate that is most populated in all-atom simulations of protein G folding [14]. The helical residue F30 is lodged between L5 of beta-strand 1, and T18 of beta-strand 2, thus organizing the entire structure of the intermediate which consists of hairpin 1 and the helix. T25 makes contacts at the hairpin-helix turn.

Table 1.2. Two-State Residues and Fits for the mutant protein G. Individual residue relaxation curves were initially fit to the following three-state kinetic model: $f(x) = d + c(a/(b-a))exp(-bx/10^9) - c(b/(b-a))exp(-ax/10^9)$. The parameter d, corresponding to fully equilibrated value of residue ordering, was obtained from long equilibrium simulation, and was not varied in the fitting process. Standard non-linear fitting was used to calculate a, b, and c. The residues listed in this table h ad one rate constant that was at least an order of magnitude faster than the other. The fits listed are therefore essentially two-state fits, and we report only the relevant slow rate constant. The three-state model was used in order to determine which residues were markedly two-state, and which ones were not. Asymptotic error on parameter a is listed as well. The table is sorted by the rate constant a.

#	a	b	c	d	err a	err b
13K	0.174	-	0.064	0.094	± 0.009	-
49T	0.190	-	0.150	0.483	0.008	-
37N	0.756	-	0.603	0.826	0.003	-
36D	0.779	-	0.168	0.326	0.01	-
15E	0.791	-	0.302	0.368	0.005	-
7I	0.941	-	0.676	0.965	0.003	-
21V	1.027	-	0.357	0.739	0.007	-
0V	1.035	-	0.379	0.813	0.01	-
24E	1.159	-	0.011	0.060	0.2	-
32Q	1.159	-	0.110	0.198	0.02	-
27E	1.283	-	0.442	0.539	0.008	-
33Y	1.308	-	0.810	0.991	0.003	-
31K	1.365	-	0.444	0.501	0.007	-
16T	1.504	-	0.606	0.958	0.005	-
3Y	1.581	-	0.779	1.000	0.003	-
17T	1.626	-	0.544	0.884	0.006	-
5L	1.724	-	0.636	1.000	0.004	-
22D	1.809	-	0.423	0.554	0.01	-
28K	1.882	-	0.026	0.057	0.00	-
18T	1.891	-	0.632	0.991	0.005	-
25T	1.931	-	0.596	0.985	0.005	-
30F	2.068	-	0.723	1.000	0.004	-

mediate formation are naturally found to be two-state, because formation of the intermediate is a purely two-state process.

The three-state residues are ones whose sidechain ordering cannot proceed normally until the intermediate has formed. They exhibit a lag phase as seen in Fig. 1.7 while the intermediate forms. In Fig. 1.9 we show the three fastest three-state residues of the mutant protein G which are significantly ordered at equilibrium: valines 6 and 54 and phenylalanine 52. Interestingly, these three residues all have the same rate of relaxation, suggesting that they become ordered together. All three are involved in bringing beta-strand 4 in hairpin 2

Table 1.3. Three-State Residues and Fits for the mutant protein G. Fits were performed as in Table 1.2. The two rate constant obtained for the residues listed here were within one order of magnitude of each other. The table is sorted by the slower of the two rate constants, which is arbitrarily designated to be parameter a.

#	a	b	c	d	err a	err b
12L	0.279	1.974	0.601	0.842	± 0.008	± 0.1
8N	0.328	2.538	0.136	0.254	0.02	0.6
46D	0.464	2.524	0.058	0.191	0.07	1.0
55T	0.710	2.158	0.384	0.710	0.02	0.2
35N	0.722	3.301	0.034	0.187	0.1	1.8
47D	0.834	1.387	0.130	0.294	0.2	0.5
1T	0.859	5.414	0.426	0.780	0.01	0.3
4K	1.018	9.252	0.178	0.210	0.01	1.1
10K	1.02	1.02	0.027	0.053	0.03	0.03
42V	1.127	1.127	0.180	0.508	0.01	0.01
50K	1.205	1.205	0.369	0.386	0.004	0.004
56E	1.319	1.319	0.610	0.666	0.003	0.003
44T	1.361	1.361	0.341	0.689	0.006	0.006
43W	1.623	3.455	0.750	1.000	0.01	0.06
39V	1.719	1.719	0.567	0.984	0.004	0.004
45Y	1.734	1.734	0.883	0.999	0.002	0.002
53T	1.749	1.749	0.517	0.850	0.005	0.005
51T	2.020	2.020	0.510	0.841	0.006	0.006
54V	2.034	2.034	0.687	1.000	0.003	0.003
6V	2.035	2.035	0.642	0.966	0.004	0.004
52F	2.039	2.039	0.858	1.000	0.002	0.002
19K	2.111	3.089	0.049	0.068	0.4	0.9
40D	2.198	2.198	0.037	0.170	0.07	0.07
2T	2.923	2.923	0.285	0.628	0.02	0.02

into contact with the rest of the protein. Valine 6 establishes contacts between beta-strands 1 and 4. Valine 54 makes contacts with valine 39 (located at the C-terminus of the helix) which hold the end of hairpin 2 against the helix. Phenyalanine 52 makes hydrophobic contacts with tyrosine 45, stabilizing hairpin 2 internally, while also making contacts with the helix.

The data obtained from the all-atom simulation is in good qualitative agreement with our lattice simulation. There is a wide distribution of residue relaxation rates, with the fast residues located in topologically important positions. The same mechanism seems to be at work here: key organizing residues form quickly holding the overall structure together, while all other residues relax more slowly toward equilibrium via fluctuations around the native fold. On the lattice we found strong overlap between the fast residues and the nucleus residues which organized the backbone transition state. In

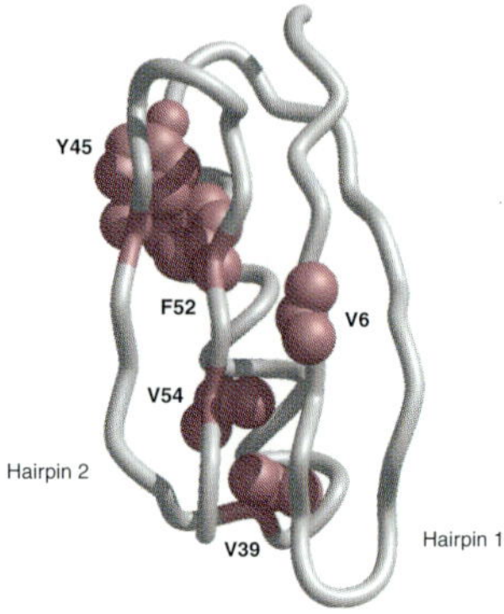

Fig. 1.9. Protein G residues exhibiting fastest three-state relaxation to highly ordered state. Residues V6, F52, and V54, shown in pink, exhibited fastest three-state relaxation, and remained highly ordered at equilibrium. All three are important post-intermediate positions: F52 and V54 secure strand 4 of hairpin 2 to the helix, while V6 makes contacts between the two hairpins.

order to make a similar comparison in the all-atom model, we proceeded to characterize its transition state ensemble.

Because the transition state ensemble lies at the top of the folding free energy landscape, its conformations are characterized by a 0.5 probability of folding (p_{fold}) during a tiny fraction of the entire folding time ("commitment time" [26]). Assuming a commitment time corresponding to 0.005% of a full folding run, we calculated the p_{fold} of at least 5 structures per trajectory. A histogram of contacts (Fig. 1.10) made by each residue for various p_{fold} ensembles reveals that phenylalanine 52 is the most important residue for the final intermediate → native folding step. Its energy contribution, which is proportional to the number of contacts it makes, appears to grow as the ensemble p_{fold} approaches one. Though less pronounced, similar increases were seen for Y3, K4, L5, V6, A23, E27, F30, W43, Y45, K50, and V54. When individual residue-residue contacts are histogrammed (Fig. 1.11), it is clear that only a handful of over 1500 possible contacts are important for stabilizing the transition state ensemble. These special contacts bring two hairpin 2 residues (F52 and V54) in contact with hairpin 1 (Y3, L5, V6) and helix residues (E27 and F30). Because of the non-local and specific nature of these contacts, folding in this model appears to be consistent with that proposed under the theory of specific nucleation [21,27–30]. Detailed comparison of these results with experimental data show excellent agreement [14].

It is clear that the nucleus characterizing the transition state ensemble under our all-atom model of folding is nicely identified by the time-series analysis of sidechain dynamics. The three fastest three-state residues - V6,

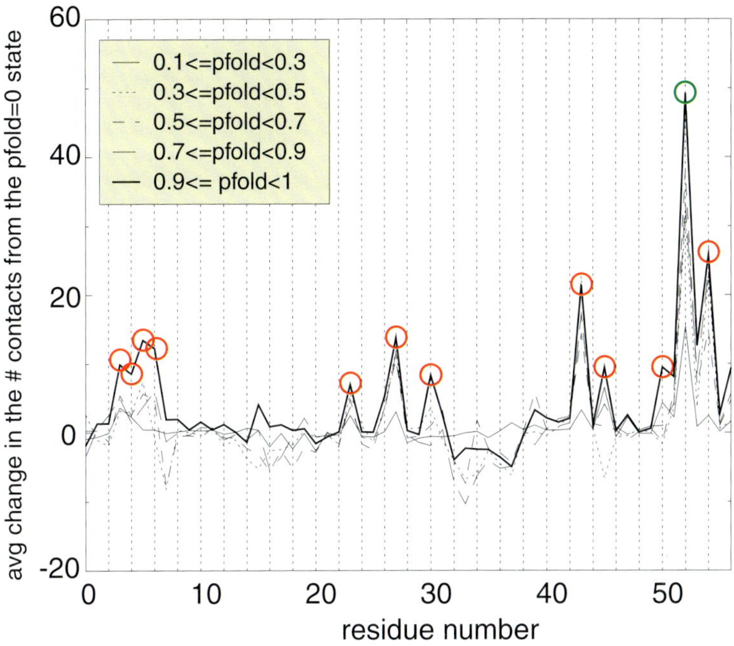

Fig. 1.10. Dependence of residue nativity upon $p_{\rm fold}$. Conformations were binned according to their $p_{\rm fold}$ values, and the average change in number of contacts, with respect to the $p_{\rm fold} = 0$ state, is plotted for each residue. Each curve corresponds to an average over all conformations within the given range of $p_{\rm fold}$.

F52, and V54 - coincide with those which are most indicative of progress along the $p_{\rm fold}$ hypersurface. Although structures with $p_{\rm fold} \approx 1$ will rapidly attain native-like backbone topologies, energies will reach equilibrium values very slowly, requiring simulations extending beyond the 2×10^9 cutoff we have used here. This is because a fairly significant amount of energy is contributed by sidechain-sidechain interactions, and the correct packing of sidechains is significantly slower once the collapse transition has occurred. In our previous study of crambin, we observed a similar phenomenon (which we referred to as the "sidechain-packing trap"; see Fig. 6E in reference [25]), where the folding of the backbone occurred on a faster timescale than that by which the native energy was fully attained. The current $p_{\rm fold}$ analysis demonstrates that, in fact, not all residues participate equally in the slow relaxation of conformations with incorrect packing. The nucleus residues (V6, F52, V54) have to attain native packing relatively early as their energy contribution is required to counterbalance the tremendous loss of backbone entropy upon collapse to a native-like conformation.

Finally, we also note a striking similarity between the thermodynamic data of the lattice model presented here and crambin obtained from all-atom

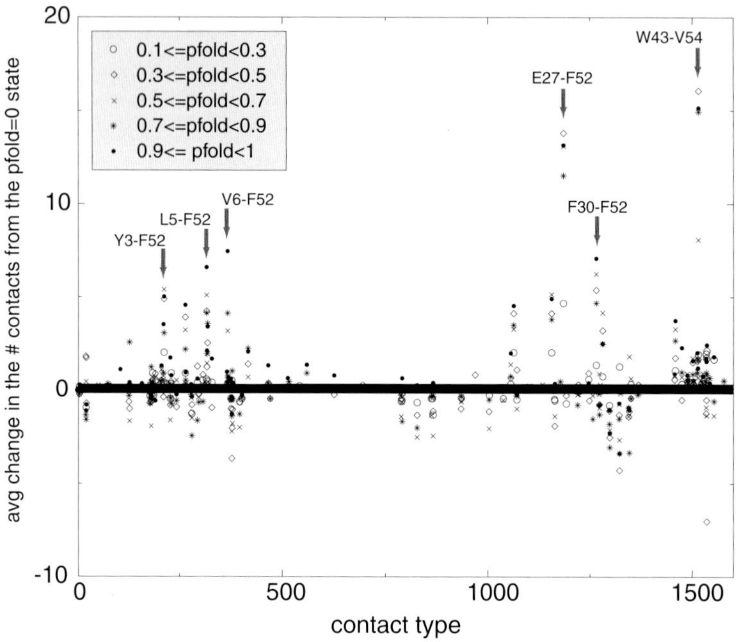

Fig. 1.11. Dependence of specific contacts upon $p_{\rm fold}$. As in Fig. 1.10, conformations were binned according to their $p_{\rm fold}$ values. For each $p_{\rm fold}$ range, the average change (with respect to the $p_{\rm fold} = 0$ state) in number of atom-atom contacts between all pairs of residues that make native contacts, is plotted for each pair of residues. Residues pairs are arbitrarily ordered in a linear fashion along the x-axis.

simulations. For crambin, we observed a rather unusual departure from a simple two state model when fitting the equilibrium energy against temperature. As temperature was lowered below the transition point, the decrease in energy was perfectly linear with temperature. As shown in Fig. 1.1, as the number of sidechain states is increased, the same linear relation between energy and temperature is observed. This suggests that sidechain degrees of freedom lead to a noticeable contribution to the heat capacity, which dominates the thermodynamic behavior at low temperatures.

1.3 Discussion

Our work demonstrates that the presence of sidechain degrees of freedom leads to a wide distribution of residue relaxation rates, even within two-state cooperative folding reactions. Figure 1.12 gives a schematic overview of the relaxation mechanism we observed. Both in lattice and in all-atom simulations, we found a small number of residues becoming fully ordered much faster than the rest of the protein. This observation is consistent with the

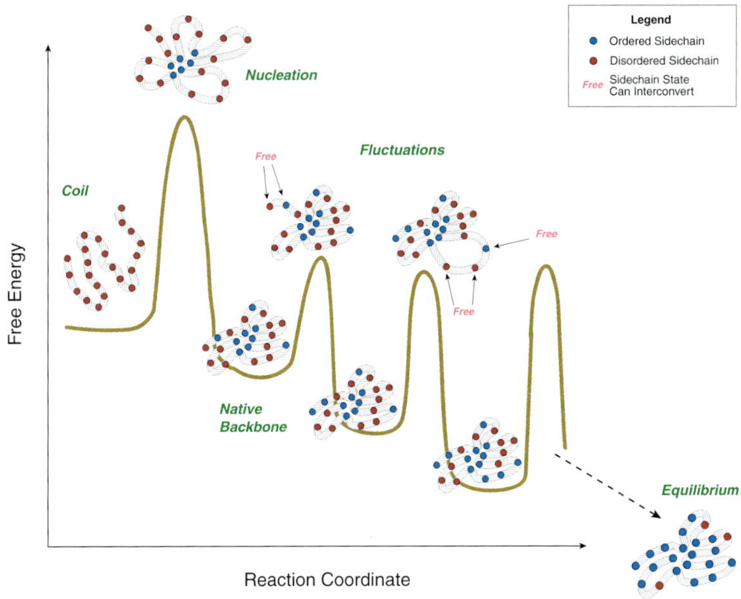

Fig. 1.12. Schematic diagram of barriers and their significance during the folding reaction. The first barrier corresponds to the nucleation event which organizes the backbone topology. Associated with this barrier is the freezing of a small group of residues - the nucleus - into their native sidechain states (blue dots). Other residues may still be partially disordered (red dots). The disordered residues become increasingly native-like via barriers corresponding to backbone fluctuations which momentarily free a few residues (see small arrows), and allow their sidechain states to interchange. Barriers become higher as chain approaches equilibrium.

nucleation-condensation view of protein folding in which the major transition state of the folding reaction involves a few residues reaching their native conformation. Importantly, in our simulations, we find that these nucleating residues are not only in correct spatial geometry with respect to each other's centers of mass, but additionally their native rotamer has been singled out and practically frozen. Once nucleation has occured, the native chain topology is strongly stabilized and certain measures, such as compactness and perhaps fluoresence, might indicate that the reaction is complete, and equilibrium has been reached (see Fig. 1.12 after nucleation barrier). This, however, is not the case as there exist many sidechains that have become partially ordered, yet have still not reached equilibrium. Because the nucleating residues have frozen and are rigid, and many other partially ordered residues are significantly stabilizing the fold, the non-equilibrated sidechains are not able to convert easily to their native rotamer. They remain in a non-native state until a backbone fluctuation momentarily allows them to interconvert. The presence of backbone breathing motions in protein globules may therefore be useful not only for function, as has been suggested before [31,32], but also in order to allow sidechain equilibrium to be achieved in a reasonable amount of time.

In simulations, a kinetic intermediate is very easily observed as a set of conformations which appears as a plateau within some range of energies during many folding trajectories [33,25]. The existence of folding intermediates in our simulation of protein G, while complicating our analysis somewhat, has one important advantage: we are able to see that the kinetics of only half of the sidechains are sensitive to the presence of the intermediate; the other residues exhibit single-exponential relaxation. In other words, a kinetic intermediate can be completely invisible if the wrong position is used to probe folding. We observed a distribution of residue relaxation rates for both the pre- and post-intermediate steps. Each of these steps is a purely two-state process as seen by the abrupt jump in rms deviation and energy. It appears, then, that a few key residues reach nativity faster than all others and propel the chain through its transition state. Further relaxation after the major event via backbone fluctuations yields a distribution of rates over the fold, the exact nature of which is governed by the extent of backbone mobility at each position in the ensemble.

At first glance this observation runs contrary to the belief that in two-state transitions all parts of the structure must reach nativity at the same rate. The argument goes that if structure is obtained gradually, with some parts folding faster than others, then there are many distinct ensembles of states for the chain to traverse. To dispell this fear, it is crucial to note that the core residues which are observed in simulation to freeze fastest also happen to be in key organizing positions. The ensemble of conformations consistent with their freezing is highly native and therefore extremely small compared with the ensemble of unfolded conformations. The major transition of protein folding occurs between these two ensembles and is a two-state transition in simulation as in reality. The entire molecule does not, however, necessarily reach equilibrium concomitantly with this barrier crossing. There can be many other smaller barriers associated with backbone fluctuations which need to be crossed in order for all sidechains to reach equilibrium (see Fig. 1.12).

It is important to note that temperature plays a key role in making sidechain relaxation possible in a reasonable amount of time. At low temperatures backbone fluctuations are small and sidechain relaxation is a very noticeable and very long process, as seen in Fig. 1.6. At optimal folding temperature, however, the energy of the post-nucleation ensemble is very close to its equilibrium value. Sidechain relaxation is still very slow, following stretched-exponential kinetics, but the product of the major transition is significantly closer energetically (and therefore structurally) to equilibrium. This suggests that under optimal conditions, the slow sidechain packing process may not be physiologically relevant because the ensemble of folded yet unequilibrated molecules is structurally close enough to the native ensemble that it may exhibit similar amounts of protection from proteolysis. The relatively small gap between mispacked and native molecules at these temperatures suggests that relevant experiments must be sensitive enough to detect such differences.

Since we could not observe full equilibration in the all-atom simulation, we return to the lattice simulations in order to discuss the relevant post-nucleation processes which establish equilibrium. In lattice simulations we found that the dynamics during short times is reminiscent of the classic nucleation mechanism that has been observed before [21]. Due to the existence of sidechain states, the nucleation-organized backbone does not reach equilibrium immediately. At long times the system can end up in traps which require some degree of backbone motion to allow sidechains to interconvert. This suggests that perhaps the energy landscape after the native fold has been acquired consists of a series of barriers, associated with backbone fluctuations, which must be crossed. As the system traverses these barriers it moves to lower and lower energies. If one makes some simple assumptions about the transition states for these barriers, the resulting relaxation process can be shown to be non-exponential in time [34–36]. Both power-law [34] and stretched-exponential [35] behavior can emerge depending on the nature of the barriers. In the lattice simulations, we observed a power-law or highly stretched exponential relaxation at long times. The landscape for slow sidechain equilibration thus seems to be one of increasingly deeper wells, rather than a single cooperative transition to nativity. This places sidechain relaxation within the set of phenomena that can be characterized as a glass. Classic lattice models without sidechain states, however, do not exhibit a glass transition at any reasonable temperature [8]. Our lattice simulations indicate that the presence of sidechain degrees of freedom may lead to glassy relaxation, but further detailed characterization of the energy landscape, as well as additional tests using more realistic models, are required to solidify this claim.

1.4 Methods

Lattice Model with Internal Monomer States. We use a standard three-dimensional lattice model in which each monomer occupies a single lattice site. A fast-folding 48-mer sequence was obtained from a lattice protein evolution study described in [20]. The standard Miyazawa-Jernigan parameter set [37] is used to compute the energy of a conformation. Two monomers are said to be in contact if they are nearest neighbors on the lattice, and are not sequence neighbors. Additionally, each monomer has n internal states, where n is a parameter of the model. We present data for 48-mer folding with $n = 1, 2$, and 4. The internal state of each monomer is stored as a number from 0 to $n - 1$. The 0 state is the native state, while the states

$1..n-1$ are non-native. If $n = 1$ then all monomers remain native throughout the simulation, and the model is equivalent to the standard lattice model. Non-native monomers do not contribute to energy. That is, two monomers in contact will contribute to energy only if they are both in their native state, the 0 state.

The standard cubic lattice move-set [38] is used to evolve the backbone conformation, and a Metropolis criterion [39] with temperature T is used to accept/reject moves. In addition to backbone moves, the internal states of the monomers must be evolved. After each backbone move is attempted, we attempt $n-1$ internal state moves. At each such move, a random monomer is chosen. If the monomer is making more than c contacts with other monomers, its internal state is not allowed to change. Otherwise, its internal state is randomly flipped to one of the other $n-1$ states, the change in energy of the conformation is computed, and the move is accepted/rejected based on the Metropolis criterion. The parameter c can take the values 0 through 4. When $c = 4$, internal states can interchange freely and are not affected by the conformation of the backbone. If $c = 0$, internal states can interchange only if the monomer makes no contacts. In this study we take $c = 0$ throughout. Folding simulations are started from random backbone conformations generated by an infinite temperature simulation. The internal state of each monomer is initialized randomly.

All-Atom Protein Folding Simulations. The all-atom Monte Carlo simulation previously described [25] was used to fold protein G (pdb code: 1IGD). By representing all sidechain and backbone heavy atoms as hard spheres, the protein was simulated as a polymer with excluded volume interactions, where chain crossings are strictly prohibited. For the D46A mutant [14], the energy of a conformation was computed as $E = E_G + E_h$, where (1) the atom-atom Gō energy $E_G = \sum C(A,B)\Delta(A,B)$, with $\Delta(A,B) = 1$ if the heavy atoms A and B were in contact and zero otherwise, $C(A,B)$ was -1 if A and B were in contact in the native conformation, 1 if they were not, and ∞ if they were sterically clashed and (2) the backbone hydrogen bonding energy $E_h = N_h h$, where N_h corresponds to the number of amide N-carbonyl O pairs in contact. h was chosen to be -0.6 in order to match experimental stabilities of the protein G helix and hairpins taken in isolation.

The torsional move set ensures that canonical bond lengths and geometries (including planar peptide bonds) are maintained throughout the entire simulation. Backbone and sidechain moves consisted of concerted random rotations of backbone ϕ/ψ and sidechain χ angles, respectively. 10 sidechain moves were completed for each backbone move in order to allow sufficient relaxation of sidechain geometries after a change in the backbone topology.

50 folding simulations were initiated from random coil conformations, obtained by simulating the mutant native state with only the excluded volume interaction turned on. The temperature was then quenched to T=1.575 and the chain was allowed to equilibrate for 2×10^9 MC steps, where 1 MC step

consisted of 1 backbone and 10 sidechain moves. Given the experimentally measured transition temperature of 360K and our simulation transition temperature of 1.95, T=1.575 corresponds to an actual temperature of $\approx$ 290 K.

From the 50 trajectories of the mutant protein G, we estimated the probability to fold [40] (p_{fold}) of conformations observed just prior to reaching the native state, by counting the number of times the native state was attained from the selected conformation in 20 separate runs of 10×10^6 MC steps.

Fitting of Residue Relaxation Curves. After collecting many long runs, we averaged the internal sidechain state of each monomer at each time step over all runs, assigning 1 if the residue was native, and 0 otherwise. For lattice simulations, 130 runs were used, and a two-state exponential fit of the form $f(x) = a_0 + a_1 \exp(-a_2 t)$ was very good for all residues. For all-atom simulations, 50 runs were used, and averaging over runs was performed by assigning 1 to each residue whose χ-angles were all within 60° of the native angles, and 0 otherwise. A value of 1 thus corresponded to observing the native rotamer. Fits to a three-state model were performed as described in Results. All fits were done using the nonlinear least-squares Marquardt-Levenberg algorithm.

Acknowledgements

We would like to thank Will Chen, Gabriel Berriz, Leonid Mirny and Phillip Geissler. Research was supported by NIH grant GM52126.

References

1. C. L. Brooks, M. Karplus, B. M. Pettitt: *Proteins: A Theoretical Perspective of Dynamics, Structure, and Thermodynamics*, John Wiley & Sons, New York (1988).
2. F. M. Richards, W. A. Lim: Quat. Rev. Biophys. **26**, 423–498 (1994).
3. J. Heringa, P. Argos: Proteins **37**, 30–43 (1999).
4. M. Levitt, M. Gerstein, E. Huang, S. Subbiah, J. Tsai: Annu. Rev. Biochem. **66**, 549–579 (1997).
5. R. L. Dunbrack, M. Karplus: Nature Struct. Biol. **1**, 334–340 (1994).
6. A. Fersht: *Structure and Mechanism in Protein Science*, W. H. Freeman and Company, New York, 1st ed. (1999).
7. E. I. Shakhnovich, A. V. Finkelstein: Biopolymers **28**, 1667–1680 (1989).
8. A. Gutin, A. Sali, V. Abkevich, M. Karplus, E. I. Shakhnovich: J. Chem. Phys. **108**, 6466–6483 (1998).
9. J. D. Bryngelson, P. G. Wolynes: J. Phys. Chem. **93**, 6902–6915 (1989).
10. P. J. Steinbach, B. R. Brooks: Chem. Phys. Let. **226**, 447–452 (1994).
11. D. Vitkup, D. Ringe, G. A. Petsko, M. Karplus: Nature Struct. Biol. **7**, 34–38 (2000).
12. A. L. Lee, J. Wand: Nature **411**, 501–504 (2001).
13. N. Gō, H. Abe: Biopolymers **20**, 991–1011 (1981).

14. J. Shimada, E. I. Shakhnovich: Nature **submitted** (2002).
15. S. Bromberg, K. A. Dill: Protein Sci. **3**, 997–1009 (1994).
16. D. K. Klimov, D. Thirumalai: Folding & Design **3**, 127–139 (1998).
17. L. Li, L. A. Mirny, E. I. Shakhnovich: Nature Struct. Biol. **7**, 336–342 (2000).
18. E. Kussell, J. Shimada, E. I. Shakhnovich: J. Mol. Biol. **311**, 183–193 (2001).
19. E. Shakhnovich, V. Abkevich, O. Ptitsyn: Nature **379**, 96–98 (1996).
20. L. A. Mirny, V. Abkevich, E. I. Shakhnovich: Proc. Natl. Acad. Sci. USA **95**, 4976–4981 (1998).
21. V. I. Abkevich, A. M. Gutin, E. I. Shakhnovich: Biochemistry **33**, 10026–10036 (1994).
22. E. L. McCallister, E. Alm, D. Baker: Nature Struct. Biol. **7**, 669–673 (2000).
23. S. Nauli, B. Kuhlman, D. Baker: Nature Struct. Biol. **8**, 602–605 (2001).
24. S. Park, M. C. R. Shastry, H. Roder: Nature Struct. Biol. **6**, 943–947 (1999).
25. J. Shimada, E. Kussell, E. I. Shakhnovich: J. Mol. Biol. **308**, 79–95 (2001).
26. P. G. Bolhuis, D. Chandler, C. Dellago, P. L. Geissler: Annu. Rev. Phys. Chem. **submitted** (2002).
27. A. R. Fersht: Proc. Natl. Acad. Sci. USA **92**, 10869–10873 (1995).
28. T. R. Sosnick, L. Mayne, S. W. Englander: Proteins **24**, 413–426 (1996).
29. E. I. Shakhnovich: Curr. Opin. Struc. Biol. **7**, 29–40 (1997).
30. V. S. Pande, A. Y. Grosberg, T. Tanaka, D. S. Rokhsar: Curr. Opin. Struc. Biol. **8**, 68–79 (1998).
31. V. Y. Alexandrov: *Cells, Molecules, and Temperature: Conformational Flexibility of Macromolecules and Ecological Adaption.*, Springer-Verlag, New York (1977).
32. P. Zavodszky, J. Kardos, A. Svingor, G. A. Petsko: Proc. Natl. Acad. Sci. USA **95**, 7406–7411 (1998).
33. V. I. Abkevich, A. M. Gutin, E. I. Shakhnovich: J. Chem. Phys. **101**, 6052–6062 (1994).
34. G. J. M. Koper, H. J. Hilhorst: Europhys. Lett. **3**, 1213–1217 (1987).
35. C. De Dominicis, H. Orland, F. Lainee: J. Physique Lett. **46**, 463–466 (1985).
36. E. I. Shakhnovich, A. M. Gutin: Europhys. Lett. **9**, 569–574 (1989).
37. S. Miyazawa, R. Jernigan: Macromolecules **18**, 534–552 (1985).
38. A. Sali, E. I. Shakhnovich, M. Karplus: J. Mol. Biol. **235**, 1614–1636 (1994).
39. N. Metropolis, A. W. Rosenbluth, M. N. Rosenbluth, A. H. Teller, E. Teller: J. Chem. Phys. **21**, 1087–1092 (1953).
40. R. Du, V. S. Pande, A. Y. Grosberg, T. Tanaka, E. I. Shakhnovich: J. Chem. Phys. **108**, 334–350 (1998).

Part II

Applications of Statistical Mechanics to Biological Systems

2 A Coarse Grain Model for Lipid Monolayer and Bilayer Studies

Steve O. Nielsen[1] and Michael L. Klein[1,2]

[1] Center for Molecular Modeling and Department of Chemistry, University of Pennsylvania, Philadelphia PA 19104-6323
[2] Address reprint requests to Michael Klein, klein@cmm.upenn.edu, www.cmm.upenn.edu

2.1 Introduction

Experimental work on complex condensed matter can address a broad range of temporal and spatial scales, from femtosecond dynamics and atomistic detail to real-time macroscopic phenomena. Simulation methods in which each atom is explicitly represented are well established but have difficulty addressing many cooperative effects of experimental and theoretical interest. There is simply too large a gap between the time and spatial scales that govern typical intramolecular events and those which are relevant for collective motions. One example is the spatial rearrangement of membrane species such as occur in the formation of a lipid raft[1] or membrane fusion. Available simulation techniques for specific time and spatial scales are illustrated schematically in Fig. 2.1. These techniques take a variety of approaches to reduce the level of detail in the representation of the system under study as the time and/or length scales grow. This will be discussed further in Sect. 2.3. Bridging these disparate scales is possible with multiscale modeling[2–4] in which the various levels of treatment are coupled and fed back into one another.

Possibly the least developed of these techniques are the ones aimed at studying events which are intermediate between the fully atomistic scale and mesoscale. Namely, events which occur on time scales of hundreds of nanoseconds to milliseconds and spatial scales of microns. Modern optical techniques[5] routinely access precisely these time and spatial scales. Dynamics occurring on time scales of nanoseconds and slower can be investigated with many different experimental methods, including nuclear magnetic resonance (NMR) spectroscopy and time-resolved X-ray diffraction. Flow cells combined with small angle X-ray scattering (SAXS) can access millisecond time scales. Spatial scales of 1 nm to 1 μm are accessably with small angle neutron scattering (SANS) and SAXS, and resolution of 20-25 Å is accessible from cyroelectron microscopy. From a theoretical point of view some phenomena in this domain can be addressed with Ginzburg-Landau, and other, expansions of the free energy[6–17]. In this chapter we present a simulation method that has ready access to events on these scales and which furthermore retains much of the specificity of atomistic models. In addition, explicitly dynamical events can be studied with simulations using molecular dynamics.

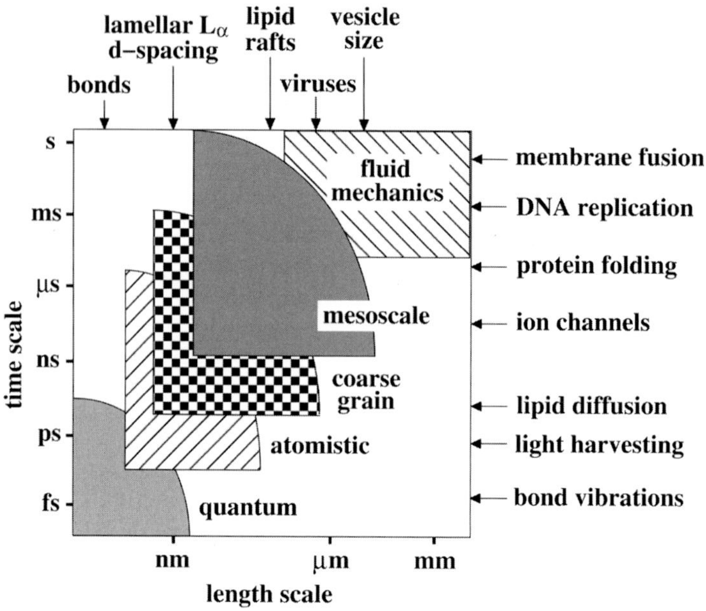

Fig. 2.1. Schematic of temporal and spatial scales accessible by simulation techniques. Also indicated are some characteristic membrane structures and events.

2.2 Challenges

There are many phenomena that lie within the mesoscopic spatio-temporal scale which may eventually be explored with coarse grain (CG) methods. In a biological context, examples of such phenomena are protein-protein interactions, lipid-protein interactions, and membrane-membrane interactions. Events that fall under these categories include membrane protein crystallization, membrane fusion, and transbilayer lipid diffusion. These are discussed below.

To understand the biological function of lipids, their physical properties must be studied in the context of membranes composed of lipid/protein mixtures. Membrane lipid composition varies widely over different organelle membranes, within a single membrane, and even across leaflets of the same bilayer membrane. These differences in membrane composition range throughout the whole spectrum of living organisms from protozoans to higher organisms such as mammals. For example, the transbilayer lipid distribution is symmetrical in the endoplasmic reticulum of mammalian cells, while it is markedly asymmetrical in the plasma membrane[18]. In the plasma membrane the majority of sphingolipids are found in the outer leaflet while most of the phosphatidylserine and phosphatidylethanolamine lipids are found in the cytosolic leaflet. Local variations in the physical properties of bilayers allow for membrane deformation and facilitate vesicle budding and fusion.

Proteins can also stimulate lipid exchange between membranes by bringing them into contact. Leaflet flips can occur by the action of protein flippases, which are thought[18] to drive vesicle budding from the plasma membrane by transferring lipids from the cytosolic to the outer leaflet. An understanding of these processes at a mesoscopic or atomic level is currently lacking.

The interaction between membranes can be accounted for by the van der Waals and electric double-layer forces which comprise the Derjaguin Landau Verwey Overbeek (DVLO) theory of colloid science and by the entropic forces due to the overlap of thermally excited surface modes[7]. Recent results[20] show that the concept of elastic deformation is relevant on lengths comparable to and even less than the bilayer thickness, involving a broad spectrum of collective modes which contribute to the forces between lipid bilayers. X-ray diffraction[21] analysis of bilayers subjected to known osmotic pressures provide information on the magnitude of both repulsive and attractive forces that exist between phospholipid and glycolipid membranes. Atomic force microscopy[22] is also a useful probe of repulsive forces. The presence of a stiff transmembrane peptide can alter the spectrum of thermally excited modes which can in turn alter the entropic forces between membranes. In Sect. 2.4.1 coarse grain (CG) and united atom simulation results[25–27] on membrane fluctuation modes will be presented.

It has been proposed[18] that lipid domains of different hydrophobic thickness and composition can aid membrane protein localization, and can influence membrane protein function. In higher organisms, membrane proteins that are destined for the plasma membrane are separated from Golgi proteins based on the length of their transmembane domains. These proteins can also be chemically sorted using address labels on their cytosolic tails to interact with a protein coat. It is clear that hydrophobic matching between the protein and its matrix is essential for protein function; this concept will be discussed in detail in Sects. 2.4.4 and 2.4.5.

Membrane protein crystallization is clearly a topic that needs a better theoretical understanding. The paucity of membrane proteins in the Protein Data Bank compared to soluble proteins reflects such a need. What is known is that bulk lipid phases and the transitions between them play a central role[19]. After protein solubilization and reconstitution in a known lipid (typically monoolein), the protein is incorporated into an inverse cubic or inverse hexagonal phase of the lipid/water mixture. The next step in crystallization is to add salt to dehydrate the water pores and provide charge screening. This dehydration can trigger either crystallization of the protein or co-crystallization of the protein with bound lipid. In Sect. 2.4.5 we will present results in which a peptide is able to induce a phase transition from a lipid lamellar bilayer phase to an inverse cubic phase.

Viral structures provide outstanding examples of macromolecular assembly and function. Structural information is obtained mainly through the use of NMR, X-ray crystallography, and electron cryomicroscopy. However, the

NMR and crystallographic techniques often provide a view of the system far removed from its physiologically active conditions. For example, some regions that are deeply buried in the X-ray structure must be exposed for virus particle interaction with the cell membrane during the process of viral infection of host cells[28,29]. Infection mechanism theories are difficult to assess from such data. On the other hand, atomistic simulations of the interactions between viruses and host cells (or even viruses and antiviral drugs) are prohibitively expensive even for the smallest viruses. There is a clear need for dynamical mesoscale simulation studies of such systems.

Another example where there is a need for an understanding of the dynamics and structure at the mesoscale level involves ion channels. The nicotinic acetylcholine receptor is a 268 kilodalton pentameric ion channel in which each subunit has four hydrophobic segments called M1–M4. The M2 segments are known to line the interior of the channel, and have been the subject of simulations isolated from their remaining M1, M3, and M4 protein parts[32,33]. A more complete description of the system, and one that would benefit fields ranging from drug design to infection control, would be to include all the atoms of the M2 segments and the pore water molecules, and to model the remaining segments (M1, M3, and M4), the membrane lipids, and the water solvent in a CG manner.

Finally, disparate time scales may be accessed by mixing levels of description in the same simulation. This is being done for mixed quantum/classical systems by QM/MM methods[30] and non-adiabatic surface hopping methods[31]. In the context of the present CG method, the force field could be adapted to embed an all-atom region inside a CG region as described above. In this chapter we focus on the development and application of a CG model that can be used for either Monte Carlo (MC) or conservative molecular dynamics (MD) simulations. We begin with an overview of existing simulation techniques which access time and length scales intermediate between atomistic and mesoscale. We then provide the methodological details of the current CG method. Next, we look at a few situations in which theoretical predictions can be examined by CG simulation methods, and at several situations in which the CG method goes beyond current theoretical and experimental reach to shed some light on novel dynamical phenomena. Lastly, we give some perspectives on future direction for research using CG models.

2.3 Models

2.3.1 Previous Work

There are many approaches in the literature for increasing the efficiency of fully atomistic simulations[34]. These methods fall into three general classes, namely Monte Carlo, conservative dynamics, and dissipative dynamics. Conservative evolution is numerically unstable and thus a small time step must be employed. The time step used is determined by the stiffest potentials in the system. The highest frequency motions are typically bond vibrations involving hydrogen. Since these fast motions are approximately decoupled from the rest of the system, they undergo many oscillations on the time scale of the remainder of the system. Hence their interaction with the remainder of the system is roughly governed by their average location, which is at the equilibrium bond length. There are algorithms, such as SHAKE and RATTLE, which constrain these high frequency bonds to remain fixed at their equilibrium extensions, thereby eliminating the stiffest motions and allowing the time step to be increased. Another common procedure to eliminate the stiff bond motion involving hydrogen is to lump hydrogen atoms together with their associated heavy atom into a single united atom site. If only static information is desired, the high frequency motions involving hydrogen can be reduced by artificially increasing the hydrogen atomic mass. Increasing the hydrogen mass by a factor of ten allows for the time step to be increased by about a factor of three.

The numerical instability inherent to conservative dynamics can be mitigated by adding dissipative terms to the system. One of the common procedures (known as the Lax method[35]) to stabilize a flux-conservative partial differential equation is to add a dissipative term. Moreover, methods have been developed in the simulation community which stabilize long time step integrators for Newtonian molecular dynamics by using very mild stochastic damping[36]. Dissipation can be accounted for with the Zwanzig-Mori projection operator formalism, which provides an exact procedure for eliminating unimportant variables from the system under study. This method leads to the generalized Langevin equation in which the total force acting on the particles of interest is composed of a colored noise term and a non-Markovian dissipative term containing a memory function satisfying the fluctuation-dissipation theorem[37]. Approximating the memory function with a delta function yields Brownian dynamics, which has been applied in studying self-assembly of amphiphiles into vesicles[38] and potassium channels[39].

The solvent is a good candidate for a less detailed treatment because it often plays a spectator role, it accounts for a sizable fraction of the system, and, especially due to the electrostatics of most water models, is very time consuming to treat in full detail. ten Wolde and Chandler[40] use a coarse grain Ising-like treatment of water to study hydrophobic polymer collapse. An exciting development by Malevanets and Kapral[41] treats the solute-solute and solute-solvent dynamics microscopically, while the solvent-solvent dynamics is treated in a mesoscale manner through multiparticle collisions which are hydrodynamically consistent. This method has been implemented for atomic solvents and will hopefully be generalized to the molecular case. The solvent can also be accounted for in an implicit manner. Continuum electrostatic models treat the solute as a low-dielectric cavity embedded in a high-dielectric medium representing the solvent. The Poisson or Poisson-Boltzmann equation is then solved numerically using either finite-difference or boundary element methods. If these numerical methods are too costly, approximations can be used such as the generalized Born approach[42].

Flekkøy and Coveney[43] introduce a procedure for deriving a coarse-grained dissipative particle dynamics (DPD) from MD. This is the first work to link DPD to the underlying microscopic dynamics. In DPD the forces between particles have fluctuating, dissipative, and conservative components. Momentum and mass conservation are imposed to produce hydrodynamical behavior at the macroscopic level. The parametrization method of Groot and Rabone[44] only requires the correct compressibilities and the correct solubilities of the various components into each other. These mutual solubilities are specified with Flory-Huggins χ-parameters.

The protein-folding community has long been interested in CG models to increase the efficiency of protein-structure predictions. Typically amino-acid residues are represented by one or two interaction sites, and the peptide backbone unit is represented as a single site[45,46]. The force field is parameterized from radial distribution functions computed from the Protein Data Bank (PDB). It seems at first that this data has little importance for determining potentials along folding pathways, and in particular for starting with unfolded sequences. But in fact even though the entire protein is in its folded state in the PDB archive, a particular short amino acid sequence appears in many structures in very different conformations, so that short sequences should be sampled in roughly a Boltzmann distribution across all structures[46].

Several of the methods discussed above are MC methods or can readily be adapted to a MC framework. However, current effort in MC is focused as much on improving the sampling efficiency as reducing the level of description. Sampling is enhanced by either using new ensembles and move sets[47] or by coupling individual MC chains to form a composite chain[48], as is done with simulated annealing and parallel tempering algorithms[49] .

2.3.2 Towards the Current CG Model

We now focus on methods and considerations closely related to the CG model that we will develop in the following section. A common motif employed by many simulation methods is to lump groups of atoms together into a CG interaction site. We give as examples the CG model of Fukunaga[50], the DPD model of Groot and Rabone[44], and the multiscale method of Goddard[4].

It is important not to overly distort the geometrical shape of the molecules when doing this grouping. As an illustrative example, for amphiphilic molecules the head group size compared to the average tail size normal to its length determines whether micelles or inverse micelles are preferred. Cone shaped molecules like phosphatidylethanolamine have small head groups and tend to form inverse micelles. Inverted cone shaped molecules like lysophosphatidylcholine tend to form micelles. Some CG models use anisotropic interaction sites to capture the underlying shape. One of the simplest anisotropic potentials is the Gay-Berne potential which is commonly used in liquid crystal simulations. The Gay-Berne sites are ellipsoids of revolution and their interaction potential has four parameters; the center of mass separation and three angles used to describe the relative orientation of the ellipsoids. It has been used in a lipid CG model[51] and in a united-residue protein folding model[45].

There is an additional consideration to take into account when deciding on the CG grouping, namely electrostatic partial charges. All-atom MD force fields typically have atom-centered partial charges to capture the electron-withdrawing capabilities of certain functional groups. Lumping atoms together into a single interaction site imbues that site with a partial charge which is the sum of all its constituent atomic charges. If it is felt that some regions of charge should remain in the CG model the interaction sites may have to be reduced in size or have their composition altered. One possibility is to associate several fictitious charges with an interaction site so as to reproduce the correct long range electrostatics. As an example of this we mention the peptide backbone structure. The peptide backbone partial charges give an α-helix its net dipole moment and its hydrogen bonding network. The backbone unit could be modeled as the electrically neutral -C_αH-CO-NH-site (possibly with two fictitious charges) or as two sites, -CO- and -NH-C_αH-, with equal positive and negative partial charges. The protein folding community often uses a single, uncharged site[45,46]. An alternative is to use a multipole expansion to capture the effect of an anisotropic charge distribution.

After selecting the interaction sites the bonded and non-bonded potentials which couple them must be determined. Short atomistic simulations can be used to attempt to include fine scale detail in a statistical manner. This is done by appealing to reverse MC simulation techniques[52,53] to implicitly capture fine structure. For example, the aqueous solvent shell structure around ionic or zwitterionic species involves a complicated hydrogen bonding network. Both all-atom MD[54] and experiments[55] show that zwitterionic phospholipid head groups and water together form such a network. The CG model cannot directly capture this network, but nonetheless the interfacial structure needs to be accounted for in order to arrive at a reasonable model.

One such CG method which captures fine detail is due to Fukunaga[50]. The bond and bend potentials are determined by taking the logarithm of the corresponding canonical distribution functions calculated from atomistic simulations. This requires that one forego simple functional forms for these potentials. The non-bonded potentials are also derived from atomistic canonical distributions. Since the canonical distributions are for a constant temperature ensemble simulation, the derived potentials are specific to that temperature. Fukunaga[50] parametrizes for a range of temperatures by running separate atomistic simulations.

An iterative method is presented in a seminal paper by Lyubartsev and Laaksonen[52], who show how to derive effective interaction potentials from all-atom simulations to capture ion solvation shell structure in an implicit manner without including any solvent in the reduced system. This is done by adjusting the CG ion-ion interaction potentials to match the all-atom ion-ion radial distribution functions from the atomistic simulation. The resulting effective potential has a complicated shape which is tabulated on a grid.

This parameter adjustment to match atomistic data warrants attention because it is difficult and time-consuming. In principle all of the parameters are coupled, so that adjusting any one of them is not an isolated act. The optimization space is hence of high dimension. The effect of the parameter adjustment must be appraised by simulating the CG system firstly to bring it to equilibrium with the updated parameter set and secondly to compute canonical averages. Furthermore, it is not obvious how to relate the change in input parameters to the change in output observables (to compute the gradients).

Although non-gradient optimization methods exist in the simulation literature[56], they work by estimating the gradient using repeated function evaluation. This is not feasible in the present context because of the cost of function evaluation. A gradient based optimization method[52] derived from statistical mechanics can be used to compute the full matrix of first partial derivatives, which relate the changes in input parameters to the changes in observables, from a single simulation. This method is presented in Sect. 2.3.3.

In the aqueous sodium chloride system studied by Lyubartsev and Laaksonen[52], a 200 point grid was used for each of the Na-Na, Cl-Cl, and Na-Cl interaction potentials, giving a fully coupled 600 by 600 matrix equation. This in turn means that 600 parameters are used to define the tabulated potentials. Optimization problems of this size are trivial to solve using the gradient method since they simply involve inverting a matrix. On the other hand, this problem is impossible to do with non-gradient methods, since 360 000 separate simulations would have to be performed to estimate the gradients.

2.3.3 A First Attempt

Water and Hydrocarbon. Parametrization of all-atom and united atom classical MD force fields is a highly developed area of study. These force fields are typically parameterized from a combination of experimental observables and quantum calculations. The present CG model uses these force fields and experimental observables as the raw data against which to fit.

We choose isotropic sites for computational simplicity and to keep the force field in a format that is easily implemented by standard MD codes. These sites must hence be small enough to not grossly distort molecular shape.

For aqueous amphiphilic systems we have adopted a hierarchical approach which we will now describe. This approach is quite general in nature and could easily be adapted to other systems. We begin with water. Actually, the treatment of water in the present model is somewhat complicated because water molecules are accounted for both implicitly and explicitly as will be described below. We only discuss the explicit representation here. With a Langmuir monolayer in mind we desire the CG water to be able to maintain a subcritical interface. Since we will, on average, be grouping about 10 atoms together to form CG units, we choose to represent a loose grouping of three water molecules as a single CG site. With no internal structure and no electrostatic partial charges, we need only specify the intermolecular potential to complete the CG water model. CG interaction sites always have softer potentials than their all-atom counterparts because the constituent atomic sites become smeared out under the spherical (isotropic) averaging. We choose a Lennard-Jones 6-4 potential[57] for this reason. With two (Lennard-Jones) intermolecular parameters we can at best hope to match two target observables. Grand canonical MC is used to optimize the two parameters to match the experimental bulk density and vapor pressure of water at room temperature.

We next consider hydrocarbons. These may be branched and/or unsaturated. We usually represent three consecutive carbon atoms and their respective hydrogens as a single site. Sites are connected by harmonic bond and bend potentials. The bond and bend force constants and equilibrium values are chosen so that the bond and bend distributions best match those of the corresponding all-atom simulation. The comparison can be made by grouping the all-atom data to correspond to the CG sites.

We now establish the change in bond distribution moments caused by a change in the bond parameters (force constant and equilibrium value). This relation allows for the parameter adjustment to be made optimally to first order. We assume the Hamiltonian

$$H = \frac{P^2}{2M} + V(R;a) \tag{2.1}$$

depends upon a parameter a, where the first term in Eq. (2.1) is the kinetic energy of the system and the second term is the potential energy. For the observable $A \equiv A(R)$ depending only on the coordinates, its expectation value in the canonical ensemble is given by

$$\langle A \rangle = \frac{\int dRAe^{-\beta H}}{\int dRe^{-\beta H}} \tag{2.2}$$

where β is the inverse of the product of Boltzmann's constant and the temperature, so that

$$\frac{\partial \langle A \rangle}{\partial a} = -\beta \left[\left\langle A \frac{\partial H}{\partial a} \right\rangle - \langle A \rangle \left\langle \frac{\partial H}{\partial a} \right\rangle \right] \tag{2.3}$$

This is the fundamental relation which links the change in input parameter to the change in output observable. We will now illustrate the gradient method with the simplest example: parameterizing an intra-molecular harmonic bond potential in the canonical ensemble. A harmonic potential has two parameters. We choose as our two observables the mean and standard deviation of the bond length, which can be obtained from the quantities $\langle r \rangle$ and $\langle r^2 \rangle$ where the angle brackets are with respect to the canonical ensemble. We write

$$V = \sum_{ij}{}' f_{\alpha\beta}\left(\left|q_i^\alpha - q_j^\beta\right|\right) + V' \tag{2.4}$$

where the prime on the sum denotes that the $\{i, j\}$ pair in question must be joined with a bond. The bond we are considering is between species α and β. V' represents all the other intra-molecular and all the inter-molecular potential energy terms in the Hamiltonian, and

$$f_{\alpha\beta}(r) = \frac{k_{\alpha\beta}}{2}\left(r - r_{eq}^{\alpha\beta}\right)^2 \tag{2.5}$$

is the harmonic potential. With our two parameters being k and r_{eq} and our two observables being the mean and standard deviation of the bond length, we need

$$\frac{\partial\langle r_{\alpha\beta}\rangle}{\partial r_{eq}^{\alpha\beta}} \quad \text{and} \quad \frac{\partial\sqrt{\langle r_{\alpha\beta}^2\rangle - \langle r_{\alpha\beta}\rangle^2}}{\partial k_{\alpha\beta}} \tag{2.6}$$

as the diagonal elements in the matrix of first partials. There must be a clear match between observables and parameters on the diagonal of this matrix for the method to work effectively. We could, for instance, have chosen to use the mean squared bond length instead of the standard deviation, but the mean squared length and the stiffness k are not obviously related. If we were using derivatives of arbitrarily high order this would not matter, but the truncation to linear order makes these choices inequivalent. The standard deviation expression of (2.6) must be expanded into terms involving $\langle r_{\alpha\beta}\rangle$ and $\langle r_{\alpha\beta}^2\rangle$, and then (2.3) used with $A = r_{\alpha\beta}$ and $A = r_{\alpha\beta}^2$. The $\partial H/\partial a$ term in (2.3) is just $\partial f_{\alpha\beta}/\partial r_{eq}^{\alpha\beta}$ and $\partial f_{\alpha\beta}/\partial k_{\alpha\beta}$ with f given in (2.5).

No torsional potentials are implemented for the hydrocarbon model. Saturated straight chain hydrocarbons are quite floppy; the torsional angle distribution over four CG sites would be featureless. If the branching and unsaturation in particular cases warrants including a torsional potential this could be done. The non-bonded potential is expected to be soft and roughly the same for different hydrocarbons. We take it to be a Lennard-Jones 9-6 potential. The well depth and contact distance are adjusted so as to reproduce the experimentally determined vapor pressure and bulk density at room temperature for a few typical bulk alkane systems.

The systematic adjustment of the parameters is not easily done for the grand canonical MC simulations, although the small number of free parameters keeps the problem from becoming too unwieldy. The difficulty is that the partial derivative of the observable (eg. vapor pressure) with respect to an adjustable parameter (eg. Lennard-Jones well depth) is difficult to derive. The iterative parameter refinement must therefore be done either with simplex optimization[56] or by trial and error.

Once the pure water and pure hydrocarbon systems have been parameterized, these parameters can be used in more complicated systems without being subject to further modification. This reduces the number of free parameters in complex systems. In this spirit, we obtain one more set of parameters: the hydrocarbon-water interaction potential. Towards this end hydrocarbon-water CG systems are simulated. The interaction potentials are chosen to be of Lennard-Jones 9-6 type and the parameters are selected to obtain phase separation and reasonable width for the hydrocarbon-water interface[57].

Amphiphilic Systems. We are now ready to parametrize an aqueous amphiphile system. We will focus on the phospholipid dimyristoylphosphatidylcholine (DMPC), but the strategy used can be applied to other systems. The DMPC molecule is coarse grained using 13 sites to represent the 118 atoms as shown in Fig. 2.2. An all-atom simulation[58] of an equilibrated DMPC bilayer in the L_α phase is used to parametrize the CG model. The CG system that we will calibrate is also prepared as a bilayer so that we are treating the same thermodynamic phase. The intra-molecular force field is parametrized with harmonic functions by matching bond and bend distributions with the corresponding distributions from the all-atom simulation. The acyl tails of the DMPC lipid are straight alkane chains and the non-bonded parameters for the tail units interacting with other tail units or water are taken from the already determined hydrocarbon-water parameters.

The lipid head groups are coarse grained into a positively charged choline site, a negatively charged phosphate site, a glycerol site, and two ester sites which have the two acyl tails attached to them (see Fig. 2.2). All combinations of non-bonded pairwise interactions among these head groups are modeled with tabulated effective potentials which aim to reproduce the radial distribution functions from the appropriately grouped all-atom simulation data (see Fig. 2.3 for examples of such effective pair potentials). These head group – head group interactions implicitly include water solvation shell structure. However, matters are not as straightforward as in the implicit solvent model of Lyubartsev and Laaksonen[52] because our solvent is also present explicitly. The explicit water serves as a momentum carrier (in MD simulations) and is desirable for dynamical studies. Furthermore, the Lennard-Jones nature of the potential for the explicit water site has an attractive well region which allows it to maintain an interface. Nonetheless, the parameterization strategy is unchanged: the tabulated potential attempts to mimic the all-atom reference radial distribution function in the fully interacting CG system. This means that the magnitude of the tabulated potentials are reduced compared to the case when no explicit solvent is present.

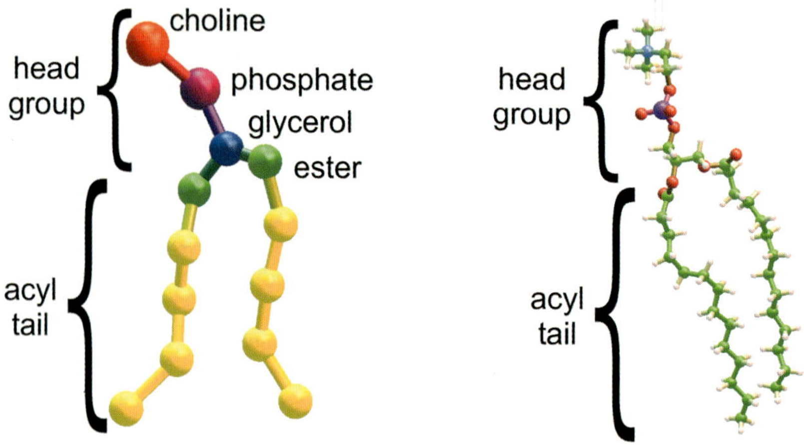

Fig. 2.2. Thirteen site model of DMPC (also denoted as $DC_{14}PC$). The choline and phosphate sites carry positive and negative electrostatic charges, respectively, of equal magnitude. The all-atom version is also shown.

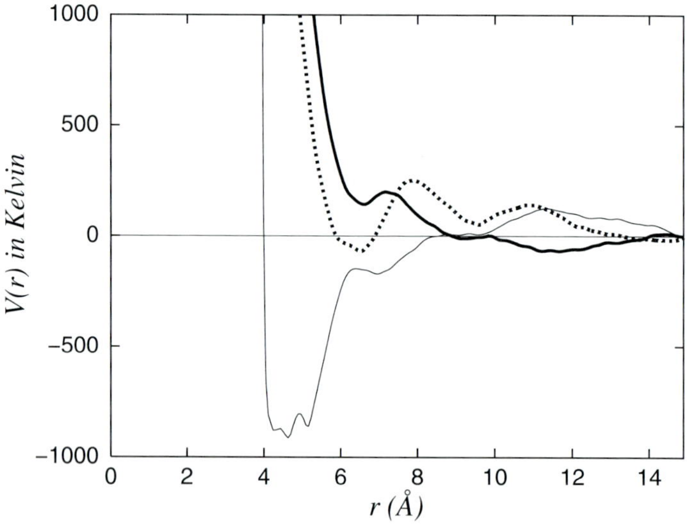

Fig. 2.3. Examples of non-bonded, tabulated, potentials in the CG force field for hydrated DMPC lipid. In thin solid line is the choline/phosphate potential, in thick solid line the choline/choline potential, and in thick broken line the phosphate/phosphate potential. The line of zero potential is drawn to guide the eye.

By using a tabulated potential each grid point is taken to be a free parameter, so that the number of parameters is large. We now show how to account for all the parameters and the coupling between them with the gradient method of Lyubartsev and Laaksonen[52]. We write

$$V = \sum V_{\alpha\beta}\left(\left|q_i^\alpha - q_j^\beta\right|\right) + V' \tag{2.7}$$

where $V_{\alpha\beta}$ is the non-bonded (non-electrostatic) potential between sites α and β and V' represents all the other potential energy terms. We choose to write

$$V_{\alpha\beta}\left(\left|q_i^\alpha - q_j^\beta\right|\right) = \int_0^\infty dr V_{\alpha\beta}(r)\delta\left(r - \left|q_i^\alpha - q_j^\beta\right|\right) \tag{2.8}$$

where δ is the Dirac delta function, and define[52]

$$S_{\alpha\beta}(r) = \sum_{ij} \delta\left(r - \left|q_i^\alpha - q_j^\beta\right|\right) \tag{2.9}$$

Note that

$$\langle S_{\alpha\beta}(r)\rangle = g_{\alpha\beta}(r)4\pi r^2 \tag{2.10}$$

where $g(r)$ is the radial distribution function which we choose as our observable. We can now write

$$V_{\alpha\beta} = \sum_{\alpha\beta} \int_0^\infty dr V_{\alpha\beta}(r) S_{\alpha\beta}(r) \tag{2.11}$$

which allows us to link the observable $g_{\alpha\beta}(r)$ with the parameter $V_{\alpha\beta}(r)$. We have

$$\frac{\partial\langle S_{\alpha\beta}(r_0)\rangle}{\partial V_{\gamma\xi}(r_1)} = -\beta\left[\langle S_{\alpha\beta}(r_0) S_{\gamma\xi}(r_1)\rangle - \langle S_{\alpha\beta}(r_0)\rangle \langle S_{\gamma\xi}(r_1)\rangle\right] \tag{2.12}$$

since it turns out that $\partial H/\partial V_{\gamma\xi}(r) = S_{\gamma\xi}(r)$

The following set of coupled linear equations is solved for $\Delta V = V^{old} - V^{new}$,

$$\langle S_{\alpha\beta}(r_0)\rangle - \langle S^*_{\alpha\beta}(r_0)\rangle = \sum_{\gamma,\xi,r_1} \frac{\partial\langle S_{\alpha\beta}(r_0)\rangle}{\partial V_{\gamma\xi}(r_1)} \Delta V_{\gamma\xi}(r_1) \tag{2.13}$$

where $\langle S^*(r)\rangle = g^*(r)4\pi r^2$ is the observable from the atomistic simulation which is our target. The r values in practice are defined on a regular grid. The matrix of derivatives needs to be inverted to obtain the solution.

For the isothermal-isobaric ensemble the partition function has an additional factor of $\int dV e^{-\beta PV}$ and the above analysis follows through in a completely analogous fashion[59].

The observables selected to correspond to parameters (namely, the diagonal entries in the matrix of first partial derivatives) should be simple enough to evaluate the derivatives explicitly. One example which would be harder to handle would be the use of c, the Ornstein-Zernike direct correlation function, instead of the full correlation function $h \equiv g - 1$. The advantage of using the direct correlation function to parameterize a non-bonded pair interaction is that it avoids the ambiguity having oscillations in the tails of the various interaction potentials which cancel each other out (and hence need not exist). The full correlation function is conceptually and pragmatically difficult to relate to the effective non-bonded potential because of many-body effects[50].

The lipid head group non-bonded interaction with both hydrocarbon and water is taken to be of Lennard-Jones form and parametrized so as to roughly match the integrated radial distribution function (in the bilayer configuration) of the two species in question out to a few selected distances[57]. This simple functional form is chosen because these interactions are of lesser importance and we only attempt to roughly capture them.

Some of the parameterization strategy just described builds specific L_α bilayer structure directly into the force field. Furthermore the derived force field is only valid for a small temperature range as discussed in Sect. 2.3.1. This is part of the tradeoff in moving to a more efficient simulation method; generality is sacrificed. However, the situation is not as limiting as it first appears. Only the lipid head group – head group interaction potentials contain explicit bilayer information since these were the only potentials that were tabulated to reproduce the thermodynamic phase-specific atomistic radial distribution function data. The rest of the interaction potentials are quite general. Furthermore, the enthalpic lipid tail and entropic changes which occur when the lipid/water system is in a different phase can partially override the structure inherent in the non-bonded potentials. Ongoing studies using this L_α bilayer derived force field include Langmuir monolayers and inverse hexagonal phases[60]; the results (eg. the surface tension of Langmuir monolayers) are encouraging and agree semi-quantitatively with experiments.

Efficiency over All-Atom MD. Work on the current CG model began by using MC as the sampling technique with a simple move set[57,61]. Large scale organization and healing were observed to be slow. For example, inverse hexagonal self-assembly did not show global structure over the MC simulation run, and defects in the self-assembly of a lipid bilayer (see Sect. 2.4.3 for more discussion) did not heal over the length of the simulation[61]. This problem would be alleviated by using a more sophisticated MC move set. Instead, MD was appraised as an alternative simulation technique. The use of MD was observed to be very efficient. Besides the cooperative global moves which occur naturally in MD, the explicit water particles act as momentum carriers and allows for material to flow in the system in a hydrodynamically consistent manner that aids global rearrangements. The efficiency of the CG MD method over all-atom MD will now be described.

The softer interaction potentials allow the use of a one order of magnitude larger propagation time-step. The reduced number of interaction sites and potentials between them yield another two orders of magnitude speedup. In the case of DMPC[57], the CG model consists of 13 sites and 24 internal potentials (12 bonds and 12 bends). The all-atom CHARMM DMPC encompasses 118 atoms and 971 internal potentials (117 bonds, 226 bends, 315 torsions, and 313 one-fours). A further two orders of magnitude efficiency gain comes from enhanced diffusion of the lipid species, for example in the plane of a bilayer or Langmuir monolayer. This is a result of the soft interaction potentials and the lack of an explicit hydrogen bonding network at the interface between lipid head groups and water. We have quantified this diffusional speedup[62,63] as follows. The two dimensional diffusion constant for L_α phase DMPC in the plane of the bilayer is 6.5×10^{-8} cm^2/s for an all-atom simulation and 6.3×10^{-6} cm^2/s for the CG model.

A drawback to the use of MD, however, is the use of effective potentials, which are related to potentials of mean force. A memory kernel should be used to properly implement the dynamics[52]. We do not use memory functions, but we do extract translational and rotational diffusion constants[62,63] for CG lipid monolayers and bilayers and compare these with all-atom results to quantify the ordering of relaxational time scales.

2.4 Applications

We begin by discussing some recent simulation work which is motivated by experimental and theoretical results on entropic forces in amphiphilic systems. We then proceed to illustrate the range of applicability of the current CG method by presenting some recent results. These results include amphiphile self-assembly, transmembrane peptide induced domain segregation and phase transitions to inverted phases, and instabilities in Langmuir monolayers.

2.4.1 Fluctuation Modes

The forces between fluid amphiphilic surfaces arise from both entropic and enthalpic factors. The enthalpic contributions arise from the two forces which comprise the Derjaguin Landau Verwey Overbeek (DVLO) theory of colloid science. These are the attractive van der Waals and repulsive electric double-layer forces, the later being present only for charged amphiphiles. The entropic contributions[7] are repulsive and arise from the overlap of thermally excited surface modes. Two of these are the undulatory and peristaltic (or squeezing) modes. These two modes are most easily visualized in the case of a hydrated bilayer. By locally orienting the bilayer to lie in the xy plane, the position of the two bilayer leaflets at a particular xy location can be taken to be h_ℓ and h_u (e.g. these locations could be the center of mass of the head groups or the center of mass of the amphiphiles). The undulatory mode is given by the fluctuations in $h_\ell + h_u$ while the peristaltic mode is given by the fluctuations in $h_\ell - h_u$ as a function of x and y. For uncharged amphiphiles the dominant interactions between amphiphilic surfaces at distances greater than 1 nm are due to the repulsive undulatory and attractive van der Waals forces[10]. Both scale as the inverse cubic distance between surfaces. The van der Waals pressure scales as[7]

$$P_{vdW} \propto -\frac{A}{6\pi D^3} \tag{2.14}$$

where A is a constant and D is the separation between surfaces. The undulatory pressure scales as[10]

$$P_{und} \propto \frac{1}{\beta^2 K_b D^3} \tag{2.15}$$

where K_b is the elastic bending modulus. We will focus on uncharged amphiphiles for the remainder of this section, but wish to point out that recent simulation work[64] has obtained results which differ qualitatively from the electrostatic predictions of DVLO theory. Moreover, it is thought[7] that electrostatic charges suppress undulatory and other fluctuations.

Simulations are beginning to evaluate the validity of these scaling predictions at the microscopic level. However, recent work has focused on the thermal fluctuations within a single surface. The best known example of these intra-surface fluctuations are capillary waves[23], which in interfacial monolayers are predicted to broaden the interface width σ due to thermal excitations as

$$\sigma^2 = \frac{1}{\beta 4\pi^2 \gamma} \int_0^\infty dq \frac{q}{q^2 + \kappa^2} \tag{2.16}$$

where q is the wavenumber, γ is the surface tension, and κ accounts for the presence of the gravitational field. This expression diverges logarithmically with system size and an upper cutoff must be invoked. Typically this cutoff is taken to be the bulk correlation length of the liquid. Logarithmic divergence is weak; for a typical system[6], a surface of extent L = 100 Å has a root mean square (rms) fluctuation of about 1.5 Å, while one with L = 1 cm has an rms fluctuation of only about 7.5 Å.

Since the capillary fluctuations are small, simulations have studied lipid bilayers in the lamellar phase, where the fluctuations are more pronounced. In this case the main entropic fluctuation force is due to the bending of the bilayer. The mean-square height fluctuation grows algebraically with the system size[6]

$$\sigma^2 \propto \frac{T}{K_b} L^2 \tag{2.17}$$

where T is the temperature, K_b is the bending modulus of the bilayer, and L is the planar spatial extent of the system. This prediction of much faster divergence is accessible to and has been tested by the simulation community using both CG and united atom models[25–27]. The elegant study of Marrink[26] examines the role of lateral tension on the undulations which has also been addressed experimentally[65]. In addition to bending fluctuations, bilayers undergo peristaltic fluctuations. The simulations of Marrink[26] clearly show the expected long-wavelength suppression of the peristaltic modes since these fluctuations cannot exceed the mean width of the bilayer.

2.4.2 Bulk Alkane and Water Surface Tension

The CG water and hydrocarbon non-bonded parameters were determined on the basis of experimental bulk density and vapor pressure data as described in Sect. 2.3.3. How accurately do these compounds reproduce experimental surface tensions at the liquid/vapor interface? This is assessed[66] by preparing a slab of liquid with vacuum at either end[67]. The surface tension is computed by averaging

$$\gamma_{inst} = \frac{L_z}{2}\left(P_{zz} - \frac{P_{xx} + P_{yy}}{2}\right) \tag{2.18}$$

over the length of the simulation, where γ_{inst} is the instantaneous surface tension. The first factor of 1/2 is included to account for the two interfaces in the simulation box, L_z is the box size in the z-direction, and P_{ij} is the ij component of the pressure tensor. The air/liquid interfacial plane is the xy plane.

The CG model gives a surface tension at the air/water interface of 72 $\pm$ 1 dyne/cm at 303.15 K. This value agrees with the experimental value of 71.18 dyne/cm. For hydrocarbons at 303.15 K, we obtain a surface tension of $\gamma = 18.0 \pm 1.5$ dyne/cm for nonane and $\gamma = 17.5 \pm 2.0$ dyne/cm for dodecane. The experimental value[68] for the surface tension of nonane is 22.01 dyne/cm; our model is reasonable but could use minor improvement in this respect. Although experimental data was used to parameterize the CG force field, atomistic simulation results could have been used since they can accurately reproduce the surface tension of water[69] and n-alkanes[70].

2.4.3 Self-assembly

Amphiphilic self-assembly is well established for generic model systems using coarse grain simulation techniques[71,72,38,73–77] and is the subject of recent atomistic studies[78,79]. We have studied many self-assembly processes from uniformly random initial conditions using the current CG model, and will present a representative selection of them to demonstrate the potential and limitations of the model.

At 303.15 K and with a water to lipid ratio of about 25:1, the thermodynamic state of DMPC is the L_α bilayer phase. Such a system with 64 lipids in

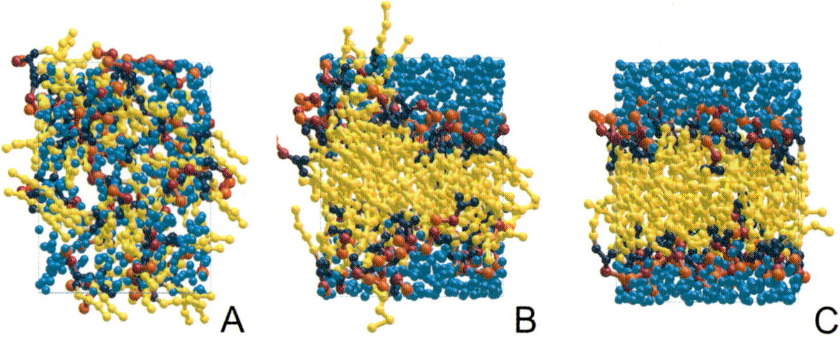

Fig. 2.4. Bilayer self assembly. A random initial condition (panel A) consisting of 64 DMPC lipids and 548 water sites (representing three water molecules each) assembles to its known thermodynamic state (panel C, the L_α phase) after passing through an intermediate state (panel B) with some defects. Water is colored light blue; lipid acyl tails yellow, lipid head groups red, blue, and purple.

the orthorhombic simulation cell is prepared randomly as shown in Fig. 2.4A. From this setup, a MC simulation with a simple move set assembles into a bilayer structure with some defects[61] as shown in Fig. 2.4B. Changing to MD causes the defects to heal quickly (within 6 ns) to the final state as shown in Fig. 2.4C.

A monolayer self assembly process is studied by randomly placing lipid and water molecules in a slab geometry with two air/liquid interfaces (Fig. 2.5A). The system self-assembles within 300 ps into two Langmuir monolayers and a cylindrical micelle in the bulk water region (Fig. 2.5B). The micelle drifts towards and fuses with one of the monolayers[63] within 1.5 ns, giving a final configuration of two unequally populated monolayers (Fig. 2.5C).

As will be discussed in Sect. 2.4.5, inverted phases present an anisotropic environment for the lipid species. To stabilize them, hydrocarbon may be added as a "filler". Such an inverse hexagonal phase was self-assembled[61] from a random initial condition consisting of 50.5 weight percent water, 33.8 weight percent diheptadecanoylphosphatidylcholine, and 15.7 weight percent nonane, as depicted in Fig. 2.6. This composition is experimentally[80] known to form the inverse hexagonal phase.

The three self-assembly processes that we have described demonstrate both the efficiency and power of our CG method. All-atom MD simulations must be prepared with the utmost care because the time scales accessible are such that it is prohibitively expensive to equilibrate a system not already almost at equilibrium. This situation is most problematic for transmembrane peptide oligomeric bundles since the orientation of the peptides with respect to one another is usually not known. The initial mutual peptide orientation

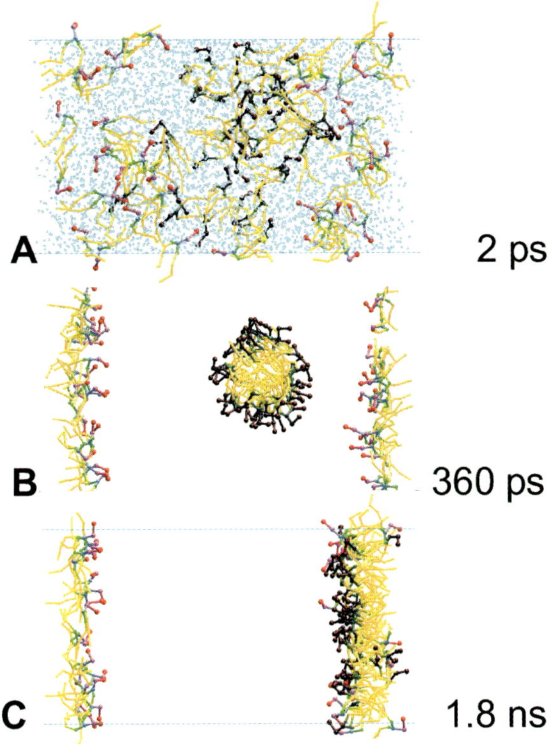

Fig. 2.5. Langmuir monolayer self assembly. A random slab initial condition (panel A) consisting of 80 DMPC lipids and 5000 water sites organizes into two Langmuir monolayers and a cylindrical micelle (panel B). The micelle then fuses with one of the monolayers (panel C). Water is shown in light blue and is omitted from panels B and C for clarity. The lipid acyl tails are colored yellow. The micellar lipids of panel B are shown with dark head groups in all three panels. The remaining head groups are colored red, purple, and green.

selected for an all-atom MD run will typically not change over the lifetime of the run because the relevant time scale is not accessible. The CG self-assembly processes presented here demonstrate that the preparation of the initial condition is largely immaterial. This both relieves the burden of system preparation and ensures that more of phase space is sampled over the course of the simulation.

However, the periodic boundary conditions and small simulation cell size may result in the stabilization of metastable structures. Care should be taken to assert whether or not the final assembled structure is the thermodynamic state. We have observed cases where an inverted phase forms when the thermodynamic phase is known to be lamellar (unpublished results).

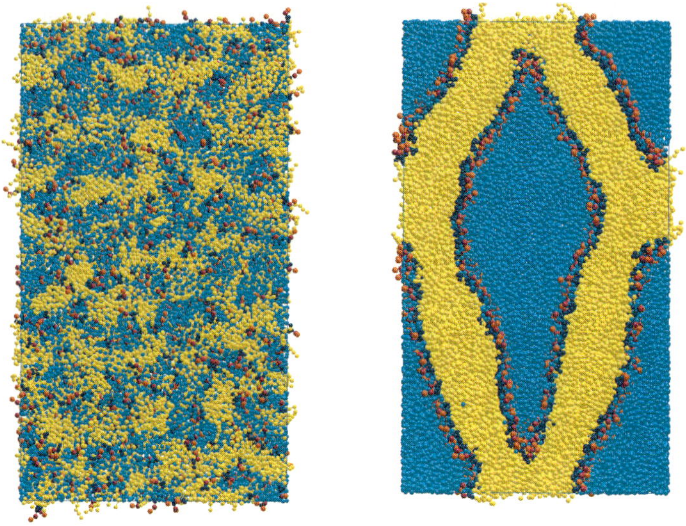

Fig. 2.6. Inverse hexagonal self assembly. A random initial condition (first panel) consisting of 738 DHPC lipids, 1968 nonane molecules, and 15060 water sites assembles into an inverted hexagonal phase (second panel). Water is colored light blue; lipid acyl tail and nonane yellow; lipid head groups red, blue, and purple.

Furthermore, the initial dynamics for the MD self assembly simulations should not be considered in detail. The instantaneous pressure is extremely high due to the large hydrocarbon-water interface and the force field was developed under equilibrium conditions. However, the gross features of the initial dynamics are correct. The dominant effect is for the system to minimize the extent of the hydrocarbon-water interface.

2.4.4 Transmembrane Peptide Induced Domain Formation

Membrane lipid composition varies widely over different organelle membranes and varies spatially within a single membrane. Such local variations in bilayers allow for membrane deformation and facilitate vesicle budding and fusion. Proteins might stimulate lipid exchange between membranes by bringing them into contact. Newly synthesized proteins and lipids destined for different organelles are laterally segregated in the Golgi membrane. It has been proposed[18] that lipid domains of different hydrophobic thickness are used to sort membrane proteins that are destined for the plasma membrane from Golgi proteins based on the length of their transmembrane domains. But this picture is misleading because it is known that proteins themselves passively induce lipid domain formation on the basis of the mismatch between the hydrophobic width of the lipid bilayer and that of the peptide[81]. What is the mechanism for this? It is related to meniscus formation, which we now discuss.

The typical picture of the bilayer-spanning part of a transmembrane peptide is of a hydrophobic alpha helix with caps that preferentially associate with the interfacial lipid head group – water region. Membrane lipids also have a hydrophobic region consisting of the conjunction of their acyl tails. Lipid bilayers are more easily deformed than alpha helical transmembrane proteins[82], and the assumption is made that the bilayer deforms to match the hydrophobic length of the protein[83], leaving the protein virtually unchanged. Theoretical considerations point to a range of effects that contribute to the free energy in the presence of such a peptide inclusion. These consist of elastic acyl chain stretching/compression, surface tension terms accounting for the average interfacial area per molecule, curvature contributions from the formation of a meniscus around the inclusion, and tilt modulus of the acyl chains[12,11,13,84,85].

Peptides whose hydrophobic core is of a different length than the hydrophobic membrane thickness induce meniscus formation. This leads, among other effects, to a lipid-mediated peptide-peptide force in the system[83,84,86–92]. Even in the absence of direct interactions between the inclusions, a fluctuation-induced tendency to aggregate exists[92]. Consider two short peptides sitting at some distance apart in the same bilayer. Each peptide will have a negative meniscus of lipids around it. There are energetic penalties the systems pays for these menisci such as curvature stress and acyl chain compression. These penalties are minimized if the two peptides come together since the system then comprises one fewer meniscus. This is the origin of the lipid-mediated force, which is attractive for this case and for the case of two long peptides. The lipid-mediated force is expected to be repulsive in the case of one long and one short peptide for the same reason[84]; the curvature associated with the menisci is greatest if the two dissimilar peptides are in close proximity.

Of current interest in the literature is the effect of a transmembrane peptide on the distribution of lipids in the bilayer in which the peptide is embedded[81]. The hydrophobic matching principle states that, in the immediate vicinity of the peptide, there is an accumulation of the lipid which is hydrophobically best matched[93].

This simple picture is complicated by the possibility of the lipid species being in phases of different fluidity. The gel to liquid crystal phase transition temperature in a phospholipid increases with increasing acyl chain length. Experimentally, then, one can choose, for a mixed lipid bilayer consisting of two different tail length lipids, to be in the gel-gel coexistence region, the gel-liquid region, or, at still higher temperature, the liquid-liquid region. Domain formation in mixed bilayers (in the absence of transmembrane peptides) is usually attributed to the formation of islands of the gel phase[94] of the longer tail lipid in the gel-liquid region of the phase diagram. We wish to study the effect of a membrane inclusion on an otherwise well-mixed system[11,12] –

hence we wish to be in the liquid-liquid region of the phase diagram where domain formation is not observed[93].

The liquid crystal to gel phase transition temperature for the phospholipid model we employ (see Sect. 2.3.3) occurs below the temperatures used in the simulations we have conducted. Thus, we are able to assess the effect of hydrophobicity in isolation from the gel/liquid crystal phase transition. Moreover, the peptide model we employ is explicitly chosen to be stiff and nonspecific, simply consisting of a hydrophobic cylinder with hydrophilic caps[95] as depicted in Fig. 2.7.

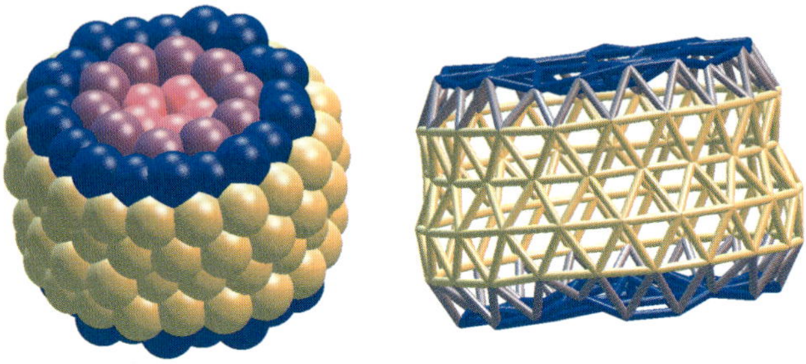

Fig. 2.7. Snapshot of the peptide model showing the van der Waals radius and the skeletal bonding structure. The peptide consists of a hydrophobic cylinder capped with hydrophilic sites. The outer, middle, and inner capping rings are composed of identical sites, but have been colored differently (blue/purple/pink) to aid the eye.

We have characterized the thickness and profile of a bilayer consisting of a mixture of lipids[95] in the absence of peptide. Will will only mention one case here, namely an equal mixture of $DC_{29}PC$ (long lipid) and $DC_{11}PC$ (short lipid), shown in Fig. 2.8. The notation is as follows. We refer to a diacylglycero-phosphatidylcholine with n carbons in each acyl chain as DC_nPC. For example, DMPC is denoted by $DC_{14}PC$. There are two features of note. The first is that the head groups all occupy the same region of space normal to the bilayer plane. This shows that the two lipids are miscible in one another. The second is the hydrocarbon tail density. The driving force for the bilayer width is the lipid tail packing. The tail density for the short lipids is depleted in the middle of the bilayer (see Fig. 2.8). The long lipids fill the hydrocarbon region up to its bulk density value.

A system consisting of 208 short lipids, 208 long lipids, and 10400 water sites is initially prepared in a patchwork fashion as shown schematically in Fig. 2.9. Since we know the head group profile normal to the bilayer plane, we characterize the extent of mixing with two-dimensional radial distribution functions in the plane of the bilayer.

The patchwork initial configuration of the two lipids melts and the two lipids mix. The time scale for this mixing process is determined by the lat-

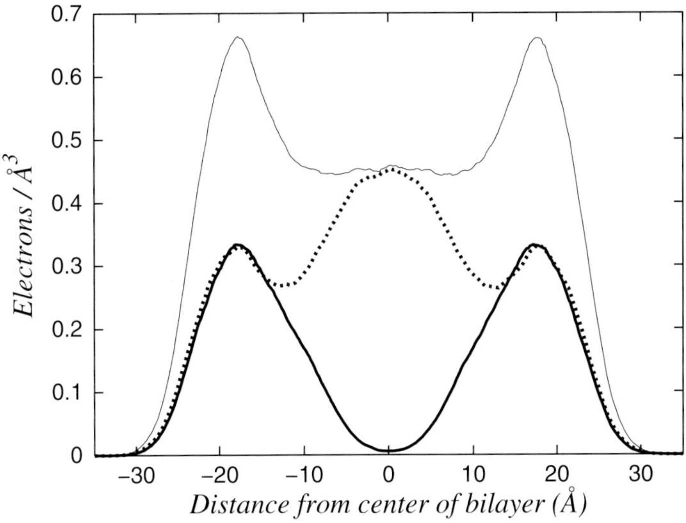

Fig. 2.8. The equilibrium lipid electron density profile normal to the bilayer plane for a mixed lipid system (shown in thin line) consisting of 80 short lipids (shown in thick solid line) and 80 long lipids (shown in thick broken line). In all three profiles, the peak around ±18 Å is due to the lipid head groups.

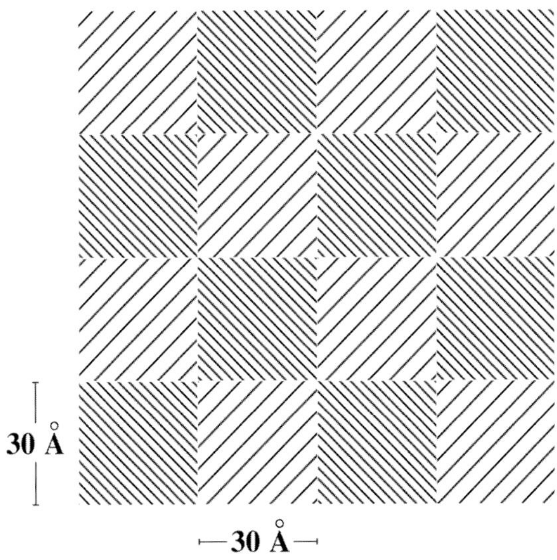

Fig. 2.9. Schematic of the patchwork initial condition for the mixed lipid bilayer simulation consisting of 208 short lipids, 208 long lipids, and 10400 water sites. Each patch contains only one of the two species of lipids. The plane of the bilayer is shown – the two leaflets are symmetrical.

eral diffusion of lipids in the plane of the bilayer. The extent of miscibility is quantified by calculating the three head group – head group radial distribution functions in the plane of the bilayer. They clearly mix as shown in Fig. 2.10. In fact, the radial distribution functions show that, if anything, the long and short lipids associate preferentially with each other rather than with themselves.

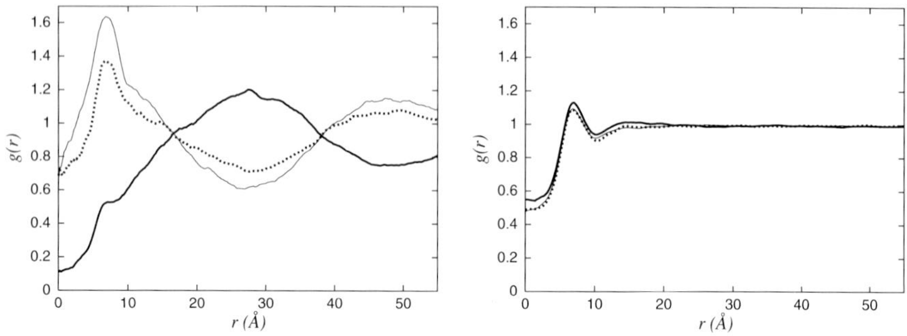

Fig. 2.10. Two-dimensional lipid-lipid radial distribution functions (rdfs) in the bilayer plane for the simulation whose initial condition is shown in Fig. 2.9. The first panel shows the initial distribution taken from the first 50 ps of the simulation. The second panel shows the equilibrated distribution taken from the last 1 ns of the MD simulation. The short lipid – short lipid rdf is shown in thin line, the long lipid – long lipid rdf in thick broken line, and the short lipid – long lipid rdf in thick solid line. The lipid location is taken to be the center of mass of the head group. Both leaflets are included, which is the reason the distributions do not go to zero at zero (projected) separation.

The short peptide depicted in Fig. 2.7 is inserted into this mixed bilayer. The hydrophobic length of the peptide is smaller than the hydrophobic thickness of the bilayer. The concentration of lipids around the tube is initially slightly enriched in the long species as shown in the first panel of Fig. 2.11. This is done deliberately so as to not bias the possible domain formation since we eventually expect the short lipid to collect around the tube. When incorporated into the membrane, a meniscus forms in the vicinity of the peptide. This meniscus is depicted in Fig. 2.12. The lipids residing next to the peptide, at a distance of roughly 15 Å from its center (see Fig. 2.12), are maximally perturbed from their equilibrium position. The lipids closer to the peptide are farther away from the bilayer center because they reside with their head groups immediately above (for the upper leaflet) the hydrophilic peptide cap, with their acyl tails bent to flank the length of the hydrophobic core of the peptide. After 10 ns of simulation, domain formation is clearly seen. The lipid species are distributed around the peptide as shown in the second panel of Fig. 2.11. The region within 30 Å of the tube center is enhanced in the short lipid species.

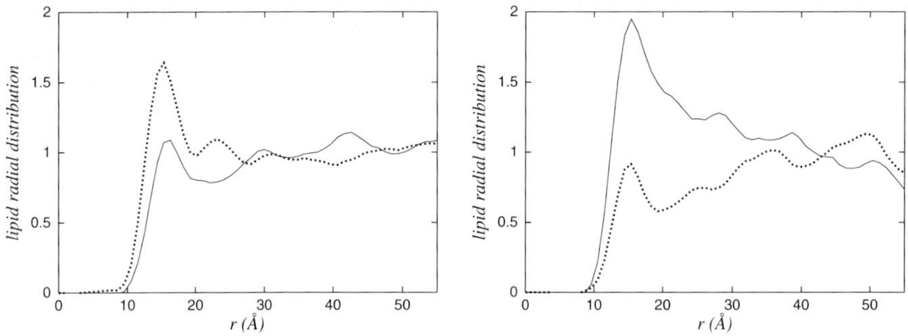

Fig. 2.11. Two-dimensional radial distribution of lipids around the transbilayer peptide. The first panel shows the distribution immediately after the the peptide is inserted. The insertion location was chosen to correspond to a region of local enhancement of the long lipid species. The second panel shows the equilibrated distribution taken from the last 1 ns of the MD simulation. The short lipid is denoted by solid line, and the long lipid by thick broken line. The lipid location is taken to be the center of mass of the head group. The peptide location is taken to be its center of mass. Both leaflets are included.

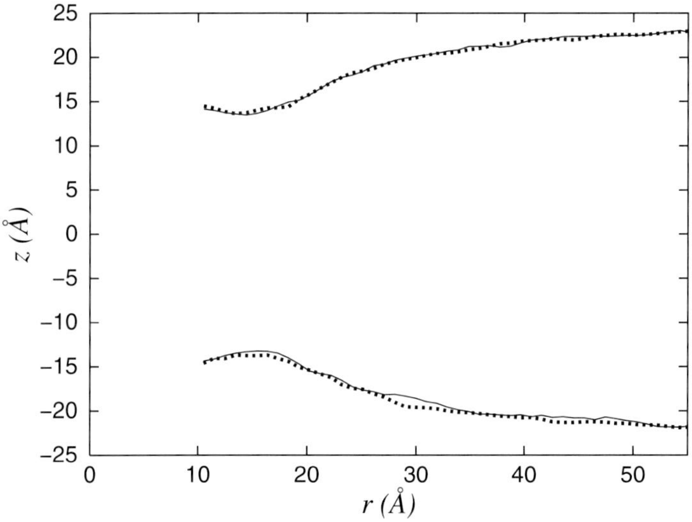

Fig. 2.12. Extent of the lipid meniscus formed around the transbilayer peptide. The radial distance r of lipids from the peptide is plotted against the distance z of lipids from the peptide in the direction normal to the bilayer plane. The short lipid is denoted with solid line and the long lipid by thick broken line. The two leaflets are shown by positive and negative values. The lipid position is taken to be the center of mass of the head group. The peptide position is taken as its center of mass.

This domain formation induces a non-homogeneous lipid concentration field around the peptide. In a lattice Monte Carlo simulation[85] it was found that if two such regions with non-homogeneous concentration fields overlap, an attractive force between the peptides results. The pair potential for this indirect force can be obtained from the free energy of the system, which depends on the distance between the peptides. We are currently pursuing this avenue by placing more than one peptide in the simulation cell.

2.4.5 Transmembrane Peptide Induced L_α to H_{II} Phase Transition

In the previous section we saw that the presence of transmembrane peptides in the lamellar lipid phase can induce meniscus and domain formation. At high concentration, transmembrane peptides can have more dramatic effects on the membrane structure. Short peptides (namely peptides whose hydrophobic length is shorter than the bilayer thickness) have been shown to induce phase transitions from the lamellar to an inverted phase (see Fig. 8 of Killian[96] for an illustration). Long peptides do not have this effect[96] – they tilt instead and the lipid remains in the lamellar phase. If the peptide hydrophobic length is grossly mismatched from the membrane thickness, the peptide may actually fail to insert[82].

For the tryptophan capped synthetic peptides used experimentally, the two dominant peptide/lipid interactions are thought to come from the hydrophobic length of the peptide and the tryptophan anchoring residues[97]. The bulky multiple tryptophan groups at each end of the peptide are thought to play a dual role in anchoring the peptide ends in the head group region of the bilayer, and in preventing peptide aggregation. In simulation work the possibility of aggregation can be entirely avoided by considering only a single peptide per unit cell. This, combined with an orthorhombic cell geometry, limits the phase transition obtainable to an inverse cubic phase, rather than than the experimentally observed inverse hexagonal (H_{II}) phase[96]. Furthermore, our peptides (see Fig. 2.7) do not have special chemistry to anchor the ends of the peptide in the head group region. We are thus able to assess the effect of hydrophobicity in isolation from these other competing effects which make the experimental work difficult to interpret.

Both the existing experimental and theoretical studies to date do not shed light on the dynamical mechanism which could drive such a phase transition. Rather, the stability of the resulting H_{II} phase is justified *a posteriori*[13,96]. We propose, by use of CG MD, a dynamical mechanism wherein the transbilayer peptide induces the formation of a meniscus (see Fig. 2.13) which causes water to move into the meniscus region. This movement depletes the water layer far from the peptide, allowing the bilayers in these regions to fuse. This fusion event pinches off the water sheets, trapping the water in cylindrical pores which are the hallmark of inverted phases (see Fig. 2.13). Figure 2.14 shows the final structure from our simulation[95]. This simulation has as its initial condition a configuration identical to that described in Sect. 2.4.4 immediately after the peptide is inserted into the bilayer. Two modifications are then made before the simulation is run. Firstly, the water to lipid ratio is reduced to 25:1 which is the normal bilayer L_α hydration level. Secondly, the unit cell size is reduced from containing roughly 400 lipids to having 200 lipids. This effectively increases the peptide to lipid ratio. A complete study of the peptide to lipid dependence is forthcoming[95].

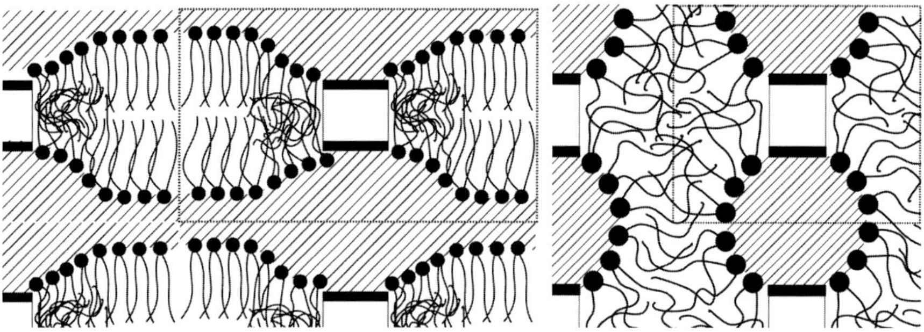

Fig. 2.13. Schematic illustration of the dynamical mechanism of the peptide-induced lamellar (first panel) to inverse cubic (second panel) phase transition. The orthorhombic simulation unit cell is shown as a striped rectangle. The hydrophobic core of the peptide is shown in white, the hydrophilic caps are shown in black, and water is indicated by diagonal stripes. See text for details.

To have the possibility of forming an inverse hexagonal phase instead of an inverse cubic phase, it is necessary to include more than one peptide in the orthorhombic unit cell. This leads to additional complications due to the possibility of a lipid-mediated peptide-peptide interaction as discussed in Sect. 2.4.4.

The inverse hexagonal and cubic unit cells are not isotropic in the sense that the position of a lipid in the cell affects how much room it has for its tail to stretch out. It has been argued that the short peptides occupy the shortest distance between water pores[13,96], relieving the longer acyl tails from having to pack into this geometry, and allowing them to occupy the more spacious "corner" regions (see Fig. 5 of May and Ben-Shaul[13] for an elegant schematic illustration).

The phase transition to an inverted phase which we observe occurs in a mixed lipid system; this transition should also occur in a single lipid system by the same mechanism. However, the transition is slightly easier in the mixed lipid case for two reasons. Firstly, the regions of the bilayer that are depleted of water gradually become enriched in the long lipid by the domain formation process discussed in Sect. 2.4.4. This facilitates the membrane fusion event since the long tail lipids would like to increase the bilayer width, thus providing another force that tends to push water out of this inter-bilayer region.

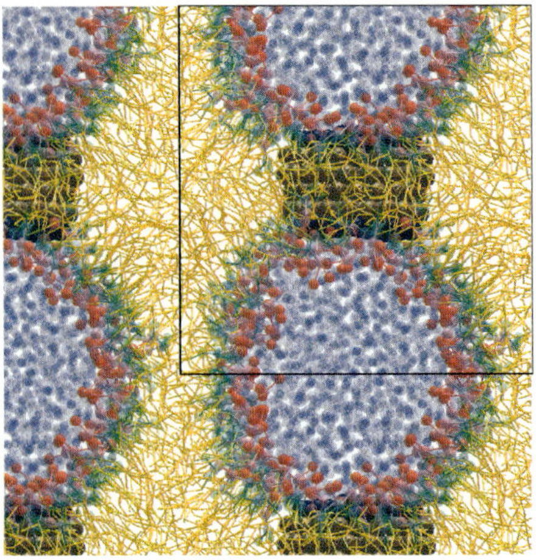

Fig. 2.14. Snapshot of the final configuration of the MD simulation consisting of one peptide, 100 short lipids, 100 long lipids, and 2000 water sites. The simulation unit cell is shown in black line. An inverse cubic phase has formed as is schematically depicted in the second panel of Fig. 2.13. Coloring is as follows. Water is colored blue; lipid head group sites are colored red, purple, and green; lipid acyl tails yellow, and peptide dark grey and blue.

Secondly, since the inverse cubic phase does not present an isotropic environment for the lipids, having different lipid species present helps to lessen the free energy penalty of this anisotropy.

2.4.6 Buckling Instabilities in Langmuir Monolayers

Langmuir monolayers (LMs) are widely used as models for studying membrane bound proteins[98,99] and as precursors for the deposition of multilayer Langmuir-Blodgett films[23]. Of interest here is their role in normal respiratory function. Premature babies who do not yet have an adequate supply of lung surfactant suffer from Respiratory Distress Syndrome (RDS), a condition characterized by lung collapse upon exhalation. The damage done to lung tissue by the cyclic collapse and re-inflation events causes respiratory failure[100]. Pulmonary surfactant supports the breathing cycle by reducing the surface tension at the air/liquid interface in the lung[101]. This surfactant is primarily a mixture of phospholipids (of which DPPC is the most abundant) and the surfactant proteins (SPs) SP-A, SP-B, and SP-C. However, the molecular mechanisms of film formation and adaption to the breathing cycle are unknown[102].

LMs are usually studied as a function of surface coverage by either changing the number of surfactant molecules at fixed surface area A, or by changing the surface area at a fixed quantity of surfactant. Experimentally one monitors the surface pressure, $\pi = \gamma_0 - \gamma$, where γ_0 is the surface tension of the pure air-water system and γ is the resultant surface tension in the presence of surfactant. Pressure-area isotherms, where π versus A is measured at constant temperature, can be mapped out and frequently show hysteresis loop behavior as some of the surfactant is lost due to irreversible monolayer collapse[103].

Surface tension is lowered by the presence of surfactant in part because the hydrogen bonding network of water is disrupted at the interface. But both all-atom MD[54] and experiments[55] show that zwitterionic head groups and water together still form a hydrogen bonding network. The DC_nPC CG lipids we employ are seen to give too low a surface tension as shown in Fig. 2.15, but even recent fully atomistic MD results[104] are far from reproducing experimental surface pressure versus area per lipid isotherms for Langmuir monolayers. In the CG model we do not have hydrogen bonds – their effect is captured in a mean-field manner by the use of effective potentials. The CG head groups are floppy due to soft potentials[63] and the head group – head group interactions might be too weak. The area per lipid is 10% to 15% too large in the CG model which results in a slight offset compared to the experimental curve (see Fig. 2.15). These discrepancies can be used to improve the CG model which has previously been parametrized only for lamellar phases.

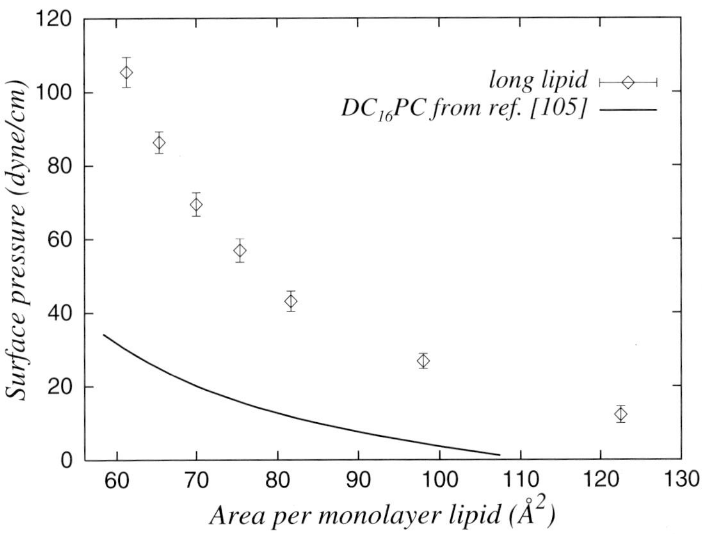

Fig. 2.15. Surface pressure in dyne/cm versus area per lipid in Å^2. The long tail lipid system of the present study is compared with the experimental result of Somerharju[105] for $DC_{16}PC$ at 321.15 K. Error bars are assigned by blocking the data into ten 50 ps windows, computing the average in each window, and then using these ten values to compute the standard deviation.

The long lipid reaches zero surface tension (see Fig. 2.15) at about 69 Å^2 area per lipid. This is close to the equilibrium lamellar phase area per lipid[106] of 75 Å^2. The smallest area per lipid systems display negative surface tension, indicating that the system is in a meta-stable state, due in part to finite size effects as discussed below.

At high coverage the system tends to increase its interfacial area, eventually leading to collapse of the monolayer[107]. In fact, the amplitude of the thermal fluctuations diverges as the surface tension approaches 0 (π approaches γ_0)[108]. Milner[14] showed that an ideal diblock copolymer monolayer develops a buckling instability only at zero surface tension. Collapse usually occurs before this limit. However, in the work by Schief[108] on DPPC monolayers, collapse is not observed until ≥ 71 dyne/cm (with $\gamma_0 = 72$). Collapse can occur into the solvent subphase or outwards on top of the monolayer.

The outward collapse is by the formation of multilayers[109]. These multilayers can be oriented in various ways. One possibility is the formation of a trilayer by a "rollover" mechanism in which a bilayer is formed on top of the monolayer[110]. If the bulk solution in equilibrium with the monolayer is at or above its critical micelle concentration (CMC), the monolayer may shed micelles readily into the bulk[14]. The collapse phenomenon is not well understood from both experimental and theoretical perspectives[110]. On the theoretical[111,8] side, however, an extensive analysis of contributions to the bending free energy has been undertaken by Hu[9].

The long tail lipid monolayers of Fig. 2.15 are destabilized[66] by adding more lipid, leading to monolayer collapse. The two mechanisms we observe whereby the monolayer relieves its high surface pressure is either through the development of curvature (see Figs. 2.16A and 2.17) to increase the interfacial area or through a loss of head groups to the exterior of the monolayer by a bridge transport mechanism, as seen in Fig. 2.16B. This bridge minimizes the energy penalty for the head groups traversing the hydrophobic region. After enough lipids are channeled to the exterior, the monolayer stabilizes and flattens (see Fig. 2.16C). For the simulation of Fig. 2.16, the instantaneous surface tension versus time is shown in Fig. 2.18.

The periodic boundary conditions of the simulations we have done limit the development of curvature in the monolayer. This geometrical constraint restricts the interfacial area increase that comes with developing a non-planar geometry, artificially stabilizing the systems down to moderate negative surface pressure values. The larger simulation cell size of Fig. 2.17 allows the extent of bending to develop to much greater curvatures than for the simulation of Fig. 2.16.

The simulation of Fig. 2.17 differs as well in that the area per lipid is decreased by compressing the simulation box to mimic experiments in which the surface pressure is controlled and to demonstrate that the observed instabilities are generic and not dependent on the method of preparation.

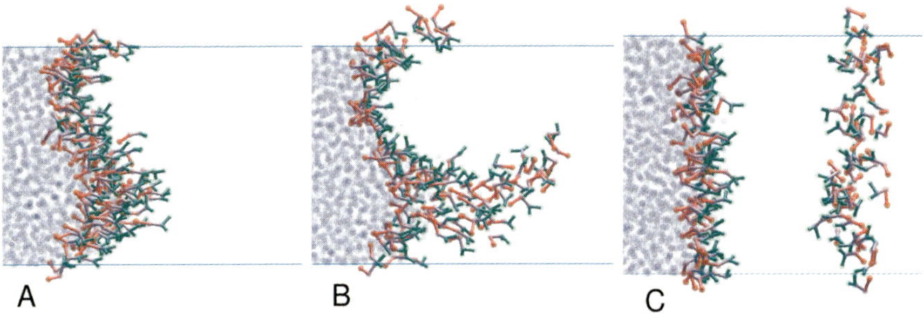

Fig. 2.16. Monolayer instability and collapse shown at one interface of the 120 long lipid per interface system. The initially flat interface (not shown) develops some curvature (panel A) and then opens a bridge to the exterior of the leaflet (panel B). This bridge transports enough material to eventually bring the system back into equilibrium with a flat monolayer interface (panel C). See Fig. 2.18 for the time and the surface tension of these snapshots. Acyl tails not shown. Coloring is as follows. Water blue, choline red, phosphate purple, glycerol blue, and ester green.

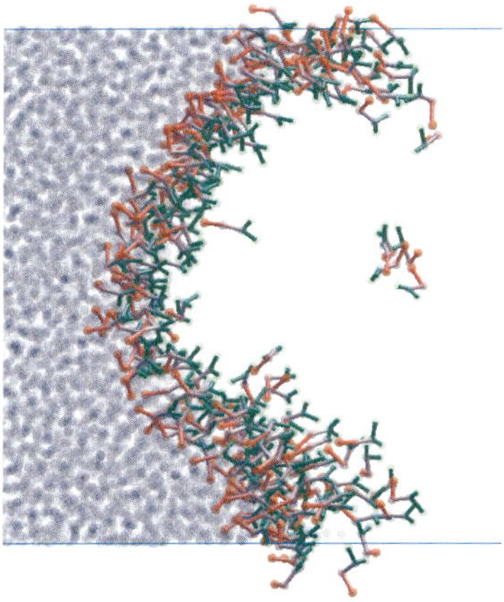

Fig. 2.17. An initially diffuse system of 250 long lipids per interface was prepared and the area per lipid was decreased by shrinking the cross-sectional box size (constrained to be square) by applying an external pressure of 4.5 atm to the system. The system is driven into the negative surface tension regime shortly after 100 ps. This snapshot is taken at 350 ps. Only one of the two interfaces is shown. Acyl tails not shown. Coloring is as follows. Water blue, choline red, phosphate purple, glycerol blue, and ester green.

It is seen that at high surface coverage the monolayer becomes unstable and sheds lipids. Experimentally, surface pressure versus area per lipid isotherms frequently show hysteresis loop behavior as surfactant is permanently lost through collapse. Lungs undergo continual expansion/compression

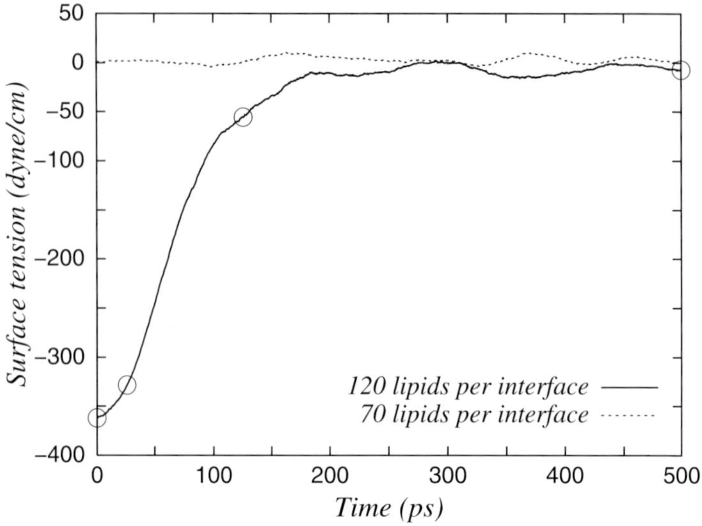

Fig. 2.18. Instantaneous surface tension (dyne/cm) versus raw simulation time (ps) for the 120 long lipid per interface unstable system shown in Fig. 2.16. The data is smoothed with a 100 ps wide symmetrical second order Savitzky-Golay filter[35]. Shown for comparison is the corresponding 70 long lipid per interface curve, which has a surface tension of roughly zero (corresponding to a surface pressure of roughly 72 dyne/cm) as shown in Fig. 2.15. The four points marked on the plot are as follows from left to right. The first point corresponds to the initially flat 120 lipid per interface monolayer. The second point corresponds to Fig. 2.16A when the interface has developed curvature. The third point corresponds to Fig. 2.16B when the monolayer is expelling lipids. The fourth point corresponds to Fig. 2.16C when the monolayer has come to equilibrium.

cycles. It is thought that the protein SP-B plays a major role in the adsorption of new material into the interface during inspiration[102]. During the subsequent expiration event, the main function of SP-C is to expel non-DPPC lipids from the interface[102]. In addition, the relative amounts of DPPC and SP-C vary with surface pressure and this variation results in changes in the elasticity and viscosity of the film[101]. These protein-induced modifications are thought to support the breathing cycle and would be interesting to pursue with CG studies.

2.5 Future Perspectives

There are three fronts along which the current CG studies can be continued. Firstly, systems can be studied whose constituent components have already been parameterized. An example of this is the study of Langmuir monolayers using existing water and phospholipid parameters as illustrated in Sect. 2.4.6.

Inaccuracies which come to light in such studies can point the way to improving the model, which has been parametrized only for lamellar phases. Secondly, existing parameters can be used as building blocks for new species. The cylindrical peptide discussed in Sects. 2.4.4 and 2.4.5 is an example of this. Thirdly, new species can be fully parameterized from a combination of all-atom MD and experimental observables.

The appeal of the second front is that a new species can be constructed very quickly. Moreover, the construction can be artificial in the sense that some interactions can be deliberately excluded or modified in order to assess their impact on the system under study. The drawbacks to this partial parameterization are that the model loses its predictive power for specific molecular systems, and the interaction parameters may not even be amenable to a crude guess based on the existing force field. As an example of this last point, the effect of unsaturation in the lipid acyl tail cannot be mimicked based on the DMPC parameter set.

There are numerous topics which are amenable to study with the CG method, some of which we now mention. Bilayers and monolayers involving a few lipid species and cholesterol are suitable to study raft formation[1]. Mixed lipid Langmuir monolayers[112] display a rich variety of microdomains of different composition and phase. It would be interesting to see how many of these are accessible with CG models. The compression/expansion cycle of the lung surfactant DPPC could be studied in the presence of the surfactant proteins (SPs) SP-A, SP-B, and SP-C, which are known to alter monolayer collapse[102]. Lipid mediated protein-protein interactions can be used to explore membrane protein crystallization[19]. Oligomeric channel protein insertion into membranes and their assembly and mutual orientation[113] across the bilayer are of interest both for antimicrobial and purely structural studies. Related to this is cyclic D,L-α-peptide self-assembly[114] and membrane insertion and disruption[115]. Monolayer structure at solid/water interfaces[116] displays novel geometry such as a hemicylindrical micelles which is being elucidated with atomic force microscopy[117]. Entropic and enthalpic interactions between amphiphile surfaces[7,6] such as micelles, lipid bilayers, microemulsion droplets, and combinations thereof can be computed as potentials of mean force. Protein alignment can be studied as a function of surface pressure in Langmuir monolayers[99]. Self-assembled vesicles from non-lipid species such as diblock copolymers[118] and surfactant-like peptides[119] offer alternatives for many applications including targeted drug delivery. In conclusion, there are clearly many possible future applications of CG models.

Acknowledgments

Discussion with Carlos F. Lopez, Ivaylo Ivanov, Eung-Gun Kim, Bin Chen, Preston B. Moore, and John C. Shelley is gratefully acknowledged. This work was supported in part by a grant from the Natural Sciences and Engineering Research Council of Canada and the National Institutes of Health.

References

1. A. Pralle, P. Keller, E.-L. Florin, K. Simons, and J. K. H. Hörber, J. Cell Biol. **148**, 997 (2000).
2. H. Rafii-Tabar and A. Chirazi, Phys. Rep. **365**, 145 (2002).
3. W. A. Goddard III, T. Cagin, M. Blanco, N. Vaidehi, S. Dasgupta, W. Floriano, M. Belmares, J. Kua, G. Zamanakos, S. Kashihara, M. Iotov, and G. Gao, Comp. Theor. Poly. Sci. **11**, 329 (2001).
4. T. Cagin, G. Wang, R. Martin, G. Zamanakos, N. Vaidehi, D. T. Mainz, and W. A. Goddard III, Comp. Theor. Poly. Sci. **11**, 345 (2001).
5. K. Svoboda and S.M. Block, Ann. Rev. Biophys. Biomol. Struct. **23** 247 (1994).
6. S. A. Safran, Statistical Thermodynamics of Surfaces, Interfaces, and Membranes, Addison-Wesley, New York, 1994.
7. J. Israelachvili and H. Wennerström, J. Phys. Chem. **96**, 520 (1992).
8. D. Marsh, Biochim. Biophys. Acta **1286**, 183 (1996).
9. J.-G. Hu and R. Granek, J. Phys. II (France) **6**, 999 (1996).
10. J. Y. Walz and E. Ruckenstein, J. Phys. Chem. B **103**, 7461 (1999).
11. P. Sens and S. A. Safran, Eur. Phys. J. E **1**, 237 (2000).
12. D. R. Fattal and A. Ben-Shaul, Biophys. J. **65**, 1795 (1993).
13. S. May and A. Ben-Shaul, Biophys. J. **76**, 751 (1999).
14. S. T. Milner, J.-F. Joanny, and P. Pincus, Europhys. Lett. **9**, 495 (1989).
15. D. Duque, X. Li, K. Katsov, and M. Schick, J. Chem. Phys. **116**, 10478 (2002).
16. C. Nielsen, M. Goulian, and O. S. Andersen, Biophys. J. **74**, 1966 (1998).
17. T. A. Harroun, W. T. Heller, T. M. Weiss, L. Yang, and H. W. Huang, Biophys. J. **76**, 3176 (1999).
18. H. Sprong, P. van der Sluijs, and G. van Meer, Nature Rev. Mol. Cell Biol. **2**, 504 (2001).
19. M. Caffrey, Curr. Opin. Struct. Biol. **10**, 486 (2000).
20. M. F. Brown, R. L. Thurmond, S. W. Dodd, D. Otten, and K. Beyer, Phys. Rev. E. **64**, 010901 (2001).
21. T. J. McIntosh, Curr. Opin. Struct. Biol. **10**, 481 (2000).
22. Y. F. Dufrêne, T. Boland, J. W. Schneider, W. R. Barger, and G. U. Lee, Faraday Discuss. **111**, 79 (1998).
23. J. K. Basu and M. K. Sanyal, Phys. Rep. **363**, 1 (2002).
24. P. Dynarowicz-Łątka, A. Dhanabalan, and O. N. Oliveira Jr., Adv. Coll. Interf. Sci., **91**, 221 (2001).
25. R. Goetz, G. Gommpper, and R. Lipowsky, Phys. Rev. Lett. **82**, 221 (1999).
26. S. J. Marrink and A. E. Mark, J. Phys. Chem. B **105**, 6122 (2001).
27. E. Lindahl and O. Edholm, Biophys. J. **79**, 426 (2000).
28. J. E. Johnson and W. Chiu, Curr. Opinion Struct. Biol. **10**, 229 (2000).
29. J. Witz and F. Brown, Arch. Virol. **146**, 2263 (2001).
30. U. Rothlisberger, P. Carloni, K. Doclo, and M. Parrinello, J. Biol. Inorg. Chem. **5**, 236 (2000).
31. See the article by R. Kapral and G. Ciccotti in this volume.
32. G. R. Smith and M. S. P. Sansom, Biophys. J. **73**, 1364 (1997).
33. Q. Zhong, D. M. Newns, P. Pattnaik, J. D. Lear, and M. L. Klein, FEBS Letters **473**, 195 (2000).
34. J. C. Shelley and M. Y. Shelley, Curr. Opin. Coll. Interf. Sci. **5**, 101 (2000).

35. Numerical Recipes in Fortran 77: The Art of Scientific Computing, 2nd edition. W. H. Press, S. A. Teukolsky, W. T. Vetterling, and B. P. Flannery, Cambridge University Press, New York, 1992.
36. J. A. Izaguirre, D. P. Catarello, J. M. Wozniak, and R. D. Skeel, J. Chem. Phys. **114**, 2090 (2001).
37. A. J. Chorin, O. H. Hald, and R. Kupferman, Proc. Natl. Acad. Sci. USA **97**, 2968 (2000).
38. H. Noguchi and M. Takasu, Phys. Rev. E **64** 041913 (2001).
39. S.-H. Chung, T. W. Allen, and S. Kuyucak, Biophys. J. **83**, 263 (2002).
40. P. R. ten Wolde and D. Chandler, Proc. Natl. Acad. Sci. USA **99**, 6539 (2002).
41. A. Malevanets and R. Kapral, J. Chem. Phys. **112**, 7260 (2000).
42. L. David, R. Luo, and M. K. Gilson, J. Comp. Chem. **21**, 295 (2000).
43. E. G. Flekkøy and P. V. Coveney, Phys. Rev. Lett. **83**, 1775 (1999).
44. R. D. Groot and K. L. Rabone, Biophys. J. **81**, 725 (2001).
45. A. Liwo, S. Ołdziej, M. R. Pincus, R. J. Wawak, S. Rackovsky, and H. A. Scheraga, J. Comp. Chem. **18**, 849 (1997).
46. A. Kolinski and J. Skolnick, J. Chem. Phys. **97**, 9412 (1992).
47. J. I. Siepmann, Adv. Chem. Phys. **105**, 443 (1999).
48. S.G. Whittington, Fields Inst. Comm. **26**, 131 (2000).
49. S. B. Opps and J. Schofield, Phys. Rev. E **63**, 056701 (2001).
50. H. Fukunaga, J. Takimoto, and M. Doi, J. Chem. Phys. **116**, 8183 (2002).
51. L. Whitehead, C. M. Edge, and J. W. Essex, J. Comp. Chem. **22**, 1622 (2001).
52. A. P. Lyubartsev and A. Laaksonen, Phys. Rev. E **52**, 3730 (1995).
53. R. L. McGreevy and L. Pusztai, Mol. Sim. **1**, 359 (1988).
54. C. F. Lopez, P. B. Moore, and M. L. Klein, manuscript in preparation.
55. G. L. Richmond, Annu. Rev. Phys. Chem. **52**, 357 (2001).
56. R. Faller, H. Schmitz, O. Biermann, and F. Müller-Plathe, J. Comp. Chem. **20**, 1009 (1999).
57. J. C. Shelley, M. Y. Shelley, R. C. Reeder, S. Bandyopadhyay, and M. L. Klein, J. Phys. Chem. B **105**, 4464 (2001).
58. S. Bandyopadhyay, J. C. Shelley, and M. L. Klein, J. Phys. Chem. B **105**, 5979 (2001).
59. J. C. Shelley, Modified coarse grain parameters for MD simulations in the NPT ensemble. Available through the web at www.cmm.upenn.edu.
60. P. E. Harper, D. A. Mannock, R. N. A. H. Lewis, R. N. McElhaney, and S. M. Gruner, Biophys. J. **81**, 2693 (2001).
61. J. C. Shelley, M. Y. Shelley, R. C. Reeder, S. Bandyopadhyay, P. B. Moore, and M. L. Klein, J. Phys. Chem. B **105**, 9785 (2001).
62. C. F. Lopez, P. B. Moore, J. C. Shelley, M. Y. Shelley, and M. L. Klein, Comp. Phys. Comm. [in press]
63. C. F. Lopez, S. O. Nielsen, P. B. Moore, J. C. Shelley, and M. L. Klein, J. Phys.: Condens. Matter [submitted].
64. P. Linse and V. Lobaskin, J. Chem. Phys. **112**, 3917 (2000).
65. R. Hirn, T. M. Bayerl, J. O. Rädler, and E. Sackmann, Faraday Discuss. **111**, 17 (1998).
66. S. O. Nielsen, C. F. Lopez, P. B. Moore, J. C. Shelley, and M. L. Klein, manuscript in preparation.
67. S. E. Feller, Y. Zhang, and R. W. Pastor, J. Chem. Phys. **103**, 10267 (1995).
68. M. M. Rudek, J. A. Fisk, V. M. Chakarov, and J. L. Katz, J. Chem. Phys. **105**, 4707 (1996).

69. J. Alejandre, D. J. Tildesley, and G. A. Chapela, J. Chem. Phys. **102**, 4574 (1995).
70. B. Chen, J. I. Siepmann, K. J. Oh, and M. L. Klein, J. Chem. Phys. **116**, 4317 (2002).
71. R. Goetz and R. Lipowsky, J. Chem. Phys. **108**, 7397 (1998).
72. J.-M. Drouffe, A. C. Maggs, and S. Leibler, Science **254**, 1353 (1991).
73. S. Karaborni, K. Esselink, P. A. J. Hilbers, B. Smit, J. Karthäuser, N. M. van Os, and R. Zana, Science **266**, 254 (1994).
74. B. Smit, K. Esselink, P. A. J. Hilbers, N. M. van Os, L. A. M. Rupert, and I. Szleifer, Langmuir **9**, 9 (1993).
75. S. Yamamoto, Y. Maruyama, and S. Hyodo, J. Chem. Phys. **116**, 5842 (2002).
76. F. K. von Gottberg, K. A. Smith, and T. A. Hatton, J. Chem. Phys. **106**, 9850 (1997).
77. F. K. von Gottberg, K. A. Smith, and T. A. Hatton, J. Chem. Phys. **108**, 2232 (1998).
78. S. J. Marrink, D. P. Tieleman, and A. E. Mark, J. Phys. Chem. B **104**, 12165 (2000).
79. S. Bogusz, R. M. Venable, and R. W. Pastor, J. Phys. Chem. B **105**, 8312 (2001).
80. M. Sjölund, L. Rilfors, and G. Lindbolm, Biochemistry **28**, 1323 (1989).
81. F. Dumas, M. C. Lebrun, and J.-F. Tocanne, FEBS Letters **458**, 271 (1999).
82. M. R. R. de Planque, E. Goormaghtigh, D. V. Greathouse, R. E. Koeppe II, J. A. W. Kruijtzer, R. M. J. Liskamp, B. de Kruijff, and J. A. Killian, Biochem. 40, 5000 (2001).
83. T. A. Harroun, W. T. Heller, T. M. Weiss, L. Yang, and H. W. Huang, Biophys. J. **76**, 937 (1999).
84. T. Gill, J. H. Ipsen, O. G. Mouritsen, M. C. Sabra, M. M. Sperotto, and M. J. Zuckermann, Biochim. Biophys. Acta **1376**, 245 (1998).
85. M. Wahab, H.-J. Mögel, and P. Schiller, Mol. Phys. **99**, 2045 (2001).
86. N. Dan, P. Pincus, and S. A. Safran, Langmuir **9**, 2768 (1993).
87. P. A. Kralchevsky and K. Nagayama, Adv. Colloid. Interf. Sci. **85**, 145 (2000).
88. P. Lagüe, M. J. Zuckermann, and B. Roux, Faraday Discuss. **111**, 165 (1998).
89. P. Lagüe, M. J. Zuckermann, and B. Roux, Biophys. J. **81**, 276 (2001).
90. S. Sharpe, K. R. Barber, C. W. M. Grant, D. Goodyear, and M. R. Morrow, Biophys. J. **83**, 345 (2002).
91. N. J. P. Ryba and D. Marsh, Biochem. **31**,7511 (1992).
92. T. R. Weikl, Europhys. Lett. **54**, 547 (2001).
93. F. Dumas, M. M. Sperotto, M.-C. Lebrun, J.-F. Tocanne, and O. G. Mouritsen, Biophys. J. **73**, 1940 (1997).
94. C. Gliss, H. Clausen-Schaumann, R. Günther, S. Odenbach, O. Randl, and T. M. Bayerl, Biophys. J. **74**, 2443 (1998).
95. S. O. Nielsen, C. F. Lopez, I. Ivanov, P. B. Moore, J. C. Shelley, and M. L. Klein, manuscript in preparation.
96. J. A. Killian, I. Salemink, M. R. R. de Planque, G. Lindblom, R. E. Koeppe II, and D. V. Greathouse, Biochem. **35** 1037 (1996).
97. H. I. Petrache, D. M. Zuckerman, J. N. Sachs, J. A. Killian, R. E. Koeppe II, and T. B. Woolf, Langmuir **18**, 1340 (2002).
98. R. Maget-Dana, Biochim. Biophys. Acta **1462**, 109 (1999).
99. A. Tronin, J. Strzalka, X. Chen, P. L. Dutton, B. M. Ocko, and J. K. Blasie, Langmuir **17**, 3061 (2001).

100. J. M. Crane and S. B. Hall, Biophys. J. **80**, 1863 (2001).
101. N. Wüstneck, R Wüstneck, V. B. Fainerman, R. Miller, and U. Pison, Colloids and Surfaces B: Biointerfaces **21**, 191 (2001).
102. E. J. A. Veldhuizen and H. P. Haagsman, Biochim. Biophys. Acta **1467** 255 (2000).
103. M. M. Lipp, K. Y. C. Lee, D. Y. Takamoto, J. A. Zasadzinski, and A. J. Waring, Phys. Rev. Lett. **81**, 1650 (1998).
104. Y. N. Kaznessis, S. Kim, and R. G. Larson, Biophys. J. **82**, 1731 (2002).
105. P. J. Somerharju, J. A. Virtanen, K. K. Eklund, P. Vainio, and P. K. J. Kinnunen, Biochem. **24**, 2773 (1985).
106. S. O. Nielsen, unpublished data.
107. R. A. Ridsdale, N. Palaniyar, F. Possmayer, and G. Harauz, J. Membrane Biol. **180**, 21 (2001).
108. W. R. Schief, L. Touryan, S. B. Hall, and V. Vogel, J. Phys. Chem. B **104**, 7388 (2000).
109. C. Ybert, W. Lu, G. Möller, and C. M. Knobler, J. Phys. Chem. B **106** 2004 (2002).
110. C. Ybert, W. Lu, G. Möller, and C. M. Knobler, J. Phys.: Condens. Matter **14**, 4753 (2002).
111. F. Schmid, C. Stadler, and H. Lange, Colloids Surf. A **149**, 301 (1999).
112. A. Gopal and K. Y. C. Lee, J. Phys. Chem. B, **105**, 10348 (2001).
113. L. Yang, Y. A. Harroun, T. M. Weiss, L. Ding, and H. W. Huang, Biophys. J. **81**, 1475 (2001).
114. J. D. Hartgerink, T. D. Clark, and M. R. Ghadiri, Chem. Eur. J. **4**, 1367 (1998).
115. S. Fernandez-Lopez, H.-S. Kim, E. C. Choi, M. Delgado, J. R. Granja, A. Khasanov, K. Kraehenbuehl, G. Long, D. A. Weinberger, K. M. Wilcoxen, and M. R. Ghadiri, Nature **412**, 452 (2001).
116. S. Bandyopadhyay, J. C. Shelley, M. Tarek, P. B. Moore, and M. L. Klein, J. Phys. Chem. B **102**, 6318 (1998).
117. L. M. Grant, F. Tiberg, and W. A. Ducker, J. Phys. Chem. B **102**, 4288 (1998).
118. B. M. Discher, Y.-Y. Won, D. S. Ege, J. C.-M. Lee, F. S. Bates, D. E. Discher, and D. A. Hammer, Science **284**, 1143 (1999).
119. S. Vauthey, S. Santoso, H. Gong, N. Watson, and S. Zhang, Proc. Natl. Acad. Sci. USA **99**, 5355 (2002).

Part III

Polymer Structure and Dynamics

3 Variable-Connectivity Monte Carlo Algorithms for the Atomistic Simulation of Long-Chain Polymer Systems

Doros N. Theodorou

Department of Materials Science and Engineering, School of Chemical Engineering, National Technical University of Athens, Zografou Campus, GR-157 80 Athens, Greece
and
Molecular Modelling of Materials Laboratory, Institute of Physical Chemistry, National Research Centre for Physical Sciences "Demokritos", GR-153 10 Ag. Paraskevi Attikis, Greece

Abstract. Recently, our ability to equilibrate atomistic models of synthetic polymers and biopolymers has been significantly enhanced through the development of Monte Carlo schemes employing moves which modify the connectivity of atoms along the chains. In this chapter, the geometric "bridging" construction underlying these moves is explained and the statistical mechanical underpinnings of Monte Carlo algorithms employing these moves to sample various, appropriately designed, ensembles are discussed. Concerted rotation, directed internal bridging, end-bridging, directed end-bridging, scission-fusion, double bridging, intramolecular double rebridging moves, and their combination with parallel tempering are developed in some detail.

Results are presented from applying the connectivity-altering Monte Carlo algorithms to predict volumetric behaviour, packing, chain conformation and entanglement properties in long-chain synthetic polymer melts (polyethylene, polypropylene, polyisoprene); melt elasticity and birefringence under conditions of steady-state flow; sorption equilibria of alkanes in polyethylene melts; and composition profiles at solid/polymer interfaces strengthened with grafted polymer chains. The molecular-level insight gained from these calculations is discussed, as is the role of the new algorithms as tools for the development of hierarchical modelling approaches to structure - processing - property - performance relations in polymer systems.

3.1 Introduction

Understanding and predicting the relations between structure, properties, processing, and performance of polymeric materials computationally is both fundamentally interesting and practically very relevant, since it can lead to faster and more economical design of materials and processes for specific applications. Today, polymer modelling is widespread and growing in both industrial and academic environments.

One serious challenge faced by polymer modellers, especially those interested in connecting macroscopic material behaviour to the chemical structure and architecture of constituent macromolecules, is that polymer structure and dynamics are characterised by extremely broad spectra of length and time scales. Intramolecular correlations and local packing of chains in the bulk exhibit features on the length scale of bond lengths and atomic radii, i.e. Å. The Kuhn statistical segment of a typical synthetic randomly coiled polymer is on the order of 10 Å and can be more than an order of magnitude higher for macromolecules with stiff backbones. The radius of gyration of entire chains in the amorphous bulk scales with the chain length X as $X^{1/2}$ and is on the order of 100 Å for typical molecular weights; the smallest dimension of microphases (lamellae, cylinders, spheres) in microphase-separated block copolymer systems is on this order of magnitude, while domain sizes in semicrystalline polymers and immiscible polymer blends may well be on the order of μm. Even broader is the range of time scales characterising molecular motion in polymers. While localized vibrational modes of chains have periods on the order of 10^{-14} s, conformational transitions of individual bonds over torsional energy barriers in the melt state have waiting times in excess of 10^{-11} s. Longer and longer sequences of segments along the backbone exhibit longer and longer correlation times. The longest relaxation time, required for a chain to diffuse by a length commensurate to its size and thus "forget" its previous conformation, is critical to the viscoelastic response of polymer melts to flow. This time scales as X^2 for low molecular weight melts in the Rouse regime and as $X^{3.4}$ for X exceeding a critical chain length that is sufficient for the development of entanglements in the reptation regime; for a C_{800} polyethylene melt at 450K this time is on the order of 3 μs, while it easily exceeds the millisecond time scale for the molecular weights encountered in typical processing operations.

Atomistic molecular dynamics (MD) is probably the most convenient simulation method for linking macroscopic properties to molecular constitution. Unfortunately, however, even when state-of-the-art developments, such as reversible multiple time step algorithms [1] are invoked, atomistic MD can typically track the evolution of systems of length scale ca. 100 Å for times no longer than a few decades of nanoseconds. While the length scale of MD can be increased significantly by use of domain decomposition strategies on parallel computers, the time scale of MD falls short of the longest relaxation times of real-life polymer systems; alternative or complementary simulation strategies are needed.

In response to this need, the polymer simulation community has moved actively in two research directions in recent years: (a) Development of multiscale modelling approaches consisting of many levels, each level addressing phenomena over a specific window of length and time scales; links between different levels can be established through systematic "coarse-graining" of the model used to represent the polymer [2,3]. (b) Development of new simulation algorithms, which can move through configuration space more efficiently than MD and achieve equilibration at all length scales; in this research direction, Monte Carlo (MC) schemes play a prominent role. This chapter focusses on a class of MC algorithms that can be used in conjunction with detailed atomistic models of dense multichain polymer systems, such as polymer melts.

MC simulations have a long history in polymer science. The earliest polymer simulations were conducted with MC using lattice models, first on single chains [4] and then on multichain systems [5]. Lattice MC work on polymers has been reviewed [6]. At melt densities, the excluded volume interactions among polymer segments, the connectivity of polymer segments along chains, and the conformational stiffness of chain backbones make it very difficult to sample configuration space efficiently. Melt MC simulations can be classified into two broad categories. The first consists of simulations aimed at extracting both structural/thermodynamic and dynamical information over some (not too short) time scales. Simulations in this category are most often conducted using coarse-grained (e.g. lattice-based) models and employ strictly local moves that emulate the actual dynamics of the polymer ("kink-jumps", "end rotations" and "crankshaft moves" in simple lattice models, single monomer displacements in bond fluctuation models [6]). Simulations in the second category dispense with the objective of following chain dynamics and aim primarily at achieving equilibration. It is in this category that the maximum power of MC simulations can be unleashed through the design of moves which, although "unphysical" from the point of view of true dynamics, induce drastic changes in the molecular configuration while at the same time satisfying the basic requirements of ergodicity and microscopic reversibility. MC schemes based on such moves may induce equilibration many orders of magnitude more efficiently than MD, for the same model of molecular geometry and energetics. An early "unphysical" move that has proved quite useful in both lattice and continuous-space simulations of dense polymer systems is the reptation, or "slithering snake" move [7–9]. This deletes a segment from one end of a (linear) chain and appends it at the other end at a randomly chosen angle, the end result being a "sliding" motion of the chain along its contour by one segment.

The 1990s have witnessed the development of several powerful bias MC algorithms appropriate for continuous-space simulations of dense polymer systems. Configurational bias (CB) MC is such an algorithm, developed by Siepmann and Frenkel [10] and by de Pablo, Laso, and Suter [11] on ideas that can be traced back to the seminal work of Rosenbluth and Rosenbluth [12]. Here one cuts off a terminal part of of chain and re-grows it segment-by segment using an energetic bias, so as to avoid overlaps with the rest of the chain and with the environment. For each added segment k, a number n_q of candidate positions q are considered, among which one is chosen according to the probability distribution

$$p_{\mathrm{CB},k}^{(q)} = \frac{\exp(-\beta \mathcal{V}_{\mathrm{CB},k}^{(q)})}{\sum_{l=1}^{n_q} \exp(-\beta \mathcal{V}_{\mathrm{CB},k}^{(l)})}. \tag{3.1}$$

Here β is $(k_B T)^{-1}$ and $\mathcal{V}_{\mathrm{CB},k}^{(l)}$ is normally chosen as the increase in system potential energy brought about by placing segment k at position l. The bias inherent in this procedure is removed at the acceptance stage. Acceptance probabilities ensure that microscopic reversibility is satisfied by incorporating Rosenbluth weights

$$W_{\mathrm{CB}} = \prod_k \sum_{l=1}^{n_q} \exp(-\beta \mathcal{V}_{\mathrm{CB},k}^{(l)}) \tag{3.2}$$

for both the forward and the reverse move; for the latter, the actual starting configuration is considered as one of the trial configurations that would be generated if the move were attempted in reverse.

For the simulation of phase equilibria, configurational bias has been used in the framework of an expanded grand canonical ensemble to alleviate problems associated with large chain molecules being inserted into/deleted from, or exchanged between dense phases [13]. In an expanded ensemble simulation, configurations are sampled according to the partition function $\sum_{y=1}^{M} Q_y \exp(w_y)$, where y is a parameter in the Hamiltonian of the system, allowed to range over M discrete values ("states"); Q_y is a conventional (e.g. grand canonical) partition function, evaluated at parameter value y; and the w_y are weighting factors modulating the probability of appearance of various y values. The expanded ensemble was originally proposed [14] for the calculation of free energies of solvation of small hydrophobic solutes and ions in water at infinite dilution, the parameter y corresponding to the strength of interaction between the solute and solvent molecules. In the implementation of [13], y modulates the length of a "tagged" chain which is allowed to fluctuate in size.

An ingeniously simple technique that can accelerate the equilibration of systems with rugged potential energy hypersurfaces, which tend to make conventional simulation schemes nonergodic at low temperatures, is parallel tempering [15]. Parallel tempering considers a larger ensemble of n systems, each equilibrated at a different temperature $T_i (i = 1, ..., M)$. The system of interest is the system of lowest temperature; the systems of higher temperature usually have the same Hamiltonian as the system of interest and are added in order to help overcome energy barriers and thereby accelerate equilibration. The systems of different temperature are considered as independent of each other, so the partition function sampled is actually the product of the individual partition functions at the different temperatures, i.e. of the form $\prod_{i=1}^{M} Q_i(N, V, T_i)$. There are two types of moves: Regular "configuration" moves, performed at each temperature, and "swapping moves", which exchange configurations between two systems i and j. A swapping move is accepted with probability $\min[1, \exp(\Delta\beta_{ij}\Delta\mathcal{V}_{ij})]$, with $\Delta\beta_{ij}$ and $\Delta\mathcal{V}_{ij}$ being the differences in reciprocal temperatures (times k_B) and energies between the two systems. A prerequisite for the scheme to work is that the energy histograms of systems adjacent in the temperature ladder should overlap. Parallel tempering has been used to sample configurations of biological molecules [16,17] and synthetic polymers and will be discussed in more detail later on in this chapter.

Reptation and configuration bias operate only at chain ends; when used with realistic continuum models, they benefit from the excess free volume available near chain ends. Their effectiveness degrades in very long-chain systems, where chain ends are scarce. Moves capable of inducing drastic reconfiguration of internal sections of chains would be highly desirable. Such moves were introduced for lattice models in the 1980s in the form of the "chain breaking", or "pseudokinetic" MC algorithms [18–20]. These algorithms alter the connectivity among polymer segments in the lattice model at the expense of introducing some polydispersity (distribution of chain lengths) in the polymer. Small alterations in the connectivity result in large jumps in the configuration space of the polymer, especially accelerating the rate of change of long-range structural features such as the end-to-end vectors and radii of gyration of the chains.

More recently, it has become possible to design connectivity-altering algorithms for continuous-space polymer models represented in atomistic detail. This chapter discusses the geometrical and statistical mechanical underpinnings of these algorithms, presents measures of their efficiency and representative applications aiming at the prediction of physical properties of long-chain polymer systems.

The chapter is organised as follows. Section 3.2 discusses the basic geometric problem of bridging, whose solution enabled the application of connectivity-altering algorithms to atomistic models. Section 3.3 presents a variety of intra- and intermolecular moves that can be designed around the bridging construction and explains a variety of ensembles in which these moves can be implemented. Finally, Sect. 3.4 presents some applications from a variety of polymer systems and property prediction problems for which the algorithms have been used so far.

3.2 The Bridging Construction

In this section we briefly examine some geometric problems whose solution allows using connectivity-altering algorithms in continuous space with detailed

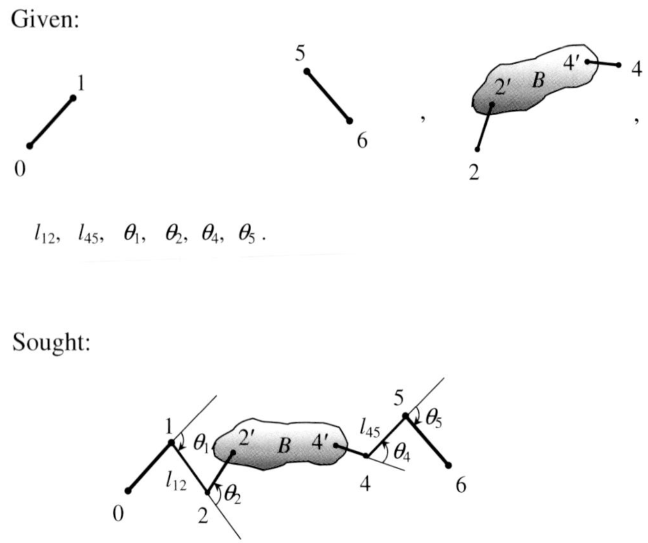

Fig. 3.1. General geometric problem of bridging: The two bonds, 01 and 56, are to be bridged through the rigid body B, leading to the formation of a chain segment with prescribed bond lengths and angles. Note that bonds $22'$ and $4'4$ are rigidly affixed to B.

atomistic models. While breaking connectivity does not pose any serious geometric problems (it can be accomplished by eliminating a sequence of one or more repeat units from the backbone, with simultaneous conversion of the units on either side into terminal units), establishing connectivity is more challenging.

The general problem of establishing connectivity, or "bridging", in detailed atomistic polymer models, can be posed as shown in Fig. 3.1. Given are two bonds, 01 and 56, in three-dimensional space. Also given is a rigid solid body of arbitrary shape, B, possessing two bonds, $22'$ and $4'4$, protruding from it; in actual applications, B is a chemical moiety of prescribed internal geometry. One seeks to bridge 01 and 56 through B by placing B at an appropriate position and orientation and constructing two new bonds 12 and 45 in such a way that the bond lengths l_{12}, l_{45} and the bond angles $\pi - \theta_1$, $\pi - \theta_2$, $\pi - \theta_4$, $\pi - \theta_5$ have prescribed values. There are 6 unknowns in the problem, namely the three translational and three orientational degrees of freedom of B in its bridging position. There are also 6 equations or constraints, arising from the requirement that $l_{12}, l_{45}, \theta_1, \theta_2, \theta_4$, and θ_5 have prescribed values. Thus, the problem is well-posed.

Particularly useful is the case where the bonds $22'$ and $4'4$, rigidly affixed to B, are coplanar, the lines on which they lie intersecting at a point 3. In this case, the bridging body B can be replaced by the triplet of points $\{2, 3, 4\}$, as far as the geometric solution is concerned. In other words, the bridging problem can be posed as follows [21] (Fig. 3.2): Given two bonds 01 and 56 in space, it is sought to bridge them with a triplet of points (atoms or pseudoatoms) $\{2,3,4\}$, such that the bond lengths $l_{12}, l_{23}, l_{34}, l_{45}$ and the bond angle supplements $\theta_1, \theta_2, \theta_3, \theta_4$, and θ_5 have prescribed values. We will call this the "trimer bridging problem".

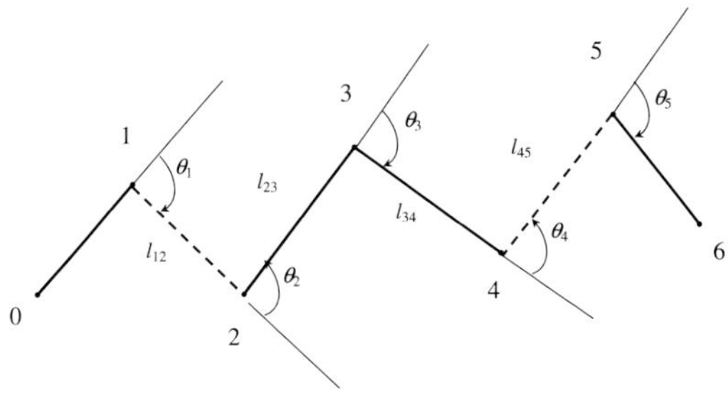

Fig. 3.2. Geometric problem of trimer bridging. The bonds 01 and 56 are to be connected through the triplet of points $\{2,3,4\}$. Bond lengths and bond angle supplements that are specified are indicated.

A solution procedure for the trimer bridging problem is given in [21]. With l_{12} and θ_1 fixed, the locus of point 2 is a circle $\mathcal{C}_2$ lying in a plane normal to the line connecting points 0 and 1, with its centre on that line. Similarly, with l_{45} and θ_5 fixed, the locus of point 4 is a circle $\mathcal{C}_4$ lying in a plane normal to the line connecting points 6 and 5, with its centre on that line. With l_{12}, l_{23} and θ_2 fixed, the distance of point 3 from point 1 is fixed, and therefore point 3 lies on a sphere of known radius centred at point 1. Similarly, with l_{34}, l_{45} and θ_4 fixed, point 3 also lies on a sphere of known radius centred at point 6. Therefore, point 3 lies on the circle $\mathcal{C}_3$ which constitutes the intersection of two known spheres (see Fig. 3.3). Thus, trimer bridging is reduced to the following geometric problem: Given three circles, $\mathcal{C}_2$, $\mathcal{C}_3$, $\mathcal{C}_4$, determine one point on each circle ($\mathbf{r}_2 \in \mathcal{C}_2, \mathbf{r}_3 \in \mathcal{C}_3, \mathbf{r}_4 \in \mathcal{C}_4$), such that the distances $|\mathbf{r}_3 - \mathbf{r}_2|, |\mathbf{r}_4 - \mathbf{r}_3|$, and $|\mathbf{r}_2 - \mathbf{r}_4|$ have prespecified values (l_{23}, l_{34}, and $(l_{23}^2 + l_{34}^2 + 2l_{23}l_{34}\cos\theta_3)^{1/2}$, respectively). This problem is cast in the form of three equations in three angular variables ϕ_L, ϕ_R, and ψ which specify the positions of $\mathbf{r}_2$, $\mathbf{r}_3$, and $\mathbf{r}_4$ along their respective circles (see Fig. 3.3). One of the equations is quadratic in $\tan(\phi_L/2)$ and another is quadratic in $\tan(\phi_R/2)$, allowing easy elimination of these variables in terms of ψ and reduction of the whole problem into a single equation of the form $F(\psi) = 0$. In general, $F(\psi)$ has four branches, because the quadratic equations contribute up to two real solutions each. Solving the trimer bridging problem amounts to finding all roots of $F(\psi) = 0$ on all four branches. The requirement of a nonnegative discriminant for the two quadratic equations leads to a couple of quartic inequalities in $\tan(\psi/2)$, which are solved analytically to determine feasibility regions for ψ, i.e. subintervals of the domain $[-\pi,\pi)$ over which the function $F(\psi)$ exists. In [21] the solution of $F(\psi) = 0$ in each of these subintervals is performed numerically, using a combined bisection and Regula Falsi algorithm.

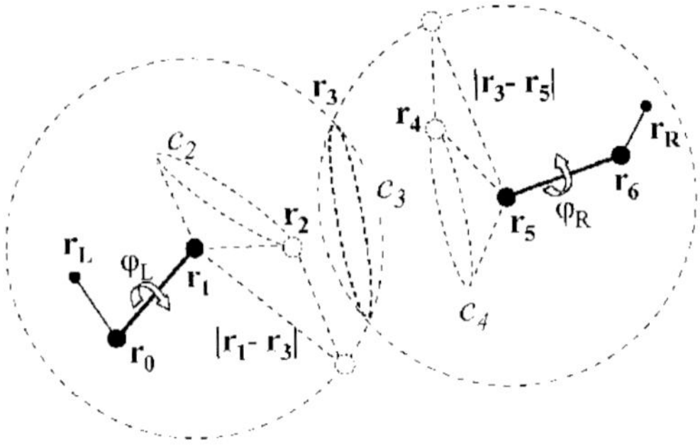

Fig. 3.3. Schematic of the trimer bridging geometric solution, according to [21]. Fixed, but otherwise arbitrary, reference points $\mathbf{r}_L$ and $\mathbf{r}_R$ are used to define the torsion angles ϕ_L and ϕ_R. Rotations of ϕ_L and ϕ_R sweep out the loci of points 2 and 4 (short-dashed circles $\mathcal{C}_2$ and $\mathcal{C}_4$, respectively). Point 3 lies on circle $\mathcal{C}_3$, which constitutes the intersection of two spheres of known radii centred at points 1 and 5. The position of point 3 along $\mathcal{C}_3$ is specified by angle ψ. (Reproduced from [21], with permission).

The original formulation of trimer bridging [22] was cast in the generalised coordinate system introduced by Flory [23] for describing the conformation of atomistically detailed chains, and followed a mathematical procedure for ring closure proposed by Gō and Scheraga [24]. Given the positions of atoms -1, 0, 1, 5, 6, and 7 and all bond lengths and bond angles, one seeks to determine the torsion angles $\phi_1, \phi_2, \phi_3, \phi_4, \phi_5, \phi_6$ in order to connect to the skeletal atoms 5, 6, and 7 and to the chain beyond in a manner that respects the bonded geometry of the chain (see Fig. 3.4). The solution strategy starts by expressing the position vector $\mathbf{r}_5$ and the unit vector $\hat{\mathbf{b}}_6$ along bond 56 in the local coordinate frame of bond 1 as functions of the torsion angles ϕ_1 through ϕ_5. Flory's transformation matrices between the local coordinate frames of skeletal bonds are used in deriving these functional expressions. The functions expressing $\mathbf{r}_5$ and $\hat{\mathbf{b}}_6$ in terms of ϕ_1 through ϕ_5 are set equal to the prespecified values of $\mathbf{r}_5$ and $\hat{\mathbf{b}}_6$, giving a set of five equations in five unknowns. By eliminating ϕ_2 through ϕ_5, this set is reduced to a single equation in one unknown, ϕ_1, which is solved numerically for all roots in $[-\pi,\pi)$ [22]. The angle ϕ_6 is computed at the end from the position of $\mathbf{r}_4$ at the solution and the known $\mathbf{r}_5$, $\mathbf{r}_6$, and $\mathbf{r}_7$.

Wu and Deem [17] discovered a very useful analogy between the rebridging problem and inverse kinematics problems involving serial chain manipulators

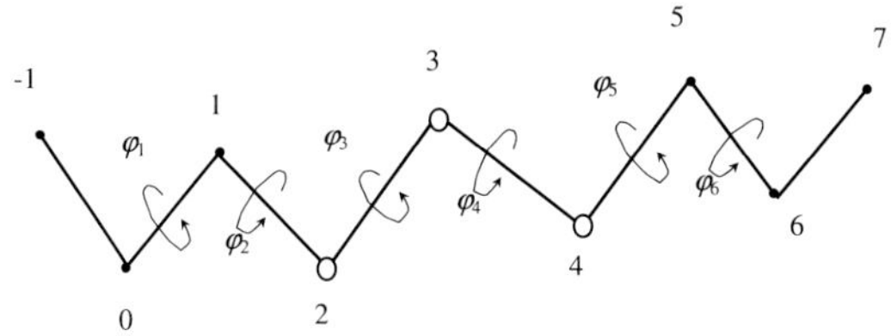

Fig. 3.4. Formulation of the trimer bridging problem in generalised coordinates, followed in [22]. The positions of atoms −1, 0, 1, 5, 6, and 7 are given. The objective is to determine the torsion angles ϕ_1 through ϕ_6 that allow bridging through the triplet of skeletal atoms $\{2, 3, 4\}$ such that all bonds lengths and bond angles are specified.

in robotics. In the latter, one is concerned with controlling the position and orientation of an end-effector through appropriate rotations of joints in a mechanism of links and joints. Figure 3.5 shows a six-degree of freedom serial chain manipulator with the notation used in robotics. For zero link lengths, the problem of controlling the end-effector can be mapped directly onto the bridging problem, as shown in part (b) of the figure. This analogy allowed Wu and Deem to adopt the analytical solution procedure already developed in the robotics literature for the general spatial 7-link 7R mechanism problem [26–28] to the general bridging problem in polymers (Fig. 3.1). In this procedure, the problem is reduced to a 16th degree polynomial equation in $tan(\phi_1/2)$, expressing the requirement that an 8×8 determinant be zero. The determinant equation is reformulated and solved for all is solutions as an eigenvalue problem. Clearly, apart from special "krankshaft" geometries admitting an infinity of solutions [22], citepanteb, the maximum number of geometric solutions to the bridging problem is limited to 16. Using the analytical rebridging algorithm in combination with parallel tempering, Wu and Deem [17] were able to achieve sufficient equilibration in order to study *cis-trans* isomerization of proline-containing cyclic peptides.

Whatever the algorithm used to solve the bridging problem, it is important to realise that using bridging to describe the local configuration of a chain system involves a non-metric preserving transformation of coordinates from Cartesian coordinates of the atoms to the generalised coordinates which are constrained, specifying the bonded geometry that must be respected during the bridging construction. Consider the trimer bridging problem, as described in Fig. 3.2 . The configuration of the trimer is ordinarily described in terms of the Cartesian coordinates of its constituent atoms, i.e. $\{\mathbf{r}_3, \mathbf{r}_4, \mathbf{r}_5\}$; we will call this coordinate set I. In using the bridging construction, however, the trimer is described indirectly through the positions of the surrounding atoms and the bond lengths and bond angles that are specified during the bridging construction. In trimer bridging, one commonly sets the values $\{l_{12}, l_{23}, l_{34}, l_{45}, \theta_1, \theta_2, \theta_3, \theta_4, \theta_5\}$ independently; we will call this coordinate

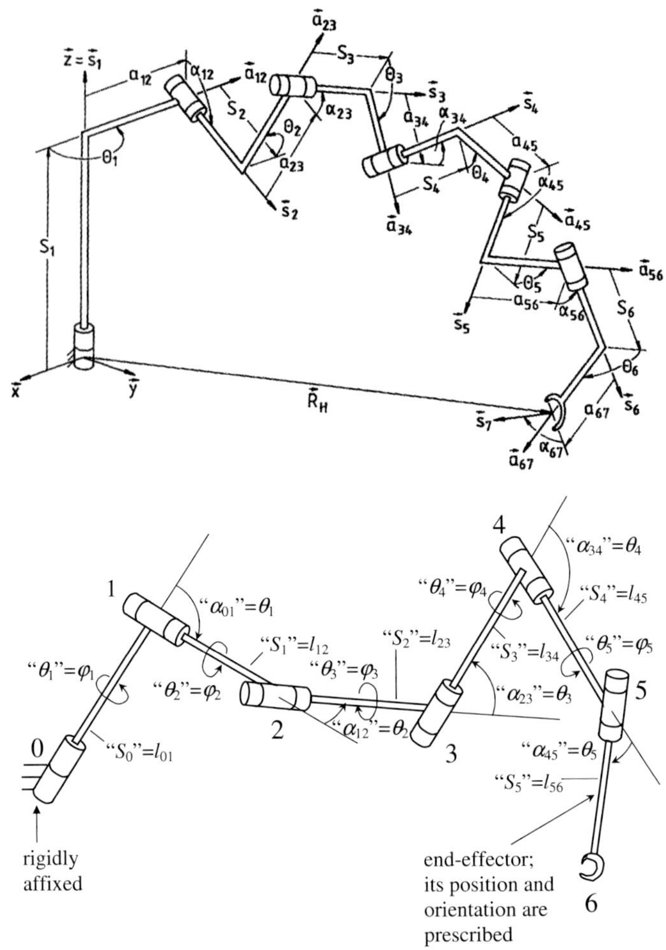

Fig. 3.5. Analogy between the bridging problem and inverse kinematics problems for serial chain manipulators in robotics. (a) A six-degree of freedom serial chain manipulator, represented in the Denavit-Hartenberg (D-H) equivalent geometry. The first pivot is rigidly affixed to the laboratory frame. The end-effector must be controlled, so as to have prescribed position and orientation. The device consists of links $\mathbf{a}_{ij}$ and joints $\mathbf{s}_i$. Each joint can rotate about its axis, thanks to the pivot on one of its ends. Pivots are rigidly affixed to the links preceding them. Successive links and joints are perpendicular. The length of joint $\mathbf{s}_i$ is denoted as S_i. Angles between the directions of successive joints (i, j) are denoted as α_{ij}. Dihedral angles of rotation around the axes of the joints are denoted as θ_i. (b) Special case of the six-degree of freedom serial chain manipulator for zero link lengths ($a_{ij} = 0$). This device is entirely analogous to the six-skeletal bond sequence considered in the trimer bridging problem, Fig. 3.2. Joint lengths S_i, interjoint angles $\alpha_{i-1,i}$ and dihedral angles θ_i of the robotics problem become bond lengths $l_{i,i+1}$, bond angle supplements θ_i, and torsion angles ϕ_i in the polymer problem, respectively. The figure indicates this mapping, with the robotics notation in quotation marks. (Part (a) is from [28], with permission).

set IV. The ratio of volume elements around a given configuration according to the latter description and according to the Cartesian coordinate description is given by the Jacobian:

$$J_{\mathrm{IV}\to\mathrm{I}} = \left| \frac{\partial(\mathbf{r}_2, \mathbf{r}_3, \mathbf{r}_5)}{\partial(l_{12}, l_{23}, l_{34}, l_{45}, \theta_1, \theta_2, \theta_3, \theta_4, \theta_5)} \right| \tag{3.3}$$

Detailed derivations useful in the calculation of $J_{\mathrm{IV}\to\mathrm{I}}$ are given in Appendix C of [21]. The result is

$$J_{\mathrm{IV}\to\mathrm{I}} = l_{12}^2 l_{23}^2 l_{34}^2 l_{45}^2 \sin\theta_1 \sin\theta_2 \sin\theta_3 \sin\theta_4 \sin\theta_5 \frac{1}{\mathcal{B}} \tag{3.4}$$

where

$$\begin{aligned}
\mathcal{B} = \Big| & -l_{12}l_{23}l_{45}[(\hat{\mathbf{b}}_1 \times \hat{\mathbf{b}}_2)\cdot\hat{\mathbf{b}}_3][(\hat{\mathbf{b}}_2 \times \hat{\mathbf{b}}_5)\cdot\hat{\mathbf{b}}_6][(\hat{\mathbf{b}}_3 \times \hat{\mathbf{b}}_4)\cdot\hat{\mathbf{b}}_5] \\
& + l_{12}l_{34}l_{45}[(\hat{\mathbf{b}}_1 \times \hat{\mathbf{b}}_2)\cdot\hat{\mathbf{b}}_5][(\hat{\mathbf{b}}_2 \times \hat{\mathbf{b}}_3)\cdot\hat{\mathbf{b}}_4][(\hat{\mathbf{b}}_4 \times \hat{\mathbf{b}}_5)\cdot\hat{\mathbf{b}}_6] \\
& + l_{23}l_{34}[(\hat{\mathbf{b}}_1 \times \hat{\mathbf{b}}_2)\cdot\hat{\mathbf{b}}_3]\{l_{12}[(\hat{\mathbf{b}}_2 \times \hat{\mathbf{b}}_3)\cdot\hat{\mathbf{b}}_4] - l_{45}[(\hat{\mathbf{b}}_3 \times \hat{\mathbf{b}}_4)\cdot\hat{\mathbf{b}}_5]\}[(\hat{\mathbf{b}}_4 \times \hat{\mathbf{b}}_5)\cdot\hat{\mathbf{b}}_6] \\
& + l_{12}l_{34}^2[(\hat{\mathbf{b}}_1 \times \hat{\mathbf{b}}_2)\cdot\hat{\mathbf{b}}_4][(\hat{\mathbf{b}}_2 \times \hat{\mathbf{b}}_3)\cdot\hat{\mathbf{b}}_4][(\hat{\mathbf{b}}_4 \times \hat{\mathbf{b}}_5)\cdot\hat{\mathbf{b}}_6] \\
& - l_{23}^2 l_{45}[(\hat{\mathbf{b}}_1 \times \hat{\mathbf{b}}_2)\cdot\hat{\mathbf{b}}_3][(\hat{\mathbf{b}}_3 \times \hat{\mathbf{b}}_4)\cdot\hat{\mathbf{b}}_5][(\hat{\mathbf{b}}_3 \times \hat{\mathbf{b}}_5)\cdot\hat{\mathbf{b}}_6] \Big|
\end{aligned} \tag{3.5}$$

In the above, $\hat{\mathbf{b}}_i$ stands for the unit vector along bond $(i-1, i)$.

The quantity $\mathcal{B}$ appearing in (3.4) and (3.5) can also be written as [21,22]

$$\mathcal{B} = \left| \frac{1}{(\hat{\mathbf{b}}_6\cdot\mathbf{e}_3)} \frac{\partial(\mathbf{r}_5, \hat{\mathbf{b}}_6)}{\partial(\phi_1, \phi_2, \phi_3, \phi_4, \phi_5)} \right| \tag{3.6}$$

in terms of the Jacobian of transformation from torsion angles to the variables set in the solution procedure of [22]. Since $\hat{\mathbf{b}}_6$ is a unit vector, it can contribute only two rows to the Jacobian of (3.6), conveniently obtained by differentiating the components of $\hat{\mathbf{b}}_6$ along two of the coordinate axes of the laboratory frame of reference; $\mathbf{e}_3$ is the unit vector along the third coordinate axis of the laboratory frame.

3.3 Monte Carlo Algorithms Based on the Bridging Construction

3.3.1 Concerted Rotation

The bridging construction is readily implemented in an intramolecular MC move aimed at rearranging the internal conformation of chains; such a move, which modifies the local conformation of an internal chain section while leaving the preceding and following parts of the chain unaffected, has come to be known as a concerted rotation (CONROT). In the single driver version of CONROT, introduced originally in 1993 [22], a chain is selected at random and a triplet of internal skeletal atoms $\{2, 3, 4\}$ is picked randomly along its contour. The triplet $\{2, 3, 4\}$ is excised and the torsion angle ϕ_0 (the driver angle) is rotated by a random amount, so as to displace atom 1, which precedes the excised triplet (see Fig. 3.6a). Skeletal atom 1 is then rebridged to atoms 5, 6 and the rest of the chain through a trimer bridging construction, leading to new positions for atoms 2, 3, and 4. The double driver (symmetric) version of CONROT [29] starts again by excising a triplet of skeletal atoms $\{2, 3, 4\}$. It then proceeds to modify two driver angles, ϕ_0 and ϕ_7, so as to displace the skeletal atoms 1 and 5, which flank the excised trimer (see Fig. 3.6b). The latter two atoms, in their new positions, are reconnected through a trimer bridging construction, leading to new positions for the triplet $\{2, 3, 4\}$. Four (five) skeletal atoms are displaced and seven (eight) torsion angles are turned in the single (double) driver version of CONROT.

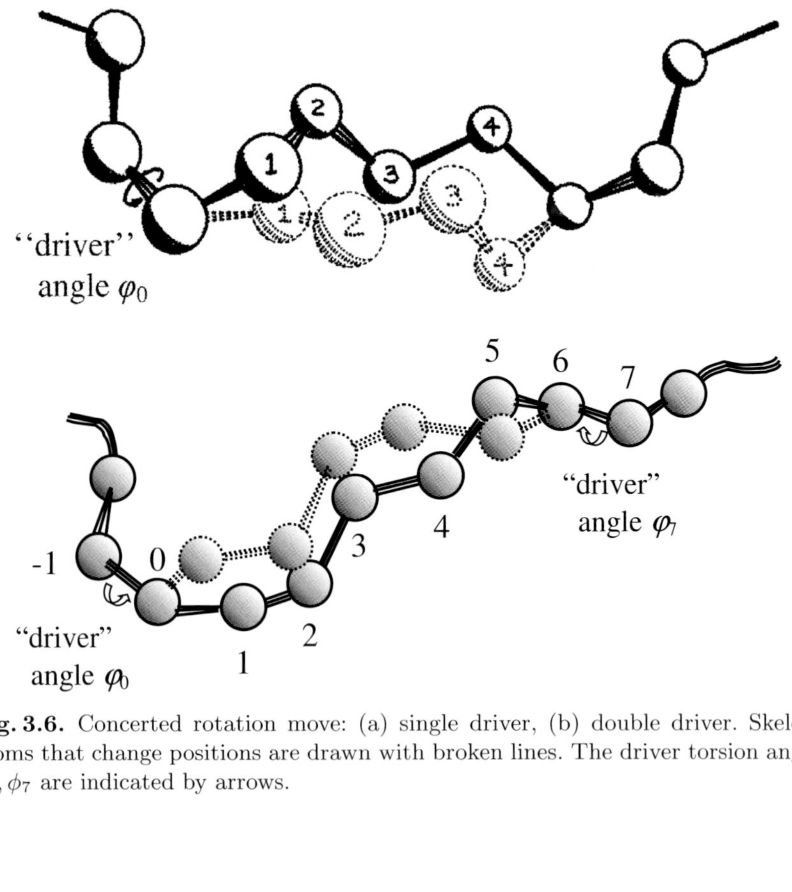

Fig. 3.6. Concerted rotation move: (a) single driver, (b) double driver. Skeletal atoms that change positions are drawn with broken lines. The driver torsion angles ϕ_0, ϕ_7 are indicated by arrows.

CONROT can be used easily with both fixed and flexible bond length and bond angle models. When flexible bond angles $\pi - \theta$ are used, each contributing a term $\mathcal{V}_\theta(\theta)$ to the Hamiltonian, one can either opt to retain the old values for all bond angles, relying on some other move (such as the "flip" [21]) to modify the bond angles, or reselect the bond angles according to the statistical weight

$$w_\theta(\theta) = \frac{\sin(\theta)\exp[-\beta\mathcal{V}_\theta(\theta)]}{\int_0^\pi \sin(\theta)\exp[-\beta\mathcal{V}_\theta(\theta)]d\theta}. \tag{3.7}$$

In the case of the double driver move, the latter option would result in a statistical weight $W_\theta = \prod_{k=1}^{5} w_\theta(\theta_k)$ associated with the bond angles selected for the new configuration.

As mentioned above, the trimer bridging construction can have up to 16 solutions. Although microscopically reversible schemes circumventing the need for calculating all solutions have been invented [29], it is recommended that all solutions be found; this enhances the ergodicity of exploring configuration space, and can be accomplished with existing very efficient algorithms [21,17]. In attempting a CONROT move, one of the solutions may be chosen at random. It is more advisable, however, to introduce a bias that avoids excluded-volume overlaps with surrounding atoms and strained conformations. In other words, one attempts a transition from the original configuration i to the configuration j dictated by one of the $N_{\text{soln}}(i \rightarrow)$ different solutions of the bridging problem according to a weight

$$W_{\text{bridge}}(i \rightarrow j) = \frac{\exp[-\beta\mathcal{V}_{\text{bridge}}(j)]}{\sum_{k=1}^{N_{\text{soln}}(i\rightarrow)} \exp[-\beta\mathcal{V}_{\text{bridge}}(k)]} \tag{3.8}$$

A good choice for the bias potential $\mathcal{V}_{\text{bridge}}$ incorporates a hard-sphere interaction, based on the assignment of a hard-sphere diameter to all atoms displaced by the move, and the torsional potential associated with all torsion angles modified [29]. An efficient strategy is to first screen all solutions for excluded-volume overlaps, then screen the remaining solutions for excessive torsional energy, and finally pick one of the surviving solutions with probability proportional to the Boltzmann factor of the torsional energy [29]. If no solution survives the excluded volume overlap and torsional screenings, i.e., if $\exp[-\beta\mathcal{V}_{\text{bridge}}(k)] \simeq 0$ for all k, then the move must be discarded. The same must be done in those rare cases where the original configuration does not pass the overlap and torsional screenings. The screening for excluded-volume overlaps can be made efficient by use of a screening overlap list [21].

To ensure microscopic reversibility, all solutions to the reverse bridging problem, starting from the trial configuration j, must be determined, corresponding to equal and opposite changes in the driver angles. The original configuration i must, of course, be found among the solutions of the inverse problem, along with an additional $N_{\text{soln}}(j \rightarrow) - 1$ "sister" solutions where, generally, $N_{\text{soln}}(j \rightarrow) \neq N_{\text{soln}}(i \rightarrow)$.

The attempted CONROT move from configuration i to configuration j is accepted with probability

$$P_{\text{accept}}(i \rightarrow j) = \min \left[1, \frac{\alpha(j \rightarrow i) \exp[-\beta \mathcal{V}(j)] J(j)}{\alpha(i \rightarrow j) \exp[-\beta \mathcal{V}(i)] J(i)}\right] \tag{3.9}$$

where $\mathcal{V}(k)$ is the total potential energy function of configuration k; $J(k)$ is the Jacobian $J_{\text{IV}\rightarrow\text{I}}$ of (3.4), evaluated at configuration k; and $\alpha(i \rightarrow j)$ is proportional to the probability of attempting a move from configuration i to configuration j. For the flexible bond angle scheme described above,

$$\alpha(i \rightarrow j) = W_\theta(j) W_{\text{bridge}}(i \rightarrow j) \tag{3.10}$$

It is advisable to test the acceptance scheme on a simple problem first, to make sure that the Jacobian and bias terms have been incorporated correctly. A system of independent "phantom chains", i.e. chains with the same bond lengths and bond angle potentials as the system of interest but devoid of any torsional potential and nonbonded interactions, constitutes a convenient test system. For such a test system, the distribution of torsion angles should come out uniform [22].

CONROT can be used in any constant-temperature ensemble (canonical, isothermal-isobaric [22,30,31], grand canonical, Gibbs, $f_1 N_2 PT$ [32]). The original application [22] involved NPT simulations of oligomeric liquids of polyethylene in a united-atom representation using a combination of CONROT and reptation moves, but since then the move has been used for many other polymers, including atactic, isotactic, and syndiotactic polypropylene [22,33], cis-1,4 polyisoprene [34,35], trans-1,4 polyisoprene [36], cis-1,4 polybutadiene and 1,2 polybutadiene [36]. The success rate of CONROT moves with a maximum change of 20° in the driver angle(s) is around 20% at usual polymer melt temperatures [22,21]. In the area of biological macromolecules, CONROT-based algorithms have been used very effectively by Deem and collaborators [25,17] in sampling equilibrium conformations of cyclic peptides, a feat very difficult to accomplish with MD.

3.3.2 Directed Internal Bridging

Uhlherr [37], Wick and Siepmann [38], and Uhlherr et al. [39] have developed combinations of the CONROT and CB algorithms, termed "internal configuration bias" (ICB), "self-adapting fixed end point configuration bias" (SAFE-CB), and "directed internal bridging" (DIB), respectively. In these algorithms, two bonds are rebridged not by a chain section of predetermined internal geometry (e.g., trimer), but through a flexible chain section started off of one of the bonds and grown towards the other in a bond–by–bond fashion through a CB scheme. Here we briefly describe Uhlherr's DIB scheme [39].

The l skeletal atoms constituting the bridging part of the chain are split into a set χ of $l-3$ atoms, which are regrown by CB, and another set ψ of 3 atoms, which are placed by the trimer bridging construction to establish the final connection. The bias potential $\mathcal{V}_{\text{CB}}(\chi)$ used in regrowing the set χ includes a fictitious potential which attracts each regrown atom k in χ towards a suitable "attractor" site m, usually chosen as the old position of the first atom in the bridging trimer. A finitely extensible nonlinear elastic (FENE) spring potential is used for this purpose, of the form

$$\mathcal{V}_{\text{FENE}}(r_{km}) = -\frac{1}{2} C R_{km}^2 \ln \left[1 - \left(\frac{r_{km}}{R_{km}} \right)^2 \right] \tag{3.11}$$

with r_{km} being the spatial distance between the regrown atom and the attractor site, and $R_{km} = 1.275(m+1-k)$ Å.

A DIB move from configuration i to configuration j is accepted with probability

$$P_{\text{accept}}(i \to j) =$$

$$\min \left[1, \frac{W_{\text{CB}}(\chi_j) \exp[\beta \mathcal{V}_{\text{CB}}(\chi_j)] \alpha(\chi_i \to \psi_i) \exp[-\beta \mathcal{V}(j)] J(\psi_j)}{W_{\text{CB}}(\chi_i) \exp[\beta \mathcal{V}_{\text{CB}}(\chi_i)] \alpha(\chi_j \to \psi_j) \exp[-\beta \mathcal{V}(i)] J(\psi_i)} \right] \tag{3.12}$$

where W_{CB} and $\mathcal{V}_{\text{CB}}$ are the Rosenbluth weights and bias potentials used in regrowing part χ of the bridging chain by configurational bias in the forward (subscript j) and backward (subscript i) direction [see (3.2)]; α are the statistical weights used in the trimer bridging construction of part ψ of the solution in the forward (subscript j) and backward (subscript i) direction [see (3.10)]; J are the Jacobian determinants associated with the bridging construction of ψ in the forward (subscript j) and backward (subscript i) direction [see (3.4)]; and $\mathcal{V}_i$, $\mathcal{V}_j$ are the total potential energies of the initial and final configuration, respectively, their Boltzmann weights in the acceptance criterion arising from the probability density of the simulated ensemble.

Combined CONROT and CB schemes have been used in simulating long cyclic alkane molecules in vacuo [37], alkane liquids [38] and polyethylene melts [39]. Uhlherr et al. [39] have presented a detailed analysis of the computational efficiency of DIB, concluding that DIB is comparable with, but not superior, to the simpler CONROT; the computational overhead of CB is not counterbalanced by better sampling of configuration space, contrary to what happens in the case of DEB, discussed below.

3.3.3 End-Bridging in the $Nn\mu^*PT$ Ensemble

CONROT is very useful in inducing local configurational rearrangements in atomistic models of chain polymers. It is not sufficient, however, for equilibrating structural features at large length scales, such as the end-to-end distance and the radius of gyration, in systems of long chains. This became obvious in the original NPT simulations of oligomeric polyethylene melts with a MC algorithm using CONROT and reptation moves; the algorithm could fully equilibrate a C_{24}, but not a C_{78} melt [30].

The need to equilibrate large length-scale structural features led to the development of a "bolder", intermolecular move based on the bridging construction, which is termed "end-bridging" (EB) [29]. As shown in Fig. 3.7, this move involves two chains, c and c'. The triplet of skeletal atoms $\{2, 3, 4\}$ is excised from the chain c, generating two new chain ends, 1 and 5. Next, the bridging construction is invoked to generate a new triplet of atoms $\{2', 3', 4'\}$, which connects the end $1'$ of chain c' to atom 5 of chain c. The end result is a dramatic change in connectivity: Chain c' (the "attacking" chain) has appended 5, 6, and the subsequent part of chain c though the newly constructed trimer $\{2', 3', 4'\}$. Chain c (the "victim" chain), on the other hand, has been shortened, now terminating at 1. Although the scheme is illustrated with a trimer, one could clearly have excised a larger internal segment and repositioned it as a bridge of predetermined internal geometry. The bridging construction guarantees that the detailed atomistic geometry of the chains is preserved.

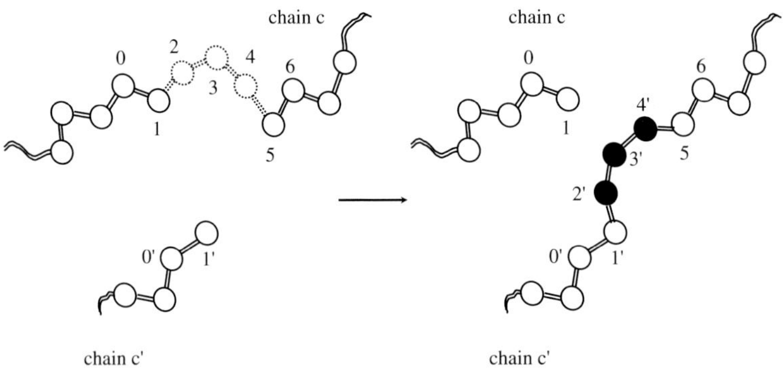

Fig. 3.7. Schematic of the end-bridging move. Initial configuration is shown on the left, and final configuration on the right. Excised atoms are outlined with broken lines and newly constructed atoms as filled circles. All other atoms (open circles) remain unchanged.

Clearly, an end-bridging move modifies the lengths of the chains participating in it. Thus, the MC simulation must be cast in an ensemble that allows chains of various lengths to be present, i.e. some polydispersity in the polymer. An appropriate "semigrand" ensemble was formulated by Pant and Theodorou [29]. They considered the polymer as a mixture of various chain species. Starting from the differential expression for the Helmholtz energy of such a mixture and introducing variable substitutions and Legendre transformations, they arrived at a fundamental equation, and therefore a partition function, describing the mixture in terms of the following thermodynamic variables: the total number of chains, N, the total number of monomer units, n, the temperature T, the pressure, P, and a set of relative chemical potentials, μ_k^*, for all chain species k but two (k_1, k_2), which are taken as reference species. Each relative chemical potential μ_k^* is defined in terms of the actual chemical potentials of the species as

$$\mu_k^* = \mu_k - \left(\frac{X_k - X_{k_1}}{X_{k_2} - X_{k_1}} \right) \mu_{k_2} - \left(\frac{X_k - X_{k_2}}{X_{k_1} - X_{k_2}} \right) \mu_{k_1} \tag{3.13}$$

with X_l being the number of monomer units in each chain of species l. The probability density of the $Nn\boldsymbol{\mu}^*PT$ ensemble in configuration space is [29]

$$\rho^{Nn\boldsymbol{\mu}^*PT}(V, \mathbf{r}; \text{connectivity}) = \tag{3.14}$$

$$\text{const} \exp \left[\beta \sum_{\substack{k=1 \\ k \neq k_1, k_2}}^{m} \mu_k^* N_k - \beta PV - \beta \mathcal{V}(\mathbf{r}; \text{connectivity}) \right],$$

where N_k stands for the number of molecules of species k, and m is the maximum number of species that may be present. V is the system volume and and $\mathbf{r}$ is the vector of coordinates of all atoms, specifying the microscopic configuration of the system.

To facilitate the implementation of the EB move, special lists are maintained of candidate atoms for bridging (e.g. atom 5 in Fig. 3.7) around each chain end (e.g. end 1′ in Fig. 3.7). The criterion for inclusion in these "end-bridging lists" is that the candidate atom must be an internal segment of another chain within a maximal "bridging" distance, beyond which solution of the bridging problem is impossible; for united-atom polymethylene, this maximal distance is taken equal to the all-trans end-to-end distance of the heptamer $\{0, 1, 2, 3, 4, 5, 6\}$ with all bond angles $\pi - \theta$ opened to a value whose equilibrium probability is less than 10^{-3} at the prevailing temperature. The number of bridgeable neighbours N_{bridge} is continuously updated for each chain end in the system.

The EB move is initiated by randomly selecting the "attacking" chain end (atom 1′ in Fig. 3.7). One of the N_{bridge} bridgeable neighbours of the attacking end is then selected at random (atom 5 in Fig. 3.7). One then proceeds to bridge $0'1'$ and 56, forming atoms $2', 3'$, and $4'$. As discussed above, the bridging construction involved in EB generally has multiple solutions. All solutions are determined and one of them is chosen after overlap screening and weighting by the Boltzmann factor of the torsional potential, as in the case of CONROT [compare (3.8), (3.10)]. The probability of accepting such an EB move from the initial configuration i to the final configuration j is then

$$P_{\text{accept}}(i \to j) = \min \left[1, \frac{\frac{1}{N_{\text{bridge}}(j \to i)} \alpha(j \to i) \rho^{Nn\boldsymbol{\mu}^* PT}(j) J(j)}{\frac{1}{N_{\text{bridge}}(i \to j)} \alpha(i \to j) \rho^{Nn\boldsymbol{\mu}^* PT}(i) J(i)} \right] \quad (3.15)$$

As in the case of CONROT, the inverse bridging problem, starting from the final configuration j, must also be solved. In the acceptance criterion of (3.15), the terms involving N_{bridge} account for the random selection of one of the bridgeable neighbours of the attacking chain end when the move is attempted in the forward (denominator) and in the reverse (numerator) direction. The remaining terms are entirely analogous to the ones appearing in the corresponding acceptance criterion for CONROT, (3.9). The α terms account for the bias in attempting the move according to a particular solution, among all solutions of the geometric bridging problem, in the forward (denominator) or reverse (numerator) direction. The quantities symbolised by ρ are the equilibrium probability densities of the ensemble being simulated in the destination (numerator) and in the origin(denominator) state, while the quantities symbolised by J are the Jacobians of transformation from the generalised coordinates used in the bridging construction to Cartesian coordinates [see (3.4)] in the destination (numerator) and origin (denominator) states.

By construction, the number-average chain degree of polymerisation of the polymer is $\bar{X} = n/N$. The profile of relative chemical potentials $\boldsymbol{\mu}^*$ controls the chain length (molecular weight) distribution at equilibrium. Pant and Theodorou [29] have considered several $\boldsymbol{\mu}^*$ profiles and calculated the expected distributions. Setting relative chemical potentials to $-\infty$ for all chain species shorter than $X_{\min}$, and equal to zero for $X_{\min}$ and longer, produces a truncated Flory number distribution function, which is zero below $X_{\min}$ and falls exponentially with X above $X_{\min}$. Setting relative chemical potentials to zero for all chain lengths within a symmetric window centred at $\bar{X}$ and to $-\infty$ outside that window produces a flat (box) number distribution within the window. On the other hand, setting relative chemical potentials equal to a parabolic function of the chain length with maximum at $\bar{X}$ within a symmetric window centred at $\bar{X}$ and to $-\infty$ outside that window, produces a truncated Gaussian number distribution within the window. As seen in Fig. 3.8, these expectations are confirmed by EBMC simulations.

The success rate of EBMC moves is normally very low (approximately 0.1% for united-atom polyethylene [21] and even lower for polymers with more complex chemical constitution). Nevertheless, the move is extremely efficient in equilibrating the long-length scale structural features of dense, long-chain systems. Assuming that chains retain their identity during an EB move as shown in Fig. 3.7 , the centre of mass of a given chain executes a random walk

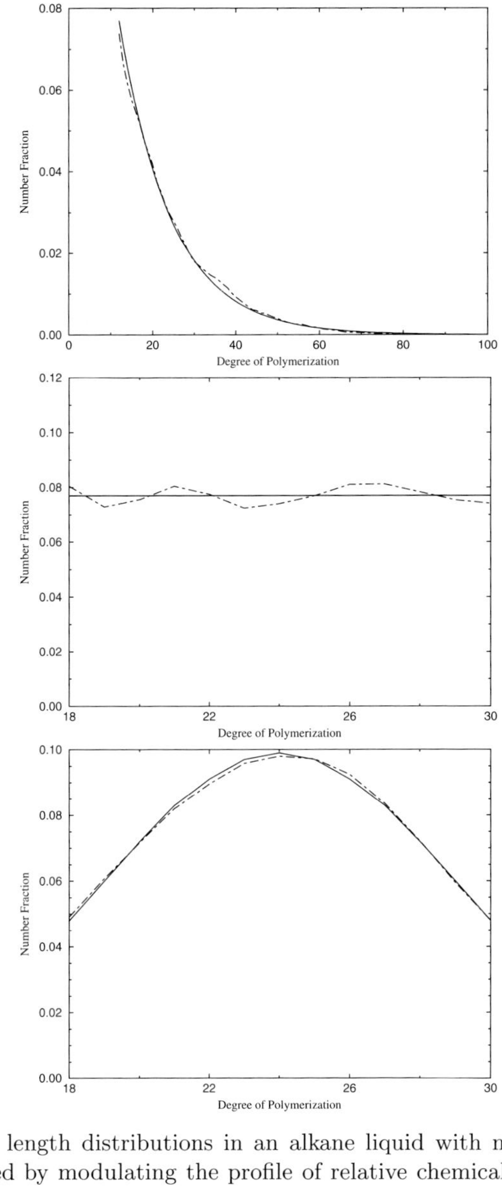

Fig. 3.8. Chain length distributions in an alkane liquid with mean chain length $\bar{X} = 24$, obtained by modulating the profile of relative chemical potentials. Lines depict theoretical expectations from the $Nn\mu^*PT$ ensemble, while dash-dotted lines are results from EBMC simulations at 450 K and 0.1 MPa. (a) truncated Flory distribution ($X_{\min} = 12$). (b) Flat distribution in the window of chain lengths [18, 30]. (c) Truncated Gaussian distribution in the window of chain lengths [18, 30]. (Reproduced from [29], with permission).

as a result of the chain appending or relinquishing mass in the successive EB moves in which it participates. Let us consider homologous melts wherein the shape of the chain length distribution is retained constant (i.e., the number distribution of chain lengths X is identical when plotted as a function of $X/\bar{X}$), but the mean chain length $\bar{X}$ is changed systematically. In a single successful EB move, the centres of mass of the two chains participating in the move will be displaced by a distance which scales as the overall mean spatial extent of chains, i.e. as the root mean square radius of gyration, hence as $\bar{X}^{1/2}$ [21]. The remaining $(N-2)$ chains will not be affected; hence, the mean square displacement of the centres of mass taken over all chains in a single move will scale as $\langle(\Delta\mathbf{r}_{\mathrm{cm}})^2\rangle_{\mathrm{EBmove}} \propto \bar{X}/N$. Within CPU time t, with τ being the mean CPU time per attempted MC move and f_{EB} the frequency of attempting EB moves, there will be $f_{\mathrm{EB}}t/\tau$ EB moves attempted, of which $f_{\mathrm{EB}}p_{\mathrm{acc}}t/\tau$ will be accepted, where p_{acc} is the rate of acceptance of EB moves. The mean square displacement of chain centres of mass during time t will be $\langle(\Delta\mathbf{r}_{\mathrm{cm}})^2\rangle \propto f_{\mathrm{EB}}p_{\mathrm{acc}}\dfrac{t}{\tau}\dfrac{\bar{X}}{N}$. Therefore, the "self-diffusivity" of chain centres of mass with respect to CPU time as a result of EB moves will be $D \propto \dfrac{\langle(\Delta\mathbf{r}_{\mathrm{cm}})^2\rangle}{t} \propto f_{\mathrm{EB}}p_{\mathrm{acc}}\dfrac{1}{\tau}\dfrac{\bar{X}}{N}$. One measure of configurational rearrangement at the level of entire chains is the CPU time t_0 required for the centre of mass of chains to be displaced by a distance commensurate with the end-to-end distance R. This is

$$t_0 \propto \frac{R^2}{D} \propto \frac{\bar{X}}{f_{\mathrm{EB}}p_{\mathrm{acc}}\dfrac{1}{\tau}\dfrac{\bar{X}}{N}} \tag{3.16}$$

or

$$t_0 \propto \frac{\tau}{f_{\mathrm{EB}}p_{\mathrm{acc}}}\frac{n}{\bar{X}} \tag{3.17}$$

A more detailed analysis, taking into account the distribution of chain lengths, leads to [21]

$$t_0 \propto \tau\frac{n}{f_{\mathrm{EB}}\bar{X}\Delta^{2.5}} \tag{3.18}$$

where Δ is the width of the chain length distribution reduced by the mean chain length.

As expected, t_0 grows linearly with the system size n. Narrowing the molecular weight distribution increases t_0, as there are less chances for an attempted end-bridging move to lead to acceptable chain lengths. The remarkable feature of (3.18) is that, for given system size and shape of the molecular weight distribution, t_0 is inversely proportional to $\bar{X}$. In other words, EBMC becomes more efficient in rearranging chain centres of mass as the mean chain length increases. This is in sharp contrast to actual chain dynamics, whereby the maximal relaxation time of the melt grows with $\bar{X}$ as $\bar{X}^2$ (Rouse regime) or $\bar{X}^{3.4}$ (reptation regime). Herein lies the power of EBMC in equilibrating long-chain systems.

Figure 3.9 displays the mean square displacement of the centre of mass as a function of CPU time in five model polyethylene melts of the same size and the same reduced molecular weight distribution (flat distribution, polydispersity index 1.08). CPU times in these figures are on a SGI 64-bit R10000 processor. Clearly, the model with $\bar{X} = 500$ displays the largest centre-of-mass displacement. In Fig. 3.10, the CPU time t_0 corresponding to the same five runs is displayed as a function of $\bar{X}$ in log-log coordinates. At large $\bar{X}$, a slope of -1 is reached, confirming the scaling analysis that led to (3.18).

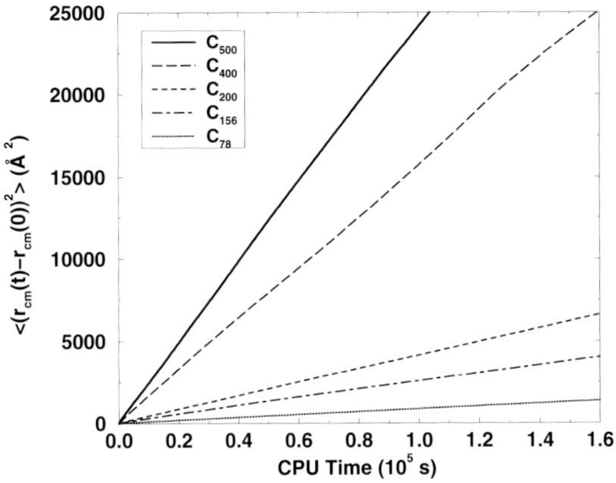

Fig. 3.9. Mean-square displacenent of the centre of mass of chains, $\langle[\mathbf{r}_{\mathrm{cm}}(t) - \mathbf{r}_{\mathrm{cm}}(0)]^2\rangle$, as a function of the CPU time t, obtained from EBMC simulations of five polyethylene melts with a flat molecular weight distribution and polydispersity index 1.08. The number of methylene and methyl units in the simulation box is $n = 4000$ in all cases and the simulation conditions are $T = 450K$ and $P = 1$ atm. The mean molecular weights $\bar{X}$ from bottom to top are $78, 156, 200, 400$, and 500, as indicated in the legend. (Reproduced from [21], with permission).

Another measure of overall conformational rearrangement is the rate at which a unit vector $\mathbf{u} = \mathbf{R}/|\mathbf{R}|$, directed along the end-to-end vector of a chain, loses memory of its original orientation. In Fig. 3.11, autocorrelation functions $\langle\mathbf{u}(t)\cdot\mathbf{u}(0)\rangle$ obtained with EBMC are shown as functions of the CPU time t for the five melt systems discussed above. The decorrelation is complete in all cases but one. Remarkably, the decorrelation is faster the larger the mean molecular length $\bar{X}$.

The equilibration of local structural features achieved by EBMC is also very good, thanks mainly to the reptation and CONROT moves accompanying EB, but also to the trimer displacements brought about by EB. This is readily seen by comparing mean square displacements of individual chain segments as functions of CPU time for EBMC, CBMC, and MD algorithms [21]. Inevitably, with the scaling of (3.18) for large-scale structural features, the

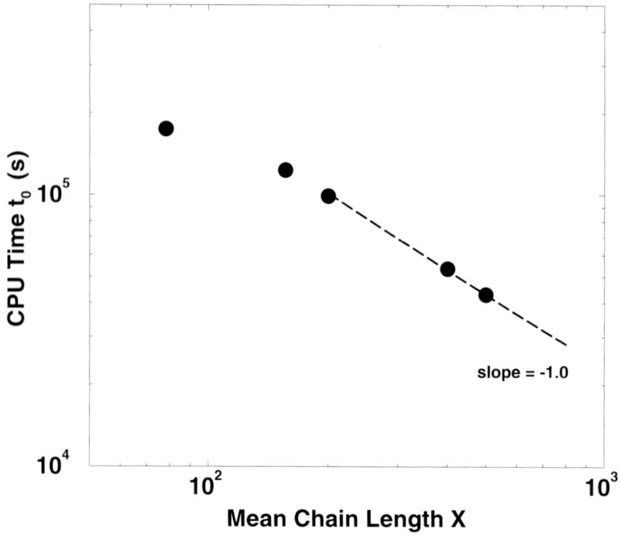

Fig. 3.10. CPU time t_0 required for the mean-square displacement of chain centres of mass in united-atom polyethylene melts simulated with EBMC to reach the equilibrium mean-square end-to-end distance $< R^2 >$ versus average chain length $\bar{X}$ on a log-log plot. The size of all model systems is $n = 4000$ and the chain length distribution is flat with polydispersity index 1.08 in all cases. The simulation conditions are $T = 450K$ and $P = 1$ atm. (Reproduced from [21], with permission).

equilibration of local structural features will become rate-controlling for high molecular weights, i.e. equilibrating the polymer becomes no more difficult than equilibrating a low-molecular weight analogue. This is seen characteristically in the simulations of C_{6000} discussed by Uhlherr et al. [40].

In long-chain, dense polymer systems where chain ends are scarce, the number of feasible candidate EB moves available at any time is relatively small. If the frequency of EB relative to other moves is high, it is possible that successive accepted EB moves may reverse each other, with little configurational rearrangement having occurred in-between. Such "shuttling" behavior is detrimental to equilibration. Introducing a good proportion of reptation moves greatly helps in avoiding this undesirable situation, as reptation induces drastic changes in the configuration of chain ends. The shuttling problem can also be overcome by resorting to directed bridging algorithms. In systems with a bulky repeat unit, where reptation and directed bridging may have very low probabilities of acceptance, shuttling can be suppressed by combining EB with parallel tempering. Double bridging and intramolecular double rebridging, which involve exclusively internal segments and no chain ends, are very effective in eliminating shuttling. All these algorithmic developments are discussed below.

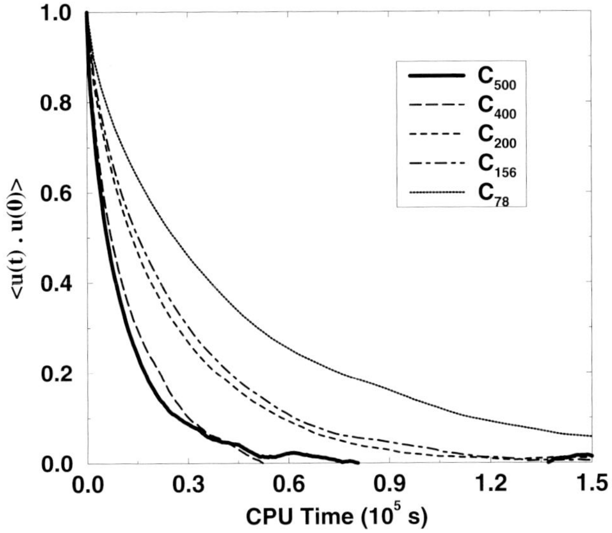

Fig. 3.11. Decay of the chain end-to-end vector orientational autocorrelation function $\langle \mathbf{u}(t) \cdot \mathbf{u}(0) \rangle$ with CPU time t as obtained from EBMC simulations of five united-atom polyethylene melts. The mean chain lengths $\bar{X}$ are indicated in the legend. The size of all model systems is $n = 4000$ and the chain length distribution is flat with polydispersity index 1.08 in all cases. The simulation conditions are $T = 450K$ and $P = 1$ atm. (Reproduced from [21], with permission).

3.3.4 Directed End-Bridging

Uhlherr et al. [39] developed a "directed end-bridging" (DEB), which combines and bridging and configuration bias. The relation of DEB to EB is identical to the relation of DIB to CONROT, discussed above. A flexible intermolecular bridge of length $l > 3$ is constructed, after elimination of a section of equal length from the victim chain. The first section χ of the bridge, containing $l-3$ skeletal atoms, is built sequentially, one atom at a time, using a CB random walk. The CB bias potential contains attractive terms which direct χ towards the stationary chain section to which it will be bridged. The last trimeric section ψ of the bridge is built using the trimer bridging construction, as described in Sect. 3.3.3. The acceptance rate from configuration i to configuration j is given by (3.12), discussed in conjunction with DIB, with the modifications that led from (3.9) to (3.15). Uhlherr [39] has conducted a detailed study of the efficiency of DEB and of its combinations with CONROT, DIB, and reptation as a function of the bridge length l. Relative to EB, each DEB move displaces a larger number of atoms and has a somewhat higher acceptance ratio, but costs more CPU time. Much more significant for equilibration, however, is that DEB suppresses "shuttling" relative to simple EB. As a result, DEB enhances performance in equilibrating long-length scale structural characteristics by an additional order of magnitude relative to EB [39].

3.3.5 Sampling of Oriented Chains: $NnbT\mu^*\alpha$ MC Simulations

MC simulations employing dynamically unphysical moves can, of course, not provide direct dynamical information. They can, however, be used to compute free energies (potentials of mean force) with respect to a chosen set of slowly evolving structural variables directly from atomistic information. Such free energies, and their constituent configurational energy and entropy functions, are valuable as input to coarse-grained theories and simulation approaches for the description of dynamical behaviour under equilibrium and nonequilibrium conditions. In the case of polymers, where relaxation times are exceedingly long by MD standards, the avenue of combining MC and coarse-grained (e.g., irreversible thermodynamics-based) formulations seems more promising than direct MD simulation.

Mavrantzas and Theodorou [42] used EBMC to calculate the free energy of oriented polymer melts, resulting from the imposition of a steady-state flow field with strain rate smaller than or equal to the longest relaxation time of chains (Deborah number ≤ 1). Considering melts with chain length below that corresponding to the onset of entanglements, they introduced the conformation tensor $\tilde{\mathbf{c}}$ as a coarse-grained descriptor of the overall shape of chains:

$$\tilde{\mathbf{c}} = 3\left\langle \frac{\mathbf{RR}}{< R^2 >_0} \right\rangle \tag{3.19}$$

where $\mathbf{R}$, as above, is the end-to-end vector of a chain, $< R^2 >_0$ is the mean square end-to-end vector under equilibrium unperturbed conditions, and the outer average is taken over all chains in the system. As defined, $\tilde{\mathbf{c}}$ is equal to the unit tensor in a quiescent melt at equilibrium; it departs from unity in a flow field of Deborah number comparable to 1, where chains do not have the time to get back to their unperturbed conformations.

Following irreversible thermodynamics-based approaches to polymer flow [41], the free energy per chain of such a melt, A/N, was postulated to be a function not only of the mass density ρ and the temperature T, but also of the conformation tensor:

$$\frac{A}{N} = \frac{A}{N}(\rho, T, \tilde{\mathbf{c}}). \tag{3.20}$$

It is convenient to introduce the conjugate thermodynamic variables associated with ρ and $\tilde{\mathbf{c}}$ in the fundamental representation of (3.20):

$$b = -\left.\frac{\partial A}{\partial V}\right|_{N,T,\tilde{\mathbf{c}}} = \rho^2 \left[\frac{\partial}{\partial \rho}\frac{A}{N}(\rho, T, \tilde{\mathbf{c}})\right]_{T,\tilde{\mathbf{c}}} \frac{N_A}{M} \tag{3.21}$$

and

$$\alpha_{\gamma\delta} = \frac{1}{k_B T}\left[\frac{\partial}{\partial \tilde{c}_{\gamma\delta}}\frac{A}{N}(\rho, T, \tilde{\mathbf{c}})\right]_{T,\rho,\tilde{c}_{[\gamma,\delta]}} \tag{3.22}$$

where M is the mean molecular weight of the polymer, N_A is Avogadro's number, the indices $\gamma, \delta \in \{1, 2, 3\}$, and the notation $\tilde{c}_{[\gamma,\delta]}$ signifies that all other components of the tensor $\tilde{\mathbf{c}}$ remain constant except $\tilde{c}_{\gamma\delta}$.

Under equilibrium conditions, the scalar field b is merely the pressure. The tensorial "orienting field" $\boldsymbol{\alpha}$ couples to the conformation tensor, inducing orientation; it is dictated by the type and strain rate of the flow.

From a general relation expressing the stress $\boldsymbol{\tau}$ in terms of the derivative of the free energy with respect to the deformation gradient tensor, one arrives at the expression

$$\boldsymbol{\tau} = -b\mathbf{I} + 2k_BT\frac{N}{V}(\tilde{\mathbf{c}}\cdot\boldsymbol{\alpha}) \tag{3.23}$$

The stress in the flowing melt is therefore readily expressed in terms of b, $\tilde{\mathbf{c}}$, and $\boldsymbol{\alpha}$.

Mavrantzas and Theodorou [42] developed a MC procedure for extracting $\tilde{\mathbf{c}}$, A (relative to that of the equilibrium quiescent polymer), and $\boldsymbol{\tau}$ under given T, b and $\boldsymbol{\alpha}$ from atomistic simulations. The polymer is allowed to have some polydispersity, which is controlled through the profile of chemical potentials $\boldsymbol{\mu}^*$, as discussed in Sect. 3.3.3. Thus, the simulation is conducted in a $NnbT\boldsymbol{\mu}^*\boldsymbol{\alpha}$ ensemble, whose probability density is

$$\rho^{NnbT\boldsymbol{\mu}^*\boldsymbol{\alpha}}(V, \mathbf{r}; \text{connectivity}) = \text{const} \quad \times \tag{3.24}$$

$$\times \exp\left[-\beta\left(-\sum_{\substack{k=1\\ k\neq k_1,k_2}}^{m} \mu_k^* N_k + bV + \mathcal{V}(\mathbf{r}; \text{connectivity}) - k_BT\sum_{k=1}^{N}\boldsymbol{\alpha}{:}\tilde{\mathbf{c}}_k\right)\right]$$

The probability density of (3.24) is very similar to that of (3.14), the only differences being the use of the scalar field b in place of P and the extra energy term $-k_BT\sum_{k=1}^{N}\boldsymbol{\alpha}{:}\tilde{\mathbf{c}}_k$, which couples the orienting field with the conformation tensors $\tilde{\mathbf{c}}_k$ of individual chains in the system. The latter are defined through (3.19) sans the outer average, with $\mathbf{R}$ replaced by the end-to-end vector $\mathbf{R}_k$ of an individual chain k.

The oriented chain simulation is conducted with EBMC, using EB, CONROT, reptation, and local "flip" and end-rotation moves. The acceptance criterion for EB moves is given by (3.15) with the probability density terms replaced by those of (3.24). For each system studied, a series of simulations is conducted with b and $\boldsymbol{\alpha}$ chosen so as to correspond to flows of increasing intensity. The conformation tensor $\tilde{\mathbf{c}}$ and, therefore, the stress tensor $\boldsymbol{\tau}$ [(3.23)] are obtained as output from the simulation. The free energy A/N relative to the equilibrium quiescent state is obtained from the whole series of simulations by thermodynamic integration [21]. Excellent numerical performance of the EBMC algorithm was noted in sampling oriented united-atom polyethylene systems [42–44]. In sampling oriented melts, one should be especially careful in establishing that the results are independent of model system size. If the simulation box is small, interactions between different images of the same chain may develop in the oriented state, artificially enhancing the orientation.

Mavrantzas and Öttinger [45] reformulated this thermodynamic approach and associated MC simulation strategy for oriented melts on the basis of Grmela and Öttinger's general equation for the nonequilibrium reversible-irreversible coupling (GENERIC). By analyzing the structure of the resulting GENERIC equation for spatially homogeneous, time-independent flows, a firm kinematic interpretation was attributed to the orienting field $\boldsymbol{\alpha}$. This interpretation depends on the specific coarse-grained viscoelastic model assumed. For the Upper-Convected Maxwell and FENE-P models, it is shown that

$$\boldsymbol{\alpha} = \frac{1}{2}\lambda_H \dot{\boldsymbol{\gamma}} \tag{3.25}$$

with λ_H being the maximum relaxation time of chains and $\dot{\boldsymbol{\gamma}}$ being the rate of strain tensor of the flow [45]. On the other hand, the stress equation (3.23), in terms of b, $\tilde{\mathbf{c}}$, and $\boldsymbol{\alpha}$, is perfectly general for all single conformation tensor models. Mavrantzas and Öttinger have also formulated GENERIC-MC approaches for mapping atomistic models to multiple-conformation tensor viscoelastic models incorporating higher modes in the coarse-grained description, such as the Rouse model. Results from these formulations will be discussed in the Applications section.

3.3.6 Scission and Fusion Algorithms for Phase Equilibria

Being able to predict phase equilibria in polymer systems is very important from the technological point of view. At the same time, it constitutes a great challenge for molecular simulation. Commonly used simulation methods (Widom test particle insertions, grand canonical and Gibbs ensemble Monte Carlo) rely upon insertions/deletions or exchanges of molecules between phases, and these result in almost certain overlap when the molecules are large and the phases are dense, even if configurational bias techniques are invoked. Good reviews of the state of the art in simulations of phase equilibria are available [47,48].

Often one is interested in the phase equilibria of systems composed of chemically similar macromolecules. In such systems, connectivity-altering moves can be used to effect pseudochemical interconversions among the macromolecular species, thereby dramatically accelerating convergence to the equilibrium composition, while at the same time entirely circumventing the insertion problem.

Zervopoulou et al. [49] illustrated this in the problem of predicting the solubilities of C_4 to C_{20} linear alkanes in molten polyethylene. Allowing some polydispersity in the polymer, which they proved to be inconsequential for the properties of interest, they introduced two new MC moves, scission and fusion (see Fig. 3.12), to generate and eliminate alkane molecules in the polymer mass. In a scission move, one randomly selects a polymer chain end and then cuts the chain near that end in order to create an oligomeric (alkane) molecule of X_1 skeletal atoms. The oligomeric molecule is created through excision and elimination of a trimer from the original chain, which is thereby reduced in length from from X_k to $X_k - X_1 - 3$. In a fusion move, one randomly selects an oligomer molecule in the system and then connects one of its ends to an end of a long chain by construction of a trimer bridge. The oligomer is thereby eliminated and the chain increases in length from X_k to $X_k + X_1 + 3$ skeletal atoms. The bridging is performed by choosing one of the N_{bridge} bridgeable neighbouring chain ends to the attacking oligomer end from a list, choosing the bond angles θ according to a weight W_θ, and selecting one of the multiple solutions of the bridging construction according to a weight W_{bridge}, exactly as described in conjunction with CONROT and EB. In addition to scission and fusion, flip, reptation, CONROT and EB moves involving both the polymer and the oligomer, and volume fluctuation moves are used to equilibrate the configuration.

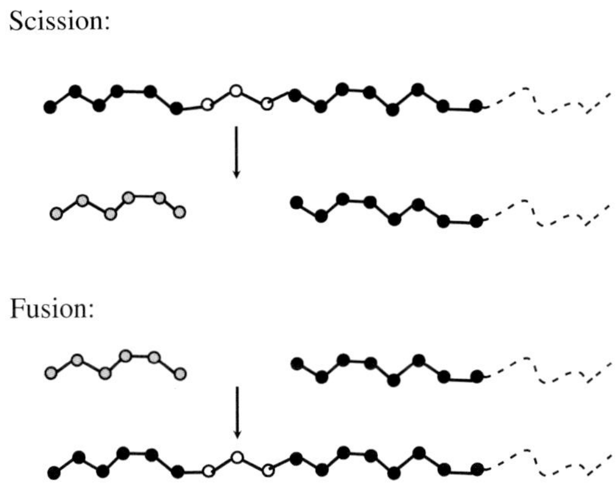

Fig. 3.12. Schematic of the scission (top) and fusion (bottom) moves. Segments of the long polymer chain and of the oligomer are shown in black and grey, respectively. The trimer bridge eliminated or formed is shown with open symbols.

Zervopoulou et al. [49] developed a new ensemble to implement scission and fusion moves. In this ensemble, dubbed the $f_1' N_\text{p} n_0 PT \boldsymbol{\mu}^*$, the pressure P, the temperature T, the total number of chains N_p, and the profile of relative chemical potentials $\boldsymbol{\mu}$ for all long-chain species but two are kept constant, as in the $NnPT\boldsymbol{\mu}^*$ ensemble discussed in Sect. 3.3.3. In addition, the number $n_0 = (X_1 + 3)N_1 + \sum_{k=2}^{m} X_k N_k$, species 1 being the oligomer and $2, \ldots, m$

being the chain species constituting the polymer, is held constant. Physically, this is the number of skeletal atoms that would result if all oligomer molecules present in a configuration were connected to polymer molecules through trimer bridges. Finally, a very important macroscopic parameter of the ensemble is the modified fugacity of the oligomer, f_1'. This is related to the oligomer fugacity f_1 and to the fugacities f, excess chemical potentials μ^{ex} and mole fractions x of the reference species k_1 and k_2 used in defining $\boldsymbol{\mu}^*$ through

$$
\begin{aligned}
f_1' &= f_1 \exp\left[-\beta(X_1+3)\frac{\mu_{k_1}^{\mathrm{ex}}-\mu_{k_2}^{\mathrm{ex}}}{X_{k_1}-X_{k_2}}\right] \times \\
&\quad \times \left(\frac{x_{k_2}}{x_{k_1}}\right)^{(X_1+3)/(X_{k_1}-X_{k_2})} \left(\frac{I_{k_1}^{\mathrm{intra}}}{I_{k_2}^{\mathrm{intra}}}\right)^{(X_1+3)/(X_{k_1}-X_{k_2})} \\
&= \frac{f_1}{\left(\dfrac{f_{k_1}/I_{k_1}^{\mathrm{intra}}}{f_{k_2}/I_{k_2}^{\mathrm{intra}}}\right)^{\frac{X_1+3}{X_{k_1}-X_{k_2}}}}
\end{aligned}
\tag{3.26}
$$

where I_k^{intra} symbolizes the integral of the Boltzmann factor of the intramolecular potential energy (excluding any bond length terms, with all bonds at their equilibrium lengths) over all orientations and internal configurations of a single chain of species k in the ideal gas state. Without loss of generality, a flexible chain model with bond stretching force constants going to infinity was used for the derivation.

Physically, $f_1'/I_1^{\mathrm{intra}}$ measures the chemical potential of oligomer molecules in relation to an equivalent mass of polymer in the (pseudo)chemically reacting system considered. If one increases (decreases) the f_1' value used in the simulation, the system at equilibrium responds by generating more (less) oligomer from the polymer. On the other hand, from the set value of f_1' one can calculate the actual fugacity f_1 of the oligomer through (3.26). The mole fractions x_{k_1} and x_{k_2} are readily obtainable from the molecular weight distribution of the polymer and the ratio of intramolecular configurational integrals is calculable by CBMC integration. It is convenient to choose the reference chain species k_1 and k_2 such that $X_{k_2} - X_{k_1} = 1$; The quantity $\dfrac{\mu_{k_1}^{\mathrm{ex}}-\mu_{k_2}^{\mathrm{ex}}}{X_{k_1}-X_{k_2}}$ then becomes the excess segmental chemical potential of species k_1 and is readily calculable through virtual augmentations of k_1-type chains in the bulk [50,32]. Thus, a series of $f_1'N_\mathrm{p}n_0PT\boldsymbol{\mu}^*$ simulations can yield the concentration of oligomer molecules as a function of the oligomer fugacity f_1, i.e., a sorption isotherm for the oligomer in the polymer.

The probability density of the $f_1'N_\mathrm{p}n_0PT\boldsymbol{\mu}^*$ ensemble has been worked out by Zervopoulou et al. [49]. With the configuration space described in terms of the generalised coordinates $\mathbf{s}_0$ (positions of all chain starts scaled with respect to the box edge length), $\boldsymbol{\psi}$ (Eulerian angles specifying the orientation of the first trimer of each chain), $\boldsymbol{\theta}$ (bond angle supplements), and $\boldsymbol{\phi}$ (torsion angles), it is

$$\rho^{\,f_1' N_{\mathrm{p}} n_0 PT \boldsymbol{\mu}^*}(N_1, \ldots, N_m, V; \mathbf{s}_0, \boldsymbol{\psi}, \boldsymbol{\theta}, \boldsymbol{\phi}) =$$

$$= \text{const.} \frac{1}{N_1!} V^{N_{\mathrm{p}}} \left[\prod_{k=1}^{m} \prod_{l=1}^{N_k} \left\{ \sin \psi_l^{(k)} \prod_{q=1}^{X_{k-2}} \left(\sin \theta_{ql}^{(k)} \right) \right\} \right] \left(\frac{\beta f_1' V}{I_1^{\mathrm{intra}}} \right)^{N_1} \times \quad (3.27)$$

$$\times \exp \left[\beta \sum_{\substack{k=2 \\ k \neq i,j}}^{m} \mu_k^* N_k - \beta P V - \beta \mathcal{V}(N_1, N_2, \ldots, N_m; \mathbf{s}_0, \boldsymbol{\psi}, \boldsymbol{\theta}, \boldsymbol{\phi}, \text{connectivity}) \right]$$

The acceptance rate of a scission move carried out on a chain of X_k skeletal atoms, leading the system from configuration i to configuration j, is

$$P_{\mathrm{accept}}(i \to j) = \min \left\{ 1, \frac{1}{N_1(i) + 1} \frac{N_{\mathrm{p}}}{N_{\mathrm{bridge}}(j \to i)} W_\theta(i) W_{\mathrm{bridge}}(j \to i) \right. \quad (3.28)$$

$$\left. \frac{\beta f_1'}{I_1^{\mathrm{intra}}} \frac{\left(l_{12}^2 l_{23}^2 l_{34}^2 l_{45}^2\right)_i}{J(i)} \exp\left[\beta(\mu^*_{X_k - X_1 - 3} - \mu^*_{X_k}) - \beta \Delta \mathcal{V}\right] \right\}$$

The acceptance rate of a fusion move carried out between a chain of X_k skeletal atoms and an oligomer, leading the system from configuration i to configuration j, is

$$P_{\mathrm{accept}}(i \to j) = \min \left\{ 1, N_1(i) \frac{N_{\mathrm{bridge}}(i \to j)}{N_{\mathrm{p}}} \frac{1}{W_\theta(j) W_{\mathrm{bridge}}(i \to j)} \right. \quad (3.29)$$

$$\left. \frac{1}{\frac{\beta f_1'}{I_1^{\mathrm{intra}}}} \frac{J(j)}{\left(l_{12}^2 l_{23}^2 l_{34}^2 l_{45}^2\right)_j} \exp\left[\beta(\mu^*_{X_k + X_1 + 3} - \mu^*_{X_k}) - \beta \Delta \mathcal{V}\right] \right\}$$

$\Delta\mathcal{V}$ is the change in potential energy (excluding bond length terms, with all bonds at their equilibrium lengths) brought about by the move. In the above equations, all symbols have the meanings attributed to them in Sects. 3.3.1 and 3.3.3. It is remarkable that the probability density and the acceptance criteria show no dependence on the energy of dissociation of bonds, on the stiffness of bond stretching constants, or on the thermal wavelength of atoms. All these contributions are lumped in the constant of (3.27).

Figure 3.13 displays the evolution of C_{10} weight fraction with CPU time in a polyethylene melt of mean molecular length C_{78} and polydispersity index

1.08 at 458 K, for comparable values of the fugacity f_1. The superiority of the scission-fusion method over the more conventional method, involving insertions and deletions of alkane molecules, in equilibrating the composition of the system is evident. For a C_{20} the conventional method is unable to provide a reliable estimate of the solubility, while the scission-fusion method continues to perform admirably well [49]. This is discussed further below, in the applications section.

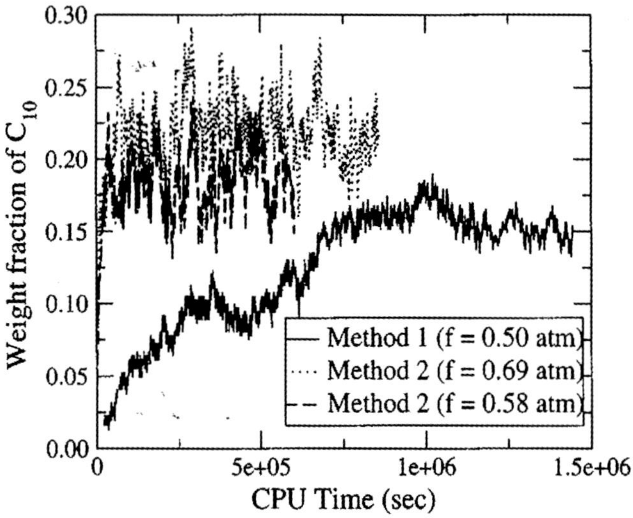

Fig. 3.13. Performance of two different methods in equilibrating the weight fraction of C_{10} dissolved in a polyethylene melt at 458 K under given values of the fugacity f_1. Method 1 is a MC simulation in the $f_1 N_p n P T \mu^*$ statistical ensemble, i.e. in a hybrid isothermal-isobaric ensemble that is grand canonical with respect to the alkane and semigrand with respect to the polymer; it employs CB insertions and deletions of alkane molecules. Method 2 is a MC simulation in the $f_1' N_p n_0 P T \mu^*$ statistical ensemble, which employs scission and fusion moves. CPU times are on an R10000 SGI workstation at 298 MHz. (Reproduced from [49], with permission).

3.3.7 Double Bridging and Intramolecular Double Rebridging

End-bridging Monte Carlo requires a finite degree of polydispersity in order to function. While this is not necessarily a shortcoming in modelling industrial polymers, which are typically polydisperse, an ability to equilibrate strictly monodisperse chain systems is desirable from the point of view of comparing simulation predictions against theory, or against experiments on anionically synthesised model systems. Also, EBMC relies on the presence of chain ends and thus does not lend itself for simulating dense phases of chains with complex nonlinear architectures.

Very recently, Karayiannis and Mavrantzas [51] developed two new connectivity-altering moves that are not subject to these shortcomings. The moves are based on the construction of *two* bridging chain sections (usually trimers) among four properly chosen *internal* skeletal atoms belonging to one or two chains in the system. The moves have been termed "double bridging" (DB) and "intramolecular double rebridging" (IDR).

Figure 3.14 presents a schematic of the DB move. An internal skeletal atom i of chain ich attacks an internal skeletal atom j of chain jch; the trimer $\{j_a, j_b, j_c\}$ adjacent to j is excised and a new trimer bridge $\{j'_a, j'_b, j'_c\}$

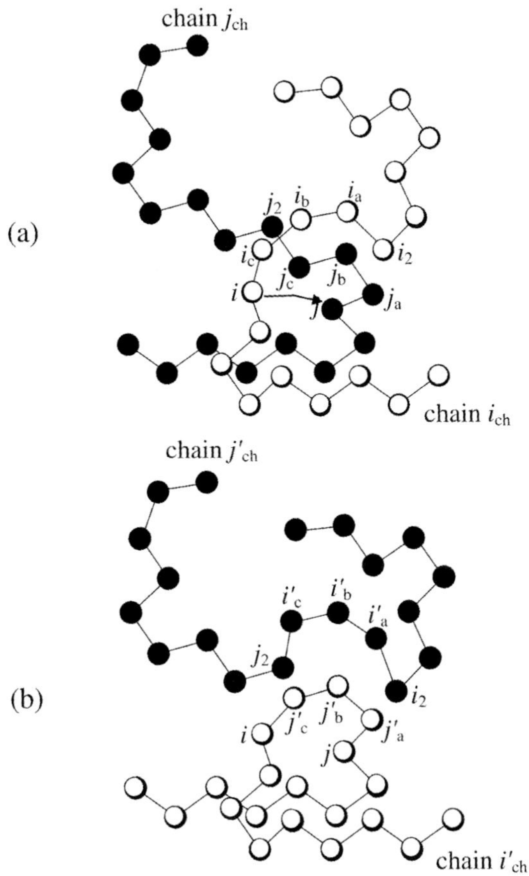

Fig. 3.14. Schematic of the double bridging (DB) move. (a): Local configuration of the two chains before the move. Trimer (j_a, j_b, j_c) is to be excised from chain j_{ch} and trimer (i_a, i_b, i_c) from chain i_{ch}. (b): Local configurations of the two new chains after the move. Trimer (j'_a, j'_b, j'_c) connects skeletal atoms i and j in chain i'_{ch}. Trimer (i'_a, i'_b, i'_c) connects skeletal atoms j_2 and i_2 in chain j'_{ch}. Chains i'_{ch}, j'_{ch}, i_{ch}, j_{ch} all have identical length.

is formed, connecting i and j. At the same time, the internal skeletal atom j_2, which is adjacent to the excised trimer on chain jch, attacks internal skeletal atom i_2, four skeletal atoms away from i on ich; the trimer $\{i_a, i_b, i_c\}$ is excised and a new trimer bridge $\{i'_a, i'_b, i'_c\}$ is built between j_2 and i_2. The chains obtained after the move, ich' and jch', have completely different conformations from those of ich and jch. Given two skeletal atoms i and j along the backbones of chains ich and jch, respectively, there are in general four different ways in which DB can be performed, depending on which two of the four trimers adjacent to j and i are excised. In a monodisperse system, if i and j are appropriately positioned relative to the ends of ich and jch, one of these four DB moves preserves monodispersity (see Fig. 3.14).

Figure 3.15 presents a schematic of the IDR move. Here, both trimer bridging constructions are performed between skeletal atoms belonging to the same chain. Given two bridgeable internal skeletal atoms on the same chain, there are two ways in which the move can be attempted. IDR modifies

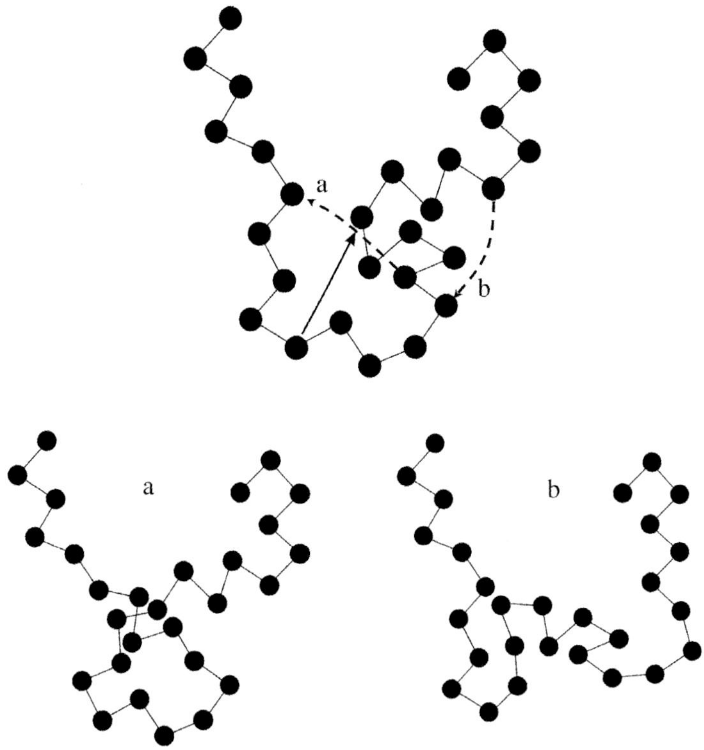

Fig. 3.15. Schematic of the intramolecular double rebridging (IDR) move. Top: Local configuration of the chain prior to the move. The attack shown as a solid arrow is combined with either of the attacks (a, b) represented by broken arrows. Bottom: trial configurations of the chain after attacks a and b, respectively.

the internal configuration of the chain, but leaves its ends unaltered. This internal reconfiguration aids in generating new possible sites for DB, so the two moves operate synergistically in long-chain systems.

The acceptance rule for DB and IDR moves is similar to (3.15). A product of two Jacobians, corresponding to the two simultaneous trimer bridging constructions, has to be used in place of each J, for both the forward and reverse moves. The weight W_θ in α accounts for all ten new bond angles generated. The weight W_{bridge} reflects the screening, for excluded volume overlaps and for excessive torsional potential, and subsequent selection, on the basis of its torsional potential, of one among all combinations of solutions of the two bridging constructions. That the two bridging constructions occur simultaneously and near each other in space generates substantial "free volume" that can accommodate some of the solutions, as a result of which the acceptance rate is much higher than the square of the acceptance rate of the EB move. This makes DB and IDR feasible.

As a measure of the rate of equilibration afforded by the DB and IDR moves, we show in Fig. 3.16 the decay of the end-to-end unit vector autocorrelation function $\langle \mathbf{u}(0) \cdot \mathbf{u}(t) \rangle$ with CPU time in the course of isothermal-

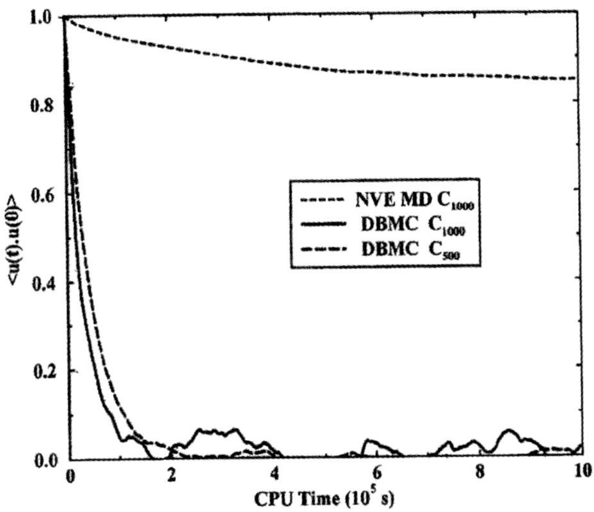

Fig. 3.16. Evolution of the autocorrelation function $\langle \mathbf{u}(t) \cdot \mathbf{u}(0) \rangle$ of the unit vector directed along the end-to-end vector of a chain with CPU time for monodisperse C_{500} and C_{1000} united-atom polyethylene melts simulated in the NPT ensemble using DB and IDR moves. The simulation conditions are $T = 450$ K and $P = 1$ atm. Both model systems contain the same number of interaction sites (8000). Also shown is the corresponding curve obtained from an NVE MD simulation of the monodisperse C_{1000} melt. CPU times are on a DEC Alpha single-processor system at 667 MHz.

isobaric MC simulations of C_{500} and C_{1000} melts with a united-atom model. (In defining the autocorrelation function, the new chain identities after a DB move are assigned as shown in Fig. 3.14.) Also shown is the corresponding decorrelation curve obtained from an NVE MD simulation of the same C_{1000} model. All model systems considered in the figure have exactly the same size. It is seen that the connectivity-altering MC is roughly two orders of magnitude more effective than MD in sampling different orientations of the end-to-end vector. Remarkably, the decay observed with MC is somewhat faster for the longer-chain system than for the shorter-chain one; this is reminiscent of the favourable scaling of EBMC efficiency with chain length, discussed above.

DB and IDR offer themselves for equilibrating systems of macromolecules of complex, precisely defined nonlinear architectures, such as H-shaped and ring polymers, grafted monodisperse polymer brushes, or bulk model systems of infinite chain length with no ends. Some of these systems, and the DB/IDR moves that can be implemented for them, are shown schematically in Fig. 3.17.

3.3.8 Connectivity-Altering Monte Carlo and Parallel Tempering

The excluded volume interactions and torsional potentials invoked in realistic models of polymers with relatively complex chemical constitutions result in extremely rugged potential energy hypersurfaces. This makes the acceptance rates of the connectivity altering moves discussed above very low at low temperatures. Under such conditions, the rate of sampling configuration space can be greatly enhanced by resorting to a parallel tempering strategy, outlined in Sect. 3.1.

Wu and Deem combined parallel tempering with intramolecular rebridging MC moves to sample the conformations of cyclic peptides [16,17] and thereby calculate the equilibrium *cis* and *trans* populations at 298 K; this would have been impossible with MD, owing to the high free energy barriers to *cis-trans* isomerisation.

Doxastakis et al. [35] combined parallel tempering and EBMC to simulate *cis*-1,4 polyisoprene (PI) in the melt state. A united atom model system of 8 chains, each containing 80 carbon atoms, was simultaneously equilibrated at 10 different temperatures, ranging from 328 K to 513 K, in the $NnPT\boldsymbol{\mu}^*$ ensemble. Histograms of the instantaneous enthalpy $\mathcal{V} + PV$ are shown in Fig. 3.18. The overlap Ω between histograms corresponding to sucessive temperatures was sufficient to ensure a good acceptance ratio for the swapping moves (see Fig. 3.19). As a result, each simulated replica visited the entire range of simulation temperatures several times, and the drastic reconfigurations induced by EBMC at high temperatures contributed to efficient sampling of the configuration space at low temperatures as well. Obtained distributions

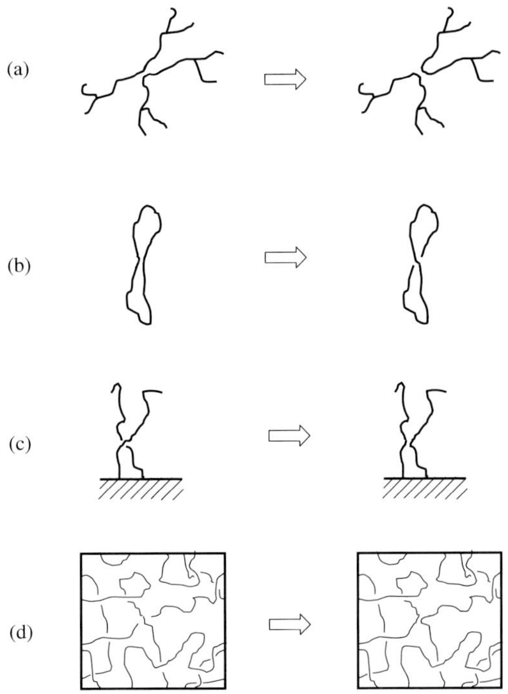

Fig. 3.17. Application of the DB and IDR moves to polymer systems of complex architecture. (a): H-shaped molecules, (b): cyclic molecules, (c): monodisperse grafted polymer brush, (d): melt of linear chains of infinite molecular weight.

of $\mathcal{V} + PV$, $\mathcal{V}$ and V and of torsion angles at various temperatures and autocorrelation functions of the chain end-to-end vectors all indicate that full equilibration was achieved at all temperatures. At the lowest temperature simulated, the rate of equilibration was dramatically enhanced relative to single-temperature EBMC runs [35]. Similarly good results were obtained for atactic, isotactic, and syndiotactic polypropylene melts simulated with a parallel tempering EBMC strategy [33].

Based on the evidence discussed above, combining connectivity-altering Monte Carlo schemes with parallel tempering constitutes an excellent strategy for enhancing equilibration under conditions (low temperatures, complex chemical constitutions) where the acceptance rate of single- temperature connectivity-altering schemes becomes too low. Furthermore, parallel tempering provides useful information over a range of temperatures and can be implemented on relatively inexpensive Beowulf clusters of personal computers with little programming effort.

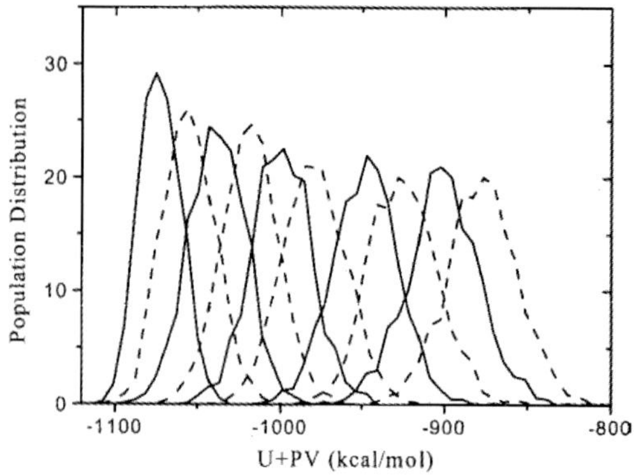

Fig. 3.18. Instantaneous enthalpy $\mathcal{V} + PV$ histograms from a parallel tempering simulation of *cis*-1,4 PI (united atom model, 8 chains of 80 carbons each). The temperature levels used are ($T_i, i = 1(1)10$ from left to right) 328, 343, 358, 373, 390, 413, 438, 463, 488, and 513 K. (Reproduced from [35], with permission).

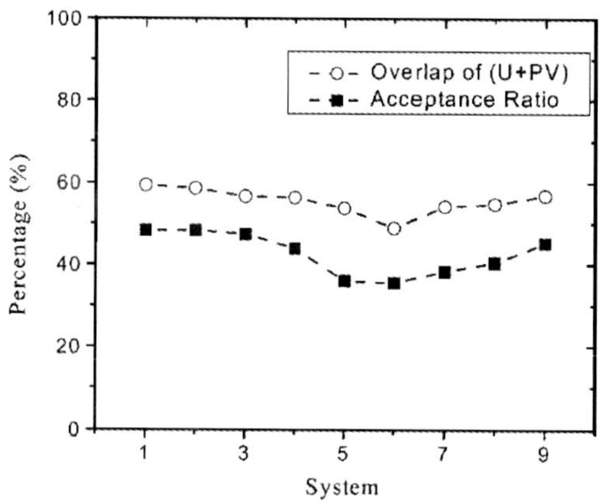

Fig. 3.19. Acceptance ratio (filled rectangles) of the swapping move between temperature states T_i and T_{i+1} as a function of i for the ten systems used in the parallel tempering simulation. Open circles show the overlap integral Ω between the corresponding instantaneous enthalpy histograms, shown in Fig. 3.18. (Reproduced from [35], with permission).

3.4 Applications

In this section we will present some results from application of the connectivity-altering schemes discussed above to specific polymer systems.

3.4.1 Structure and Volumetric Properties of Long-Chain Polyethylene Melts

Linear polyethylene (PE) has been used as a test case throughout the development of connectivity-altering MC algorithms [22,29,21,39]. Recently, Uhlherr et al. [40] simulated a sample with uniform (rectangular) distribution of chain lengths ranging from 2400 to 9600 carbon atoms (mean chain length C_{6000}, mean molecular weight 84000 g mol^{-1}) in the $NnPT\boldsymbol{\mu}^*$ ensemble on a parallel computer, using CONROT, DEB, reptation, flip, end rotation, and volume fluctuation moves. For the first time, the direct atomistic simulation of such a high molecular weight polymer, similar to the commercial grades used for injection moulded articles, has become possible, thanks to the excellent equilibration properties of the DEB move.

Figure 3.20 displays the mean square end-to-end distance $\langle R^2 \rangle$ and the mean square radius of gyration $\langle s^2 \rangle$ as functions of the chain length (number of skeletal carbons) from a C_{6000} melt simulation at 450 K [40]. The functional dependence is linear, indicating that the system is well-equilibrated and that perturbations of the chain dimensions due to the finite size of the simulation

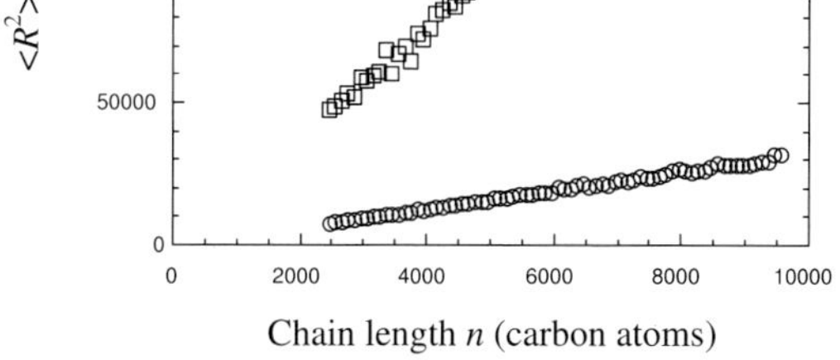

Fig. 3.20. Mean square end-to-end distance $\langle R^2 \rangle$ and mean square radius of gyration $\langle s^2 \rangle$ as functions of the chain length X from a $NnPT\mu^*$ simulation of a C_{6000} melt, where chain lengths range from 2400 to 9600. The temperature of the simulation was 450 K and the pressure 1 bar.

box are absent. The characteristic ratio $C_\infty = \lim_{X\to\infty} \frac{\langle R^2 \rangle}{X l^2}$, with l being the skeletal bond length, is evaluated as 8.4 ± 0.2. This is somewhat larger than the commonly accepted value $C_\infty = 7.8$ obtained via SANS measurements [52,53], a fact attributable to the torsional potential used in the simulation.

According to Flory's "random coil hypothesis," chain conformations in the melt should be unperturbed, i.e. similar to those of isolated single chains subject only to local interactions along their backbones. Computer simulations [30,29,21] were instrumental in showing that, for PE, this hypothesis is true down to the atomistic level, if "local interactions" are appropriately defined. Torsion angle distributions, intramolecular pair density functions, and chain dimensions are practically indistinguishable between atomistically simulated PE melt chains and single chains subject to the same potentials but devoid of any nonbonded interactions between methylene or methyl segments separated by more than four bonds along their backbone. Figure 3.21 provides evidence for this in the case of a C_{500} melt with uniform chain length distribution and polydispersity index 1.09 at 450 K and 1 atm. The

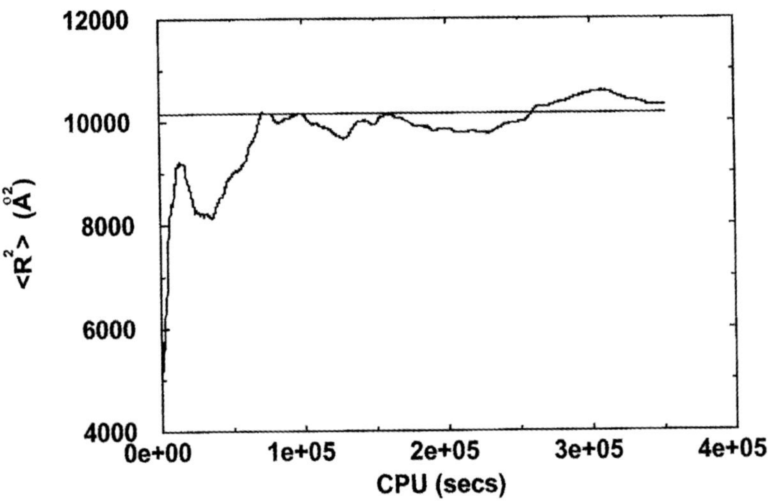

Fig. 3.21. Mean square end-to-end distance as a function of CPU time in a C_{500} PE melt of polydispersity index 1.09, simulated with EBMC at 450 K and 1 bar. The horizontal line indicates the corresponding value of $< R^2 >$ for a set of isolated unperturbed chains with the same length distribution. The isolated unperturbed chains are subject to the same potentials as the melt chains, with the exception that there are no intermolecular interactions and no intramolecular nonbonded interactions between segments separated by more than 4 bonds along the backbone; they have been sampled in continuous space with a MC algorithm [21]. CPU times are on an SGI 64-bit R10000 processor for a melt system containing 4000 skeletal (methylene and methyl) segments.

mean-square end-to-end distance $< R^2 >$ from EBMC quickly stabilises at a constant value, which is indistinguishable from the corresponding $< R^2 >$ of single unperturbed chains.

Figure 3.22 displays the variation of specific volume v of strictly monodisperse linear PE melts as a function of chain length X in the range C_{24} - C_{1000} at 450 K and 1 atm. The simulation results have been obtained in the NPT ensemble using DB and IDR moves, using a newly proposed united-atom model for PE [51]. They are being compared with accurate experimental results, available in [54]. One observes the characteristic hyperbolic dependence of v on X, attributed to the higher volume of end-segments in relation to internal segments in free volume theories [55]. The prediction is within 1% of the experimental value for all chain lengths examined. This is a significant improvement over earlier molecular models, which tended to overestimate the density by *ca.* 4%. The characteristic ratio using this model turns out to be $C_\infty = 8.0$, quite close to the commonly accepted experimental value of 7.8 in the melt state [52,53].

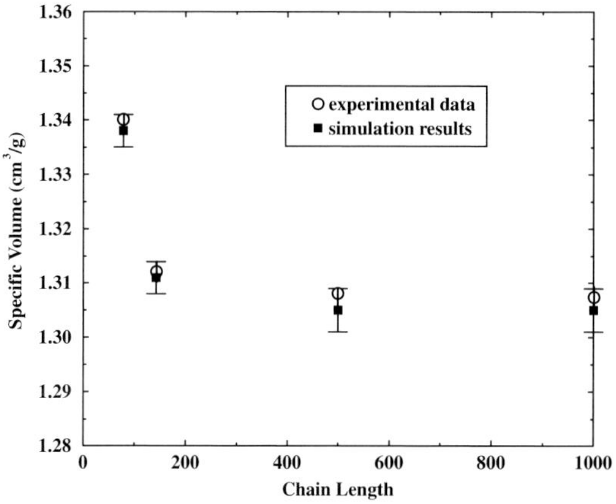

Fig. 3.22. Dependence of the specific volume v on chain length X in monodisperse linear PE at 450 K and 1 atm, as predicted through NPT MC simulations using DB and IDR moves, and as measured experimentally. (Reproduced from [51], with permission).

Local packing in molten PE can be reproduced very well by united-atom simulations. Figure 3.23 compares the static structure factor $S(k)$ predicted from a monodisperse C_{1000} simulation [51] to high-quality experimental data from X-ray diffraction [56], accumulated under comparable conditions. Agreement is seen to be excellent. The first sharp diffraction peak at $k = 1.35$ Å^{-1} stems mainly from intermolecular structural correlations; the intermolecular pair distribution function $g(r)$ exhibits damped oscillations with wavelength $\Delta r = 4.65$ Å$\simeq 2\pi/k$, corresponding to the distance between nearest-neighbour carbon atoms belonging to different chains. This first sharp diffraction peak has a 17% contribution from extended-range density fluctuations appearing in $g(r)$ beyond $r = 10$ Å [40]. The peaks in $S(k)$ beyond $k = 3.5$ Å^{-1} are of intramolecular origin, due to the bonding and conformational preferences along the chains, and are more or less indistinguishable from the corresponding peaks of liquid n-butane [40].

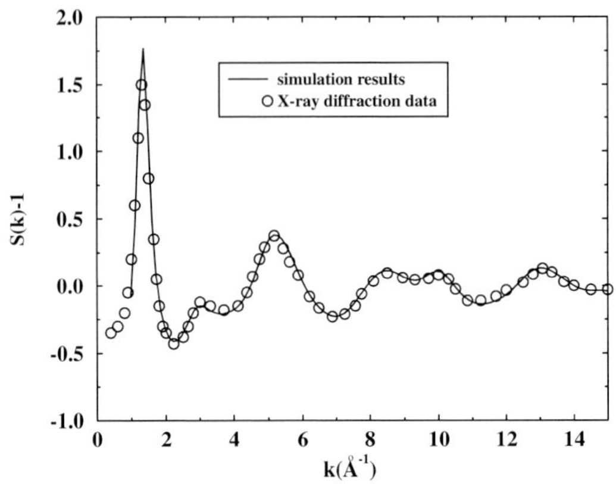

Fig. 3.23. Static structure factor $S(k)$ of a monodisperse linear C_{1000} PE, as obtained from NPT MC simulations using DB and IDR moves, being compared to an experimental X-ray diffraction pattern from PE. The simulation is at 450 K and 1 atm; the experimental data are at 430 K and 1 atm. (Reproduced from [51], with permission).

Long-chain atomistic model configurations of PE melts, thoroughly equilibrated through connectivity-altering MC algorithms, offer the unique opportunity of performing direct topological analysis for the identification of entanglements, which are of crucial importance to the transport and viscoelastic properties of polymer melts, but also to the ultimate mechanical properties of solid polymers. Although entanglements have been extremely useful in explaining experimentally observed behaviour, their direct definition in terms of the topology of chains has been elusive. Uhlherr *et al.* [40] have analysed their model C_{6000} melt configurations using the formalism proposed by Wool [57] for determining the critical entanglement molecular weight M_c, which marks the crossover from Rouse to reptation dynamics in the chain-length dependence of the zero-shear viscosity. Wool defines entanglements in terms of the ratio p/n, where n is the number of chains intersecting a plane drawn randomly through the system and p is the number of crossings of the plane. A chain that crosses the plane three times forms a "bridge", which is able to interlock with chains on either side of the plane, and therefore to transmit forces across the plane. Thus, the critical length $M = M_c$ for the onset of an entanglement network in the melt, according to Wool, occurs when $p/n = 3$. As a result of the unperturbed conformations adopted by chains in the melt [57], $p/n \propto M^{0.5}$, leading to the simple expression $M_c = 9M(p/n)^{-2}$, whereby M_c can be determined through a count of p and n for chains of molecular weight M in an entangled melt.

The Wool approach was implemented on the C_{6000} model PE configurations by drawing a large number ($\simeq 100$) of regularly spaced planes parallel to each face of the simulation cell in each configuration. The estimated M_c was found to be essentially independent of M, as expected for a fully entangled melt. The calculated average was $M_c = 3400 \pm 450$ g mol^{-1}, corresponding to a chain length of approximately 240 backbone atoms, or 120 monomers. While reported experimental values range from 3000 to more than 5000 g mol^{-1}, this result agrees quite well with the commonly cited value [58] of 3800 g mol^{-1} at 423 K. More detailed topological analyses of the entanglement structure are now underway.

3.4.2 Simulations of Polypropylene Melts of Various Tacticities

Polypropylene (PP) is a very important polymer from the technological point of view. The presence of pseudoasymmetric carbon atoms along its backbone gives rise to the possibility of different stereochemical sequences. Three types of PP are commonly produced: isotactic (iPP), where chains are sequences of meso (m) dyads, syndiotactic (sPP), where chains are sequences of racemic (r) dyads, and atactic, where chains are more or less random sequences of m and r dyads. Equilibrium epimerized atactic PP, henceforth denoted as aPP, has a well-defined stereostructure; its chains are Bernoullian sequences of m and r dyads, with a fraction of m dyads equal to 48% [59] (see Fig. 3.24). The three types of PP have very different properties; iPP and sPP are semicrystalline, while aPP remains completely amorphous at all temperatures, having a transparent, leathery appearance at room temperature.

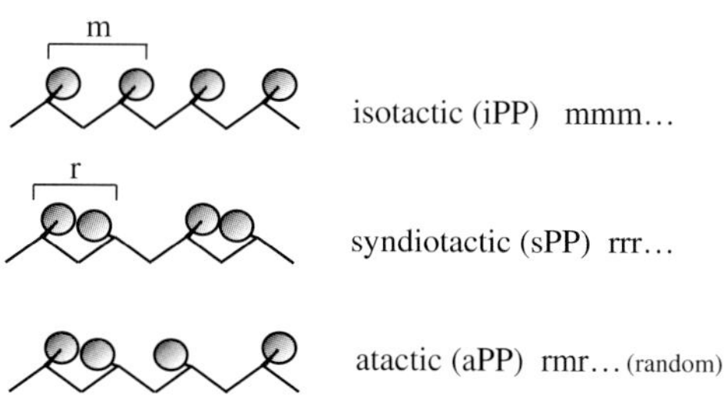

Fig. 3.24. Diffent tacticities of polypropylene. Chains are considered in an all-*trans* conformation. The spheres are methyl substituents. Hydrogens have been omitted, for clarity.

Samara [33] developed a parallel tempering EBMC strategy to simulate PP in the melt state. In the case of iPP and sPP, EB moves were designed so as to maintain the regular stereochemical sequence of chains. In the case of aPP, the semigrand ensemble sampled by the EBMC method was generalised to allow variations not only in the chain length, but also in the stereochemical sequence, subject to a prescribed Bernoullian distribution; four possible types of trimer bridges were considered in each EB construction [33].

For the first time, the parallel tempering EBMC simulation was able to give estimates of the equilibrium chain dimensions in the melt (molecular weight 3200 g mol^{-1}). Calculated characteristic ratios C_∞, as obtained from the simulated melts and from sampling single continuous unperturbed chains (CUCs), are shown in Table 3.1. In the same table appear experimental values, obtained by small angle neutron scattering measurements from iPP and aPP melts [60], values estimated for unperturbed chains through the Rotational Isomeric State (RIS) model of Suter and Flory [61], as well as values from an MD simulation [62] of single chains in a Θ solvent, where they adopt unperturbed conformations. Results from the parallel tempering EBMC melt simulations are in very good agreement with J.-P. Ryckaert's results for Θ chains and with available experimental evidence. Clearly, sPP chains are significantly stiffer than aPP and iPP chains, which are very similar in dimensions. Reliable SANS data from sPP melts would be highly desirable to validate this finding.

Table 3.1. Characteristic ratios of polypropylenes

PP type	*PT EBMC*		Experiment [60]	RIS model [61]	Θ chains [62]
	Melt	CUC			
aPP	6.2 ± 1.0	5.2 ± 0.4	5.5	5.5	6.1
sPP	8.5 ± 1.1	9.6 ± 0.1		11.0	8.0
iPP		6.6 ± 0.3	6.2	4.2	6.1

Despite these differences in chain conformation, the volumetric properties of iPP, sPP, and aPP are very similar in the melt state. This is seen characteristically in Fig. 3.25. Both simulations [63] and experiment [64] indicate negligible differences in density between the three different tacticities. Given the disparity in conformational and the similarity in volumetric behaviour, an interesting question is how the segmental dynamics compares between iPP, sPP and aPP. This question was addressed with MD simulations [63], leading to the conclusion that segmental dynamics is slowest in sPP and fastest in iPP, with aPP lying in the middle. The reason for this dynamical difference is that conformational isomerisations between the dominant *trans* and *gauche* states are more frequent in iPP, where the relative populations of these states

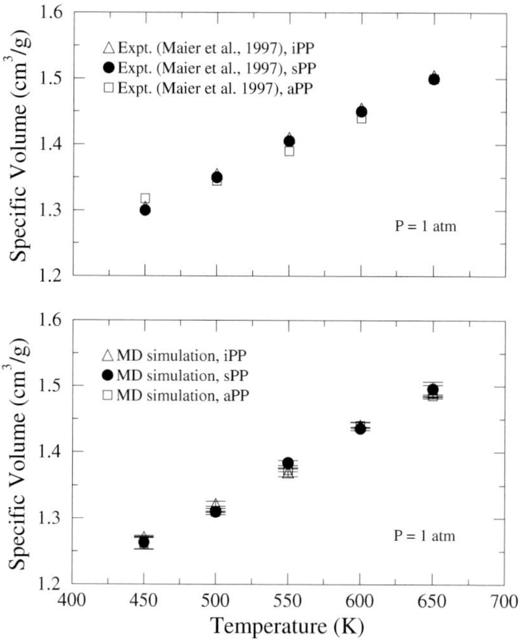

Fig. 3.25. Specific volume of isotactic, syndiotactic, and atactic polypropylene melts as a function of temperature at 1 atm, as obtained from experiment [64] (top) and from molecular dynamics simulations [63] (bottom). (Reproduced from [63], with permission).

are closer together [63]. This prediction was a posteriori confirmed by NMR relaxation experiments [65].

Another interesting question that was elucidated with the help of parallel tempering EBMC is how exactly local interactions, leading to unperturbed conformational behaviour, should be defined in the case of PP. As mentioned above, in the case of PE an appropriate definition of single unperturbed chains in continuous space is to include all bonded and torsional interactions, plus nonbonded interactions between pairs of atoms separated by four bonds or less. Such a definition does not work for PP, however; it leads to torsion angle distributions which are very different from those of melt chains, and to chain dimensions which are much smaller than in melt simulations or in experiment.

A systematic study was undertaken to determine the effects of changing the range of what is considered as local interactions on unperturbed single chain dimensions [33]. Figure 3.26 displays four models that were tried, with progressively increasing range. The torsion angle distributions obtained by sampling single chains of aPP according to these models are shown in Fig. 3.27. Clearly, the "4 bond" and "4 C bond" models produce results very far from the melt distribution, while the "5 C bond" and "6 C bond" models

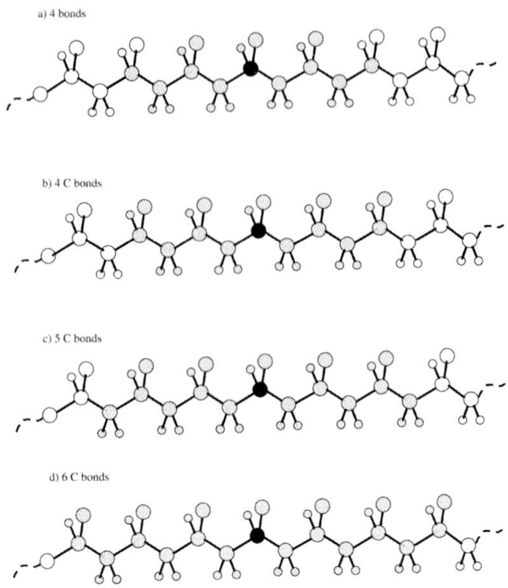

Fig. 3.26. Four models of local interactions considered in sampling single unperturbed chains of PP. In each case, the segment whose skeletal carbon is marked as a black sphere interacts with all cross-hatched atoms. Open symbols mark atoms with which interactions are not included.

produce results which are practically indistinguishable and very close to the melt results. Single chain dimensions obtained from the "5 C bond" and "6 C bond" models are indistinguishable from those of bulk melt chains [33]. The conclusion is that, in order correctly to sample unperturbed chains of PP, one must incorporate nonbonded interactions of each atom with atoms of segments lying no less than 5 skeletal bonds away from the considered atom. Clearly, one must exercise great care in defining and sampling unperturbed chains to determine conformational stiffness in the case of polymers with more complex chemical constitution than PE.

3.4.3 Simulation of Polydienes

Polydienes, such as polyisoprene (PI) and polybutadiene (PB), are of great technological importance, as they constitute main components of rubbers. Chain packing and conformation in these polymers depend sensitively on their stereochemical configuration (1,4 or diene versus 1,2 or vinyl addition during polymerisation, *cis* versus *trans* configuration around the double bond in the 1,4 units) and in turn dictate their rheological properties in the melt (important in processing), their segmental dynamics and permeability by gases (important, e.g. in making tyres airtight) and their miscibility in the melt state.

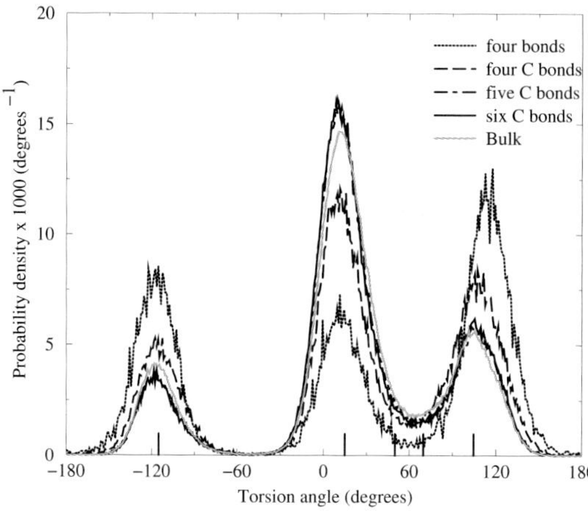

Fig. 3.27. Torsion angle distributions in model atactic PP at 600 K. Distributions for single unperturbed chains sampled according to the "four bond", "four C bond", "five C bond", and "six C bond" models are shown with different line types. The "five C bond" and "six C bond" results are practically coincident. Also shown, as a grey line, is the distribution obtained from bulk EBMC simulations of an aPP melt. The short, bold vertical lines on the abscissa mark values of the torsion angles in the five rotational isomeric states considered in the RIS model of Suter and Flory [61].

A combination of EBMC and parallel tempering has been used to study the properties of *cis*-1,4 PI melts of mean chain length 40 to 200 carbon atoms and polydispersity index 1.08 in the temperature range 328 to 513 K [34,35]. Use of this strategy enabled, for the first time, the full equilibration of PI melts of this chain length. The united atom model employed is described in detail in [34]. The torsion angle distributions it produces are consistent with those obtained in earlier simulations of the polymer. Figure 3.28 displays the characteristic ratio at 413 K, defined in terms of the average square skeletal bond length $l_{\mathrm{ave}}^2 = 2.18$ Å^2. The quantity n_{mer} is the degree of polymerisation (average number of repeat units per chain, each repeat unit containing four skeletal bonds). An extrapolation of the results shown in Fig. 3.28 to infinite chain length gives a value of C_∞ of 4.5 to 4.8. The value of C_∞ at 413 K, based on experimental measurements, is 4.8 to 5.5. The apparent slight underestimation in the simulation may be partly due to the presence of some *trans*-1,4 linkages in real PI, which were absent from the model polymer. At lower temperatures, agreement with reported experimental values improves [35].

The specific volume at 413 K and 1 atm is shown as a function of mean chain length in Fig. 3.29. The dependence is parabolic, as already seen in the case of PE. The dashed line represents a fit through the simulation points

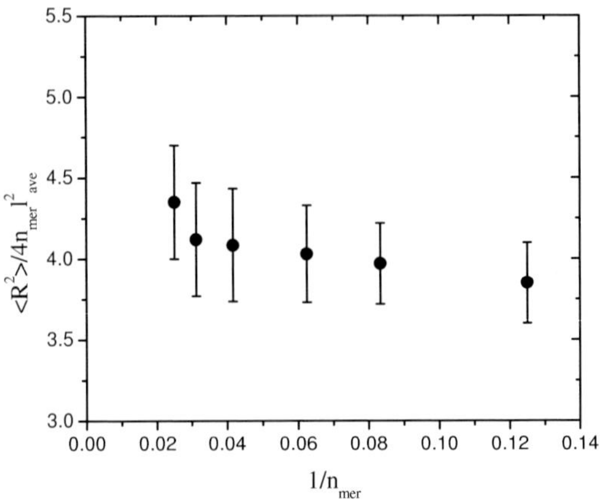

Fig. 3.28. Characteristic ratio of *cis*-1,4 polyisoprene as a function of the inverse degree of polymerisation, as obtained from EBMC simulations at 413 K (reproduced from [34], with permission).

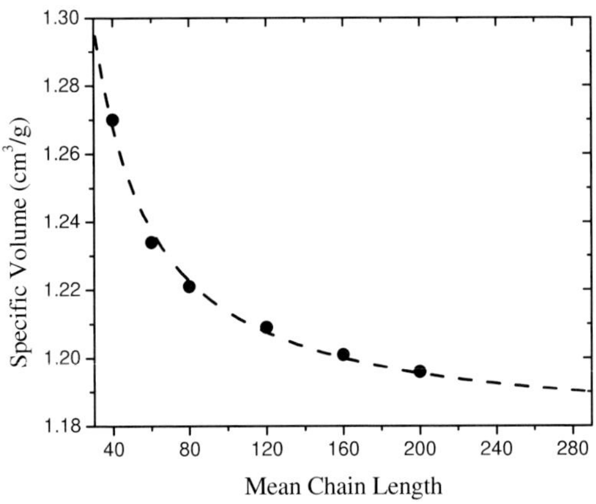

Fig. 3.29. Dependence of the specific volume v on mean chain length $\bar{X}$ (number of carbon atoms per chain) in *cis*-1,4 PI, as predicted by the EBMC simulation runs at 413 K and 1 atm. The dashed line shows a hyperbolic fit according to (3.30).

[34] with the equation

$$v = v_\infty + v_0/\bar{X} \tag{3.30}$$

From this fit, the specific volume of PI of very large molecular weight at 413 K is estimated as $v_\infty = 1.178$ cm^3/g. Nemoto et al. [66] and Han et al. [67] have correlated experimental data on the specific volume of PI as a function of temperature. According to the first correlation, v_∞ at 413 K is

1.196 cm^3/g; according to the second, 1.183 cm^3/g. Clearly, the simulation estimate is in very good agreement with experimental evidence.

Apart from dramatically accelerating equilibration at low temperatures, parallel tempering offers the opportunity to study thermodynamic and structural properties systematically as functions of temperature. Figure 3.30 shows the temperature dependence of the specific volume of a *cis*-1,4 PI melt of mean chain length 80 carbons (open circles). The two filled circles represent single-temperature EBMC results at 373 K and 413 K using an eight times larger simulation box; they ensure that the parallel tempering run is free of system size effects. The dashed and straight lines display the Nemoto et al. [66] and Han et al. [67] correlations, respectively. Clearly, the parallel tempering EBMC simulation does an excellent job capturing the thermal expansion of the polymer melt.

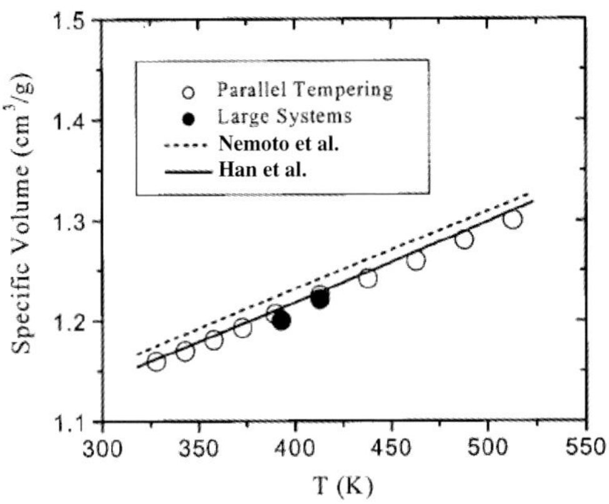

Fig. 3.30. Open symbols: Dependence of the specific volume of *cis*-1,4 PI on temperature as predicted from parallel tempering EBMC simulations of a model melt consisting of 8 chains with average length 80 carbons. Filled symbols: Results from two single-temperature EBMC runs carried out at 373 K and 413 K using a simulation containing 64 chains, to ensure that system size effects are absent from the parallel tempering simulations. Dashed line: Nemoto et al. correlation for the volumetric properties; Solid line: Han et al. correlation for the volumetric properties (reproduced from [35], with permission).

3.4.4 Prediction of Melt Elasticity

Let us consider a polymer melt subjected to a steady-state flow. If the strain rate imposed by the flow is high, in comparison to the longest relaxation time of chains, then the chains will orient, i.e. the clouds of segments constituting the chains will depart from an isotropic orientation distribution. If the flow is very strong, it may even "unravel" the polymer coils, causing changes in their intrinsic shape. This is the molecular origin of viscoelastic behaviour in polymer melts, which is particularly pronounced at high molecular weights.

To describe these effects in a very coarse-grained sense, one may introduce a "single conformation tensor" model, wherein the mean chain orientation and shape are described in terms of the quantity $\tilde{\mathbf{c}}$ defined in (3.19). As described in Sect. 3.3.5, Mavrantzas and Theodorou [42] have introduced and implemented a computational methodology, involving EBMC simulations in the $NnbT\boldsymbol{\mu}^*\boldsymbol{\alpha}$ ensemble, to capture flow-induced changes in the melt structure and compute the free energy, energy, entropy, and stress of unentangled melts subjected to steady-state flows.

Figure 3.31 depicts the average shape of chains in a C_{78} melt (a) at equilibrium; (b) subject to a steady-state uniaxial elongational flow along the x-direction with $\alpha_{xx} = 0.3$ (anisotropy of the stress tensor $\tau_{xx} - \tau_{yy} \simeq 28$ atm). In both (a) and (b) the average chain segment cloud has been visualised in the laboratory frame of reference. Clearly, it is isotropic (spherical) in the equilibrium (quiescent) melt case but assumes the shape of a prolate ellipsoid of revolution in the presence of the flow. Chain segment clouds, as whole objects, orient along the x-direction.

Figure 3.32 shows again the average shape of chains in the C_{78} melt, but this time in the frame of the principal axes of the instantaneous radius of gyration tensor. The average intrinsic shape of the chain segment cloud is thus revealed under (a) equilibrium and (b) steady-state uniaxial elongational flow ($\alpha_{xx} = 0.3$) conditions. One sees that, under the relatively gentle flow conditions considered, the intrinsic shape of chains is not significantly affected; it remains similar to that of a "cake of soap", as pointed out long ago for random walks by Šolc and Stockmayer [68] and subsequently seen for atomistically represented unperturbed chains sampled according to the Rotational Isomeric State model [69]. Under the conditions considered in Figs. 3.31 and 3.32, the flow brings about an overall orientation, but not "unravelling" of the chains; this is also seen from a direct comparison of the torsion angle distributions under equilibrium and flow conditions [42]. On the other hand, for stronger flow fields (larger α_{xx} values) the simulations reveal a progressive elongation of the intrinsic shape of chains in the x direction, accompanied by an enhancement in the population of *trans* torsion angles [42].

Figure 3.33 shows the free energy, ΔA, relative to the equilibrium (undeformed) state for the same C_{78} melt subjected to steady-state uniaxial elongational flows of different strengths α_{xx} at $T = 450$ K and $b = 1$ atm. ΔA values have been obtained by thermodynamic integration of the $\tilde{\mathbf{c}}$ versus $\boldsymbol{\alpha}$ relation obtained from the simulations [42]. Also shown is the variation of the internal energy of the melt, ΔU, as a function of α_{xx}, obtained directly by averaging the potential energy in the course of the simulations. The entropic contribution $T\Delta S$, calculated as $\Delta U - \Delta A$, is also shown. ΔA provides a direct measure of the elastic response of the melt to the steady-state flow.

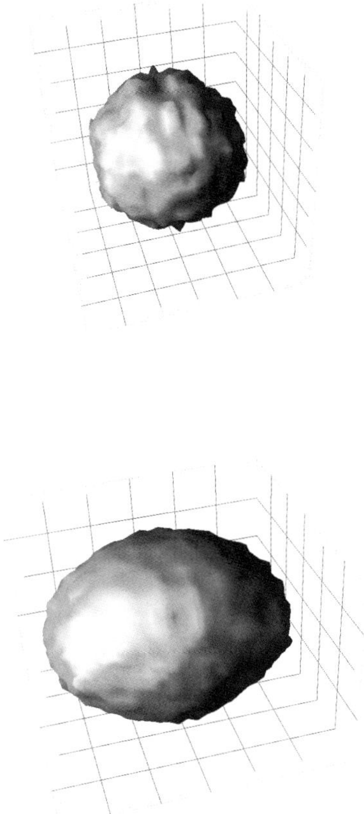

Fig. 3.31. Average shape of the segment cloud of chains in a C_{78} melt with flat chain length distribution and polydispersity index 1.08 (a, top) at rest; (b, bottom) under steady-state elongational flow with $\alpha_{xx} = 0.3$. To create the isosurfaces depicted in the figure, all chains in the sampled configurations were translated in the laboratory frame of reference so as to bring their centres of mass at a common point. The isosurfaces are drawn for a segment density of 0.35 mers/nm^3 and contain 87.5 % and 83.2 % of the total segments in cases (a) and (b), respectively. The distance between successive grid points on the axes represents 12.45 Å of real length. T is 450 K and b is 1 atm. (Reproduced from [42], with permission).

Clearly, for the conditions considered in Fig. 3.33, this response is purely entropic, originating in the orientation effect discussed above. This ceases being true under stronger flow conditions; for a C_{24} melt and values of α_{xx} higher than 0.3, the simulations reveal a significant energetic component ΔU which is negative and increases in absolute value with increasing α_{xx}. This energetic component is linked to the changes in intinsic chain shape discussed above. Detailed analysis attributes it to the development of more attractive lateral, "nematic type" nonbonded interactions between the oriented chains, which

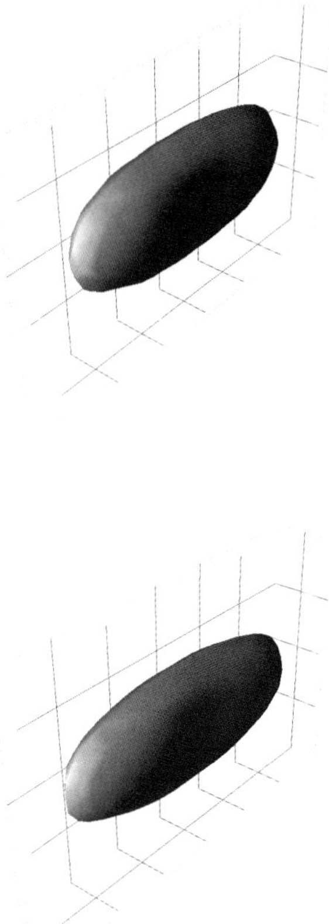

Fig. 3.32. Average shape of the segment cloud of chains in the coordinate frame of their principal axes in a C_{78} melt with flat chain length distribution and polydispersity index 1.08 (a, top) at rest; (b, bottom) under steady-state elongational flow with $\alpha_{xx} = 0.3$. To create the isosurfaces depicted in the figure, all chains in the sampled configurations were translated in the laboratory frame of reference so as to bring their centres of mass at a common point and rotated so as to make their longest, intermediate, and shortest principal axes coincide with three preset orthonormal directions. The isosurfaces are drawn for a segment density of 5 mers/nm^3 and contain 59.2 % and 56.3 % of the total segments in cases (a) and (b), respectively. The distance between successive grid points on the axes represents 9.96 Å of real length. T is 450 K and b is 1 atm. (Reproduced from [42], with permission).

are partly unravelled in the x-direction, and also to a reduction in intramolecular torsional energy reflecting the shift from *gauche* to *trans* torsional states which accompanies the unravelling [42]. It is, however, remarkable that, under the conditions of small α_{xx} representative of common polymer processing operations, the melt response is purely entropic (Fig. 3.33), as postulated by most mesoscopic theories of melt viscoelasticity.

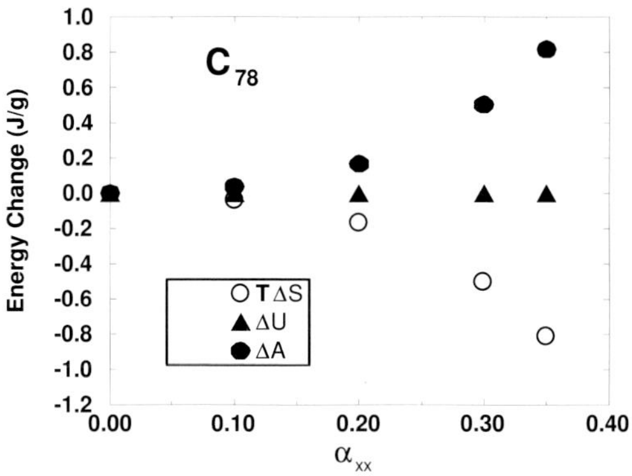

Fig. 3.33. Changes in the Helmholtz free energy ΔA, internal energy ΔU, and entropy multiplied by the temperature, $T\Delta S$, relative to the equilibrium undeformed state for a C_{78} polymer melt subjected to steady-state uniaxial elongational flow, as functions of the orienting field α_{xx} ($T = 450$ K, $b = 1$ atm).

The free energy function of Fig. 3.33 defines a "spring force law" for the single-conformation tensor model of (3.20), derived directly from atomistic information. It can be compared against various "dumbbell" models used in the mesoscopic modelling of polymer flows. The comparison can be made either at the level of ΔA itself [45] or at the level of its first derivative (functional dependence of $\tilde{\mathbf{c}}$ on $\boldsymbol{\alpha}$) [21]. It is found that the atomistically computed elastic response is quite close to that predicted on the basis of the finitely extensible nonlinear elastic (FENE) and preaveraged FENE (FENE-P) models with parameter values of these models computed from the equilibrium mean square end-to-end distance of the atomistic chains and their contour length at full extension. On the other hand, the Maxwell dumbell model is in poor agreement with the simulation results, because it does not take into account the finite extensibility of the chains.

Equation (3.23) leads to estimates of the stress tensor τ which are in excellent agreement with, but much less subject to statistical error than, estimates based on the virial theorem [42]. Within the context of the GENERIC formalism [45] one can associate the orienting field $\boldsymbol{\alpha}$ with the strain rate tensor $\dot{\boldsymbol{\epsilon}}$. Values of $\dot{\epsilon}_{xx}$ for the elongational flow situation considered here can thereby be calculated from α_{xx} and from the Rouse time of chains, which is extracted from equilibrium molecular dynamics simulations of the unentangled melt. Estimates of the elongational viscosity η_E obtained as $\frac{\tau_{xx}}{\dot{\epsilon}_{xx}}$ at low α_{xx} values are consistent with Trouton's rule $\eta_E = 3\eta_0$ [44].

Experimentally, stress in flowing melts is measured through birefringence measurements. By ascribing polarisability tensors to all repeat units in the sampled configurations and using the Clausius-Mossotti equation, one can calculate the index of refraction tensor, $\mathbf{n}$, as an ensemble average [43]. Figure 3.34 displays the anisotropy of the index of refraction, $n_{xx} - n_{yy}$, as a function of the anisotropy of stress, $\tau_{xx} - \tau_{yy}$, for two polyethylene melts subjected to steady-state uniaxial elongational flow, as computed from Monte Carlo simulations in the $NnbT\boldsymbol{\mu}^*\boldsymbol{\alpha}$ ensemble. The relation between $n_{xx} - n_{yy}$ and $\tau_{xx} - \tau_{yy}$ is linear, as observed experimentally ("stress optical law"). Furthermore, the coefficient of proportionality in this linear relationship, $C_{\rm simul} = (2.35 \pm 0.10) \times 10^{-9}$ Pa^{-1}, predicted for the C_{200} melt, is in excellent agreement with the experimental value reported by Janeschitz-Kriegl for high-molecular weight polyethylene subjected to elongational flow [43], $C_{\rm expt} = 2.20 \times 10^{-9}$ Pa^{-1}. At very high stress anisotropies (corresponding to very high strain rates), some nonlinearity is observed in the predicted birefringence, the $n_{xx} - n_{yy}$ versus $\tau_{xx} - \tau_{yy}$ curve becoming concave downwards [43,44].

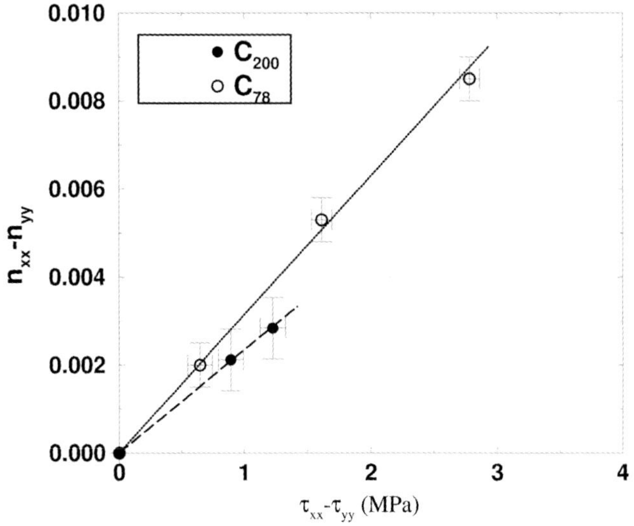

Fig. 3.34. Anisotropy of the index of refraction, $n_{xx} - n_{yy}$, as a function of the anisotropy of stress, $\tau_{xx} - \tau_{yy}$, for C_{78} and C_{200} melts subjected to steady-state uniaxial elongational flow along the x-direction (T=450 K, b = 1 atm).

Pre-oriented melt configurations generated under conditions of steady-state uniaxial elongational flow, using EBMC in the $NnbT\boldsymbol{\mu}^*\boldsymbol{\alpha}$ ensemble, are excellent starting points for conducting nonequilibrium molecular dynamics (NEMD) simulations to track the relaxation of melt structure and stress to equilibrium upon cessation of the flow. Such NEMD computer experiments have been performed on unentangled PE melts [70], leading to estimates of the relaxation time spectrum which are fully consistent with those obtained through equilibrium MD [71].

3.4.5 Sorption Equilibria of Alkanes in Polyethylene

An accurate knowledge of phase equilibria in oligomer-polymer systems is important for many applications. The effective design of polymer production processes requires knowledge of the conditions that ensure full miscibility of the polymer with the monomer and any solvent present during polymerisation, and partial miscibility during separation and collection of the final product. On the other hand, devolatilisation, i.e., removal of residual monomer and solvent from a polymer product under low pressure and high temperature conditions, is of critical importance for the protection of the environment and for the elimination of health and safety hazards from toxic monomers and solvents.

Phase equilibria in oligomer-polymer systems have been studied extensively with experiments, equation of state-based thermodynamic analysis, and molecular simulation. As already mentioned, traditional Monte Carlo–based simulation methods [47] require insertions/deletions of oligomer molecules within, or exchanges of oligomer molecules between, dense phases; thus, they become exceedingly time consuming when large oligomer molecules are of interest, even when configurational bias strategies are used.

Zervopoulou et al. [46] have employed Monte Carlo simulations in the $f_1' N_\mathrm{p} n_0 PT \boldsymbol{\mu}^*$ statistical ensemble to predict the sorption thermodynamics of C_4 to C_{20} alkanes in molten polyethylene around 450 K over a wide range of alkane fugacities. By introducing scission and fusion moves, as explained in Sect. 3.3.6, these simulations obviate the need to perform insertions/deletions or exchanges of molecules, and thus circumvent the associated excluded volume overlap problems. Zervopoulou et al. have compared their results against those obtained from a more conventional $f_1 N_\mathrm{p} n PT \boldsymbol{\mu}^*$ simulation [46], i.e., a hybrid isothermal isobaric–grand canonical simulation employing insertions/deletions of oligomer molecules in a polymer matrix being simulated with EBMC. In the following we will refer to the insertion and deletion-based $f_1 N_\mathrm{p} n PT \boldsymbol{\mu}^*$ MC simulation as "Method 1" and to the scission and fusion-based Monte Carlo simulation in the $f_1' N_\mathrm{p} n_0 PT \boldsymbol{\mu}^*$ ensemble as "Method 2".

Figure 3.35 presents predictions of the sorption isotherm of n-decane (C_{10}) in molten linear PE at 458 K, as obtained from the two simulation methods. Over the range of molecular weight distributions studied (mean PE chain length C_{78} to C_{200}, flat chain length distribution with polydispersity index $I = 1.08$) the sorption equilibria have been found to be insensitive to the exact molecular weight distribution of PE [46]. Predictions are based on the NERD force field. Also shown in Fig. 3.35 are experimental measurements from Bonner *et al.* [72] and a calculation based on the self-associating fluid theory (SAFT) equation of state. All sets of data agree very well. The isotherm has a characteristic upward-curving "Flory-Huggins" shape; as more and more alkane is dissolved in the polymer with increasing fugacity, the "fractional free volume" of the polymer phase increases, making it easier for the alkane to dissolve. Although Methods 1 and 2 give the same results, the CPU time required for the scission and fusion-based Method 2 to equilibrate the composition is shorter by a factor larger than 20 than for the insertion and deletion-based Method 1 [46]. The horizontal error bars on the points from Method 2 in Fig. 3.35 arise during the process of converting the relative fugacity f_1' of the alkane to an absolute fugacity f_1; this is done by virtually incrementing the PE chains present in the simulated system by one segment [46].

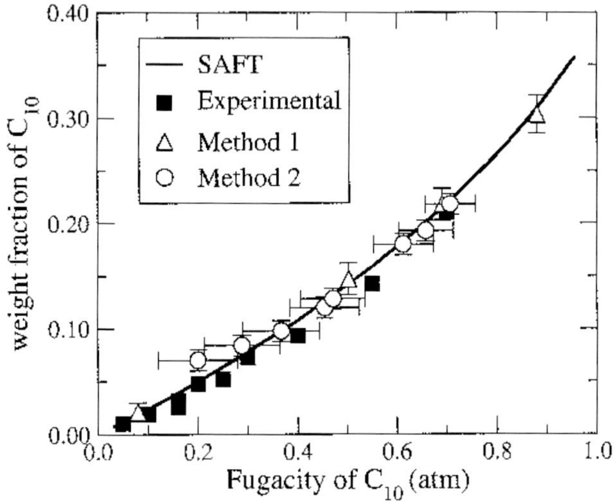

Fig. 3.35. Sorption isotherm of C_{10} in molten polyethylene at $T = 458$ K. Predictions are shown from the insertion and deletion-based MC Method 1 and from the scission and fusion-based MC Method 2, using the NERD force field. Also shown are experimental data and a calculation based on the SAFT equation of state.

Figure 3.36 shows calculated sorption isotherms for n-eicosane (C_{20}) in molten PE at 474 K. Predictions from the scission and fusion-based Method 2 only are shown, obtained with the NERD force field [46]. Method 1 cannot converge for C_{20} under the conditions considered here, despite the use of configurational bias in the insertion/deletion moves. Predictions from Method 2 are in excellent agreement with the SAFT calculation, which is expected to be very accurate for this system. Unfortunately, direct experimental evidence is unavailable. This alludes to the difficulty of performing accurate experimental measurements of phase equilibria in oligomer-polymer systems, a difficulty which can be alleviated through the development of reliable molecular simulation and theoretical calculation schemes.

Volumetric changes in the polymer phase as a result of sorbing the oligomer can readily be captured with the MC schemes of Zervopoulou et al. [46]. Fig-

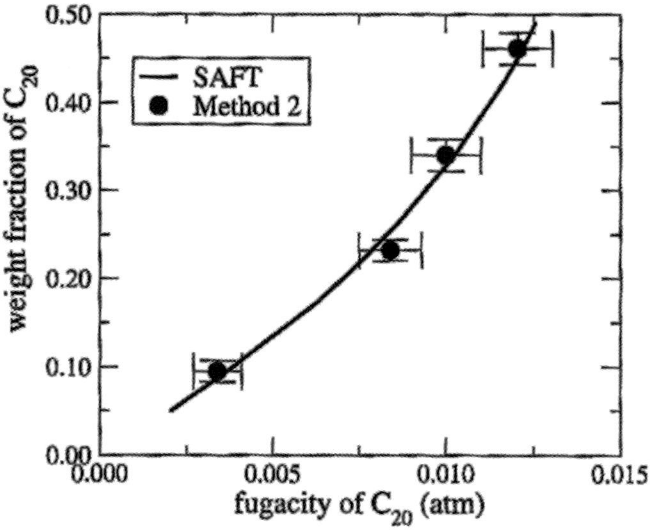

Fig. 3.36. Sorption isotherm of C_{20} in molten polyethylene at $T = 474$ K. Predictions are shown from the scission and fusion-based MC Method 2, using the NERD force field, and from the SAFT equation of state.

ure 3.37 displays the percent equilibrium swelling of a PE melt at 474 K as a result of sorbing n-pentane at various pressures. The points displayed have been obtained with Method 1. Method 2 leads to identical results. Clearly, the swelling is substantial. Results from the simulations are in very good agreement with predictions from the SAFT equation of state, especially when the TraPPE model is employed in the simulations. Interestingly, the mass density of the alkane-swollen polymer phase is found to be quite insensitive to pressure [46]. This results from a compromise between two opposing tendencies: a tendency to densify in response to the application of higher and higher pressures, and a tendency of "free volume" to increase as a result of the increased concentration of the lower-density sorbed alkane. Over a wide range of pressures, the mixed phase behaves as if it were incompressible; this may explain the success of simple incompressible theories, such as the Flory-Huggins theory, in qualitatively capturing sorption behaviour in these systems.

3.4.6 Polymers at Interfaces

Connectivity-altering MC schemes can be used for the rapid equilibration of interfacial systems containing long-chain polymers.

Daoulas et al. [73] used EBMC simulations in the $Nn\mu^*PT$ ensemble in order to study the thermodynamic and conformational properties of polyethylene melt/graphite interfaces, wherein some of the polyethylene chains are

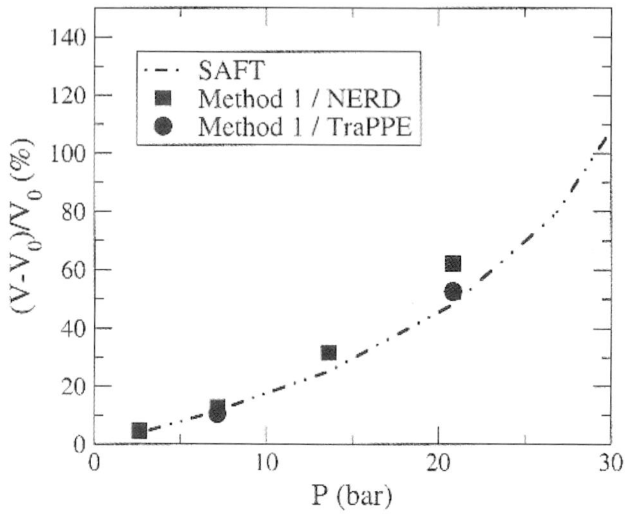

Fig. 3.37. Percentage of equilibrium swelling (volume increase) of a molten polyethylene phase as a result of sorbing pentane from a pure pentane phase at various pressures. The temperature is 474 K. Predictions are shown from the insertion and deletion-based MC Method 1, using the NERD and TraPPE force fields, as well as from the SAFT equation of state approach.

terminally grafted to the graphite substrate. Such use of terminally grafted chains is technologically very important for enhancing adhesion between a polymer and a solid substrate.

The Daoulas *et al.* simulations employed a realistic united-atom PE model and an atomistic representation of the graphite substrate. Grafted and free chains of mean length C_{78} to C_{250} and polydispersity index 1.08 to 1.14 were considered. The surface density of grafted chains was 0.54 to 2.62 chains/nm^2 and the temperature was 450 K. Thin film systems consisting exclusively of PE chains terminally grafted to the graphite (with no free chains) were also simulated. End-bridging moves involving grafted and free chains were designed so as not to perturb the topology of grafting by, e.g., creating chains grafted on both ends [73]. Excellent performance of the equilibration algorithm was noted.

Figure 3.38 displays the volume fraction profiles of segments belonging to the grafted ($\phi_g(z)$, profiles decaying with increasing z) and segments belonging to the free ($\phi_f(z)$, profiles rising with increasing z) as functions of the distance z from the graphite surface for a C_{156} system at two surface grafting densities. Also shown are results from a lattice-based self-consistent field (SCF) approach to the same problem [74]. The chain representation in the SCF model has been designed so as to respect the mass density, contour length, and conformational stiffness of the atomistic chains; SCF calculations have been conducted using the same chain length distribution, surface grafting density, and temperature as in the atomistic simulations. On the basis of this and similar figures, it is seen the SCF calculation can capture the overall spatial distribution and conformation of grafted and free chains with remarkable success. Only on the length scale of individual methylene units, close to the solid substrate, are there some atomistic layering effects which cannot be captured by the more coarse-grained SCF calculation.

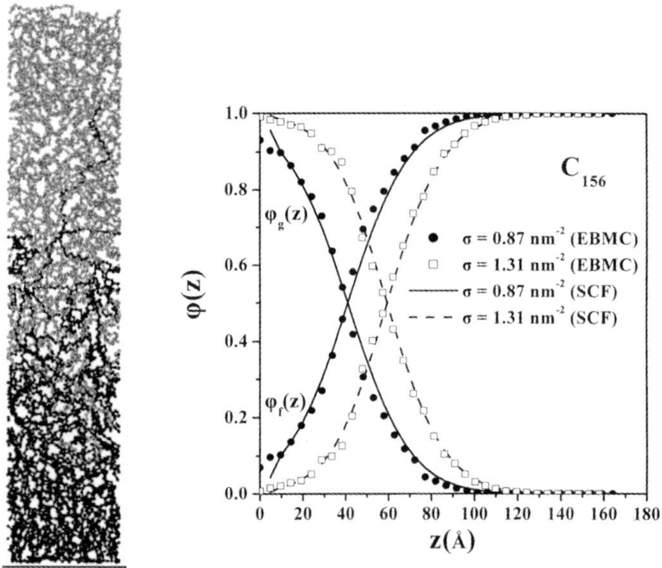

Fig. 3.38. Volume fraction profiles of free and grafted chains obtained through EBMC simulation of a C_{156} melt, consisting of free and grafted chains, next to a graphite surface. Two different interfacial systems, with grafting densities 0.87 chains/nm^2 (solid circles) and 1.31 chains/nm^2 (open squares), are considered. The solid and dashed curves show the corresponding predictions of the SCF model of [74]. A configuration of the interfacial system with surface grafting density 1.31 chains/nm^2 is shown on the left, with grafted chains in black and free chains in grey lines.

An interesting feature of these interfaces, with profound consequences for adhesion [74,76], is that, with increasing surface grafting density, a layer consisting almost exclusively of surface grafted chains builds up next to the solid substrate, pushing the free chains towards the bulk melt. Thus, the width of the region over which grafted and free chains interpenetrate does not necessarily increase upon increasing the surface grafting density. For the grafted chains, this phenomenon is sometimes referred to as a transition from a "wet brush" to a "dry brush" regime; it was first predicted on the basis of scaling arguments by de Gennes [77].

3.5 Conclusions and Outlook

We have discussed a family of connectivity-altering Monte Carlo algorithms which permit the rapid equilibration of dense polymer systems consisting of long chains. Thanks to their excellent scaling with chain length, these algorithms can, for the first time, fully equilibrate melts of molecular weight 10^4 - 10^5, comparable to that encountered in plastics processing operations, at all length scales. Relying upon the solution of a well-understood geometric bridging problem, they have been adapted to deal with a variety of monomer constitutions. Furthermore, there is excellent synergy between these algorithms and state-of-the-art techniques for overcoming barriers in rugged potential energy landscapes, such as parallel tempering.

Application of the connectivity-altering algorithms has yielded predictions of the structure, volumetric properties, chain conformational characteristics, and sorption equilibria in several polyolefin and polydiene systems, in excellent agreement with experiment. Moreover, the algorithms have been used in order to derive potentials of mean force with respect to selected, slowly varying degrees of freedom (e.g., the conformation tensor) for use within nonequilibrium thermodynamic approaches (e.g., GENERIC) to flow and transport phenomena. More generally, the algorithms constitute useful tools for deriving "coarse-grained" model representations, cast in terms of a few, slow degrees of freedom or "order parameters", directly from atomistic information.

The well-equilibrated configurations sampled by the connectivity-altering Monte Carlo algorithms serve as excellent starting points for molecular dynamics simulations aimed at elucidating and predicting dynamical properties in polymer melts or polymer solutions; also for forming glassy configurations and studying their properties.

Research on the applications discussed in Sect. 3.4 is continuing. Current efforts pursue four additional directions: (a) equilibration of melts consisting of chains with highly nonlinear architectures, such as star and comb polymers; (b) prediction of mixing thermodynamics in polymer blends (e.g., blends of polyisoprene and polybutadiene) from atomistic simulations; (c) combination with schemes for the pretabulation of interactions between inflexible multiatom moieties (e.g., aromatic groups) for the equilibration of stiff polymers consisting of such moieties (e.g., polyimides); (d) study of phase transitions of melts under conditions of steady-state flow, e.g. of the shift in melting point expected when molten polyethylene is subjected to uniaxial or biaxial elongational flow.

On the algorithmic side, efforts are made to enhance the acceptance rate of connectivity-altering moves and to reduce the "shuttling" tendency in runs employing these moves. Modified moves employing local relaxation of the environment around the point where a bridging event takes place are currently under investigation.

Acknowledgements

I am grateful to my collaborators Vlasis Mavrantzas, Krishna Pant, Larry Dodd, Manolis Doxastakis, Alfred Uhlherr, Nikos Karayiannis, Loukas Peristeras, Christina Samara, Travis Boone, Vagelis Harmandaris, Vanessa Zervopoulou, Andreas Terzis, Ioannis Economou, Erwan Nicol, Patricia Gestoso, and Angeliki Giannousaki, without whom the work presented herein would not have been possible. The European Commission is thanked for valuable financial support through the Brite-EuRam project MPFLOW, the GROWTH project DEFSAM, and the TMR research network NEWRUP. Additional support provided by the Greek General Secretariat for Research and Technology through two PENED programmes (contracts 218-95 EΔ, 95-99 EΔ) is deeply appreciated. Finally, CECAM and SIMU are thanked for enabling my participation in many stimulating and enlightening research workshops and conferences, where some of the ideas that led to the results presented herein initially arose.

References

1. M. Tuckerman, B.J. Berne, G.J. Martyna: J. Chem. Phys. **97**, 1990 (1992). G.J. Martyna, M.E. Tuckerman, D.J. Tobias, M.L. Klein: Mol. Phys. **87**, 1117 (1996).
2. D.N. Theodorou: SIMU Newsletter **1**, 19 (2000) *http://simu.ulb.ac.be/newsletters/newsletter.html*
3. A. Uhlherr, D.N. Theodorou: Curr. Opin. Solid State and Mat. Sci. **3**, 544 (1998)
4. F.T. Wall, A. Hiller, D.J. Wheeler: J. Chem. Phys. **22**, 1036 (1954)
5. F.T. Wall, W.A. Seitz: J. Chem. Phys. **54**, 5338 (1971)
6. K. Binder (ed.): *Monte Carlo and Molecular Dynamics Simulations in Polymer Science* (Oxford University Press, New York 1995)
7. A.K. Kron: Polymer Sci. USSR **7**, 1361 (1965)
8. M. Vacatello, G. Avitabile, P. Corradini, A. Tuzi: J.Chem.Phys. **73**, 543 (1980)
9. R. H. Boyd: Macromolecules **22**, 2477 (1989)
10. J.I. Siepmann, D. Frenkel: Mol. Phys. **75**, 59 (1992)
11. J.J. de Pablo, M. Laso, U.W. Suter: J. Chem. Phys. **96**, 2395 (1992)
12. M.N. Rosenbluth, A.W. Rosenbluth: J. Chem. Phys. **23**, 356 (1955)
13. J.J. de Pablo, F.A. Escobedo: J. Chem. Phys. **105**, 4391 (1996)
14. A.P. Lyubartsev, A.A. Martsinovski, S.V. Shevkunov, P.N. Vorontsov-Velyaminov: J. Chem. Phys. **96**, 1776 (1991)
15. C.J. Geyer: 'Markov Chain Monte Carlo Maximum Likelihood'. In: *Computing Science and Statistics, Proceedings of 23rd Symposium on the Interface* (American Statistical Association, New York 1991) pp 156-163. M.C. Tesi, E.J.J. van Rensburg, E. Orlandini, S.G. Whittington: J. Stat. Phys. **82**, 155 (1996)

16. M.G. Wu, M.W. Deem: Mol. Phys. **97**, 559 (1999)
17. M.G. Wu, M.W. Deem: J. Chem. Phys. **111**, 6625 (1999)
18. O.F. Olaj, W. Lantschbauer: Makromol. Chem., Rapid Commun. **3**, 847 (1982)
19. M.L. Mansfield: J. Chem. Phys. **77**, 1554 (1982)
20. W.G. Madden: J. Chem. Phys. **87**, 1405 (1987); J. Chem. Phys. **88**, 3934 (1988)
21. V.G. Mavrantzas, T.D. Boone, E. Zervopoulou, D.N. Theodorou: Macromolecules **32**, 5072 (1999)
22. L.R. Dodd, T.D. Boone, D.N. Theodorou: Mol. Phys. **78**, 961 (1993)
23. P.J. Flory: *Statistical Mechanics of Chain Molecules* (Wiley Interscience, New York 1969)
24. N. Gō, H.A. Scheraga: Macromolecules **3**, 178 (1970)
25. M.W. Deem, J.S. Bader: Mol. Phys. **87**, 1245 (1996)
26. H.-Y. Lee, C.-G. Liang: Mech. Mach. Theory **23**, 209 (1988)
27. H.-Y. Lee, C.-G. Liang: Mech. Mach. Theory **23**, 219 (1988)
28. H.-Y. Lee, C. F. Reinholtz: ASME J. Mech. Design **118**, 296 (2000)
29. P.V.K. Pant, D.N. Theodorou: Macromolecules **28**, 7224 (1995)
30. L.R. Dodd, D.N. Theodorou: Advan. Polym. Sci. **116**, 249 (1994)
31. T. Spyriouni, I.G. Economou, D.N. Theodorou: Macromolecules **31**, 1430 (1998)
32. T. Spyriouni, I.G. Economou, D.N. Theodorou: Phys. Rev. Lett. **80**, 4466 (1998)
33. C.T. Samara: Simulation of polypropylene of various tacticities with the Monte Carlo method. Ph.D. Thesis, University of Patras, Patras, Greece (2000)
34. M. Doxastakis, V.G. Mavrantzas, D.N. Theodorou: J. Chem. Phys. **115**, 11339 (2001)
35. M. Doxastakis, V.G. Mavrantzas, D.N. Theodorou: J. Chem. Phys. **115**, 11352 (2001)
36. E. Nicol, P. Gestoso, M. Doxastakis, D.N. Theodorou, in preparation.
37. A. Uhlherr: Macromolecules **33**, 1351 (2000)
38. C.D. Wick, J.I. Siepmann: Macromolecules **33**, 7207 (2000)
39. A. Uhlherr, V.G. Mavrantzas, M. Doxastakis, D.N. Theodorou: Macromolecules **34**, 8554 (2001)
40. A. Uhlherr, M. Doxastakis, V.G. Mavrantzas, D.N. Theodorou, S.J. Leak, N.E. Adam, P.E. Nyberg: Europhys. Lett. **57**, 506 (2002)
41. A.N. Beris, E.J. Edwards: *Thermodynamics of flowing systems with internal microstructure* (Oxford University Press, Oxford 1994)
42. V.G. Mavrantzas, D.N. Theodorou: Macromolecules **31**, 6310 (1998)
43. V.G. Mavrantzas, D.N. Theodorou: Comp.Theor.Polym.Sci. **10**, 1 (2000)
44. V.G. Mavrantzas, D.N. Theodorou: Macromol. Theory Simul. **9**, 500 (2000)
45. V.G. Mavrantzas, H.C. Öttinger: Macromolecules **35**, 960 (2002)
46. E. Zervopoulou, V.G. Mavrantzas, D.N. Theodorou: Macromolecules **115**, 2860 (2001)
47. J.J. de Pablo, Q. Yan, F.A. Escobedo: Ann. Rev. Phys. Chem. **50**, 377 (1999)
48. A.Z. Panagiotopoulos: J. Phys: Condens. Matter **12**, R25 (2000)
49. E. Zervopoulou, V.G. Mavrantzas, D.N. Theodorou: J. Chem. Phys. **115**, 2860 (2001)
50. T. Spyriouni, I.G. Economou, D.N. Theodorou: Macromolecules **30**, 4744 (1997)
51. N.Ch. Karayiannis, V.G. Mavrantzas, D.N. Theodorou: Phys. Rev. Lett. **88**, 105503 (2002)

52. J.C. Horton: Macromolecules **22**, 681 (1989)
53. L.J. Fetters, W.W. Graessley, R. Krishnamoorti, D.J. Lohse: Macromolecules **30**, 4973 (1997)
54. G.T. Dee, T. Ougizawa, D.J. Walsh: Polymer **33**, 3462 (1992)
55. V.A. Harmandaris, M. Doxastakis, V.G. Mavrantzas, D.N. Theodorou: J. Chem. Phys. **116**, 436 (2002)
56. K.G. Honnell, J.D. McCoy, J.G. Curro, K.S. Schweizer, A.H. Narten, A. Habenschuss: J. Chem. Phys. **94**, 4659 (1991).
57. R. P. Wool: Macromolecules **26**, 1564 (1993).
58. J.D. Ferry: *Viscoelastic Properties of Polymers* (Wiley, New York 1980)
59. U.W. Suter, P. Neuenschwander: Macromolecules **14**, 528 (1981)
60. D.G.H. Ballard, P. Cheshire, G.W. Longmann, J. Schelten: Polymer **19**, 379 (1978)
61. U.W. Suter, P.J. Flory: Macromolecules **8**, 765 (1975)
62. J.-P. Ryckaert: 'Simulation of polymers using realistic potentials'. In: *Monte Carlo and molecular dynamics of condensed matter systems, Conference Proceedings Vol. 49*, ed. by K. Binder, G. Ciccotti (Società Italiana di Fisica, Bologna 1996) Chap. 27, pp 729-746
63. S.J. Antoniadis, C.T. Samara, D.N. Theodorou: Macromolecules **32**, 8635 (1999)
64. R.D. Maier, R. Thomann, J. Kressler, R. Mülhaupt: J. Polym. Sci. Polym. Phys. **35**, 1135 (1997)
65. S.M. Lippow, X.-H. Qiu, M.D. Ediger: personal communication.
66. N. Nemoto, M. Moriwaki, H. Odani, M. Kurata: Macromolecules **4**, 215 (1971)
67. C.D. Han, J. Kim, J.K. Kim: Macromolecules **22**, 383 (1989)
68. K. Šolc and W.H. Stockmayer: Macromolecules **54**, 2756 (1971)
69. D.N. Theodorou, U.W. Suter: Macromolecules **18**, 1206 (1985)
70. V.A. Harmandaris, V.G. Mavrantzas, D.N. Theodorou: Macromolecules **33**, 8062 (2000)
71. V.A. Harmandaris, V.G. Mavrantzas, D.N. Theodorou: Macromolecules **31**, 7934 (1998)
72. D.C. Bonner, N.F. Brockmeier, Y.I. Cheng: Ind. Eng. Chem. Process Des. Dev. **4**, 437 (1974)
73. K.Ch. Daoulas, A.F. Terzis, V.G. Mavrantzas: J. Chem. Phys., in press (2002)
74. A. Terzis, D.N. Theodorou, A. Stroeks: Macromolecules **33**, 1385 (2000)
75. A. Terzis, D.N. Theodorou, Stroeks: Macromolecules **33**, 1397 (2000)
76. A. Terzis, D.N. Theodorou, and A. Stroeks: Macromolecules **35**, 508 (2002)
77. P.-G. de Gennes: Macromolecules **13**, 1069 (1980)

4 Bridging the Time Scale Gap: How Does Foldable Polymer Navigate Its Conformation Space?

Alexander Grosberg[1,2]

[1] Department of Physics, University of Minnesota, 116 Church Street SE, Minneapolis, MN 55455, USA
[2] Institute of Biochemical Physics, Russian Academy of Sciences, Moscow 117119, Russia

Abstract. We analyze the physical ideas used to analyze the computational Monte Carlo experiments on dynamics of protein folding and dynamics of other complex systems. We show that the concept of reaction coordinate can be formulated in a systematic way using the concept of commitor. This quantity can be usefully described by associating certain system of resistors with the space of conformations of the protein, in which case the flow of current and the distribution of potentials, governed by the Kirchoff rules, give full description of the topology of folding pathways.

4.1 Introducing the Characters

Most novels are based on a love triangle. Science in this regard is becoming competitive to novels: in addition to experiment and theory, computation is now the third party to the scientific triangle. Scientists of the past lived a quiet family life, maintaining satisfactory relations of the two sides, theory and experiment. As long as theories were in a reasonable agreement with experiments, everything was in order. Because of the triangle, our life now is potentially frustrated: harmonizing two sides of the triangle, we risk to frustrate the third. This concentrates exclusively on the relation between theory and computation. How can we understand theoretically certain computer experiments?

Specifically, the subject matter has to do with the ways to understand the mobility of foldable polymers, and it is of course motivated by the protein folding problem. The hope is to show that theoretical understanding of computational models provides an insight useful beyond the theory-computation side of the scientific triangle.

Dynamics of protein folding is by no means a new problem. There is now vast literature on this subject, including experimental, theoretical, and computational works [1]. One of the central aspects is known under the nickname of "an adequate reaction coordinate, or order parameter, problem." The issue has a long history and, in the view of this author, a confusing name. In fact, the issue is not so much about the coordinate, but about the potential profile along this coordinate. In this article, I want to discuss a simple physical analogy which may shed a new light on this problem.

The problem under consideration is quite general in spirit. Similar difficulties exist not only in various models of protein folding, but also in other complex systems. They arise whenever relevant space of configurations is not only multidimensional, but also geometrically complex (curved, restricted, etc). In such cases, the simple idea of barrier crossing is at least insufficient, and sometimes outright confusing. There is enormous literature on the subject (the reader may want to consult the references in the recent review article [2]).

One popular language used in discussions of similar problems is that of energy (or free energy) landscapes. It is hard to say who was the first to introduce this approach. The very term "energy barrier" suggests that some landscape-related images were around for a very long time. In recent years, this terminology has become very popular thanks mainly to the works by Wolynes and his co-authors on energy landscapes in protein folding context (see recent review [3] and references therein).

4.2 Setting Up the Stage: Conformation Space and Reaction Coordinate

4.2.1 Conformation Space: Lattice Polymer

Computationally most tractable polymer model is the lattice one. It represents the polymer as a self avoiding walk on a (cubic) lattice. An appropriate set of local moves is usually defined to allow the dynamics realized as a succession of these moves. In the most common Stockmayer-Verdier model [4], these moves include end flip, corner flip, and crankshaft flip. The only requirement for the set of local moves is that they have to provide for ergodicity: every conformation should be possible to transform into every other conformation. Importantly, ergodicity does not require that the path from one conformation to another be short or simple, it only requires that such path exists.

We now have to make one step further in the direction of abstraction and ask: what is the space of conformations for such lattice polymer? This space is in fact a graph [5,6]. Each vertex of the conformational graph represents one particular conformation. Two vertices of the graph are connected by a bond if and only if the corresponding conformations can be transformed into one another by a *single* elementary move. When all conformations are included, this graph is connected, which is to say that the system is ergodic. In general, this graph, or certain regions of it, may have non-trivial fractal dimension [7]. In addition, certain parts of conformation graph are shown to be of a small world network type [8]. Finally, if we restrict consideration with only compact conformations, then ergodicity may be broken [9]; controlling, say, maximal gyration radius of the polymer in real space we can observe percolation transition in the graph of conformations.

4.2.2 Conformation Space: Off-lattice Polymer

In the simplest off-lattice model, the chain conformation is fully described by the set of monomer position vectors, $\boldsymbol{r}_i$. Naively, this suggests that the space of conformations for the N-mer is a $3N$-dimensional Euclidean space. This is wrong. To see why and how this is wrong, it is enough to consider a "polymer" with two monomers ($N = 2$) with fixed bond length: apart from translational freedom of the first monomer, the "space of conformations" is, of course, a sphere - curved surface with non-Euclidean internal geometry. This very simple argument can be pulled quite far to show that conformation space is very much curved for long polymers, and its internal geometry is not at all Euclidean [10]. One should not be distracted by the fact that conformation *can* be fully characterized with the Euclidean position vectors $\boldsymbol{r}_i$ - this only means that the curved space of conformations can be *embedded* in a bigger Euclidean space - just like we usually imagine a sphere in a $3D$ space. This embedding in no way cancels or downplays the importance of the fact that internal geometry of conformation space is not Euclidean, because of course all the dynamical trajectories of the systems are in this space, not in the bigger flat one.

Clearly, the difference between on-lattice and off-lattice cases is neither deep nor important in this context. To understand it, let us begin with neglecting the excluded volume constrains. In this case, lattice model has just 6 (on the cubic lattice in $3D$) possible positions for every bond. Similarly, in an off-lattice model based on rotational isomers, there are a few permissible rotational states for every bond. With 2 possible states for every one of N monomers, the graph of conformations would have been an N-dimensional hypercube; similar figure with 6 or other finite number of vertices along each of the N axis has no special name, so, for the lack of a better word, let us call it a "cube." Of course, imposition of the excluded volume constraints erases many (in fact, even the majority) of the "cube" vertices, the ones which correspond to conformations with overlapping monomers. Nevertheless, the remaining part, which is the real conformations graph, is still a part of a cube, which obviously has nothing to do with the regular lattice - discrete analog of a flat Euclidean space. Similarly, for the off-lattice model with continuous set of rotations, conformation space is compact (in the strict mathematical sense of the word) and curved.

Thus, the stage of the protein folding drama is a compact space, some part of a "cube" or its continuous analog. It is terribly curved and restricted, it has complex topological and fractal properties. How can we understand the motion of a protein in such space?

4.2.3 Reaction Coordinate Problem

Speaking about protein folding kinetics, it is fashionable to resort to the so-called landscape theory. This theory invokes an intuitively appealing image of a rugged terrain in a mountain country. Everyone is aware about one difficulty of such approach, and everyone is prepared to brave it: While vertical coordinate in the landscape picture is energy, there are great many "horizontal" coordinates; underlying "geography" is multidimensional. The bigger problem, however, is the above mentioned complicated geometry of the underlying space of conformations.

The problem is frequently formulated in terms of the choice of proper reaction coordinate, or, in other words, dynamic order parameter. This is supposed to be a quantity suitable to monitor the progress of a molecule from somewhere in the sea of unfolded states and all the way into the correctly folded state. The choice of a "good" reaction coordinate is mixed with the issue of a preferred path of folding in conformation space. In fact, these two issues are very different.

First, as regards the preferred path, it can be only understood in terms of some projection of a full conformation space. Clearly, there are always great many "inessential" small scale movements which are not important for the overall folding progress. In order to see a pathway, we must imagine a projection along all these "inessential" coordinates on the smaller space of only "essential" ones. The choice or identification of the latter is a difficult problem, but the necessary one if we want to discuss the folding pathway. Assuming that this problem is somehow resolved, we are immediately tempted to say that the coordinate along the pathway is the reaction coordinate.

In fact, the issue of a pathway is difficult and open one (see [11] and reference therein), while the question of the reaction coordinate is independent of it. To understand this point, it is useful to recognize the difference between the question of the reaction coordinate and that of the relevant (free) energy profile [12].

Consider an elementary example, kinetics of a first order phase transition, such as gas-liquid. As it is well known, the transition occurs via nucleation and growth mechanism. In the usual theory [13], nucleus is assumed to be spherical, and its radius serves as an order parameter or reaction coordinate. What does this mean? This means, we assume the free energy associated with the nucleus of radius R to include two contributions, proportional to the nucleus volume and surface, respectively. The former has to do with the deviation from the point of thermodynamic phase equilibrium (say, overcooling) Δ, the latter is interpreted as a surface tension, γ. This yields the R-dependent free energy

$$F(R) = -\Delta R^3 + \gamma R^2 \ , \tag{4.1}$$

which has a barrier at finite R. Then, what one does is one writes the diffusion equation in "the space of nuclei sizes R" and obtains the qualitatively reasonable description of the nuclei development [13]. As a matter of fact, recent simulations [14] indicate that this theory is less than perfect quantitatively, but this is not so important for the present discussion. We should concentrate on the choice of R and the derivation of the free energy profile, Eq. (4.1).

What happens if we choose some other variable, instead of R? In this context, averaged density is sometimes viewed as an example of totally inappropriate reaction coordinate. Indeed, thinking about gas-liquid transition in terms of changing density, one imagines a succession of more and more dense spatially uniform states. Along such pathway, the barrier is enormous, it is proportional to the total volume of the system. However, this conclusion is only due to the assumption of spatial uniformity. In fact, as we know well, when averaged density increases very slightly from its gas value, it is getting overwhelmingly more favorable for the system to break spatial uniformity and to have one small nucleus with the density of a liquid, leaving density in other places unaffected. Therefore, we *can* describe the transition in terms of density, but we should be prepared to pay dearly - we should defy our intuition and to realize that the state with the given averaged density may not be, and mostly likely is not, the state with uniformly distributed density.

In this sense, it is trivial to derive a one-to-one relation between R and the averaged density (assuming for simplicity that there is only one nucleus in the system): $\rho = \rho_g / \left(1 - \frac{4\pi R^3 (\rho_l - \rho_g)}{3N}\right)$, where N is the total number of particles, while ρ_g and ρ_l are the gas and liquid densities, respectively. Therefore, we can easily re-formulate the free energy profile $F(R)$ (4.1) taking ρ as another independent variable.

The message from the discussion above is simple. It is the free energy profile, not the coordinate itself, that is important to determine properly if we want to have an adequate description of the transition kinetics.

What is the standard method to determine the free energy profile as a function of the chosen reaction coordinate? There is only one particular limit in which the answer is simple. Specifically, let us assume that there is one particular variable in the system which changes much slower than any other variable; let us call this slow coordinate p. In this assumption, p is "slow" in the sense that all other degrees of freedom, which we call q for simplicity, have enough time to arrive to the conditional thermodynamic equilibrium while p does not change appreciably. This gives rise to the effective free energy profile

$$F(p) = -k_B T \ln \left[\int \exp\left(-\frac{\mathcal{H}(p,q)}{k_B T} dq \right) \right] , \qquad (4.2)$$

where $\mathcal{H}(p,q)$ is the corresponding energy.

Is this simple case generic? Could one generally define reaction coordinate as the slowest degree of freedom in the system? The answer is certainly no. Indeed, for the formula like (4.2) to apply, it is not enough to have strongly differing time scales in the system; it is necessary to have just one degree of freedom much slower than all others. There is no ground to expect such situation for a generic system, nor is there any hope to find such special situation for the proteins. What can one do then?

4.3 Unfolding the Drama: Commitor, p_{fold}, and the Reaction Coordinate

4.3.1 Commitor

The practical solution to the problem was suggested in the work [15] and, independently, in [2]. This solution is based on the quantity called p-fold [15] or "commitor" [2]. We shall use here both terms interchangeably.

Let us first discuss this quantity for a lattice model [15]. To understand what is p-fold, we assume that there are two well defined states of the system, in case of lattice toy proteins those are folded and unfolded states. For the present discussion, it is not important that folded state is usually represented by just a single compact conformation, while unfolded state is a big ensemble of coil-like conformations. What is important is that in the conformation space graph certain vortices are labeled as belonging to unfolded state, while certain others are labeled as belonging to the folded state.

Now, we pick an arbitrary conformation, or an arbitrary vertex of the conformation space graph, and define p-fold for this conformation in the following way. Imagine that we perform many Monte Carlo runs using the chosen conformation as an initial one. Every time, we run Monte Carlo dynamics for as long as it takes for the system to arrive for the first time into either folded or unfolded state. Then, we have to collect the statistics over many runs, and determine the probability, p, that a run, or randomly chosen trajectory, will first arrive into the folded state before ever touching the unfolded state. This probability is p-fold.

Quite similarly, commitor can be defined for an off-lattice system [2]. In this case, we choose a particular configuration (or conformation), and then initiate many trajectories going out of this point by randomly choosing the full set of momenta (or velocities). Once again, we determine the probability that the trajectory first hits the folded state before ever touching the unfolded state.

What is the meaning of this quantity, whether one prefers calling it p-fold or commitor? It is simple. It measures the commitment of the system for folding. If p is large, close to unity, that means he system is very close to the folded state, its descend back to the unfolded state is unlikely. Vice versa, if p is small, that means the system is close to the unfolded state.

It was conjectured in the work [15] that the value of p-fold, although time consuming for computer experiment. represents the best possible approximation for the reaction coordinate. Indeed, if there were the slowest degree of freedom in the system, then p would be clearly a well defined monotonic function of this coordinate. As the system progresses along the selected singly degree of freedom from an unfolded to the folded state, its value of p necessarily increases. Thus, if the reaction coordinate in the traditional sense existed, it would be basically the same as commitor. However, p can be computed by a constructive algorithm, albeit time consuming one, while reaction coordinate is never known. And, most importantly, p-fold, or commitor, remains well defined even for the generic system, with no special slow degree of freedom.

The concept of commitor, or p-fold is currently widely used in protein folding simulations (see, for instance, [11]). Nevertheless, there is a feeling that better intuition about this quantity is necessary. For instance, the works on the statistical mechanics of folding trajectories [16] lead to the questions like the ones about local extrema of p-fold. Is it possible to find a state such that all surrounding states have p-fold smaller (or larger) than the given one? If the answer were to be positive, that would render the concept of commitor useless at the very least. Luckily, the intuition suggests that it should be impossible. Below, we shall prove that it is impossible indeed.

4.3.2 Direct Current Analogy

In this section, we shall describe the physical analogy which helps intuitive understanding of the commitor, and also allows to prove their general properties, like monotonous behavior mentioned in the previous paragraph. This analogy is presented in terms of direct currents governed by the Kirchoff rules. To formulate it, let us be a little more specific about the model under study.

Suppose our polymer has certain available conformations, and let us call the conformations with letter $\mathcal{C}$. There is potential energy landscape, which means that there is certain energy $U(\mathcal{C})$ associated with every conformation $\mathcal{C}$. Consider now a pair of conformations, say $\mathcal{C}$ and $\mathcal{C}'$, such that they are connected on the conformation graph; in other words, they can be transformed one into another via a single Monte Carlo move. Consider the flow between $\mathcal{C}$ and $\mathcal{C}'$. Of course, this means, we should imagine a large number (an ensemble) of computers independently running Monte Carlo simulation of our system, and then we should ask what is number of copies switching from $\mathcal{C}$ to $\mathcal{C}'$ at the given time? Equivalently, we can speak of a probability to find a single Monte Carlo process jumping from $\mathcal{C}$ to $\mathcal{C}'$ at the given time. To be specific, let us assume that $\mathcal{C}$ is higher in energy than $\mathcal{C}'$: $U(\mathcal{C}) > U(\mathcal{C}')$. Then, Metropolis criteria [17] accepts the move from $\mathcal{C}$ to $\mathcal{C}'$ every time the move is offered, we get that the flow from $\mathcal{C}$ to $\mathcal{C}'$ is simply proportional to the occupation number of $\mathcal{C}$ at the present time; apart from units and proportionality coefficient,

$$I_{\mathcal{C}\to\mathcal{C}'} = n_{\mathcal{C}} \ . \tag{4.3}$$

As regards the opposite moves from $\mathcal{C}'$ to $\mathcal{C}$, they are accepted with probability smaller than unity, namely $\exp\left[U(\mathcal{C}') - U(\mathcal{C})\right]$, so that

$$I_{\mathcal{C}' \to \mathcal{C}} = n_{\mathcal{C}'} e^{U(\mathcal{C}') - U(\mathcal{C})} \ . \tag{4.4}$$

Here and below, to save some writing, we omit the temperature factor, assuming $k_B T = 1$; in other words, we pretend that potential energy $U(\mathcal{C})$ is measured in the units of $k_B T$: $U(\mathcal{C}) \to U(\mathcal{C})/k_B T$.

It is convenient to rewrite equations (4.3) and (4.4) by introducing the following notations. For every conformation $\mathcal{C}$, we define the quantity

$$\phi_{\mathcal{C}} = n_{\mathcal{C}} e^{U(\mathcal{C})} \ , \tag{4.5}$$

and for every connection between $\mathcal{C}$ and $\mathcal{C}'$, we define another quantity

$$R_{\mathcal{C}\mathcal{C}'} = e^{\max\left\{U(\mathcal{C}),\ U(\mathcal{C}')\right\}} \equiv e^{U(\mathcal{C})} \ . \tag{4.6}$$

The nice thing about quantities $\phi_{\mathcal{C}}$ is that in equilibrium, when occupation numbers $n_{\mathcal{C}}$ are governed by Boltzmann distribution $n_{\mathcal{C}} \propto e^{-U(\mathcal{C})}$, the values of $\phi_{\mathcal{C}}$ become independent of $\mathcal{C}$, a constant all over the conformation space. Using $\phi_{\mathcal{C}}$ and $R_{\mathcal{C}\mathcal{C}'}$, the master equations (4.3) and (4.4) can be presented in the form

$$I_{\mathcal{C}\mathcal{C}'} \equiv I_{\mathcal{C} \to \mathcal{C}'} - I_{\mathcal{C}' \to \mathcal{C}} = \frac{\phi_{\mathcal{C}} - \phi_{\mathcal{C}'}}{R_{\mathcal{C}\mathcal{C}'}} \ . \tag{4.7}$$

In this formula, it is now easy to recognize the Ohm's law for direct currents, which leads to the following physical interpretation.

We imagine that the conformational graph of the system is a network of resistors. The resistance between nodes $\mathcal{C}$ and $\mathcal{C}'$ is $R_{\mathcal{C}\mathcal{C}'}$. Governed by the Kirchoff rules, there are potentials on each node, $\phi_{\mathcal{C}}$, and currents between the nodes, $I_{\mathcal{C}\mathcal{C}'}$. The knowledge of these potentials and the current distribution is equivalent to the full knowledge of the dynamics of the system in question.

As we understand perfectly well, the distribution of currents in the resistor network is determined by the externally applied voltage. We have to understand now the way to apply the external voltage such as to get the results relevant for the determination of p-fold or commitor. Before doing that, it is useful to discuss one more equivalent formulation.

4.3.3 Diffusion Equation and Continuous (Off-lattice) Models

Given the flow I, we can write down the rate of change for the occupation number at any site $\mathcal{C}$. This is the sum over all neighboring sites $\mathcal{C}'$ of the difference between the flow out of $\mathcal{C}$ and into $\mathcal{C}$:

$$\frac{dn_{\mathcal{C}}(t)}{dt} = \sum_{\text{neighbors},\ \mathcal{C}'} \left[-I_{\mathcal{C} \to \mathcal{C}'} + I_{\mathcal{C}' \to \mathcal{C}}\right] \ . \tag{4.8}$$

This can be conveniently rewritten taking into account equations (4.3) and (4.4):

$$\frac{dn_{\mathcal{C}}(t)}{dt} = -\mu_{\mathcal{C}} n_{\mathcal{C}} + \sum_{\text{neighbors, } \mathcal{C}'} n_{\mathcal{C}'} e^{U(\mathcal{C}')-U(\mathcal{C})} , \tag{4.9}$$

where $\mu_{\mathcal{C}}$ is the connectivity (the number of connections) of the site $\mathcal{C}$. Remember now that the Laplace operator Δ can be defined on an arbitrary graph; its action on the arbitrary function $f_{\mathcal{C}}$ is defined as

$$\Delta f = -f_{\mathcal{C}} + \frac{1}{\mu_{\mathcal{C}}} \sum_{\text{neighbors, } \mathcal{C}'} f_{\mathcal{C}'} . \tag{4.10}$$

In terms of this operator, we finally obtain

$$\frac{d}{dt} f_{\mathcal{C}}(t) = \sqrt{\mu_{\mathcal{C}}} \Delta \left[\sqrt{\mu_{\mathcal{C}}} f_{\mathcal{C}}(t) \right] , \quad \text{where } f_{\mathcal{C}}(t) = \left(\frac{n_{\mathcal{C}}(t)}{\sqrt{\mu_{\mathcal{C}}}} e^{U(\mathcal{C})} \right) . \tag{4.11}$$

Without going too deeply into this subject, we point out here that this form of equation allows to perform easily the generalization for the continuous off-lattice limit. In this case, $\sqrt{\mu}$ is related to the local curvature of the conformation space.

4.3.4 Stationary and Transient Regimes

In most cases, the computer experiment on protein folding or similar complex dynamics is performed by running simulations starting from certain initial site $\mathcal{C}_0$. This can be described by adding proper δ-function in the diffusion equation (which we write here for simplicity in terms of n instead of f):

$$\frac{d}{dt} n_{\mathcal{C}}(t) = \mu_{\mathcal{C}} e^{-U(\mathcal{C})} \Delta \left[n_{\mathcal{C}}(t) e^{U(\mathcal{C})} \right] + \delta(t) \delta\left(\mathcal{C}_0\right) . \tag{4.12}$$

It is instructive to imagine another formulation in which there are (infinitely) many simulations running simultaneously, so that we start certain number of simulations per unit time, Q. In this case, the system as a whole is in stationary state, with occupation numbers $n_{\mathcal{C}}$ independent of time and satisfying the equation

$$0 = e^{-U(\mathcal{C})} \Delta \left[n_{\mathcal{C}} e^{U(\mathcal{C})} \right] + Q \delta\left(\mathcal{C}_0\right) . \tag{4.13}$$

Importantly, this is not the equation for equilibrium, its solution is not Boltzmann distribution. The solution of this stationary diffusion equation is equivalent to the determination of direct time-independent currents and potentials in the resistor network.

Note, the quantity $\mu_{\mathcal{C}}$ does not enter the equations for the stationary (time independent) conditions, but does play important role in the non-stationary transient regimes. This may seem to lead us into a trouble, given that our goal is to study the commitor, which seems to be essentially non-stationary quantity. This latter impression turns out wrong, as we show now.

4.3.5 Direct Current Formulation of the First Return Problem: Casino Problem and Its Easy Solution

The concept of commitor, as it is formulated above, is based on the classical probabilistic "first return" problem. Indeed, commitor of the state $\mathcal{C}$ is the probability that the trajectory of the random walk on the graph, starting from $\mathcal{C}$ will arrive *for the first time* into the folded state $\mathcal{F}$ before ever hitting the unfolded state $\mathcal{U}$. Let us consider the simple classical example [18].

Suppose a gambler arrives at the casino with certain amount of money x_0 and keeps gambling indefinitely unless arriving at the desperate zero money state ($x = 0$), in which case he/she is discarded. The question is this: what is the probability that the player looses money (arrives for the first time at $x = 0$) at the time t? The standard way to solve this problem is to realize that the time-dependent probability distribution of the money at possession of the gambler satisfies the diffusion equation:

$$\frac{\partial n_x(t)}{\partial t} = D\frac{\partial^2 n_x(t)}{\partial x^2} + \delta(t)\delta(x - x_0) \ , \tag{4.14}$$

where D is an appropriate diffusion coefficient (which can be absorbed into the time measurement units), and δ-functions describe the initial condition. The central idea of this approach is to realize that the concept of *first return* is embodied in the absorbing boundary condition:

$$n_x(t)|_{x=0} = 0 \ . \tag{4.15}$$

This boundary condition ensures that in the path integral solution of the diffusion equation, all trajectories are discarded which ever visit the left half-line ($x < 0$). Now, assuming $n_x(t)$ is found, the probability to arrive at $x = 0$ at time t *for the first time* is given by the flux into $x = 0$: $W(t) = -D\partial n_x(t)/\partial x|_{x=0}$. The corresponding solution is not difficult to find:

$$W(t) = \frac{x_0}{2\sqrt{\pi D}t^{3/2}} e^{-x_0^2/4Dt} \ , \tag{4.16}$$

and the classical theorem [18] says that $\int_0^\infty W(t)dt = 1$: sooner or later, gambler looses all money with probability one.

Here is now the simple solution of this problem based on the direct current analogy (see also [19]). Consider semi-infinite line of resistors (or just a wire of uniform resistance) going along the x-axis. In terms of analogy, the absorbing boundary condition (4.15) means that the boundary is grounded, i.e., is kept at the zero potential (4.5). Suppose now we feed a direct current into the point x_0 (of course, applying a constant voltage to the point x_0). Then, the whole statement of the theorem is that the current leaving the system through the grounded terminal at $x = 0$ is equal to the current entering the system through the terminal at x_0 - which is trivial.

Of course, this simple solution comes from the use of stationary equation (4.13).

4.3.6 Direct Current Formulation of the Commitor

Now we are prepared to formulate the way to determine the commitor of any given state $\mathcal{C}$ using the direct current analogy. We understand that the first arrival condition can be imposed by absorbing boundary conditions, or grounding the corresponding sites. Therefore, the formulation is as follows.

Suppose all the sites $\mathcal{U}$ corresponding to the unfolded state are grounded, as well as the site $\mathcal{F}$ corresponding to the folded state. Suppose further that we feed a direct (stationary, time independent) current I into the site $\mathcal{C}$. This current flows partly to $\mathcal{U}$ and partly to $\mathcal{F}$; in obvious notations, $I = I_{\mathcal{U}} + I_{\mathcal{F}}$. Then, the commitor is nothing but the fraction of current going into the $\mathcal{F}$ terminal. According to the Ohm's law, this can also be written in terms of the corresponding resistances:

$$p_{\text{fold}}(\mathcal{C}) = \frac{I_{\mathcal{F}}}{I} = \frac{R_{\mathcal{CU}}}{R_{\mathcal{CU}} + R_{\mathcal{CF}}} \; . \tag{4.17}$$

It is instructive to consider here a simple example, which is the diffusion in one-dimensional potential landscape $U(x)$. Of course, in this case the choice of reaction coordinate is trivial, there nothing but x. Nevertheless, we can imagine that there are two ends of the diffusion interval, which we call points $\mathcal{U}$ and $\mathcal{F}$, and we want to know what is the probability to start from some point $\mathcal{C}$ between $\mathcal{U}$ and $\mathcal{F}$ and arrive at $\mathcal{F}$ for the first time without ever touching $\mathcal{U}$. In this case, the resistor network is one-dimensional, all resistors are connected in series. Remembering the expression (4.6), we see momentarily that formula (4.17) yields

$$p_{\text{fold}}(\mathcal{C}) = \frac{\int_{\mathcal{C}}^{\mathcal{U}} e^{U(x)} dx}{\int_{\mathcal{F}}^{\mathcal{U}} e^{U(x)} dx} \; . \tag{4.18}$$

It is clearly seen both in the example (4.18) and in the general formula (4.17) that the commitor is always positive (non-negative) and never exceeds unity - as it must be for the probability.

4.3.7 Direct Current Formulation of the Landscape

Our result (4.17) generates an insight into the general meaning of commitor and its properties. For instance, we can proceed in the following way. Let us ask what are the surfaces of $p_{\text{fold}} = \text{const}$, and what is their topology.

To understand it, let us imagine let another experiment with the direct current. Let us ground this time only the folded state site $\mathcal{F}$, and let us feed the current into the unfolded state sites $\mathcal{U}$. In other words, the potentials of the terminals at the folded and unfolded states are equal to $\phi_{\mathcal{F}} = 0$ and $\phi_{\mathcal{U}} = V$, respectively. In this case, according to the Ohm's law, the potential at the arbitrary site $\mathcal{C}$ should be equal to

$$\frac{\phi_{\mathcal{C}}}{V} = \frac{R_{\mathcal{CF}}}{R_{\mathcal{CU}} + R_{\mathcal{CF}}} = 1 - p_{\text{fold}} \; . \tag{4.19}$$

Thus, this potential essentially gives us the value of the commitor p_{fold}. Of course, this realization of the direct current model corresponds very directly to the most common computer experiment in which we start from somewhere in the unfolded region and look for the first arrival into the folding state.

The result (4.19) indicates that the topology of every surface of constant commitor is such that it separates folded state from unfolded state; moreover, and more general, every surface $p_{\text{fold}} = \text{const} = p_0$ separates the regions with $p < p_0$ and with $p > p_0$. In particular, the commitor has no local maximum or local minimum. Its only minimum is the global one at the unfolded state, where it is equal to zero, and its only maximum is that at the folded state where it is equal to unity. This confirms the intuition according to which the commitor can be used as a reaction coordinate.

Now, as we mentioned in the beginning of this article, the central question arises: what is the relevant free energy profile if we choose to use p_{fold} as the reaction coordinate? It turns out possible to address this question in quite general form. IN this argument, we shall disregard the discrete character of conformation space graph and pretend to work with the continuous conformations space.

Consider our latest direct current formulation, in which terminals are at the folded and the unfolded states, and equipotential surfaces are simultaneously the surfaces of constant commitor. Since the current does not flow along the equipotential surface, we can ask what is the resistance of the conformation space layer sitting between the surfaces $p_{\text{fold}} = p$ and $p_{\text{fold}} = p+dp$. Since these surfaces are equipotential, all resistors connecting them may be viewed as connected in parallel. For them, the conductivities must be summed together, which yields

$$\frac{1}{R(p)} = \sum_{\mathcal{C},p=\text{const}} \frac{1}{R_{\mathcal{C}\mathcal{C}'}} \; . \tag{4.20}$$

Remembering the formula (4.6), it is natural to define $U_{\text{eff}}(p)$ according to $R(p) = e^{U_{\text{eff}}(p)}$, and then we obtain

$$U_{\text{eff}}(p) = -\ln\left[\sum_{\mathcal{C},p=\text{const}} e^{-U(\mathcal{C})}\right] \; . \tag{4.21}$$

Strikingly, this formula coincides with (4.2), except it does not involve any assumptions regarding slow degree of freedom and the like. Instead of being the definition of the statistical mechanics partition function as (4.2), formula (4.21) describes the parallel connection of the resistors. What this formula proves is that one can use the commitor as the reaction coordinate, with the relevant free energy profile given by the formula (4.21). In particular, the concept of the folding barrier is clearly formulated as the place with highest resistance.

It should be born in mind that equation (4.21) deserves further attention. It may be possible that for some types of conformational space graphs our approach here is too cavalier. For instance, the relation between the discrete and continuous versions may deserve further attention. We may return to this question in the future.

4.4 Culmination: So What?

The outcome of our analysis is an insight. We were trying to understand the computer experiments in which commitor, or p_{fold} is employed to study the dynamics of protein folding. We found that, using the direct current analogy, we can convince ourselves that p_{fold} does indeed give rise to the reasonable reaction coordinate, whose topological properties are reliably understood. We gave also plausible arguments suggesting the effective free energy profile along the direction in the conformation space which is characterized by the quantity of p_{fold} as a generalized coordinate. What remains to be a challenge is to find an explicit expression of p_{fold} in terms of microscopic coordinates. This problem remains unsolved even for the lattice toy models, not to mention real systems.

References

1. *Protein Folding, Evolution, and Design, Course CXLV, International School of Physics "Enrico Fermi", Italy, July 11–21, 2000*, ed. by R.Broglia, E.Shakhnovich, and G.Tiana (Italian Physical Society, Bologna, 2001)
2. P. Bolhuis, C. Dellago, P. Geissler, D. Chandler: J. Phys. - Condensed Matter **12**, A147-A152 (2000)
3. H. Frauenfelder, R. Austin, P. Wolynes: Rev. Mod. Phys. **71**, 419-430 (1999)
4. P. H. Verdier, W. H. Stockmmayer: J. Chem. Phys., **36**, 227 (1962)
5. J.D. Bryngelson, J.N. Onuchic, N.D. Socci, P.G. Wolynes: Proteins, **21**, 167-195 (1995)
6. M. R. Betancourt, J.N. Onuchic: J. Chem. Phys., **103**, 773-787 (1995)
7. R. Du, A. Grosberg, T. Tanaka: Phys. Rev. Letters **84**, 1828-1831 (2000)
8. A. Scala, L. A. Nunes Amaral, M Barthèlèmy: Europhys. Lett., **55**, 594-600 (2001)
9. R. Du, A. Grosberg, T. Tanaka, M. Rubinstein: Phys. Rev. Letters **84**, 2417-2420 (2000)
10. E. Shnol, preprint, Pouschino (1980)
11. V. Pande, A. Grosberg, T. Tanaka, D. Rokhsar: Current Opinion in Structural Biology **8**, 68-79 (1998)
12. P. Hanggi, P. Talkner, M. Borkovec: Rev. Mod. Phys. **62**, 251 (1990)
13. E. Lifshitz, L. Pitaevskii: *Physical Kinetics*, reprint edition (Butterworth-Heinemann, 1981)
14. S. Auer, D. Frenkel: Nature **409**, 1020-1023 (2001) sphere colloids
15. R. Du, V. Pande, A. Grosberg, T. Tanaka, E. Shakhnovich: J. Chem. Phys. **108**, 334-350 (1998)

16. P. Eastman, N. Grönbech-Jensen, S. Doniach: J. Chem. Phys. **114**, 3823 (2001) by reaction path annealing
17. N. Metropolis, *et al*: J. Chem. Phys. **21** 1087 (1953)
18. W. Feller: *An introduction to probability theory and its applications*, 2nd edn. (J. Wiley & Sons, New York 1971)
19. S. Redner: *A guide to first passage processes* (Cambridge Univ. Press, 2001)

5 Multiscale Computer Simulations for Polymeric Materials in Bulk and Near Surfaces

Cameron Abrams, Luigi Delle Site, and Kurt Kremer

Max-Planck-Institute for Polymer Research, Postfach 3148, Ackermannweg 10, D-55021 Mainz, Germany

Abstract. We review some recent approaches to simulate polymer melts on different levels of description. The methods aim at a link between these approaches, in order to faithfully simulate specific materials and eventually predict material properties. In the present contribution this is explained in some detail for different coarse grained bead spring models and different micro-meso mapping schemes for bulk properties. In the case of polycarbonate near surfaces a combined approach of mesoscale model simulations and QM-DFT calculations is presented.

5.1 Introduction

Polymer materials can be crystalline, amorphous (glasses, melts, rubber, gel) or in solution. Especially polymer melts in the glassy state are standard materials for many applications (yoghurt cups, compact discs, housings of technical equipment etc.). They combine relatively low specific weight and ductitility with processing at moderate temperatures. In the melt state, polymers are viscoelastic liquids where the crossover from elastic to viscous behaviour can be adjusted by the chain length. Added to a solvent, polymers can be used as viscosity modifiers and, depending on parameters, be either shear thickening or shear thinning, as used e.g. for drag reduction. Crosslinking chains into a disordered network results in gels or rubber. Applications range from gels in (low fat) food, hydrogels in modern body care (nappies ...) via biological systems (cytoskeleton, DNA) all the way to classical elastomers (e.g. car tyres), to name very few. Here, the interplay of connectivity, chain length and chain stiffness determines the properties.

This wide range of applications is made possible by the variability of physical properties, which is based on the many different chemical molecular building blocks as well as on various molecular architectures and huge differences in molecular weights of polymers. It is the combination and the rather delicate interplay of local chemical with more global architectural and size properties, which makes macromolecules so versatile and interesting. This means that many different length and time scales are relevant, and that understanding the properties on one scale is not at all sufficient to understand the material properties.

The simplest polymers are chain molecules with identical chain segments, repeat units or monomers. To give a very few examples, there are:

PE	$(CH_2)_N$	polyethylene
PS	$(CH_2(CH(C_6H_5))_N$	polystyrene
PEO	$((CH_2)_2O)_N$	polyethylene oxide
BPA-PC	$((C_6H_4)C(CH_3)_2(C_6H_4)CO_3)_N$	bisphenyl A polycarbonate

from the widely used PE (e. g. plastic bags) to the more complicated, but technically very relevant polycarbonate BPA-PC (compact discs). Many other cases exist, which can become very complex as especially biological examples demonstrate (DNA, proteins), where several different building blocks are present in one huge molecule. While most polymers are not water soluble, PEO has the exceptional property, that it is both water and oil soluble. Other important water soluble polymers are polyelectrolytes, which in water dissociate into ions and due to this are soluble even though their backbone is hydrophobic in most cases. Those are currently in the center of research along the border line between physics and biology, which is beyond the scope of the present contribution. The typical simulation approaches for polyelectrolytes however are conceptually very similar to the ones discussed here [2–5].

Understanding the relationship between atomistic structure and material properties is the long-standing goal of fundamental materials research. Polymeric materials offer many particularly challenging hurdles to this understanding, because of a unique dispersion of mechanisms of structure/property relations over many decades in length and time scales, from (sub)atomic to macroscopic. This feature of polymers has led to enormous success in describing *generic* aspects of their behavior, through scaling relations and dimensional arguments [6,7] as well as simulations based on simplified models such as bead spring chains or even lattice models [1,4]. However, within the large class of materials generically known as "polymers" exists an astounding variety of material properties which must depend on specific differences in molecular structure. A fundamental understanding of the mechanisms leading to this important variety of properties requires the construction of theories and models that can account for the atomic-scale differences among polymers.

5.2 Length and Time Scales for Polymer Simulations

To illustrate the importance of different time and length scales, we mention a technically important quantity, the shear viscosity η of a polymer melt. It contains the whole complexity of the problem. If one changes the process temperature of a BPA-PC melt from 500K to 470K, the viscosity rises by a factor of ten. This is a direct result of the local interactions on the atomistic level as it could -in principle- also have been achieved by an equivalent change of the chemical structure of the monomer. (The glass transition temperature T_G of BPA-PC is around 420K and different polymers display huge differences in T_G.) On the other hand, increasing the chain length by a factor of 2 also shifts the viscosity by a factor of ten, since for melts of long chains , one observes $\eta \propto N^{3.4}$. This power law is a universal property of linear polymers, and holds for all known polymers independent of the chemical structure of the backbone. Thus, both material-specific as well as universal properties produce a comparable variation of the macroscopic (dynamic) properties. This variation can easily extend over several orders of magnitude.

At first sight, it might be tempting to perform an all-atom computer simulation of a melt of polymers in order to determine properties like viscosity, morphology etc. However, there are two major complications. The first stems from the choice of interaction potentials, while the second is related to the many scales involved. An all atom simulation necessarily has to use an empirical force field. All quantum simulations (Car Parinello density functional simulations, path integral quantum Monte Carlo or combinations thereof) are still confined to very small systems and orders of magnitude slower than force field approaches [1,2]. The precondition for such an approach is an empirical energy function for the interaction of all atoms in the system. This determines the force field to solve Newton's equations of motion for the system. Though conceptually straight forward, such an attempt contains a number of unsolved problems and complications. First, though usually not considered, are quantum effects. One might think that typical temperatures for macromolecular systems (room temperature and higher) are well above the Debye temperature of the relevant atoms. This is true for the carbon atoms, however, not necessarily for the many hydrogens present.Their thermal de Broglie wave length at room temperature is about $1\AA$. In a paper by Martonak et al. [8] employing path integral quantum Monte Carlo simulations of a PE crystal revealed, that even at room temperature quantum effects are crucial to understand their anisotropic thermal expansion. This is obviously a rather special example. However, it should be kept in mind as a general sign of caution. Secondly, for the force field the intra-molecular interactions can be derived from a proper parameterization of quantum calculations on chain fragments. However to parameterize the inter-molecular or non bonded interactions, usually experimental quantities like the heat of vaporization of low-molecular weight liquids are used. There, additional difficulties can arise from the quality and availability of experimental data. Moreover, it is often impossible to optimize all properties to the same degree of accuracy and confidence. Thus, one has to be very careful and there is no single force field for a system which, without further verification, can be used at significantly different temperatures or compositions. Hence, there is NO universal force field. In principle, a new parameterization has to be performed for every new set of simulation parameters. Keeping this in mind, force field simulations can be very useful and have provided important insight into microscopic properties. Typical examples can be found in the overviews [3,5].

Whether such a fully atomistic simulation, if possible, would be useful at all, is questionable, since it would provide an enormous amount of data. Almost all the generated information would be irrelevant for the questions under consideration, e.g. the above mentioned viscosity η. In order to make suggestions for material improvements, or to qualitatively and quantitatively understand certain properties, it is crucial to structure and properly interpret the results rather than just collect data. This also is often easier with simplified models.

This leads us to the second problem field, the main topic of this article. Polymers can be characterized by a hierarchy of different length and especially time scales, which span a wide range. Fig. 5.1 illustrates this and shows the typical range needed.

On the microscopic level the properties are dominated by the local vibrations of bond angles and lengths. The typical time is of $10^{-13}sec$ as given in the figure resulting in a simulation time step to integrate Newton's equations of motion of about $10^{-15}sec$. This $\mathring{A}$-regime is well characterized by the bond angles and bond lengths resulting from the quantum mechanical energy levels. Up to torsional degrees of freedom, chains are in their vibrational ground states, at typical experimental temperatures. The properties on this level are solely determined by the detailed chemical structure of the molecules involved.

On a more coarse grained level, one cannot resolve all the atomistic details of the chains any more. The chain looks like a more or less flexible thread. This is the universal, entropy-dominated coil regime. The many possible conformations of the chains and the many ways to pack chains in a melt determine the morphology. In many questions, the intra-chain entropy plays the most important role. However, once two states of similar intra-chain entropy are available (e.g. chains in a mixture of two different polymer species A and B and in the two phase regime of an A-rich and a B-rich phase) tiny energy differences in the interaction, originating from the microscopic structure, of order $O(k_BT/N)$ are sufficient to change the morphology completely and to drive the phase separation. A striking illustration of this fact is, that even protonated and deuterated polystyrene phase separate, if the chains are long enough [9]. On the mesoscopic level many properties can be understood on the basis of simple coarse grained (bead spring) models. Characteristic time and length scales, as they are present in typical experiments are indicated in Fig.5.1. On the even coarser level, the semi-macroscopic level the behavior is dominated by the overall relaxation of conformations of the objects. The typical overall relaxation time, depending on chain length varies between N^2 for short and $N^{3.4}$ for longer chains. As explained in the beginning, prefactors originating from the microscopic interaction of the monomers, cause an equally large variation of scales. The resulting times can easily reach seconds. If one approaches the glass transition temperature even much longer characteristic times can be observed. Thus a satisfactory numerical description of material properties needs a combination of both aspects.

This illustrates the long standing challenge within the modeling of complex polymer materials, to develop methods which allow to cover the range from microscopic to mesoscopic and then on to the semi-macroscopic regime.

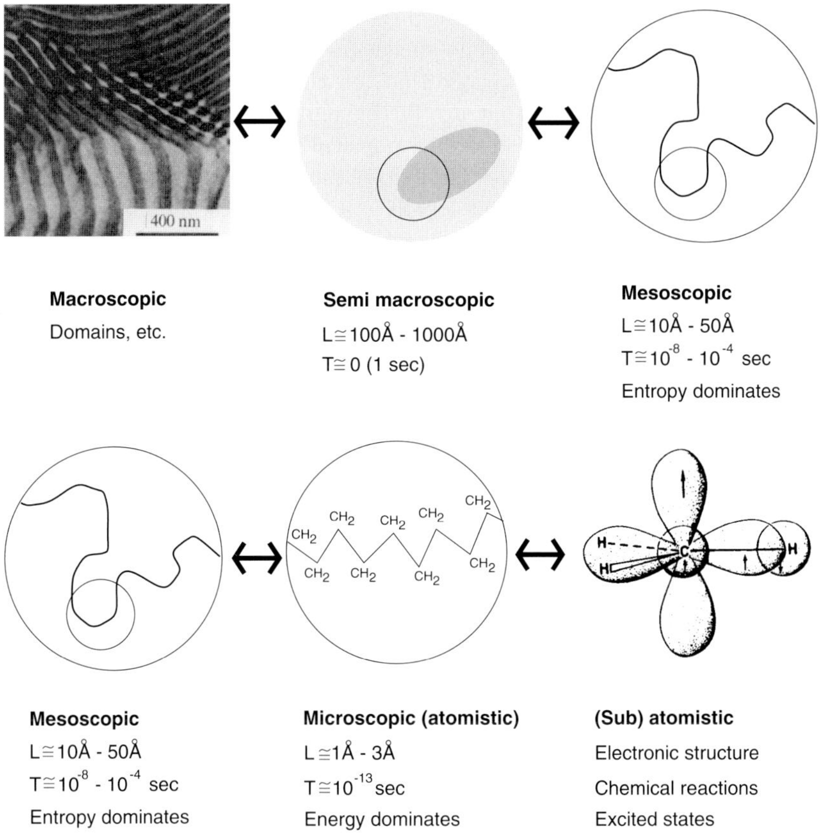

Fig. 5.1. Polymers exhibit phenomena on many length scales (from entire devices down to electrons) and associated time scales (from years to femtoseconds). Starting from the top left side, one can observe the overall morphology of a polymer material. Looking a little bit more closely, e.g. by marking a chain in a melt ore dense solution of otherwise identical chains, the individual polymer can be observed only as a very pale shadow. A typical spatial extension of the shadow is given by the overall coil diameter, as indicated. The characteristic time for this picture to change can vary dramatically depending on chain length and temperature, starting at about $10^{-4}s$ for short chains and 'high temperatures', with essentially no upper limit. Looking again more closely, more of the polymer structure is revealed. This is the universal, entropy dominated coil regime. Again the variation in time can be very large, cf. text. Typical times, as they are present in many experiments, are indicated. Only if the objects are examined much more closely, chemical details of the polymers can be identified. There local chemical details govern the properties and all bond lengths, angles etc. are determined by the energy levels, originating from quantum mechanics. The lower time limit is determined by the highest frequency of oscillations, which depending on the model used are the bond length or bond angle oscillations. To study excited states or reactions, the electronic structure is to be considered and quantum methods are required. Methods for treating individual scales are well developed. The challenge is to connect them systematically.

Thus one has to construct suitable links or mapping schemes, which connect the different approaches. The successful mapping of an atomistic to a mesoscale model has the advantage that, on the mesoscopic level, time and length scales are accessible that are far beyond the reach of atomistic simulations. Thus, qualitatively different physical problems can be treated. A good mesoscale model preserves enough of the original chemical identity of the atomistic model, that it does locally not behave like a generic Gaussian coil, which it has to do on huge length scale. It contains just enough information to reproduce certain aspects of, say, polystyrene or polypropylene chains under the corresponding conditions. It is no surprise that in recent years, a number of atomistic-to-mesoscopic mappings have been published; a review encompassing scale bridging from electronic to macroscopic degrees of freedom, cf. Fig.5.1, is presented in Ref [3]).

Moreover, it has turned out that coarse-grained models are useful not only in their own right to study large scale phenomena. They are also a reliable tool for the generation of well-equilibrated atomistic structures, provided one can perform an inverse mapping from the mesoscopic model back to a fully atomistic model which can then be analyzed. In many cases, the fully atomistic model is needed for comparison with experiment because the experimental information often necessarily involves atoms, like in nuclear-magnetic-resonance (NMR) spectroscopy [10], neutron scattering [11]or positron annihilation spectroscopy [12].

5.3 Dual-Scale Modelling Ansatz

5.3.1 Mesoscopic Models in Bulk and Near Surfaces

Perhaps the simplest particle-based continuous-space representation of a polymer molecule is the 'bead-spring' model, (Fig. 5.2), in which point-mass beads are tethered to one another by massless springs. It has served for twenty years as the prototypical polymer molecule in computer simulations of single macromolecules [4,13–15]. More recently, it has gained popularity specifically for simulation studies of a wide variety of dense polymer systems, including melts, [14–16] gels, [17–20] brushes, [21–23] to name a few examples. (By 'dense', we mean an excluded volume fraction greater than about 0.4, typical for polymer melts.) The model's appeal for such studies stems not only from its simplicity, but also the way it allows for an approximation of the polymer molecule excluded volume. Realistic simulations of dense systems must include excluded volume effects, because they dominate the static and dynamic liquid properties [7]. Moreover, the bead-spring model serves today as a basis for systematic coarse-graining, which seeks to develop computationally tractable representations of atomistically detailed polymer molecules [3].

Realizing that a faithful representation of excluded volume is essential in dense liquid simulations, a systematic study of these models has been

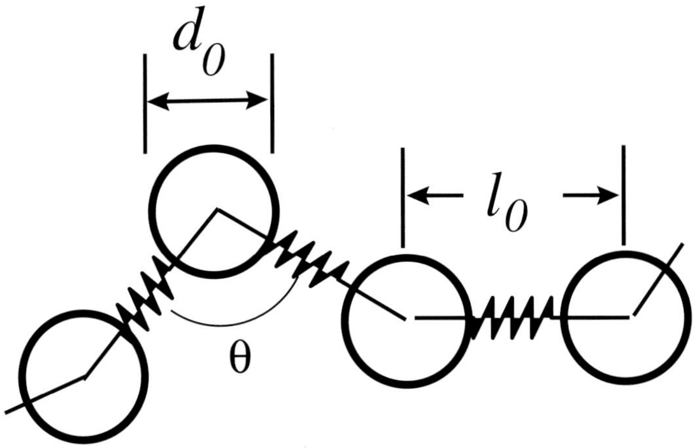

Fig. 5.2. Schematic of a bead-spring polymer chain. d_0 is the effective bead diameter, l_0 is the average bond length , and θ is a bond angle. From [24].

performed by Abrams and Kremer [24,25]. For this they considered the relatively simple case of linear chains with a single bead diameter, d_0, and a single prescribed *average* bond length, l_0, as depicted in Fig. 5.2. The understanding of how the physical properties derived from the simulations depend on the choice of these parameters is essential for the choice of a particular coarse-grained model. The consequences turned out to be crucial for both static and dynamic aspects. We here shortly review results, which were obtained for bulk melts and thin confined melt films, both composed of simple, freely-jointed, bead-spring polymer chains. l_0/d_0 was chosen to vary between 0.73 and 1.34.

The polymer chains are modeled as sequences of N identical particles, connected sequentially by massless springs. All particles in the system, except bonded neighbors along a chain, interact according to a purely repulsive shifted 12–6 Lennard-Jones potential, cut off at $r = 2^{1/6}\,\sigma$:

$$U_{LJ}\left(r_{ij}\right) = \begin{cases} 4\epsilon\left[\left(\frac{\sigma}{r_{ij}}\right)^{12} - \left(\frac{\sigma}{r_{ij}}\right)^{6} + \frac{1}{4}\right] & r_{ij} < 2^{1/6}\sigma; \\ 0 & r_{ij} \geq 2^{1/6}\sigma, \end{cases} \tag{5.1}$$

with the standard LJ units $[\epsilon] = kT$. All beads have the same excluded volume diameter, d_0, defined as the separation at which the Lennard-Jones potential has a value of $1kT$. With this choice of potential, $d_0 = 1\ \sigma$. Bonds between adjacent neighbors $(i,i+1)$ along a chain are enforced by a stiff harmonic potential with a given average bond length, l_0.

The systems were simulated in the NVT ensemble, using a Langevin thermostat, [14] with friction $\Gamma = 0.5$ at constant $T = 1.0$. For this they ran systems of $M = 80$ chains of length $N = 50$ and a number density of $\rho = 0.85\sigma$.

For the case of thin films or bulk systems with a solid surface a standard potential $U_w(z)$ models flat repulsive walls at $z = 0$ and $z = L_z$, the z-boundaries of the simulation cell:

$$U_w(z) = U_{10-4}(z) + U_{10-4}(L_z - z); \tag{5.2}$$

$$U_{10-4}(z) = \begin{cases} 2\pi\epsilon_w\sigma_w^2 \left[\frac{2}{5}\left(\frac{\sigma_w}{z}\right)^{10} - \left(\frac{\sigma_w}{z}\right)^4 + \frac{3}{5}\right] & z < \sigma_w; \\ 0 & z \geq \sigma_w. \end{cases} \tag{5.3}$$

We here only report the variation of the polymer liquid structure. For the resulting changes in the dynamics we refer to [25]. In Fig. 5.3,

the intermolecular monomer radial distribution functions, $g(r)$, are shown for the various melts, along with $g^*(r)$ for the corresponding monomeric fluid. It is clear from these data that the liquid structure is quite sensitive to l_0/d_0. This holds for both cases, whether one considers systems of fixed number density ρ, or excluded volume fraction ϕ. One can define a correlation length λ as the radius at which amplitudes in the oscillations of $g(r)$ become less than 1%, with $\lambda^* \approx 5\sigma$ being that of the simple monomer liquid. One observes that $\lambda \approx \lambda^*$ for the melts with $l_0/d_0 > 1$, while when $l_0/d_0 < 1$, $\lambda < \lambda^*$ and it decreases with l_0/d_0. When the bond length is below the excluded volume diameter, monomer-monomer correlations weaken, whereas these correlations are little if at all affected when $l_0/d_0 > 1$. The manner in which chains pack gives rise to the less ordered liquid structure for $l_0/d_0 < 1$, and the similarly ordered liquid structure for $l_0/d_0 > 1$, relative to the simple fluid. Hence, by simply changing l_0/d_0 from about 1.3 to 0.7, one finds a gradual transition from polymeric liquids whose structure is determined primarily by monomer packing to polymeric liquids whose structure is determined by intramolecular configurational entropy of the chains.

In the case of a polymer melt near a flat hard wall the effects are even more pronounced. In Fig. 5.4 the resulting monomer density profiles, $\rho(z)$, scaled by the bulk number density, $\langle\rho\rangle$, for each of the l_0 cases, as well as $\rho^*(z)$ for a monomeric fluid at $\rho = 0.85\ \sigma^{-3}$ are given.

The wall potentials were placed at $z = 0$ and $z = L_z$, and the data in Fig. 5.4 are symmetrically averaged with respect to the $z = L_z/2$ center plane. In all cases, $\rho(z)$ reveals wall-induced layering in the monomer density. Roughly speaking, relative to the simple monomeric fluid, layering is strongly suppressed in melts for which $l_0/d_0 < 1$. This suppression has been observed previously only for systems of lower number density and $l_0/d_0 = 0.97$ [26]. What is even more striking is that, judging from peak intensities, the layering appears to be slightly *enhanced*, relative to the simple fluid, in the melts for which $l_0/d_0 > 1$.

In a similar way the chain diffusion is very sensitive to the choice of l_0/d_0. A commensurate ratio leads to a significantely slowed down diffusion constant, as local cages in the liquid can nicely be created by the system.

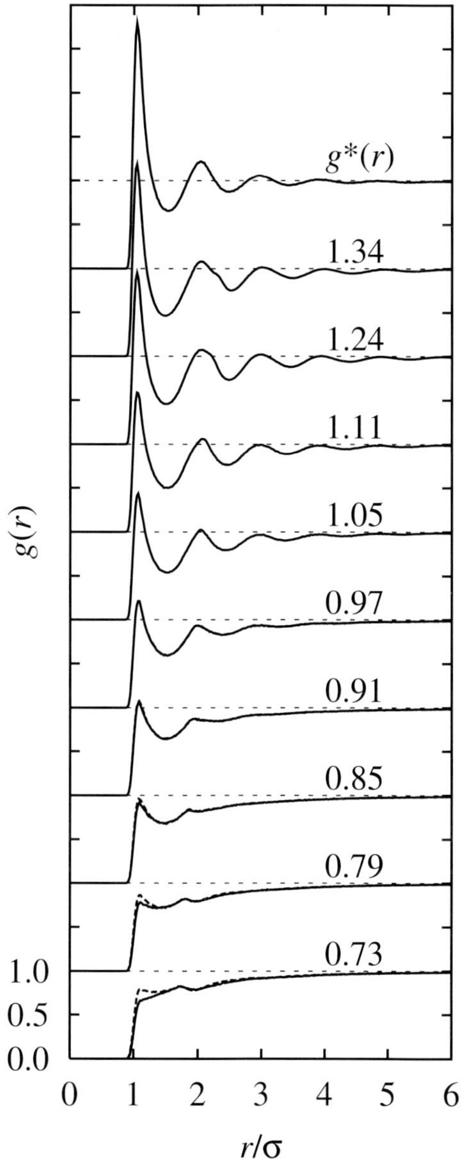

Fig. 5.3. Intermolecular radial distribution functions, $g(r)$, for bulk bead-spring melts. Solid curves correspond to samples with $\rho = 0.85\ \sigma^{-3}$, and broken curves to those melts for which $l_0/d_0 < 1$ and fixed $\phi = 0.445$. Labels denote bond length, l_0. $g^*(r)$ was computed from a simple repulsive monomeric fluid at $\rho = 0.85\ \sigma^{-3}$. From [24].

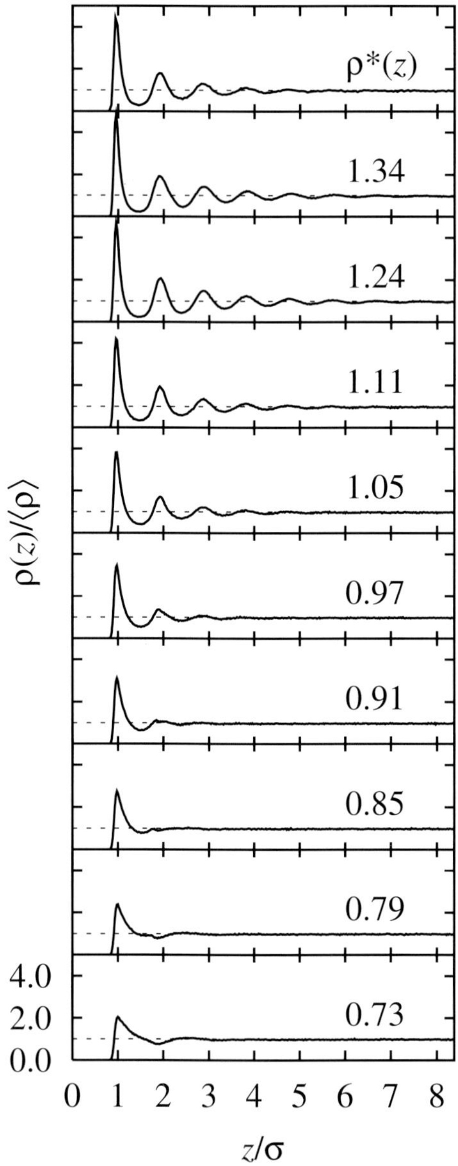

Fig. 5.4. z-dimension scaled density profiles, $\rho(z)/\langle\rho\rangle$, for confined bead spring polymer melts, plotted over the domain $0 \leq z \leq L_z/2$. Labels denote bond length, l_0. $\rho^*(z)$ was computed from a simple repulsive monomeric fluid at $\langle\rho\rangle = 0.85\ \sigma^{-3}$, where $\langle\rho\rangle$ is $\rho^*(z)$ averaged over $0.8 < z < L_z - 0.8$. From [24].

This maximises the bead entropy, but at the same time creates significant barriers for diffusion [25].

5.3.2 Systematic Molecular Coarse-Graining

5.3.2.1 Mapping Schemes

We here compare schemes of coarse graining for polycarbonate melts which takes the above considerations into account. The basic original idea was presented in detail in Refs. [27] and [28]. These papers describe a mapping scheme for bisphenol-derived polycarbonates in which each repeat unit is replaced by two spherical beads; referred to as a "2:1" mapping. Here we focus on a related scheme in which the repeat unit is replaced by four beads, a "4:1" mapping. This is represented schematically in Fig. 5.5.

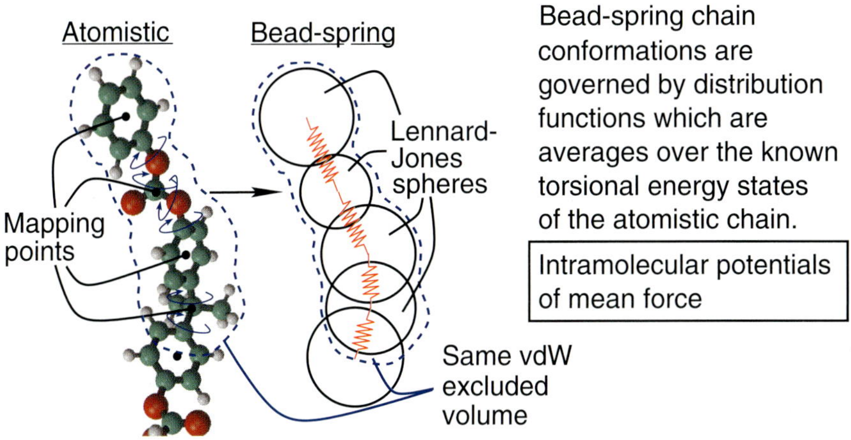

Fig. 5.5. A schematic representation of the 4:1 mapping scheme for coarse-graining of bisphenol-A-polycarbonate (BPA-PC). A chain-ending repeat unit is shown. From [29].

All results are obtained via standard molecular dynamics simulations of bisphenol-A-polycarbonate liquid samples containing 80 chains, each chain bearing 10 chemical repeat units, at a mass density of 1.055 g/cc and a temperature of 570 K. Details on the simulation technique can be found elsewhere [27,30].

The key features of this model are as follows. The bead centers correspond to specific groups of atoms in the molecular structure. Bead centers are mapped to the carbonate carbon, the isopropylidene backbone carbon, and one to each of the phenyl ring centers of mass. With these definitions of the mapping points, the molecular structure of the repeat unit constrains allowable bead-bead distances and bead-bead-bead angles along the coarsened backbone. One important exception is the phenylene-carbonate-phenylene angle, which is constrained to satisfy a *distribution* of angles which reflects an average over the atomic-scale torsional states "within" this bead. These constraints are translated into intramolecular potentials used directly in molecular dynamics simulations of these coarsened bead-spring objects. Furthermore, the bead diameters are chosen such that the overall excluded volume of the repeat unit as closely as possible matches that known for BPA-PC from equation-of-state predictions [31]. This prescribes *intermolecular* potentials which are also included in the MD simulations.

5.3.2.2 Coarse Grained Liquid Structure

We first examine the structure of the simulated BPA-PC liquids. Of particular interest in this discussion are as before the radial distribution functions, $g(r)$, obtained from bulk simulations, and depth profiles obtained from simulations of the bead-spring liquids confined to slit-pore geometries. In Fig. 5.6, we show comonomer-specific $g(r)$'s for the "4:1" and "2:1" liquids. These liquids were composed of chains of ten repeat units each, which corresponds to chains of 23 and 43 coarse grained beads for the 2:1 and 4:1 mapping schemes, respectively. From this data, it is evident that the liquid structure for the different mapping schemes is quite different, as to be expected from the results of the simple bead spring models discussed before. Most notable are the long-ranged decaying oscillations in $g(r)$ for the 2:1 system. This starkly resembles the $g(r)$ of simple dense liquids of spherical particles [32]. In contrast, such long-lived correlations are not apparent in the $g(r)$ from the 4:1 simulations. The differences in the liquid structure are a direct result of the imposed *intramolecular* structure of the two mapping schemes.

This difference in liquid structure is again evident if we consider how the bead-spring objects pack next to a flat wall. In Fig. 5.7, we show the depth profiles of carbonate and isopropylidene beads obtained from sampling over liquids equilibrated in confining slit pores. Both the density of the adsorbed layer and the decay length of the density oscillations are much higher for the 2:1 liquid than the 4:1 liquid. This arises again because of the commensurability between bond length (1.24 σ) and bead diameter (1 σ) in the 2:1 model. The above discussed systematic studies of the effects of l_0/d_0 on chain packing in dense melts of flexible bead-spring chains clearly demonstrates that the coarse-grained liquid structure is extremely sensitive to the relationship between these two lengths. [24,25].

The resulting different packing also in the present case has significant influence on the dynamics as well. It turns out, as expected from [25], that the 4:1 chains move more quickly than the 2:1 chains as a result of the much lower sphere-packing efficiency in the 4:1 melts *vis-à-vis* the 2:1 melts. This high packing efficiency, demonstrated by the large nearest-neighbor peak in $g(r)$, means that each bead in the 2:1 melt has many more excluded volume interaction partners on average than do beads in the 4:1 melt. Hence, the bead-specific friction, or resistance to accumulation of mean-squared dis-

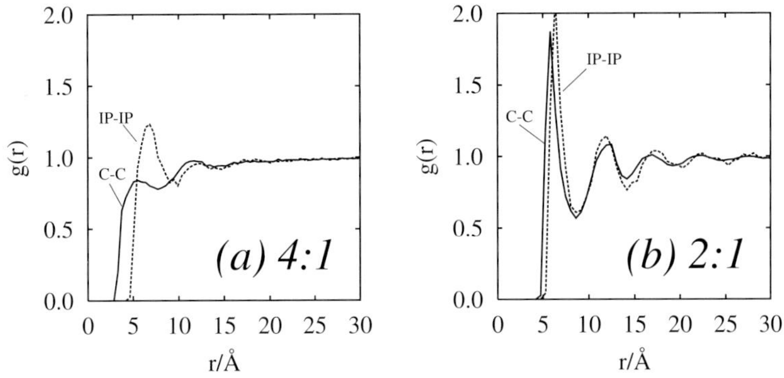

Fig. 5.6. Carbonate-carbonate ("C-C") and isopropylidene-isopropylidene ("IP-IP") intermolecular radial distribution functions from MD simulations of N=10 BPA-PC: comparison of (a) "4:1" and (b) "2:1" molecular coarse-graining resolutions. From [29].

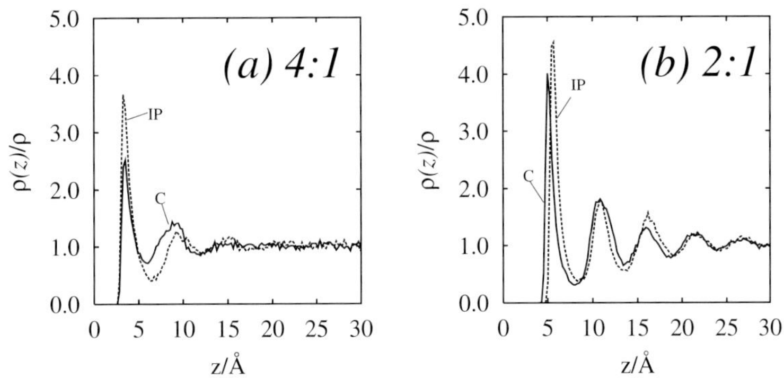

Fig. 5.7. Density depth profiles of carbonate ("C") and isopropylidene ("IP") beads from MD simulations of N=10 BPA-PC confined in a slit pore: comparison of (a) "4:1" and (b) "2:1" molecular coarse-graining resolutions, from [29].

placement, is higher in the 2:1 liquids. This slowdown in the bead motion for the 2:1 melt means that more simulation time steps are required to produce the necessary amount of average molecular rearrangement for equilibration, relative to the 4:1 melt. As a result, though more complex, the simulations with the 4:1 mapping scheme use much less computer time than those with the 2:1 mapping scheme. Despite these differences, both models carry the essential microscopic information. When they are used to reconstruct an all atom "sample" the resulting structures from both models are very similar (for $T = 570K$) [29].

5.4 Specific Surface Effects: BPA-PC Near a Ni Surface

For many applications the understanding of organic materials near metal surfaces is technically very importnant. Polycarbonates near a nickel surface is such an example. The idea of combining coarse-graining models for polymers with *ab initio* Density Functional Theory (DFT) calculations for small molecules analogues of polymer chemical subunits, can result particularly useful when polymers interact with interfaces where local chemistry may influence global properties of polymer systems. Such an approach would represent a reasonable compromise between a local quantum-based, but computationally expensive, description of the relevant polymer-surface interaction and simplified bead-spring models which allows simulations of large polymer systems. Within this schematic framework, recently, the particular case [30] of BPA-PC (Fig.5.8) near a Ni {111} surface (see Fig.5.9) was treated.

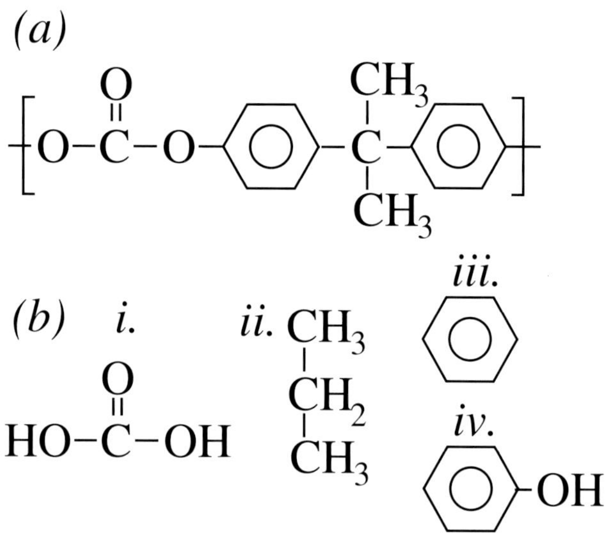

Fig. 5.8. (a) Chemical structure of the repeat unit BPA-PC. (b) Analogous molecules used in the *ab initio* studies: (i) carbonic acid, (ii) propane, (iii) benzene, and (iv) phenol, from [30].

The general strategy consists in cutting the chain into comonomeric molecules, small enough to study the interaction with the surface and examine the adsorption of each on a Ni {111} surface. In this case we considered three molecules analogous to the comonomeric subunits of BPA-PC [Fig.5.8(b)]: carbonic acid (i), propane (ii) and benzene (iii) (phenol (iv)), representing carbonate, isopropylidene, and phenylene, respectively. The first important result to emerge from the *ab-initio* calculations, is that the carbonic acid and propane molecules do not stick to the surface. In fact the adsorption energy

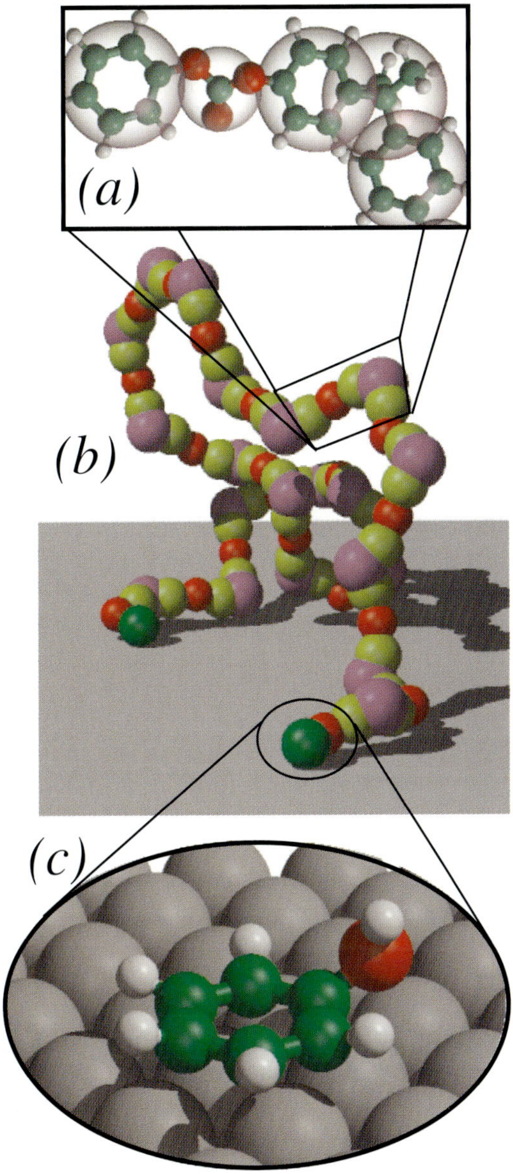

Fig. 5.9. The multiscale model of BPA-PC on nickel. (a) The coarse-grained representation of a BPA-PC segment; the coarse-grained beads are transparent spheres, superimposed on the underlying chemical structure, where the carbon atoms are green, the oxygens red, and the hydrogens white; (b) Coarse-grained model of an $N = 20$ BPA-PC molecule, with ends adsorbed on a flat surface; configuration from a 160-chain liquid simulation. (c) A phenol molecule adsorbed on the bridge site of a (111) nickel surface; configuration computed via CPMD simulation, from [30].

in both cases is of the order $0.01eV$ and is independent from the location of the molecule above the different surface sites, i.e. the molecule experiences an effective uniform hard wall. Benzene, which from our calculations shows a fairly strong adsorption in isolation, $1.05eV$, is sterically hindered to adsorb when incorporated into a BPA-PC chain, due to neighboring carbonate and isopropylidene groups. However, phenoxy end groups are not in this way sterically hindered, and hence may adsorb strongly to the surface. The latter aspect can be treated in a more complete way by considering phenol, which, due to the presence of the oxygen, better represents the phenoxy end groups. From the extensive DFT calculations emerges a strong chemisorption of phenol on nickel surface, the energy being $0.91eV$, this leads us to the conclusion that indeed the probability that there is chemisorption of end-chains on the surface is extremely high. The quality of the DFT calculations is assured by comparing recent DFT work on the Ni(111)/benzene system[33–35], with experiment and noticing that it gives us an adequate description of the strength of the adsorbate-substrate bond, the error being 20%, which is sufficiently small for our present purposes. The next step consists in incorporating these results properly within the coarse-grained model described in previous sections. This approach allowed us to show how the interplay of entropic and energetic contributions can alter the structure of a polymer melt near a metal surface. In particular, we predict that polycarbonate chain ends adsorb strongly to a nickel surface next to a polycarbonate liquid. This fact may have important consequences and applications in different technological fields. For the technical part of the *ab initio* calculations we used the plane-wave pseudopotential CPMD code, [36], implemented with finite-temperature density functional theory (FEMD) [37,38], using PBE[39] generalized gradient approximation (GGA). Several geometry optimizations, starting from plausible structures compatible with possible orientations of the respective comonomers in a polymer chain, were performed at each of the four high-symmetry sites of the {111} surface. The adsorption energy (E_{ad}), defined as the energy of the adsorption system relative to the clean surface and isolated molecule, characterizes the strength of the interaction of each submolecule with the surface. DFT calculations also provided a basic understanding of the possible mechanism of adsorption of the end-chains. We found that in the most energetically favourable configuration the center of mass of benzene and phenol lies over the bridge site and the molecular orientation is parallel with respect to the surface. The strength of the interaction rapidly decays as a function of the incidence angle ($0.05eV$ beyond 40 degrees) as well as of the distance from the surface ($0.03eV$ beyond $3.0Å$). This causes near the surface to obtain rather unusual conformations, which best can be characterized by the similarities to two-end adsorbing brushes in contact to a surface on one side and to a bulk melt on the other side [30].

5.5 Other Approaches: Automatic Coarse-Graining

As shown above developing a reliable mesoscopic polymer model is a cumbersome process. A number of different steps, which may vary between approaches have to be taken: (i) The degree of coarse graining, how many real atoms per coarse-grained bead, has to be chosen and the positions of the coarse-grained (CG) beads in relation to the atoms have to be determined. (ii) The form of the intra-chain and inter-chain potentials need to be chosen, if they are not directly derived from the distributions. (iii) Their free parameters, especially for the nonbonded interactions, have to be optimised in a way that the CG model reproduces the structure of the model. While (i) and (ii) are intellectual challenges, (iii) often is a menial task which should be automated to the extent possible. This chapter describes a recent approach by Müller-Plathe and coworkers to systematically and automatically parameterize interaction parameters of mesoscale models for polymers in solution. Its purpose is to be able to carry out the parameterisation (iii) for a given degree of coarse graining (i) and form of the potential (ii) quickly and reproducibly, so that every physical model can be evaluated with a set of optimum parameters. First, reference data have to be obtained, such as structural properties of the polymer of interest. In the present study they concentrated on the sodium salt of poly(acrylic acid) (PAA) as an aqueous solution of about 2 wt.%. They were obtained by performing an atomistic simulation of an oligomer (23 monomers) solvated by about 3200 water molecules. The coarse-grained model contained one bead per monomer centered at the centre of mass of the atomistic monomer, reducing the number of polymer atoms by 8. More importantly, the coarse-grained model disposed of the explicit solvent, so that the total number of sites was reduced from approximately 3350 to 23, Fig.5.10.

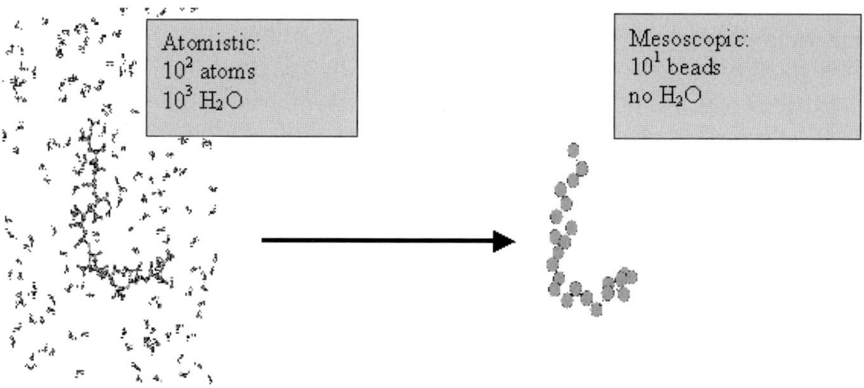

Fig. 5.10. An example for mapping between atomistic and mesoscopic models: The sodium salt of poly(acrylic acid) (2 wt.%) in aqueous solution. From [40].

The coarse-grained intrachain interactions consisted of a harmonic bond potential, a Gaussian expansion of the multiple-minima bond-angle potential and a short cosine expansion of the dihedral angle potential. The piecewise connected non bonded potential between monomers is of the form

$$V(r) = \begin{cases} \epsilon_1[(\sigma_1/r)^8 - (\sigma_1/r)^6] & r \leq \sigma_1 \\ \epsilon_2[sin(\frac{(\sigma_1 - r)\pi}{(\sigma_2-\sigma_1)2}] & \sigma_1 < r \leq \sigma_2 \\ \epsilon_3[cos(\frac{(r-\sigma_2)\pi}{(\sigma_3-\sigma_2)} - 1] - \epsilon_2 & \sigma_2 < r \leq \sigma_3 \\ \epsilon_4[-cos(\frac{(r-\sigma_3)\pi}{(\sigma_4-\sigma_3)} + 1] - \epsilon_2 - 2\epsilon_3] & \sigma_3 < r \leq \sigma_4 = r_{cutoff} \end{cases}$$

This choice at first sight appears complicated and arbitrary, however it has proven useful [41], and it contains enough flexibility to encompass an effective description of the solvent. For a more detailed discussion see [41]. The reference data include distributions of bond lengths and bond angles as well as radial distribution functions $RDF_{target}(r)$ obtained from the atomistic simulation but calculated for the coarse-grained beads.

The 23-mer of PAA with the CG model was simulated as an isolated molecule in space, in other words the same oligomer as with the atomistic model. The CG parameters were adjusted, until the target RDFs and other distributions were reproduced satisfactorily. We defined a least-squares merit function $f(p_1, p_2, ...)$ which was to be minimised

$$f(\{p_1, p_2, ...\}) = \int_0^{cutoff} w(r)[RDF(r, \{p_1, p_2, ...\}) - RDF_{target}(r)]^2 dr \tag{5.4}$$

where $p_1, p_2, ...$ are the parameters in the optimisation set and $w(r)$ is an optional weighting function. For the minimisation, we use a standard amoeba simplex scheme [42]. Note, that every evaluation of f involves an entire molecular dynamics (or Brownian dynamics) simulation of the CG system, including equilibration, check for convergence etc. At this point, the apparently straight forward scheme can become technically tricky and computationally expensive, which they noted already in the first application of the automatic parameterisation method, namely the development of atomistic force fields from experimental data [43]. The typical agreement of the target RDF of PAA and two CG RDFs, which can be obtained after parameter optimisation is shown in Fig.5.11.

With the CG model parameterised, the simulation was extended to much longer chains of PAA in aqueous solution. The results for the calculated hydrodynamic radius of such chains match well the results from dynamic light scattering (Fig. 5.12).

This shows that the CG model retains enough of the true identity of PAA to reproduce its structure on a scale much larger than that of the atomistic model from which it was developed. In the meantime this approach has been extended not only to other polymers in solution but also to bulk polymer

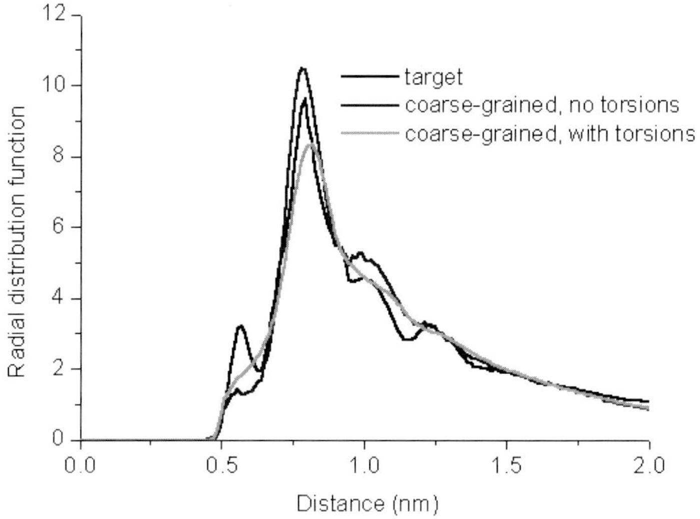

Fig. 5.11. Monomer-monomer radial distribution function for poly(acrylic acid), first and second neighbours excluded. The target function from the reference atomistic simulation is shown as well as two different coarse-grained models after convergence. From [40].

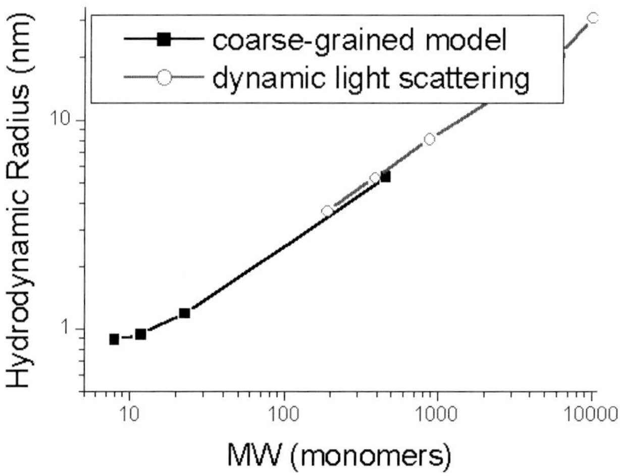

Fig. 5.12. The hydrodynamic radius of poly(acrylic acid) in aqueous solution as a function of molecular weight. From [44].

melts which allow for an all atom simulation of oligomers [43]. Compared to the previous ansatz, this is a more pragmatic way, which in special cases leads more directly to the desired results.

5.6 Conclusions, Outlook

In spite of all progress made over the last years a number of challenges remain. On each level of description, new and improved methods have been developed. Still better models are needed. Most important, however, is the controlled and systematic improvement of links between the different simulation schemes, i.e.:

- Systematic coarse graining procedures, including the inverse mapping step are to be improved and developed. Steps must cover the Micro (many atoms) $< - >$ Meso (many monomers) $< - >$ Macro (many chains) regimes and link to quantum simulations at the low end and to self consistent field calculations and finite element like approaches at the upper end.

In order to achieve the longstanding goal of predicting macroscopic material properties from first principles, much effort is needed over the coming years. In particular, a few key ingredients have to be developed, such as

- quantum simulations of reasonably sized systems (Car Parinello techniques, path integral quantum Monte Carlo, combinations of both): coupling electronic and conformational degrees of freedom (beyond the ansatz of [30])
- improved methods to parameterize and validate force fields for atomistic classical molecular simulations, especially suitably parameterized intermolecular interactions (for polar molecules)
- new methods for static and dynamic studies on the semimacroscopic to macroscopic level, such as dissipative particle dynamics for composite materials based on the microstructure of the polymers

This needs a significant interdisciplinary coordinated effort of researchers coming from different fields of expertise. Sizeable projects are underway or have been performed at a number of places in the world and range from more software development oriented activities, such as the NEDO project in Japan and European activities within the Training and Mobility program to more basic method development oriented activities like the materials simulation competence center funded by the German ministry of science and technology which is coordinated by our group.

Acknowledgements

The work described here is supported by the BMBF Kompetenzzentrum Materialsimulation (Förd. Kennz. 03N6015) and the TMR grant (ERB-4061-PL97-0661) "New Routes to Understanding of Polymer Materials (NEWRUP)". We would like to thank the many colleagues (especially F. Müller-Plathe) and students who were involved in various aspects of the described examples over the past years.

References

1. K.Binder and G.Ciccotti (eds.): *Monte Carlo and Molecular Dynamics of Condensed Matter Systems.* Oxford University Press, Oxford, 1996.
2. K.Kremer in [1]
3. J.Baschnagel, K.Binber, P.Doruker, A.A.Gusev, O.Hahn, K.Kremer, W.L.Mattice, F. Müller-Plathe, M.Murat, W.Paul, S.Santos, U.W.Suter and V.Tries: Adv.Polym.Sci. **152**, 41 (2000).
4. K.Binder (ed.): *Monte Carlo and Molecular Dynamics Simulations in Polymer Science.* Oxford Univ.Press, Oxford, 1995
5. L.Monnerie and U.W.Suter (eds.):*Atomistic Modelling of Physical Properties*, in Advances in Polymer Science **116**, Springer, Heidelberg (1994).
6. P.G.de Gennes: *Scaling Concept in Polymer Physics.* Cornell University Press, Ithaca, New York, 1979.
7. M.Doi and S.F.Edwards:*The Theory of Polymer Dynamics.* Oxford Scientific Publications, Oxford, 1986.
8. R.Martonak, W.Paul and K.Binder: J.Chem.Phys. **106**, 8918 (1997)
9. M.D.Gehlsen, J.Rosendale, F.S.Bates, G.D.Wingall, L.Hansen, K.Almdal: Phys.Rev.Lett **68**, 2452 (1992).
10. R.Faller, F. Müller-Plathe and A.Heuer: Macromolecules **33**, 6602 (2000)
11. J.Eilhard, A.Zirkel, W.Tschöp, O.Hahn, K.Kremer, O.Schärpf, D.Richter and U.Buchenau: J.Chem.Phys. **110**, 1819 (1999)
12. H.Schmitz and F. Müller-Plathe: J.Chem.Phys **112**, 1040 (2000).
13. M.P.Allen and D.J.Tildeslesy, *Computer Simulations of Liquids.* Oxford University Press, New York, 1987.
14. G.S.Grest and K.Kremer: Phys.Rev.A **33**, 3628, (1986).
15. K.Kremer and G.S.Grest: J.Chem.Phys. **92**, 5057 (1990).
16. S.Barsky: J.Chem.Phys. **112**, 3450 (1999).
17. E.R.During, K.Kremer and G.S.Grest: Macromolecules **26**, 3421 (1993).
18. E.R.During, K.Kremer and G.S.Grest: J.Chem.Phys **101**, 8169 (1994).
19. M.Pütz, K.Kremer and R.Everaers: Phys.Rev.Lett **84**, 298 (2000).
20. E.M.Aydt and R.Hentschke: J.Chem.Phys. **112**, 5480 (2000).
21. M.Murat and G.S.Grest: Macromolecules **27**, 418 (1989).
22. G.S.Grest: Macromolecules **22**, 418 (1994).
23. F.S.Csajka and C.Seidel: Macromolecules **33**, 2728 (2000).
24. C.F.Abrams and K.Kremer: J.Chem.Phys. **115**, 2776 (2001).
25. C.F.Abrams and K.Kremer: J.Chem.Phys. **116**, 3162 (2002).
26. I.Bitsanis and G.Hadziioannou: J.Chem.Phys. **92**, 3827 (1990).

27. W.Tschöp, K.Kremer, J.Batoulis, T.Bürger and O.Hahn: Acta Polym. **49**, 61 (1998).
28. W.Tschöp, K.Kremer, J.Batoulis, T.Bürger and O.Hahn: Acta Polym. **49**, 75 (1998).
29. C.F.Abrams and K.Kremer: Submitted to Macromolecules (2002).
30. L. Delle Site, C. Abrams, A. Alavi, K. Kremer: Phys. Rev. Lett. (2002) in press.
31. I.C.Sanchez and J.Cho: Polymer **36**, 2929 (1995).
32. L.Verlet: Phys.Rev. **159**, 98 (1967).
33. F. Mittendorfer and J. Hafner: Surface Science, **472**, 133, (2001).
34. S. Yamagishi, S. J. Jenkins, and D. A. King: J. Chem. Phys. **114**, 5765, (2001).
35. G. Held, W. Braun, H.-P. Steinrück, S. Yamagishi, S. J. Jenkins, and D. A. King: Phys. Rev. Lett. **87**, 216102, (2001).
36. CPMD v. 3.4.1, J. Hutter, A. Alavi, T. Deutsch, M. Bernasconi, S. Goedecker, D. Marx, M. Tuckerman, M. Parrinello. Max-Planck-Institut für Festkörperforschung and IBM Zurich Research Laboratory, 1995-1999.
37. A. Alavi, J. Kohanoff, M. Parrinello, and D. Frenkel: Phys. Rev. Lett. **73**, 2599, (1994).
38. A.Alavi in [1], chapter 25, page 649.
39. J. P. Perdew, K. Burke and M. Ernzerhof. Phys.Rev. Lett. **77**, 3865, 1996.
40. D.Reith, H.Meyer and F. Müller-Plathe: Macromol. **34**, 2235 (2001).
41. H.Meyer, O.Biermann, R.Faller, D.Reith and F. Müller-Plathe: J.Chem.Phys. **113**, 6264 (2000)
42. W.H.Press, S.A.Teukolsky, W.T.Vetterling and B.P.Flannery: *Numerical Recipes in C* (2nd Edition, Cambridge University Press, New York (1992).
43. R.Faller, H.Schmitz, O.Biermann and F. Müller-Plathe: J.Comp.Chem. **20**, 1009 (1999).
44. D.Reith, B.Müller, F. Müller-Plathe and S.Wiegand: Macromolecules (2002) in press.

Part IV

Complex and Mesoscopic Fluids

6 Effective Interactions for Large-Scale Simulations of Complex Fluids

Jean-Pierre Hansen[1] and Hartmut Löwen[2]

[1] Department of Chemistry, Lensfield Road, Cambridge CB2 1EW, UK
[2] Institut für Theoretische Physik II, Heinrich-Heine-Universität Düsseldorf, Universitätsstraße 1, D-40225 Düsseldorf, Germany

Abstract. The simulation of complex fluids naturally involves widely different length scales. Integrating out parts of the microscopic degrees of freedom leads to the concept of effective interactions and provides a "coarse-grained" picture which can be simulated much more efficiently than a full microscopic model. This approach bridges length scales in complex fluids. In this chapter, we justify this procedure on a Statistical Mechanics level and apply it to a variety of different systems ranging from charged colloidal dispersions and polymer solutions (including star polymers and dendrimers) to mixtures of colloids and polymers and binary colloidal mixtures. Problems arising when this concept is transferred to nano-scales are pointed out. Finally the much harder problem of bridging different time scales in complex fluids is briefly discussed.

6.1 Introduction

While early Monte Carlo (MC) and Molecular Dynamics (MD) methods were historically designed to simulate simple fluids, represented, e.g., by the hard sphere or Lennard-Jones models, these agorithms are ill adapted to deal with fluids of increasing complexity, or "soft matter", because of the simultaneous presence of widely different length and time scales. Consider for example dispersions of spherical (e.g. polystyrene balls), rod-like (e.g. the "tobacco mosaic virus", TMV) or lamellar (e.g. clay) charge-stabilized colloidal particles in water. There are at least three length scales: the microscale of the solvent molecules, the nanoscale of the width of the electric double-layers formed by the co- and counterions (approximately equal to the Debye screening length), and the mesoscale of the colloidal particles (typically hundreds of nanometers). There are at least as many relevant time scales (although they are not always as clearly separated), and direct simulations based on a femtosecond time scale, or MC trial moves over molecular distances, are clearly inadequate to describe large scale phenomena over times of milliseconds and over distances of several microns. Returning to the case of charged colloids, the practical limit of "brute force" approaches, even neglecting the molecular nature of water (replaced by a dielectric continuum) is a total charge of less than $100e$ per polyion [1].

In this chapter we describe systematic coarse-graining procedures which lead to effective interactions between the largest, mesoscopic particles in multicomponent, multiscale fluid mixtures. These effective interactions follow from a rigorous "tracing out" of microscopic degress of freedom, and can be used in standard MC or MD simulations of samples involving only the large particles which now play the role of molecules in atomistic simulations. After a formal Statistical Mechanics justification of the coarse-graining procedure in Sect. 6.2, it will be successively applied to interacting electric double-layers (Sect. 6.3), to polarizable dielectric media (Sect. 6.4), to solutions of linear polymers (Sect. 6.5), star polymers and dendrimers (Sect. 6.6), to colloid-polymer (Sect. 6.7) and binary colloid mixtures (Sect. 6.8). In Sect. 6.9 the new challenges of coarse-graining nanoscale rather than mesoscale colloidal systems will be briefly considered, with biomolecular (e.g. protein) solutions in mind. Some conclusions will be drawn in Sect. 6.10.

It is worth stressing that this chapter will deal almost exclusively with the bridging of length scale gaps in complex fluids. The problem of how to cope with widely different time scales is far more difficult for supramolecular systems, and the corresponding methodology is still in its infancy.

6.2 Efficient Coarse-Graining Through Effective Interactions

An efficient statistical description of multi-component systems involving particles of widely different sizes requires a controlled-coarse-graining which may be achieved by integrating ("tracing") out the degrees of freedom of the majority components of "small" particles, which may be solvent molecules, microscopic ions ("micro-ions") or monomers of macro-molecules. For the sake of simplicity, consider an asymmetric binary "mixture" of N_1 "large" spherical particles, with centres of positions $\{\boldsymbol{R}_i\}$ $(1 \leq i \leq N_1)$, and $N_2 \gg N_1$ "small" particles at positions $\{\boldsymbol{r}_j\}$ $(1 \leq j \leq N_2)$. Restriction will be made to thermodynamic equilibrium states. If classical statistics apply, integration over momenta is trivial, and the focus will be on configurational averages. The total potential energy of the mixture may be conveniently split ino three terms:

$$U(\{\boldsymbol{R}_i\}, \{\boldsymbol{r}_j\}) = U_{11}(\{\boldsymbol{R}_i\}) + U_{22}(\{\boldsymbol{r}_j\}) + U_{12}(\{\boldsymbol{R}_i\}, \{\boldsymbol{r}_j\}) \tag{6.1}$$

At a fixed inverse temperature $\beta = 1 / k_B T$, the configurational part of the Helmholtz free energy F of the two-component system may be formally expressed as:

$$\begin{aligned} \exp(-\beta F) &= Tr_1 \, Tr_2 \, \exp(-\beta U) \\ &= Tr_1 \, \exp(-\beta U_{11}) \, Tr_2 \, \exp(-\beta(U_{12} + U_{22})) \\ &= Tr_1 \, \exp(-\beta U_{11}) \, \exp(-\beta F_2(\{\boldsymbol{R}_i\})) \\ &= Tr_1 \, \exp(-\beta V_{11}(\{\boldsymbol{R}_i\})) \end{aligned} \tag{6.2}$$

where the short-hand trace notation implies integration over the configuration space of species 1 or 2, i.e.

$$Tr_\alpha = \frac{1}{N_\alpha!} \int d^{3N_\alpha} r$$

$V_{11}(\{\boldsymbol{R}_i\})$, the effective interaction energy of the large particles, is the sum of their direct (or bare) interaction energy U_{11}, and of the configurational free energy of the fluid of small particles in the "external" field of the large particles F_2; the latter depends parametrically on the configuration $\{\boldsymbol{R}_i\}$ of the large particles:

$$V_{11}(\{\boldsymbol{R}_i\}) = U_{11}(\{\boldsymbol{R}_i\}) + F_2(\{\boldsymbol{R}_i\}) \tag{6.3}$$

Up to now, no approximation has been made. If the large particles are fixed in a given configuration, (6.3) provides the exact energy from which the direct-configurational forces between the large particles can be derived. Once these particles move under the action of these effective forces, friction forces due to exchange of momentum with the bath of small particles will set in, and hydrodynamic forces due to induced flow of the latter will come into play. This difficult problem of hydrodynamic interactions [2] is not the subject of the present paper, which is restricted to static equilibrium properties independent of velocity-dependent forces.

Two key aspects of the effective interaction V_{11} must be underlined. Firstly, due to the presence of a free energy, F_2, V_{11} is obviously state-dependent, and has an entropic contribution of the small particles ($F_2 = U_2 - T S_2$). Secondly, although the direct interaction U_{11} may be pair-wise additive, this is no longer true of V_{11}. The free energy $F_2(\{\boldsymbol{R}_i\})$ generally has many-body-contributions, so that V_{11} will be of the more general form (with the change of notation $N_1 \to N$ and $V_{11} \to V_N$):

$$V_N(\{\boldsymbol{R}_i\}) = V_N^{(0)} + \sum\sum_{i \leq j} v_2(\boldsymbol{R}_i, \boldsymbol{R}_j) + \sum\sum\sum_{i \leq j \leq k} v_3(\boldsymbol{R}_i, \boldsymbol{R}_j, \boldsymbol{R}_k) + \ldots . \tag{6.4}$$

$V_N^{(0)}$ is a state dependent but configuration-independent "volume" term, which has no bearing on the local structure of the large particles, but through its contribution to the thermodynamic properties, it can, in some cases, strongly influence their phase behaviour [3].

Expression (6.3) for the effective interaction, or potential of mean force, was derived in the canonical ensemble, where the total numbers of small and large particles are fixed (closed system). In many practical situations the binary system is in osmotic equilibrium with a pure phase of the small particles (e.g the solvent), and the appropriate ensemble for such an open system is the semi-grand canonical ensemble where N_1 and the chemical potential μ_2 of the small particles (rather than N_2) are fixed. The corresponding thermodynamic potential is the semi-grand potential $\Omega_2 = \Omega_2(T, N_1, \mu_2; \boldsymbol{R}_i)$, and the effective interaction energy of the large particles will then be:

$$V_{11}(\boldsymbol{R}_i) = U_{11}(\boldsymbol{R}_i) + \Omega_2(\{\boldsymbol{R}_i\}) \tag{6.5}$$

which will again be state-dependent, a function of temperature, volume V and μ_2 (rather than $\rho_2 = N_2/V$).

In summary, the initial two-component system, involving a large number of microscopic degrees of freedom, has been reduced to an effective one-component system involving only the degrees of freedom of the mesoscopic particles. The price to pay is that the effective interaction energy is state-dependent and generally involves many-body terms. Approximations must now be invoked to calculate the highly non-trivial F_2 or Ω_2 term, i.e. the part of the interaction energy between the large particles induced by the small particles. Three different strategies have so far been used in practical implementations:

a) For any given configuration $\{\boldsymbol{R}_i\}$ of the large particles, the small particles are subjected to the "external" potential $U_{12}(\{\boldsymbol{R}_i\}, \{\boldsymbol{r}_j\})$, and hence form an inhomogeneous fluid, characterized by a local density $\rho(\boldsymbol{r}; \{\boldsymbol{R}_i\})$. The thermodynamic potentials F_2 or Ω_2 are functionals of $\rho(\boldsymbol{r})$, and full use can be made of the classical density functional theory (DFT) of non-uniform fluids to obtain tractable forms of F_2 or Ω_2 [4]. Given a form of $\Omega_2[\rho^*(\boldsymbol{r})]$, where $\rho^*(\boldsymbol{r})$ is a properly parametrized trial density, the equilibrium density follows from the variational principle:

$$\left.\frac{\delta \Omega_2[\rho^*(\boldsymbol{r})]}{\delta \rho^*(\boldsymbol{r})}\right|_{\rho^*=\rho} = 0 \tag{6.6}$$

Substitution of the optimum $\rho(\boldsymbol{r})$ into Ω_2 yields the equilibrium grand potential for any configuration $\{\boldsymbol{R}_i\}$. The optimization (6.6) may be implemented by steepest descent or conjugate gradient techniques, and the resulting effective potential energy between large particles can then be used directly in standard MC or MD simulations [5]. In the latter case, the forces $\boldsymbol{F}_i$ acting on the large particles may be directly calculated from a classical version of the Hellmann-Feynman theorem:

$$\begin{aligned} \boldsymbol{F}_i &= -\boldsymbol{\nabla}_i V_{11}(\{\boldsymbol{R}_j\}) \\ &= -\boldsymbol{\nabla}_i U_{11}(\{\boldsymbol{R}_j\} - \langle \boldsymbol{\nabla}_i U_{12}(\{\boldsymbol{R}_j\}, \{\boldsymbol{r}_l\} \rangle_{\{\boldsymbol{R}_j\}} \end{aligned} \tag{6.7}$$

where the angular bracket denotes an equilibrium average over the degrees of freedom of the small particles, for a fixed configuration $\{\boldsymbol{R}_i\}$ of the large ones. If the interaction energy U_{12} between the two species is pairwise additive ($U_{12} = \sum_{i=1}^{N_1} \sum_{j=1}^{N_2} u_{12}(|\boldsymbol{r}_i - \boldsymbol{R}_j|)$), the force $\boldsymbol{F}_i$ is directly expressible in terms of the local density $\rho(\boldsymbol{r})$:

$$\boldsymbol{F}_i = -\boldsymbol{\nabla}_i U_{11}(\{\boldsymbol{R}_j\}) - \int \rho(\boldsymbol{r}) \, \boldsymbol{\nabla}_i u_{12}(\boldsymbol{r} - \boldsymbol{R}_i) \, d\boldsymbol{r} \tag{6.8}$$

The optimization can also be achieved "on the fly", along lines directly inspired by the Car-Parrinello method for ion-electron systems [6]. Successive minimization and large particle updating steps are replaced by a single dynamical evolution, which involves the physical motion of the large particles and fictitious dynamics of the local density of small particles, parametrized by a plane wave expansion [7].

b) The previous DFT optimization method calculates directly the total effective energy of interaction between the large particles, or the resulting forces acting on each of these particles, without dividing V_N up into pair, triplet and higher order interactions, as written in (6.4). Another strategy is to attempt to compute these various contributions separately. At very low concentration of large particles, the effective pairwise interaction v_2 is expected to be dominant. In order to map out v_2 as a function of the distance r between two large particles, one may use standard MC or MD algorithms to simulate a bath of small particles in the field of two fixed large particles. Equation (6.7) may then be used to calculate the mean forces acting on the two mesoparticles (which are opposite if the latter are identical) for each distance $r = |\boldsymbol{R}_1 - \boldsymbol{R}_2|$. The effective pair potential $v_2(r)$ finally follows from an integration of the forces. This procedure must be repeated for each distance r, but there are no time-scale or ergodicity problems, since the two large particles are fixed. The same goal can be achieved by appealing once more to DFT for the inhomogeneous fluid of small particles, subjected to the force field of two fixed large particles. The optimization may be carried out in r-space, using an adequate Eucledian or non-Eucledian [8,9] grid on which the local density of small particles is defined. For two identical large particles, the local density has obvious cylindrical symmetry, but under favourable conditions, a considerable simplification occurs by fixing one of the large particles and considering an infinitely dilute solution of large particles in a bath of small particles around the fixed large particle. The density profile of the large particles in the zero concentration limit is directly related to the effective pair potential between two large particles in a bath of small particles [10], i.e.,

$$v_2(r) = -k_B T \lim_{\rho_1 \to 0} \ln\left(\frac{\rho_1(r)}{\rho_1(r \to \infty)}\right) \tag{6.9}$$

The advantage is that the two density profiles $\rho_1(r)$ and $\rho_2(r)$ are now spherically symmetric, but the method requires the prior knowledge of an accurate density functional for an asymmetric binary mixture. This strategy may be generalized to the calculation of three-body and higher order effective interactions, by considering the density profiles of large and small particles around two or more fixed large particles [11]. Applications of this strategy will be discussed in Sects. 6.5-6.8.

c) Although the effective interaction energy (6.3) or (6.5) is not, in general, pairwise additive at finite concentrations of the large particles, it would be very convenient, for computational purposes, to reduce it, at least approximately, to a pairwise additive form. Contrarily to the two-body potential $v_2(r)$ discussed in the previous paragraph, which is only valid in the low density limit of large particles, the effective pair potential corresponding to finite concentrations is expected to be density-dependent, and will, in some average sense, incorporate the contributions of higher order terms in (6.4). Such effective density-dependent pair potentials can, in some cases, be derived from approximate functionals or from inversion procedures, examples of which will be described in Sect. 6.5.

6.3 Electric Double-Layers

Electric double-layers around mesoscopic colloidal particles of various shapes (spheres, rods, platelets, ...) or around polyelectrolytes make the generally dominant contribution to the effective interaction between highly-charged particles, which will be referred to as polyions [12,13]. Most simulations are based on a primitive model, whereby the discrete nature of the aqueous solvent is neglected, and a macroscopic value of the dielectric permittivity ϵ is assumed. At very low polyion concentration, strategy b) of the previous section may be adopted to compute an effective pair interaction between two polyions, which is screened by microscopic counterions of opposite sign, as well as coions in the presence of added salt. The resulting effective pair potential turns out to be invariably repulsive, of the screened Coulomb form predicted a long time ago by Derjaguin, Landau, Verwey and Overbeek (DLVO) [14] as long as the microions are monovalent. However if divalent counterions are present, they are more strongly correlated, and this may lead to a short-range attraction between equally-charged polyions, due to an overscreening effect [15]. Although most of the work on effective pair interactions has focussed so far on spherical polyions, some recent MC simulations have investigated the case of parallel lamellar colloids [16], and this work has very recently been extended to charged discs of various relative orientations [17]. The triplet interaction between spherical polyions has similarly been calculated by MD simulations of co and counterions in the field of three fixed polyions [18], and turns out to be attractive under most circumstances. In the opposite limit of high concentrations, each polyion is confined to a cage of neighbouring polyions, so that many-body interactions are expected to be important, and pairwise additivity of the effective interaction is expected to break down. It is then reasonable to consider a Wigner-Seiz cell model, where a cell of geometry adapted to the shape of the polyions (e.g. a spherical cell for spherical polyions) contains one polyion at its centre, surrounded by co and counterions, such that overall charge neutrality is ensured, and with appropriate boundary conditions for the electric field on the surface of the cell. A physically reasonable boundary condition is to impose that the normal component of the electric field vanishes on the surface. The initial problem involving many polyions is thus approximately reduced to the much simpler problem of a single polyion surrounded by its electric double-layer. Although all information on correlations between polyions is lost, the cell model allows a calculation of the thermodynamic properties of concentrated suspension, from MC or MD simulations of the inhomogeneous fluid of microions contained in the cell, as well as an estimate of the effective polyions charge, taking into account the phenomenon of counterion "condensation" [19,20]. Such simulations provide stringent tests for approximate DFT calculations, including Poisson-Boltzmann (PB) theory.

At moderate polyion concentrations, the two previous strategies break down. Strategy a) of the previous section, based on the step by step or "on the fly" optimization of an appropriate free energy functional of the microion density profiles, is the most appropriate [7]. The free energy functional $F_2[\rho_+(\boldsymbol{r}), \rho_-(\boldsymbol{r}), \{\boldsymbol{R}_i\}]$ of the co- and counterion densities is conveniently split into ideal, Coulomb, external and correlation parts:

$$\begin{aligned} F_2[\rho_+, \rho_-] = \; & F_{id}[\rho_+] + F_{id}[\rho_-] + F_{Coul}[\rho_c] \\ & + F_{ext}[\rho_+] + F_{ext}[\rho_-] + F_{corr}[\rho_+, \rho_-] \end{aligned} \tag{6.10}$$

where:

$$F_{id}[\rho_\alpha] = k_B T \int \rho_\alpha(\boldsymbol{r}) \left[\ln(\Lambda_\alpha^3 \, \rho_\alpha(\boldsymbol{r})) - 1\right] d\boldsymbol{r} \tag{6.11}$$

$$F_{Coul}[\rho_\alpha] = \frac{e^2}{2} \int d\boldsymbol{r} \int d\boldsymbol{r}' \frac{\rho_c(\boldsymbol{r}) \, \rho_c(\boldsymbol{r}')}{|\boldsymbol{r} - \boldsymbol{r}'|} \tag{6.12}$$

$$\begin{aligned} F_{ext}[\rho_\alpha] &= \int \varphi_{ext}(\boldsymbol{r}) \, \rho_\alpha(\boldsymbol{r}) \, d\boldsymbol{r} \\ &= \sum_{i=1}^{N_1} \int u_{1\alpha}(\boldsymbol{r} - \boldsymbol{R}_i) \, \rho_\alpha(\boldsymbol{r}) \, d\boldsymbol{r} \end{aligned} \tag{6.13}$$

In (6.12), $\rho_c(\boldsymbol{r}) = z_+\rho_+(\boldsymbol{r}) + z_-\rho_-(\boldsymbol{r})$ is the charge density of the microions (of valences z_α). The polyion-microion potentials $u_{1\alpha}$ in (6.13) contain a hard core repulsion and a long-range Coulomb attraction (counterions) or repulsion (coions). Rapid variations of the densities profiles $\rho_\alpha(\boldsymbol{r})$ near the surfaces of the polyions, which would pose numerical problems in r-space (grid) or k-space (large $\boldsymbol{k}$ Fourier components) may be avoided by the use of appropriate classical polyion-microion pseudopotentials [7]. The correlation term F_{corr} may be expressed within the local density approximation (LDA) [7]. If it is neglected, the functional (6.10) reduces to the mean-field Poisson-Boltzmann (PB) form. Optimization based on the functional (6.10-6.13) has been achieved with the "on the fly" MD strategy for spherical polyions with counterions only (no salt) [7], and in the presence of salt (i.e. with co and counterions) [21]. The effective forces between colloids are reasonably well represented by a pair-wise additive screened-Coulomb form provided the (effective) polyion charge and the screening length are treated as adjustable parameters. Other applications include rigid rod-like polyions [22], and flexible polyelectrolytes [5], the latter being investigated by MC simulations coupled with steepest descent optimization, to allow a more efficient exploration of polyelectrolyte configuration space. If F_{corr} is neglected in the functional (6.10), and the ideal terms are replaced by their quadratic expansion in powers of $\Delta\rho_\alpha(\boldsymbol{r}) = \rho_\alpha(\boldsymbol{r}) - \rho_\alpha$ (where ρ_α is the bulk concentration of microions), the total functional is quadratic in the $\rho_\alpha(\boldsymbol{r})$, and the Euler-Lagrange equations resulting from the extremum conditions (6.6) can be solved analytically [7]. The resulting total effective energy of the polyions is then strictly pair-wise additive, and the effective pair potentials are of the linearly screened DLVO form. The entire procedure is justified only for relatively weak microion inhomogeneities (i.e. $|\Delta\rho_\alpha(\boldsymbol{r})|/\rho_\alpha < 1$), i.e. for low absolute polyion valence $|Z_p|$. If the polyion charge is distributed over a number ν of interaction sites, each carrying a charge $Z_p\, e/\nu$, linear screening may be an adequate approximation for each interaction site. The resulting "Yukawa site" model, where all sites on neighbouring particles interact via a screened Coulomb (or Yukawa) pair potential, has been used to simulate charged rods [22] or charged discs representing clay particles [23].

An excellent literature survey of the recent simulation work on charged-stabilized colloidal suspensions is provided by the review of M. Dijkstra [24].

6.4 Simulating the Polarization of Dielectric Media

The coarse-graining methods developed for poly-ionic systems may be extended to take into account the polarization of dielectric media. This is important when dealing with mesoscopic interfaces, or the solvation of highly charged macromolecules of biological interest, like DNA or proteins, by water [25]. Since a full molecular description of the solvent surrounding the macromolecules would be computationally prohibitive, water is generally treated as a dielectric continuum which is polarized by the charge distribution on the macromolecules. The key problem is to determine the spatially varying polarization $\boldsymbol{P}(\boldsymbol{r})$ induced in the dielectric, for any configuration of the "external" charges carried by the macromolecules and counter- and coions and to calculate the resulting electrostatic potential $\Psi(\boldsymbol{r})$, due to the external and induced polarization charges.

Electrostatic problems involving dielectric polarization can be solved variationally, as in the case of electric double-layers considered in the previous section, by minimizing an appropriate functional F of the polarization density $\boldsymbol{P}(\boldsymbol{r})$ [26] or of the polarization charge [27]

$$\rho_{pol}(\boldsymbol{r}) = -\boldsymbol{\nabla}\cdot\boldsymbol{P}(\boldsymbol{r}) = \boldsymbol{\nabla}\cdot[\chi(\boldsymbol{r})\boldsymbol{\nabla}\Psi(\boldsymbol{r})] \tag{6.14}$$

where $\chi(\boldsymbol{r})$ is the local dielectric susceptibility. The non-linear nature of the problem is immediately apparent from the self-consistency requirement which links $\Psi(\boldsymbol{r})$ to $\rho_{pol}(\boldsymbol{r})$:

$$\Psi(\boldsymbol{r}) = \Psi_0(\boldsymbol{r}) - \int \frac{\rho_{pol}(\boldsymbol{r}')}{|\boldsymbol{r}-\boldsymbol{r}'|} d\boldsymbol{r} \tag{6.15}$$

where $\Psi_0(\boldsymbol{r})$ is the "external" electrostatic potential, due to the charge distribution on the macromolecules. The advantage of using $\boldsymbol{P}(\boldsymbol{r})$ as a variational field is that an electrostatic free energy functional can be constructed in the form [28,29]

$$\begin{aligned} F_{el}[\boldsymbol{P}(\boldsymbol{r})] = \; & \frac{1}{2}\int \Psi_0(\boldsymbol{r})\rho(\boldsymbol{r})d\boldsymbol{r} \\ & +\frac{1}{2}\int |\boldsymbol{P}(\boldsymbol{r})|^2/\chi(\boldsymbol{r})d\boldsymbol{r} - \int \Psi_0(\boldsymbol{r})\boldsymbol{\nabla}\cdot\boldsymbol{P}(\boldsymbol{r})d\boldsymbol{r} \\ & +\frac{1}{2}\int\int \frac{(\boldsymbol{\nabla}\cdot\boldsymbol{P}(\boldsymbol{r}))(\boldsymbol{\nabla}'\cdot\boldsymbol{P}(\boldsymbol{r}'))}{|\boldsymbol{r}-\boldsymbol{r}'|} d\boldsymbol{r}\, d\boldsymbol{r}' \end{aligned} \tag{6.16}$$

Minimization of (6.16) with respect to $\boldsymbol{P}(\boldsymbol{r})$, for a fixed macromolecular configuration (and hence $\Psi_0(\boldsymbol{r})$), leads back to the usual constitutive equations of electrostatics, and the value of F_{el} at the minimum coincides with the standard expression for the electrostatic energy of a polarizable medium in an external field. This functional has been used by Marchi et al. in MD simulations of polypeptides, with an "on the fly" optimization strategy [26].

A functional of the polarization charge (6.14), rather than of the polarization itself can be constructed, which upon minimization, leads back to the standard relations of electrostatics [30]. At its minimum, this functional reduces to minus the electrostatic energy, so that it cannot be used in an obvious way in dynamical optimization algorithms, but working with $\rho_{pol}(\boldsymbol{r})$, rather than with $\boldsymbol{P}(\boldsymbol{r})$, has two advantages. First a scalar rather than a vectorial field is to be handled on a grid. But more importantly, if the interface between dielectrics is sharp, so that the susceptibility is essentially a step function, the polarization charge (6.14) reduces to a surface charge which may be defined on a 2d (rather than full 3d) grid, resulting in considerable computational savings. In MD or MC simulations of macromolecular charge distributions near interfaces, the polarization charge is efficiently calculated by a steepest descent algorithm on a step by step basis. The method has been successfully tested for simple model systems [30], and is at present being applied to the simulation of ion channels through membranes, where the channel protein and the embedding lipid bilayer are treated as a dielectric continuum responding to the moving electric charges on the cations and on the water molecules [31]. Note that the role of the macromolecule and of water are inverted compared to the macromolecular hydration problem [27,26].

6.5 Coarse-Graining Linear Polymer Solutions

A solution of linear polymers involves many different length scales ranging from microscopic bond length over the persistence length and the radius of gyration (coil size) to the mean-inter coil distance, see Fig. 6.1. Therefore, the direct simulation of on or off-lattice models of polymer solutions or melts are very computer-intensive [32], because even the simplest linear polymers involve thousands of monomers. Even if the latter are grouped into Kuhn segments, corresponding to one persistence length, the system will involve a very large number of linearly connected, interacting particles. If N is the number of polymer coils, each made up of M segments, the total number of degrees of freedom is $3NM$ which is a factor of M ($\gg 1$) larger than for simple fluids or rigid colloidal particles, assuming that the organic or aqueous (for polyelectrolytes) solvent is replaced by a continuum. The question hence naturally arises of how to coarse-grain the initial, fully microscopic model involving M monomers or segments per chain. The situation is somewhat different from the previously examined cases, involving large and small particles. The polymer case is more "democratic" in that all monomers play identical roles, at least in the scaling limit $M \to \infty$, where end effects become negligible. An old idea, which goes back at least to Flory [33] is to derive an effective interaction between the centres of mass (CM) of neighbouring polymer coils, by integrating over the individual monomer degrees of freedom of two or more coils, for fixed relative positions of their CM's. Consider first the case of two isolated polymer coils with monomer coordinates $\{\boldsymbol{r}_{i\alpha}\}_M$ ($i = 1, 2; 1 \leq \alpha \leq M$) and CM's:

$$\boldsymbol{R}_i = \sum_{\alpha=1}^{M} \boldsymbol{r}_{i\alpha} \tag{6.17}$$

If $U(\{\boldsymbol{r}_{i\alpha}\})$ is the total potential energy of interaction of all monomers, the probability distribution of the CM's is.

$$P(\boldsymbol{R}_1, \boldsymbol{R}_2) = \frac{1}{Q_{2M}} \int e^{-\beta U(\{\boldsymbol{r}_{i\alpha}\})} \prod_{i=1,2} \delta(\boldsymbol{R}_i - \sum_{\alpha} \boldsymbol{r}_{i\alpha}) \prod_{\alpha} d\boldsymbol{r}_{i\alpha} \tag{6.18}$$

where Q_{2M} is the corresponding configurational partition function (equal to the $6M$-dimensional integral in (6.18), without the δ-functions). By analogy with (6.2), the effective pair interaction between the CM's is then given by:

$$v_2(\boldsymbol{R}_1 \boldsymbol{R}_2) = -k_B T \ln[P(\boldsymbol{R}_1 \boldsymbol{R}_2)] \tag{6.19}$$

The effective potential will only depend on $r = |\boldsymbol{R}_1 - \boldsymbol{R}_2|$. $v_2(r)$ is expected to be of the order of the radius of gyration R_g of the polymers, since for $r \geq R_g$, there will be little overlap between two coils. Swollen polymers in good solvent (where $R_g \sim M^\nu$, with $\nu \simeq 0.6$, the Flory exponent) are highly fractal objects, i.e. the mean monomer density inside a coil $\sim M^{1-3\nu}$ goes to zero in the scaling limit. An immediate consequence is that $v_2(r)$ is a "soft" potential, and that $v_2(r = 0)$ is finite [34], i.e. the polymer coils may be modelled as penetrable spheres. On and off-lattice simulations of self-avoiding walk (SAW) polymers [35,36], as well as renormalization group (RG) calculations for the continuous "thread" model [37], show that the pair potential $v_2(r)$ is well approximated by a single Gaussian:

$$\frac{v_2(r)}{k_B T} = \xi \exp(-J(r/R_g)^2) \tag{6.20}$$

where $J \simeq 1$; the simulations yield [36] for $M \to \infty$:

$$\xi \simeq 1.80 \pm 0.05 \tag{6.21}$$

while the RG yields the following $\epsilon = 4 - d$ expansion.

$$\xi = 0.94\epsilon + 0.62\epsilon^2 + O(\epsilon^3) \tag{6.22}$$

These results are independent of molecular detail, so that the simplest lattice models and most efficient MC sampling (e.g. the pivot algorithm) can be used to determine the effective interactions. Note that for self-avoiding walk (SAW) polymers (which involve only excluded volume interactions between monomers), the effective interaction is purely entropic in nature as signalled by the scaling with $k_B T$.

When nearest-neighbour attractions between monomers are included to allow for solvent conditions (strong attractions correspond to poor solvent), the effective pair potential between the CM's becomes less repulsive, and develops an attractive part as Θ-conditions are approached [35,38]. If ϵ_0 denotes the depth of the attraction, ergodicity problems become more and more severe in the simulations when $\epsilon_0 / k_B T > 1$, but can be overcome by using Bennett's overlapping distribution method [39,40]. Returning to the SAW model, appropriate for good solvent conditions, the method for determining the effective pair potential can be extended to effective three- and more-body interactions, by simulating three or more polymers for various configurations of their CM's [41]. The main qualitative results are that more-than-two body interactions alternate in sign (the three-body potential being mostly attractive), and that the absolute amplitudes of higher order interactions do appear to decrease with increasing order in line with scaling theory [42].

However the strategy of adding higher order effective interactions in simulations of polymer solutions of finite concentration is computationally inefficient. A much more efficient strategy is to determine state-dependent effective pair interactions by a systematic inversion procedure [36]. The pair distribution function $g(r)$ of the CM's of systems of SAW polymers at finite concentration is calculated by direct simulations of a few hundred polymers on a lattice, using efficient MC algorithms [40]. An effective concentration-dependent effective pair potential between the CM's is then determined by Ornstein-Zernike (OZ) inversion, assuming some adequate closure relation, like the HNC closure [43]. In view of the softness of the resulting pair potential, HNC theory becomes asymptotically exact in the high concentration limit, and is extremely accurate at all concentrations [44,45]. The inversion is a noniterative, one-step procedure, and it has been proven that there is a one-to-one correspondence between any given $g(r)$ and a $v(r)$ (uniqueness theorem [46]). The resulting effective pair potentials turn out to be weakly dependent on concentration [36]. They may be used in large-scale simulations of polymer solutions, of polymers at interfaces or of colloid-polymer mixtures [47], to study, in particular, the effect of polymer interactions on the depletion force between colloidal particles [36,48]. In its original formulation, this coarse-graining strategy has one obvious drawback, namely that simulations of the full monomeric representation of polymer solutions are initially required to determine the CM pair contribution function $g(r)$, for each polymer concentration, a rather formidable task, even if the resulting effective potentials may then be used to explore a range of different large-scale phenomena. However even this drawback can be overcome by calculating the monomer-monomer pair distribution function within the accurate PRISM theory [49], and then extracting the CM pair distribution function from its monomeric counterpart, together with the form factor (or internal structure factor) of a single polymer coil, using a recently proposed, accurate relation between these three correlation functions [50].

6.6 Star Polymers and Dendrimers

The ideas of coarse-graining, as applied to solutions of linear polymer chains in the previous section, can be generalized to polymers with a more complicated architecture. We shall discuss solutions of star polymers and dendrimers in more detail. Star polymers [51] consist of f linear polymer chains which are chemically anchored to a common centre (f is called functionality or arm number). Obviously, linear polymers are a special case of star polymers when $f = 1, 2$ depending whether the end or middle segment is taken as "centre". Dendrimers, on the other hand, can be viewed as iterated star polymers: periodically, any linear chain branches off into n additional chains (n is called degree of branching) which is repeated g times (g is called generation number). For $f > 3$, in contrast to linear chains, star polymers and dendrimers possess a natural centre which serves as an appropriate statistical degree of freedom.

Let us first focus on *star polymers in a good solvent.* A full monomer-resolved computer simulation is completely out of reach of present-day computers: If N is the number of stars and M the number of monomers per chain, a total number of NfM particles has to be simulated, f times more than for a solution of linear chains and fM times more than for simple fluids. The strategy b) of Sect. 6.2, however, can be efficiently used to make progress. First consider only two stars at fixed separation r and average the force acting

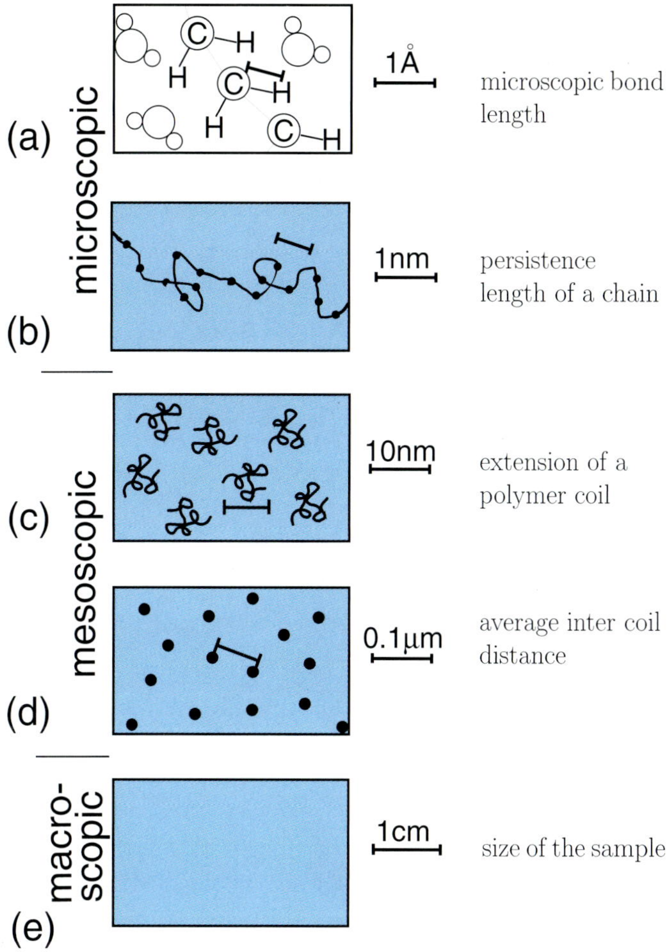

Fig. 6.1. Polymer solution on different length scales. (a): microscopic picture, water and hydrocarbon chains are shown, the chemical bonds have a range of typically 1Å. (b): On a larger scale, the persistence length of the chains is relevant. (c): the spatial extension σ of a single polymer coil. (d): all the coils are point particles on this scale governed by the mean intercoil distance (e): size of the macroscopic sample.

on their centres during an ordinary MC or MD simulation of the monomers. Such a simulation involves $2fM$ particles only. A typical simulation snapshot is shown in Fig. 6.2. This is repeated for different r. By integrating the distance-resolved data for the force, the effective interaction potential $v(r)$ is obtained. This interaction is repulsive, since the presence of another star reduces the number of configurations available to the chains. For small arm numbers $f \leq 10$, the simulation results confirm an effective pair potential of

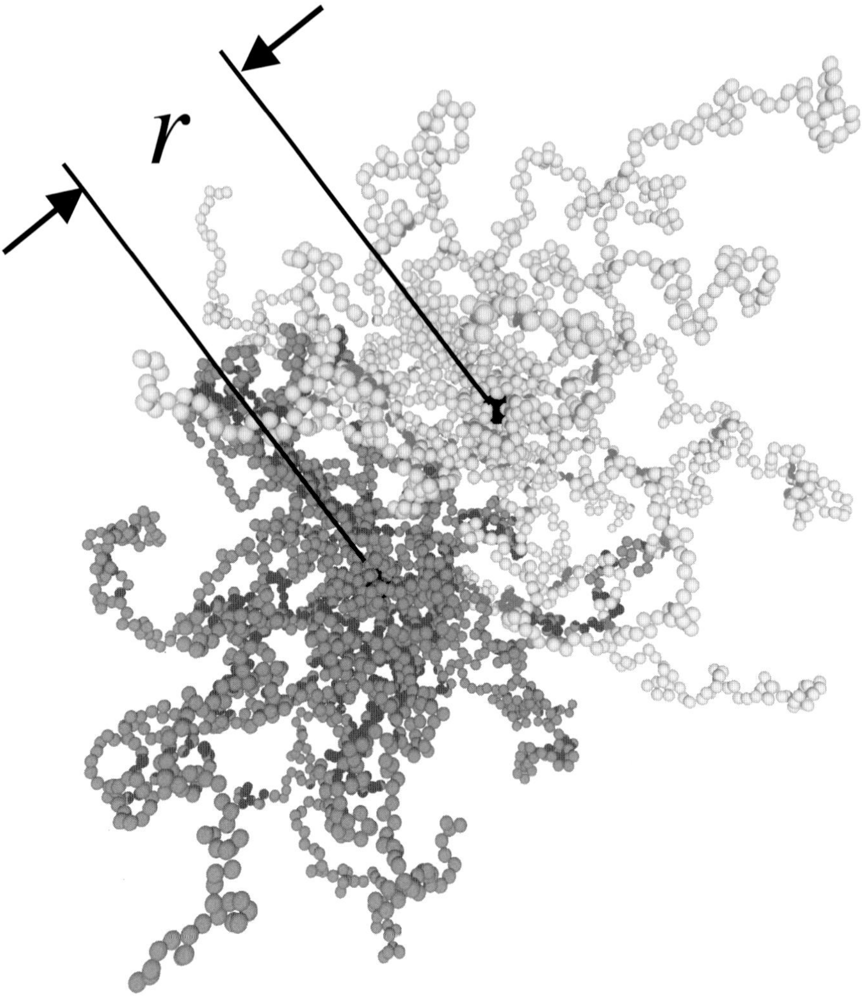

Fig. 6.2. Typical configuration for two stars with $f = 18$ and $M = 50$ monomers per chain as obtained from a snapshot during a Molecular Dynamics simulation with r denoting the distance between their centres. By courtesy of A. Jusufi.

the log-Gauss form:

$$v(r) = \frac{5}{18} k_{\mathrm{B}} T f^{3/2} \begin{cases} -\ln(\frac{r}{\sigma}) + \frac{1}{2\tau^2\sigma^2} & \text{for } r \leq \sigma; \\ \frac{1}{2\tau^2\sigma^2} \exp\left(-\tau^2 \frac{r^2-\sigma^2}{\sigma^2}\right) & \text{for } r > \sigma, \end{cases} \tag{6.23}$$

where σ is the corona diameter of a single star measuring the spatial extent of the monomeric density. For large distances r, the interaction is Gaussian as for linear chains. It then crosses over, at the corona diameter of the star, to a logarithmic behaviour for overlapping coronae as predicted by scaling theory

[52] which implies a very mild divergence as $r \to 0^+$. The matching at $r = \sigma$ is done such that the force $-dv/dr$ is continuous. In (6.23) , $\tau(f)$ is known from a fit to computer simulation results; for $f = 2$ we obtain $\tau = 1.03$ in line with the Gaussian potential (6.20) used for linear chains.

For larger arm numbers, $f > 10$, on the other hand, a geometric blob picture of f cones around the star centre, each containing one linear chain is justified [53]. The effective force for nearly touching coronae decays exponentially with r, the associated decay length is the outermost blob-diameter $2\sigma/\sqrt{f}$. This motivates a log-Yukawa form of $v(r)$ [54]:

$$v(r) = \frac{5}{18} k_B T f^{3/2} \begin{cases} -\ln(\frac{r}{\sigma}) + \frac{1}{1+\sqrt{f}/2} & \text{for } r \leq \sigma \\ \frac{\sigma}{1+\sqrt{f}/2} \frac{\exp(-\sqrt{f}(r-\sigma)/2\sigma)}{r} & \text{for } r > \sigma \end{cases} \tag{6.24}$$

again matched at the corona diameter $r = \sigma$ such that the force is continuous. This potential was verified in monomer-resolved simulations [55] for a large range of arm numbers.

Using scaling theory and monomer-resolved simulations of a triangular configuration of three stars [42], triplet interactions were shown to be negligibly small outside the corona and at most 11 percent of the pairwise forces for penetrating triplets inside the corona; consequently the effective pair-wise description for the many-body system is adequate provided the number density ρ_s of the stars is not much higher than the overlap density $1/\sigma^3$. Large scale simulations involving many stars were performed using the pair potential of (6.24) [56,57]. Due the crossover of $v(r)$ at $r = \sigma$ from a harsh Yukawa to a soft logarithmic behaviour, uncommon structural and thermodynamical properties were obtained. First, the main peak of the liquid structure factor changes non-monotonically with increasing density [57]. Secondly, the bulk phase diagram exhibits [56] a reentrant melting behaviour for $34 < f < 44$ and stable anisotropic crystal lattices. The latter finding has been supported by experiments [58].

Next let us briefly discuss *star polymers in a poor solvent*. The only work in this direction is close to the Θ-point where the chains are weakly interacting. Consequently the resulting effective repulsion is weaker than in good solvent. More quantitatively, an effective potential between two plates is available within a self-consistent field approach for polymers grafted on flat plates where the grafting density is high and the self-avoidance is weak [59]. This was extended to spherical particles by employing the Derjaguin approximation [60,61] providing an analytical expression for the effective pair potential $v(r)$. In the limit of small core sizes, this expression has been successfully tested against scattering data for $f = 64$ arm stars in a solvent close to Θ conditions [62]. What is still unexplored is a systematic approach for arbitrary solvent quality which continously switches between good solvent quality to the Θ point and beyond.

Much more stretched configurations are achieved for *polyelectrolyte stars* ("porcupines") due to the strong Coulomb repulsion of the charged monomers along the chains. If one brings two polyelectrolyte stars together they hardly interdigitate but retract. A variational analysis [63] for the effective force, which includes Coulomb interactions and entropies of the counterions, reveals that the entropy of the counterions which are inside the coronae of the two polyelectrolyte stars dominates the interaction, confirming an old idea of Pincus [64]. The analytical theory was quantitatively verified by computer simulations with explicit monomers and counterions [63]. Inside the corona, the resulting effective force could be fitted by an inverse-power law $\propto r^{-\gamma}$ where the exponent γ slightly depends on the actual charging conditions but is always around $0.7 - 0.8$. By integration, an effective potential is obtained which stays finite at the origin and behaves inside the corona as $v(r) = v(0) - Cr^{1-\gamma}$ with a positive constant C. However, the actual value $v(0)$ for completely overlapping stars is much larger than k_BT so that significant overlap is rare. Due to the softness of the interaction, similar structural anomalies as obtained for star polymers are expected including a non-monotonic variation of the first peak in the structure factor for increasing density and reentrant melting.

Finally, *dendrimers* in a good solvent have been addressed. For a branching degree $b = 2$ and a generation number $g = 4$ an effective Gaussian potential can be derived theoretically [65], provided the centre-to-centre distance r is larger than the corona diameter σ. In formal analogy to linear polymer chains one obtains:

$$v(r) = B\exp(-r^2/\sigma^2) \tag{6.25}$$

The important difference from the case of linear polymer chains is that the prefactor

$$\frac{B}{k_BT} = \frac{M^2 v_0}{\pi^{3/2}\sigma^3} \tag{6.26}$$

is much larger than 1, with M denoting the total number of monomers per dendrimer and v_0 the excluded volume parameter per monomer. The Gaussian interaction (6.25) was confirmed quantitatively by scattering experiments [65]. A detailed comparison for $b = 2$ dendrimers with higher generation number $g = 5$ reveals that the effective potential $v(r)$ is well described by a sum of two Gaussians [66] comprising the effect of the stiff inner region and the floppy outer region of a dendrimer. The effect of increasing the degree of branching b is much less explored. In principle a Gaussian pair potential (6.25) (or a superposition of them) is again expected but the prefactor B/k_BT should grow with increasing b. Thereby it should be possible to tune the prefactor B/k_BT to larger values, where freezing is expected [67].

Coarse-graining star polymers and dendrimers thus maps them onto simple liquids with soft interactions (socalled mean-field fluids) which in turn implies peculiar properties. Hence the concept of effective interactions not only allows for efficient simulation, but also provides insight into the physical behaviour.

6.7 Colloids and Polymers: Depletion Interactions

If a sterically-stabilized colloidal particle is brought into a non-adsorbing polymer solution, the latter are depleted in a zone around the colloidal surfaces due to the colloid-polymer repulsion. The width of this zone is of the order of the radius of gyration $d_p/2$ of the polymers. If one now brings two colloidal particles close to each other, the two depletion zones overlap, which brings about a free energy gain of the polymers relative to a situation of non-overlapping zones, resulting in an effective attraction between the colloids, the so-called depletion attraction. Alternatively one can view the attraction arising from an unbalanced osmotic pressure exerted on the colloidal particles by the surrounding polymers.

The simplest model for colloid-polymer mixtures including the depletion effect is the so-called Asakura-Oosawa (AO) model [68] which assumes hard core interactions between the colloids of diameter d_c, further hard-core interactions between the polymers and the colloids with a range $(d_c + d_p)/2$, but no interaction at all between polymers. The ideality of the polymers is a crucial approximation which is fulfilled only for dilute polymer solutions, but it allows to investigate many of the statistical properties of the AO model analytically. For instance, the effective interaction $v(r)$ between a colloidal pair can be calculated to be the product of the polymer osmotic pressure $P_p = k_B T \rho_p$ and the overlap volume of the two depletion zones consisting of two spherical half-caps. Explicitly it reads

$$\frac{v(r)}{k_B T} = \begin{cases} \infty & \text{for } r \leq d_c \\ \rho_p \frac{\pi}{6}(d_c + d_p)^3 \left[1 - \frac{3r}{2(d_c+d_p)} + \frac{1}{2}\frac{r^3}{(d_c+d_p)^3}\right] & \text{for } d_c < r \leq d_c + d_p \\ 0 & \text{for } r \geq d_c + d_p \end{cases} \quad (6.27)$$

Furthermore, by a simple geometric consideration, it can be shown that effective triplet and higher-order many-body forces vanish provided the size ratio between colloids and polymers $q = d_p/d_c$ is smaller than 0.154. In this case, the AO model is formally equivalent to an effective one-component system with a short ranged attraction, which immediately opens the way for large-scale simulations.

The phase diagram of the AO model was explored by computer simulations on three different levels: first, one-component calculations using the effective pair potential (6.27) have been performed [69], which are exact for $q < 0.154$. Secondly, more recently, Dijkstra has simulated the full effective Hamiltonian including effective many-body forces to arbitrary order for $q = 1$ [70]. Finally, brute force simulation with explicit ideal-gas polymers have been carried out [71]. The emerging phase diagram involves three phases: gas (i.e. colloidal poor), liquid (i.e. colloidal rich) and an fcc colloidal crystal. A liquid phase is stable if the ratio q is larger than $q_c \approx 0.5$.

On the other hand, theoretical progress was made by constructing a free volume theory for the fluid bulk free energies [72] which provides a reliable estimate for the gas-liquid transition. A free-energy density functional for the AO colloid-polymer mixtures, valid for arbitrary inhomogeneous situations, was constructed [73] in the spirit of Rosenfeld's fundamental measure approach [74], which reproduces the effective interaction (6.27) for a colloid pair and the free volume theory of Ref. [72]. This density functional was applied to wetting phenomena of planar walls. A novel type of wetting involving growth of only few colloidal liquid layers on top of the wall as liquid-gas coexistence is approached was predicted by density functional theory [75] and confirmed by computer simulations [70,71]. This wetting scenario only shows up for ratios larger than q_c, so that one can speculate that it is produced by the intrinsic many-body nature of the effective forces.

Obviously, the AO model has the short-coming of idealized interactions. More realistic models involve a non-zero polymer-polymer interaction and a softer polymer-wall interaction [76]. On the other hand, full two-component simulations of colloids and polymers were performed [77,78] where the polymers are defined on a lattice. Clearly these include any effective many-body interactions. A second computationally less demanding technique is to calculate effective pair interactions between a colloid and a polymer first by a monomer-resolved reference simulation. This strategy was followed in the more general context of mixtures of colloids and star polymers for small size ratios q. Supported by theoretical scaling arguments the following pair interaction between a hard-sphere colloid and a star polymer was obtained [79,47]:

$$v_{cp}(r) = k_B T \Lambda f^{3/2} \left(\frac{d_c}{2r+d_c}\right) \tag{6.28}$$

$$\times \begin{cases} -\ln(\frac{2r-d_c}{\sigma}) + (\frac{(2r-d_c)^2}{\sigma^2} - 1)((1+4\kappa)/(1+2\kappa)) + \zeta & \text{for } r \le \frac{d_c+\sigma}{2}; \\ \zeta \,\mathrm{erfc}(\kappa(2r-d_c)/\sigma)/\mathrm{erfc}(\kappa) & \text{else,} \end{cases}$$

Here, Λ and κ are known parameters depending on the functionality f of the star, $\zeta = \sqrt{\pi}\mathrm{erfc}(\kappa)\exp(\kappa^2)/(\kappa(1+2\kappa^2)$, σ denotes the corona diameter of the star and $\mathrm{erfc}(x)$ is the complementary error function. For $r \to d_c/2$ the potential diverges logarithmically as for the star-star interaction (6.23). Linear polymer chains are obtained as the special case $f = 2$ where $\Lambda = 0.46$ and $\kappa = 0.58$. The two-component system with effective pair interactions was investigated in detail by further simulation and liquid integral equation theory. For different arm numbers f, the fluid-fluid demixing transition was calculated [79] in good agreement with experimental data. Furthermore, the freezing transitions was discussed. Above a critical arm number of $f_c \approx 10$, fluid-fluid demixing was preempted by freezing [80].

In case of polymer size comparable or larger than the colloidal diameter d_c, effective many-body forces play a significant role. Complementary methods such as monomer-resolved liquid intergral equations methods combined with the PRISM approach [81] or field-theoretic calculations [82] have provided valuable insight into the structure of colloid-polymer mixtures. The limit of large q contains completely different physics, since the colloidal spheres represent then small perturbations for the long polymer chains [78].

6.8 Binary Colloidal "Alloys"

Binary mixtures of large and small sterically-stabilized colloids exhibit many unexpected phases. Such a binary colloidal "alloy" can be modelled as a two-component hard sphere system involving two colloidal diameters d_1 and d_2 ($d_2 < d_1$). If the size ratio $q = d_2/d_1$ is larger than 0.4, a large variety of stable phases involving different superlattice crystals are predicted by theory [83] and simulation [84] and observed experimentally [85] . We shall focus here more on the case of small q, where a depletion picture, similar to that discussed for colloid and polymer mixtures, should hold. In contrast to the AO-model, however, many-body forces are present for any q in the hard sphere mixture, but these can be shown to be small with respect to the pairwise contribution [9,11].

The effective depletion potential $v(r)$ between a pair of big spheres in a sea of small spheres has been efficiently calculated and analytically parameterized, based on the procedure of Roth et al. [10] as applied to the two-component hard sphere density functional of Rosenfeld [74]. As a function of r, $v(r)$ involves a short-ranged attraction followed by an oscillatory behaviour decaying exponentially with the bulk correlation length of the small spheres. The density functional predictions were confirmed by computer simulations [86,87] and by experiments [88].

One key question addressed during the last decade was whether the depletion attraction is strong enough to drive fluid-fluid phase separation (for a recent review see [24]). The problem was finally solved by computer simulation using the effective one-component model with the depletion pair potential [89,90]: a fluid-fluid demixing is obtained for $q < 0.1$ but this is always metastable with respect to the freezing transition into an fcc solid made up by the big spheres. The simulations of the effective one-component model were confirmed by full simulations of the true binary system [89] showing once more that the influence of triplet forces is negligible [9,11]. Simulation results for the phase diagram, as presented in terms of the packing fractions $\eta_1 = \pi\rho_1 d_1^3/6$ of the big particles and η_2^r of a reservoir of small particles in coexistence with the whole system, are shown in Fig. 6.3 for $q = 0.2, 0.1, 0.05$. Besides the fluid-fluid demixing which remains metastable, an isostructural solid-solid transition occurs at high η_1 for $q < 0.05$, as familiar for one-component systems with a short-ranged attraction [91,92]. As q decreases, the fluid-solid coexistence line becomes more and more horizontal until the sticky hard-sphere limit ($q \to 0^+$) is achieved where a vacuum coexists with a close-packed crystal. This example shows again that the effective interaction

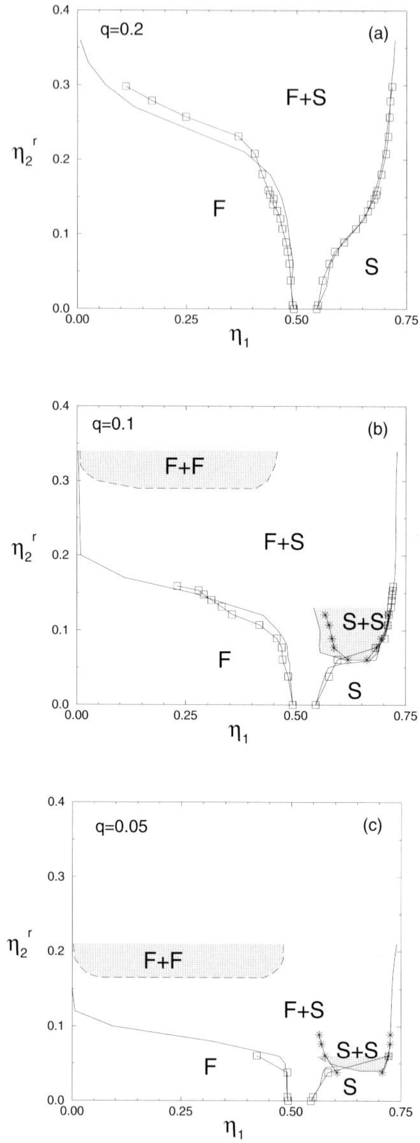

Fig. 6.3. Phase diagram of binary hard-sphere mixtures with size ratios (a) $q = 0.2$, (b) $q = 0.1$, and (c) $q = 0.05$ as a function of the large-sphere packing fraction η_1 and the small-sphere reservoir packing fraction η_2^r. F and S denote the stable fluid and solid (fcc) phase. $F + S$, $F + F$, and $S + S$ denote, respectively, the stable fluid-solid, the metastable fluid-fluid, and the (meta)stable solid-solid coexistence regions. The solid and dashed lines are the effective one-component results; the squares and the asterisks (joined by lines to guide the eye) denote, respectively, the fluid-solid and the solid-solid transition obtained from direct simulations of the true binary mixtures. Reproduced from Ref. [89] with permission.

picture allows qualitative and quantitative understanding of the topology of phase diagrams.

Further current research is focused on the sensitivity of the depletion potential to polydispersity of the small particles [93], and to small-small and big-small interactions beyond the hard sphere model [94]. Polydispersity smears out the oscillatory behaviour, while keeping the deep attraction near contact unchanged. Different small-small and big-small interaction have a significant impact on the effective big-big interaction: for instance, an attraction between the small particles or the big and small particles leads to repulsive effective interactions due to an accumulation of small particles on the surfaces of the big ones. This "inverse depletion effect" may be called accumulation repulsion. These recent findings imply that the effective interactions can be systematically tuned via the basic interactions and polydispersity.

6.9 From Colloidal to Nanoscales

Although the concept of effective interactions as described in Sect. 6.2 is exact in principle, there are limitations in applying it in practice. These limitations become more and more important if one considers smaller and smaller macroparticles, such that molecular details become more relevant. In fact, on nanoscales, chemical specificity is starting to become crucial, resulting in many different phenomena like solvation effects, hydration, hydrophilicity, hydrophobicity, forces determined by chemical bonding etc. In fact, these effects are essential to explain the structure and function of biological macromolecules in solution, e.g. of proteins. Under these circumstances, it is clear that one cannot get away with relatively simple effective interactions, characterized by few parameters, like those discussed previously in the context of colloidal length scales. The immediate question arising is when and where does the simple coarse-graining concept break down if one crosses over from the colloidal to the nanoscale. Basically there are *two major caveats*: The first concerns the choice of the microscopic degrees of freedom which are to be integrated out; the second concerns internal degrees of freedom and modelling of the big particles themselves. In the sequel we shall discuss these two points in detail and illustrate them using two examples.

Regarding the first point, even on nanoscales there remains an enormous number of microscopic degrees of freedom. The relevant question is which of these have to be included explicitly in the starting Hamiltonian or may be ignored or replaced by effective parameters. This is a tricky question when length scales are less clearly separated than in colloidal systems. For long polymer chains, scaling theory implies that molecular details are unimportant for most purposes, such that one can get away with simple lattice models (as discussed for linear polymer chains) or with a simple monomeric description of beads (as discussed for star polymers). For charged colloids, in the primitive approach, the solvent molecules are not considered explicitly but only enter via the dielectric constant. The charged microions, on the other hand, are included explicitly, since their Coulomb interaction is stronger than the dipolar forces acting between the solvent. While this seems to be justified for micron-sized colloidal particles with typical interparticle spacing of microns (provided molecular details on the colloidal surface are encaptured by an effective colloidal charge), it is questionable when the colloidal diameter or the intercolloidal distance is becoming comparable to the correlation lenth or interaction range of the microcopic degrees of freedom, i.e. for nanoparticles.

As an illustration, we consider the effective interaction between two nano-sized charged colloidal particles in a *hard sphere solvent*. A systematic comparison between primitive-model calculations where the solvent is neglected and the full system including the hard sphere solvent was performed recently by Allahyarov and one of us [95,87]. On an intermediate level, one can formally integrate out the solvent, ending up with effective interactions between the charged particles. If these are approximated to be pairwise, one obtains the so-called solvent averaged primitive model [96]. In this model, the interaction between charged species comprises the bare Coulomb interaction and the effective depletion interaction between hard spheres as discussed in Sect. 6.8. Extensive computer simulations [95,87] have shown that the total effective force between charged colloidal particles does depend on the presence of the discrete solvent. Even the sign of the effective interaction can be different in the primitive model as compared to the full solvent result. An example for divalent counterions and a charge asymmetry of $q_p : q_c = 64 : 2$ is shown in Fig. 6.4 where the ratio of the three hard-core diameters of colloidal particles, counterions and solvent particles is $d_p : d_c : d_s = 14 : 2 : 1$. In fact, interpreting the solvent size as a microscopic scale, the colloidal diameter is 14 times larger and thus falls into the nano-regime. While the primitive model reference calculations yield a repulsive effective force, the simulations including a hard-sphere solvent result in an attractive force.

The solvent-averaged primitive model, on the other hand, reproduces the data of the full solvent simulation rather well. The simulation time for the solvent-averaged primitive model is similar to that of the primitive model, since the number of particles simulated is the same, while the full simulation requires the inclusion of many solvent particles. Hence the concept of effective interactions as applied to the solvent degrees of freedom alone, makes simulations feasible in the spirit of McMillan-Mayer theory. But even more importantly, the solvent-averaged model also provides insight into the basic physics: the depletion attraction between a colloidal sphere and counterions favours an accumulation of the latter on the colloidal surfaces, thus enhancing the screening. It is this effect which, together with strong Coulomb correlations, leads to the attraction evident in Fig. 6.4 which is completely missing in the primitive model. For large colloidal spheres, on the other hand, it was shown in Ref. [87] that the effect of a discrete hard sphere solvent can completely be accounted for by taking a different (effective) colloidal charge as input in the

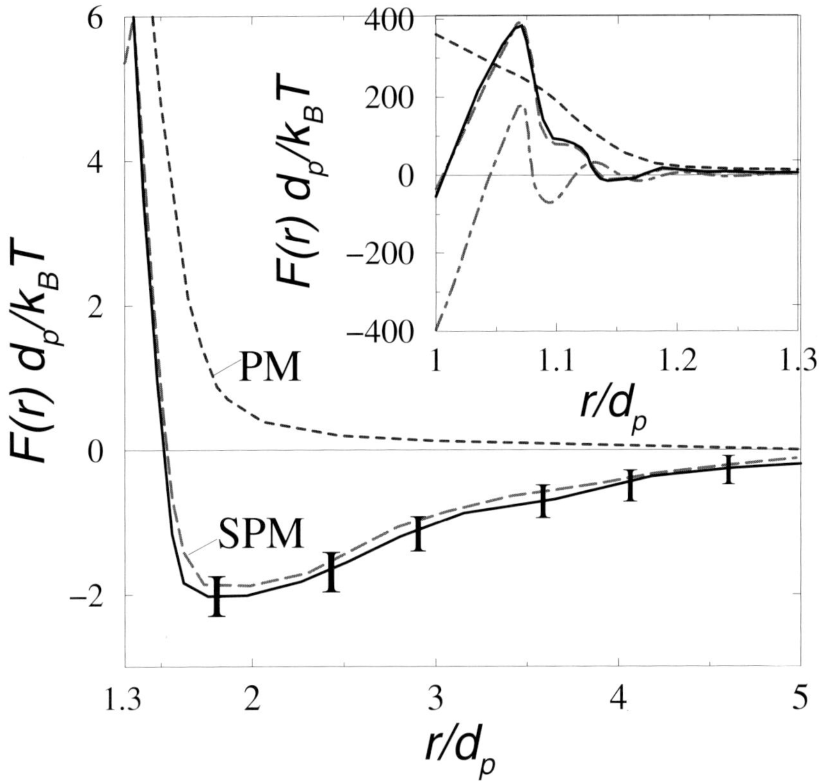

Fig. 6.4. Reduced distance-resolved force $F(r)d_p/k_BT$ versus reduced distance r/d_p between two charged colloids as obtained by computer simulations. The inset shows the same for nearly touching polyions of molecular distances. The parameters are: $q_c = 2$, $q_p = -64$, $d_p : d_c : d_s = 14 : 2 : 1$, $\epsilon = 81$, the volume fraction of the polyions in the periodically repeated box is $\eta_p = 5.8 \times 10^{-3}$. Solid line with error bars: full simulation including the hard sphere solvent; long-dashed line: solvent averaged primitive model; short-dashed line: primitive model; dot-dashed line in inset: solvent depletion force alone (for comparison).

primitive model calculations. Consequently simple coarse-graining is justified provided an effective colloidal charge is used.

The second caveat for coarse-graining concerns the description of the big colloidal particles. Up to now we have modelled them as homogeneously charged hard spheres in the context of charged suspensions. Coming down to nanoscales, the molecular details and internal degrees of freedom of the large particles themselves, which are neglected in statistical descriptions of colloidal dispersions, are becoming more and more relevant. We shall illustrate this using an example where the *discreteness of the charge pattern* on the colloidal surfaces turns out to be crucial provided the particles are nano-sized. This is of particular importance for proteins which are characterized by nanoscale electrostatic patches.

In a recent computer simulation [97], the effective interaction between two nano-sized colloidal spheres (modelling globular proteins) with a discrete charge pattern was calculated within the primitive model as a function of added salt concentration. A snapshot of a protein pair is shown in Fig. 6.5. Discrete elementary point charges were placed on the protein surface with a finite depth. Consequently the effective interaction not only depends on the centre-to-centre distance r but also on the relative orientations of the two proteins. These represent additional statistical degrees of freedom associated with the macroparticles. The data were compared to the standard description with the charge uniformly smeared over the particle surface. A key quantity controlling protein crystallization [98] is the second osmotic virial coefficient B_2 which can readily be measured by scattering methods in dilute protein solutions. B_2 can be shown [97] to be related to the effective interaction by

$$B_2 = \frac{1}{2} \int d^3 r [1 - \exp(-v(r)/k_B T)] \tag{6.29}$$

in formal analogy to the orientation-independent case of the smeared charged model. Here $v(r)$ is the integral of the canonical orientational average of the distance-resolved effective force projected onto the separation vector between the two proteins [97].

A detailed calculation of B_2 as a function of added salt reveals that it behaves non-monotonically as a function of added salt concentration, in agreement with several experimental studies [99]. This non-monotonicity, however, disappears when the surface charge uniformly smeared out. It can thus be traced back to strong Coulomb correlations induced by the discrete binding centers near the surface. This example shows that important effects are lost even qualitatively, when the coarse graining of nanoparticles is pushed too far.

6.10 Conclusions

In summary, we have demonstrated that the concept of effective interactions allows large-scale simulations and provides additional insight into the physical mechanisms governing colloidal dispersions and polymer solutions. We conclude with a discussion about multiple *time* scales in the context of colloidal dispersions.

The dynamics of colloidal particles embedded in a solvent involves many different time scales ranging from the collision time $\tau_s \approx 10^{-14} sec$ of the solvent molecules, over the relaxation time $\tau_B \approx 10^{-9} sec$ of the total colloid momentum and the propagation time $\tau_H \approx 10^{-9} sec$ of hydrodynamic interactions, to the Brownian time $\tau_0 \approx 10^{-6} sec$ on which diffusive motion of the

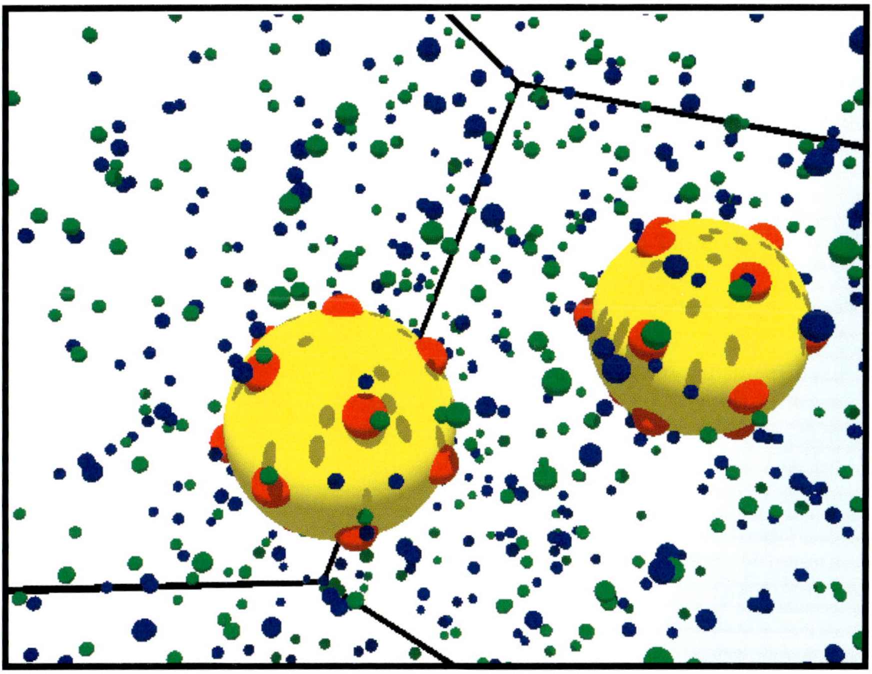

Fig. 6.5. Simulation snapshot of a microion configuration around two model proteins separated by $r = 1.7d_p$, d_p denoting the protein hard core diameter. The proteins carry 15 discrete charges e, monovalent salt molarity is $c_s = 0.206 Mol/\ell$. The globular protein molecules are shown as two large grey spheres. The embedded small dark spheres on their surface mimic the discrete protein charges. The small grey spheres are counterions, while the black spheres are coions.

colloidal particles is observed. Consequently there is almost complete time scale separation

$$\tau_s << \tau_B \simeq \tau_H << \tau_0 . \tag{6.30}$$

It is a challenging question whether - in analogy with bridging length scale gaps - one can "integrate out" the fast dynamical processes which happen on the time scale τ_s in order to arrive at an "effective dynamics" on larger time scales. The traditional approach is a stochastic one, as embodied in Langevin and Fokker-Planck formulations [114,115], but a rigorous derivation of the Fokker-Planck equation from the initial full Liouville equation for dilute colloidal suspensions exposes the intrinsic limitations of the Fokker-Planck equation, due to the similarity of the time scales τ_B and τ_H [116]. Clearly a dynamical counterpart of the effective interaction concept governing the thermodynamics and statics is missing. From a simulation point of view however, relying on an (almost) complete time scale separation between solvent and colloidal dynamics, one typically describes the motion of the colloidal particles by a completely overdamped Langevin equation with stochastic forces exerted onto the colloids modelling the random solvent kicks [100,101]. Within such a Brownian approach operating on a time scale τ_0, the hydrodynamic interactions act instantaneously. Much recent effort was spent to treat these hydrodynamic interactions approximatively [102] by using different computational schemes such as lattice-Boltzmann techniques [103], Stokesian Dynamics [104], dissipative particle dynamics [105], and fluidizing the solid colloidal particles [106]. Although these algorithms are powerful in different applications, it is fair to say that all of these approaches lack a rigorous theoretical justification.

Having established a stochastic Brownian approach for the colloidal particles resulting from the solvent dynamics, a much more modest question concerns the dynamical utility of the effective interaction concept originating from integrating out small particles different from the solvent (counter- and salt ions, polymers, small colloidal particles etc). One may conjecture that it is only in the case of a complete time scale separation between the big and small particle dynamics that the effective interaction has a true dynamical meaning. This is the reason why the effective interaction potential is frequently combined with Brownian dynamics simulations for the colloids. For instance, the effective DLVO-potential has been combined with Brownian dynamics simulations in order to investigate the glass transition [107], long-time self-diffusion [108], linear shear flow [109–111], and phase transitions in driven colloidal mixtures [112,113]. It would be very interesting to test and study systematically how far one can get with the coarse-graining approach as far as dynamical questions are concerned.

Acknowledgements

We thank E. Allahyarov, M. Dijkstra, J. Dzubiella, A. Jusufi, C. N. Likos, A. A. Louis and R. Pierre for help and for valuable comments.

References

1. V. Lobaskin, A. Lyubartsev, P. Linse, Phys. Rev. E **63**, 020401 (2001).
2. C. W. J. Beenakker, P. Mazur, Physica **126A**, 349 (1984).
3. See e.g R. van Roij, M. Dijkstra, J.P. Hansen, Phys. Rev. E **59**, 2010 (1999).
4. For an extensive review of classical DFT see R. Evans in "Fundamentals of Inhomogeneous Fluids", edited by D. Henderson (Marcel Decker, New York, 1992).
5. J. P. Hansen and E. Smargiassi, in "Monte Carlo and Molecular Dynamics of Condensed Matter Systems", edited by K. Binder and G. Ciccotti (Societa Italiana di Fisica, Bologna, 1996).
6. See e.g. G. Galli and M. Parrinello, in "Computer Simulations in Materials Science", P. 282, edited by M. Meyer and V. Pontikis (Kluwer, Dordrecht, 1991)

7. H. Löwen, J. P. Hansen, P. A. Madden, J. Chem. Phys. **98**, 3275 (1993).
8. F. Gygi, G. Galli, Phys. Rev. B **52**, R 2229 (1995).
9. S. Melchionna, J. P. Hansen, Phys. Chem. Chem. Phys. **2**, 3465 (2000).
10. R. Roth, R. Evans, S. Dietrich, Phys. Rev. E **62**, 5360 (2000).
11. D. Goulding, S. Melchionna, Phys. Rev. E **64**, 011403 (2001).
12. J. P. Hansen, H. Löwen, Ann. Rev. Phys. Chem. **51**, 209 (2000).
13. L. Belloni, J. Phys. Condens. Matter **12**, R549 (2000).
14. B. V. Derjaguin, L. D. Landau, Acta Physicochim. USSR **14**, 633 (1941); E. J. W. Verwey and J. T. G. Overbeek, *Theory of the Stability of Lyophobic Colloids.* (Elsevier, Amsterdam, 1948).
15. A. Delville, R. J. M. Pellenq, Molecular Simulation **24**, 1 (2000); R. Messina, C. Holm, K. Kremer, Phys. Rev. Lett. **85**, 872 (2000); T. Terao, T. Nakayama, Phys. Rev. E **63**, 041401 (2001); B. Hribar, V. Vlachy, Langmuir **17**, 2043 (2001).
16. A. Delville, J. Phys. Chem. B **103**, 8296 (1999); A. Delville, P. Levitz, J. Phys. Chem. B **105**, 663 (2001)
17. S. Meyer, P. Levitz, A. Delville, J. Phys. Chem. B **105**, 10684 (2001).
18. H. Löwen, E. Allahyarov, J. Phys.: Condensed Matter **10**, 4147 (1998).
19. R. D. Groot, J. Phys. **95**, 9191 (1991).
20. M. J. Stevens, M. L. Falk, M. O. Robbins, J. Chem. Phys. **104**, 5209 (1996).
21. H. Löwen, G. Kramposthuber, Europhys. Lett. **23**, 673 (1993).
22. H. Löwen, Phys. Rev. Lett. **72**, 424 (1994). J. Chem. Phys. **100**, 6738 (1994).
23. S. Kutter, J. P. Hansen, M. Sprik, E. Boek, J. Chem. Phys. **112**, 311 (2001).
24. M. Dijkstra, Current Opinion in Colloid and Interface Science **6**, 372 (2001).
25. For a review, see T. Tomasi , M. Perico, Chem. Rev. **94**, 2027 (1994).
26. M. Marchi, D. Borgis, N. Levy, P. Ballone, J. Chem. Phys. **114**, 4377 (2001).
27. D. M. York, M. Karplus, J. Phys. Chem. A **103**, 11060 (1999).
28. R. A. Marcus, J. Chem. Phys. **24**, 966, 979 (1956).
29. B. U. Felderhof, J. Chem. Phys. **67**, 493 (1977).
30. R. Allen, J. P. Hansen, S. Melchionna, Phys. Chem. Chem. Phys. **3**, 4177 (2001).
31. R. Allen, S. Melchionna, J. P. Hansen to be published.
32. C. Mischler, J. Baschnagel, K. Binder, Advances in Colloid and Interface Science **94**, 197 (2001).
33. P. J. Flory, W. R. Krigbaum, J. Chem. Phys **18**, 1086 (1950).
34. A. Y. Grosberg, P. G. Khalatur, A. R. Khokhlov, Makromol. Chem. Rapid Comm. **3**, 709 (1982).
35. J. Dautenhahn, C. K. Hall, Macromolecules **27**, 5399 (1994), and references therein.
36. A. A. Louis, P. G. Bolhuis, J. P. Hansen, E. J. Meijer, Phys. Rev Letters **85**, 2522 (2000) P. G. Bolhuis, A. A. Louis, J. P. Hansen, E. J. Meijer, J. Chem. Phys. **114**, 4296 (2001).
37. B. Krüger, L. Schäfer, A. Baumgärtner, J. Phys. (Paris) **50**, 319 (1989).
38. V. Krakoviack, A. A. Louis, J. P. Hansen, to be published.
39. C. H. Bennett, J. Comp. Phys. **22**, 245 (1976).
40. D. Frenkel, B. Smit, "Understanding Molecular Simulation", 2^d Edition (Academic Press, San Diego, 2001).
41. P. G. Bolhuis, A. A. Louis, J. P. Hansen, Phys. Rev. E **64**, 021801 (2001).
42. C. von Ferber, A. Jusufi, C. N. Likos, H. Löwen, M. Watzlawek, Europhys. Journal E **2**, 311 (2000).

43. See e.g. J. P. Hansen, I. R. McDonald, "Theory of Simple Liquids" 2^d edition (Academic Press, London, 1986).
44. A. A. Louis, P. G. Bolhuis, J. P. Hansen, Phys. Rev. E **62**, 7961 (2000).
45. C. N. Likos, Physics Reports **348**, 267 (2001).
46. R. L. Henderson, Phys. Lett. **49A**, 197 (1974).
47. A. Jusufi, J. Dzubiella, C. N. Likos, C. von Ferber, H. Löwen, J. Phys.: Condensed Matter **13** 6177 (2001).
48. A. A. Louis, P. G. Bolhuis, E. J. Meyer, J. P. Hansen, cond-mat (011518).
49. For a review see K. S. Schweizer, J. G. Curro, Adv. Chem. Phys. **98**, 1 (1997).
50. V. Krakoviack, J. P. Hansen, A. A. Louis, cond-mat / 0110387
51. For a review see: G. S. Grest, L. J. Fetters, J. S. Huang, D. Richter, Advances in Chemical Physics, Volume XCIV, 67 (1996)).
52. T. A. Witten, P. A. Pincus, Macromolecules **19**, 2509 (1986).
53. M. Dauod, J. P. Cotton, J. Phys. (Paris) **38**, 983 (1977).
54. C. N. Likos, H. Löwen, M. Watzlawek, B. Abbas, O. Jucknischke, J. Allgaier, D. Richter, Phys. Rev. Letters **80**, 4450 (1998).
55. A. Jusufi, M. Watzlawek, H. Löwen, Macromolecules **32**, 4470 (1999).
56. M. Watzlawek, C. N. Likos, H. Löwen, Phys. Rev. Letters. **82**, 5289 (1999).
57. M. Watzlawek, H. Löwen, C. N. Likos, J. Phys. Condensed Matter **10**, 8189 (1998).
58. G. A. McConnell, A. P. Gast, Macromolecules **30**, 435 (1997).
59. S. T. Milner, T. A. Witten, M. E. Cates, Macromolecules **21**, 2610 (1988).
60. J. Mewis, W. J. Frith, T. A. Strivens, W. B. Russel, A. I. Ch. E. J. **35**, 415 (1989).
61. U. Genz, B. D'Aguanno, J. Mewis, R. Klein, Langmuir **10**, 2206 (1994).
62. C. N. Likos, H. Löwen, A. Poppe, L. Willner, J. Roovers, B. Cubitt, D. Richter, Phys. Rev. E **58**, 6299 (1998).
63. A. Jusufi, C. N. Likos, H. Löwen, Phys. Rev. Letters (in press).
64. P. Pincus, Macromolecules **24**, 2912 (1991).
65. C. N. Likos, M. Schmidt, H. Löwen, M. Ballauff, D. Pötschke, P. Lindner, Macromolecules **34**, 2914 (2001).
66. C. N. Likos, private communication.
67. A. Lang, C. N. Likos, M. Watzlawek, H. Löwen, J. Phys. Condensed Matter **12**, 5087 (2000).
68. S. Asakura, F. Oosawa, J. Chem. Phys. **22**, 1255 (1954).
69. M. Dijkstra, J. M. Brader, R. Evans, J. Phys.: Condensed Matter **11**, 10079 (1999).
70. M. Dijkstra, to be published.
71. I. Götze, Diploma Thesis, University of Düsseldorf, 2002.
72. H. N. W. Lekkerkerker, W. C. K. Poon, P. N. Pusey, A. Stroobants, P. B. Warren, Europhys. Lett. **20**, 559 (1992).
73. M. Schmidt, H. Löwen, J. M. Brader, R. Evans, Phys. Rev. Letters **85**, 1934 (2000).
74. Y. Rosenfeld, M. Schmidt, H. Löwen, P. Tarazona, Phys. Rev. E **55**, 4245 (1997).
75. M. Schmidt, H. Löwen, J. M. Brader, R. Evans, J. Phys.: Condensed Matter **14**, L1 (2002).
76. A. A. Louis, R. Finken, J. P. Hansen, Phys. Rev. E **61**, R1028 (2000).
77. E. J. Meijer, D. Frenkel, Physica A **213**, 130 (1995).

78. A. Johner, J. F. Joanny, S. Diez Orrite, J. BonetAvalos, Europhys. Letters **56**, 549 (2001).
79. J. Dzubiella, A. Jusufi, C. N. Likos, C. von Ferber, H. Löwen, J. Stellbrink, J. Allgaier, D. Richter, A. B. Schofield, P. A. Smith, W. C. K. Poon, P. N. Pusey, Phys. Rev. E **64**, 01040 (2001).
80. J. Dzubiella, C. N. Likos, H. Löwen, to be published.
81. M. Fuchs, K. S. Schweizer, Phys. Rev. E **64**, 021514 (2001).
82. A. Hanke, E. Eisenriegler, S. Dietrich, Phys. Rev. E **59**, 6853 (1999).
83. H. Xu, M. Baus, J. Phys.: Condensed Matter **4**, L663 (1992).
84. M. D. Eldridge, P. A. Madden, D. Frenkel, Mol. Phys. **79**, 105 (1993).
85. P. Bartlett, R. H. Ottewill, P. N. Pusey, Phys. Rev. Lett. **68**, 3801 (1992).
86. R. Dickman, P. Attard, V. Simonian, J. Chem. Phys. **107**, 205 (1997).
87. E. Allahyarov, H. Löwen, Phys. Rev. E **63**, 041403 (2001).
88. J. C. Crocker, J. A. Matteo, A. D. Dinsmore, A. G. Yodh, Phys. Rev. Letters **82**, 4352 (1999).
89. M. Dijkstra, R. van Roij, R. Evans, Phys. Rev. E **59**, 5744 (1999).
90. N. G. Almarza, E. Enciso, Phys. Rev. E **59**, 4426 (1999).
91. P. Bolhuis, M. Hagen, D. Frenkel, Phys. Rev. E **50**, 4880 (1994).
92. C. N. Likos, Z. T. Németh, H. Löwen, J. Phys. Condensed Matter **6**, 10965 (1994).
93. D. Goulding, J. P. Hansen, Mol. Phys. **99**, 865 (2001).
94. A. A. Louis, R. Roth, E. Allahyarov, H. Löwen, to be published.
95. E. Allahyarov, H. Löwen, J. Phys. Condensed Matter **13**, L277 (2001).
96. H. Löwen, E. Allahyarov, J. Dzubiella, C. von Ferber, A. Jusufi, C. N. Likos, M. Heni, Philos. Trans. R. Soc. Lond. A **359**, 909 (2001).
97. E. Allahyarov, H. Löwen, A. A. Louis, J. P. Hansen, *Discrete charge patterns, Coulomb correlations and interactions in protein solutions,* Europhys. Letters (in press).
98. A. George, W. W. Wilson, Acta Cryst. D **50**, 361 (1994); G. A. Vliegenthart, H. N. W. Lekkerkerker, J. Chem. Phys. **112**, 5364 (2000).
99. B. Guo, S. Kao, H. McDonald, A. Asanov, L. L. Combs, W. W. Wilson, J. Cryst. Growth **196**, 424 (1999); D. N. Petsev, B. R. Thomas, S. T. Yau, P. G. Vekilov, Biophysical Journal **78**, 2060 (2000).
100. For a review, see P. N. Pusey, in "Liquids, Freezing and the Glass Transition", edited by J. P. Hansen, D. Levesque and J. Zinn-Justin (North Holland, Amsterdam, 1991).
101. J. K. G. Dhont, "An Introduction to Dynamics of Colloids", Elsevier, Amsterdam, 1996.
102. G. Nägele, Physics Reports, **272**, 215 (1996).
103. A. J. C. Ladd, R. Verberg, J. Stat. Phys. **104**, 1191 (2001).
104. See e.g.: R. Pesché, G. Nägele, Europhys. Lett. **51**, 584 (2000); H. M. Schaink, P. A. Nommensen, R. J. J. Jongschaap, J. Mellema, J. Chem. Phys. **113**, 2484 (2000).
105. P. B. Warren, Current Opinion in Colloid and Interface Science **3**, 620 (1998).
106. H. Tanaka, T. Araki, Phys. Rev. Lett. **85**, 1338 (2000).
107. H. Löwen, J. P. Hansen, J. N. Roux, Phys. Rev. A **44**, 1169 (1991).
108. H. Löwen, G. Szamel, J. Phys. Condensed Matter **5**, 2295 (1993).
109. S. Butler, P. Harrowell, J. Chem. Phys. **103**, 4653 (1995); J. Chem. Phys. **105**, 605 (1996).

110. S. R. Rastogi, N. Wagner, S. R. Lustig, J. Chem. Phys. **104**, 9234 (1996).
111. N. Olivi-Tran, R. Botet, B. Cabane, Phys. Rev. E **57**, 1997 (1998).
112. H. Löwen, G. Hoffmann, Phys. Rev. E **60**, 3009 (1999).
113. J. Dzubiella, G. P. Hoffmann, H. Löwen, *Lane formation in binary colloidal mixtures driven by an external field,* Phys. Rev. E (in press).
114. N. G. van Kampen, "Stochastic Processes in Physics and Chemistry (North Holland, Amsterdam, 1990).
115. W. Hess and R. Klein, Adv. Phys. **32**, 173 (1983).
116. For a review, see L. Bocquet and J. P. Hansen, in "Dynamics: Models and Kinetic Methods for Non-Equilibrium Many-Body Systems", edited by J. Karkheck (Kluwer Academic, Dordrecht, 2000).

Part V

Slow Dynamics and Reactivity

7 Simulation of Models for the Glass Transition: Is There Progress?

Kurt Binder[1], Jörg Baschnagel[2], Walter Kob[3], and Wolfgang Paul[1]

[1] Institut für Physik, Johannes-Gutenberg-Universität Mainz, D-55099 Mainz, Germany
[2] Institut Charles Sadron, Université Louis Pasteur, 6, rue Bossingault, F-67083 Strasbourg, France
[3] Laboratoire des Verres, Université Montpellier II, F-34095 Montpellier, France

Abstract. The glass transition of supercooled fluids is a particular challenge for computer simulation, because the (longest) relaxation times increase by about 15 decades upon approaching the transition temperature T_g. Brute-force molecular dynamics simulations, as presented here for molten SiO_2 and coarse-grained bead-spring models of polymer chains, can yield very useful insight about the first few decades of this slowing down. Hence this allows to access the temperature range around T_c of the so-called mode coupling theory, whereas the dynamics around the experimental glass transition is completely out of reach. While methods such as "parallel tempering" improve the situation somewhat, a method that allows to span a significant part of the region $T_g \leq T \leq T_c$ is still lacking. Only for abstract models such as the infinite range 10-state Potts glass with a few hundred spins this region can be explored. However this model suffers from very strong finite size effects thus making it difficult to extrapolate the results obtained for the finite system sizes to the thermodynamic limit.

For the case of polymer melts, two different strategies to use lattice models instead of continuum models are discussed: In the first approach, a mapping of an atomistically realistic model of polyethylene to the bond fluctuation model with suitable effective potentials and a temperature-dependent time rescaling factor is attempted. In the second approach, devoted to a test of the entropy theory, moves that are artificial but which lead to a faster relaxation ("slithering snake" algorithm) are used, to get at least static properties at somewhat lower temperatures than possible with a "realistic" dynamics. The merits and shortcomings of all these approaches are discussed.

7.1 Introduction

The reason for the slowing down of the dynamics of supercooled liquids and the resulting glass transition to an amorphous solid is one of the biggest unsolved problems in the physics of condensed matter [1–5] and it is also a particular challenge for computer simulation [6–12]. The present introductory section intends to remind the reader on the main experimental facts and some theoretical ideas about the glass transition, and will also serve to make clear why in this problem there exists a gap of time-scales that simulations need to bridge.

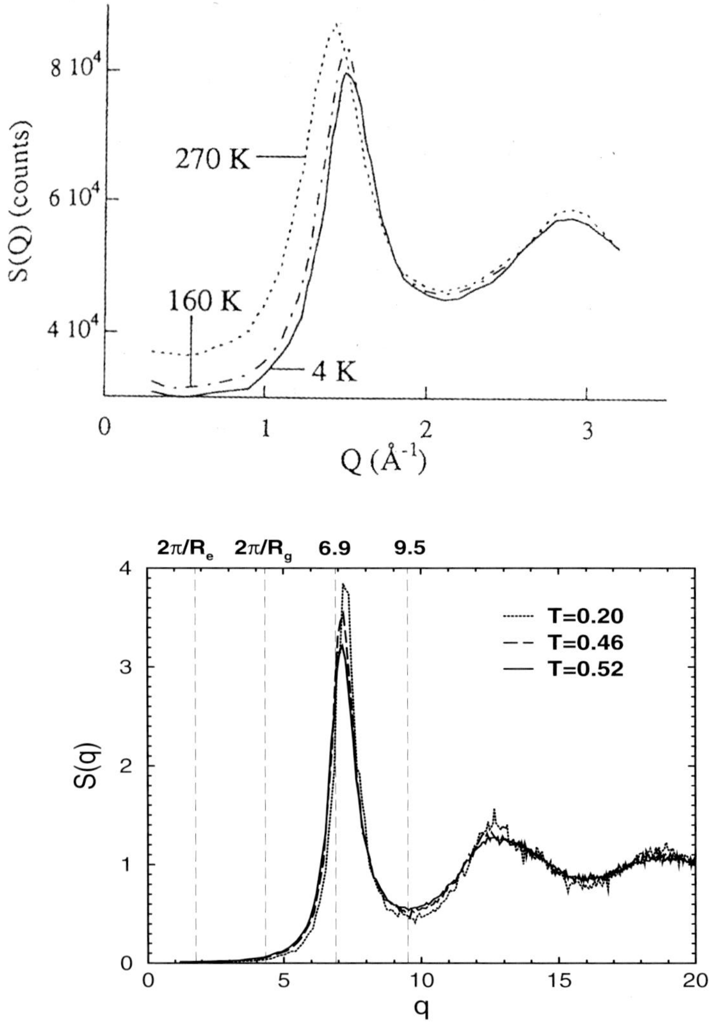

Fig. 7.1. a) Static collective structure factor of polybutadiene at temperatures $T = 4\text{K}$, $T = 16\text{K}$, and $T = 270\text{K}$. Note that for this system the glass transition temperature is $T_g = 180K$ and the critical temperature of mode coupling theory [3] is $T_c = 220\text{K}$. The scattering background is not subtracted here, thus the zero of the ordinate axis is not known precisely, and the ordinate units are just measuring absolute scattering intensities. From Arbe *et al.* [13]. b) Static collective structure factor $S(q)$ plotted versus wave-vector q, for a bead-spring model of flexible polymer chains with chain length $N = 10$. Beads interact with the potential given in Eqs. (7.11)-(7.13). and lengths are measured in units of σ, temperatures in units of ε. Three temperatures $T = 0.2$, 0.46 and 0.52 are shown (note that $T_g \approx 0.41$ and $T_c \approx 0.45$ for this model). The vertical lines highlight characteristic inverse length scales (related to the end-to-end distance R_e and radius of gyration R_g of the chains as well as the first maximum and minimum of $S(q)$). From Baschnagel el at. [14].

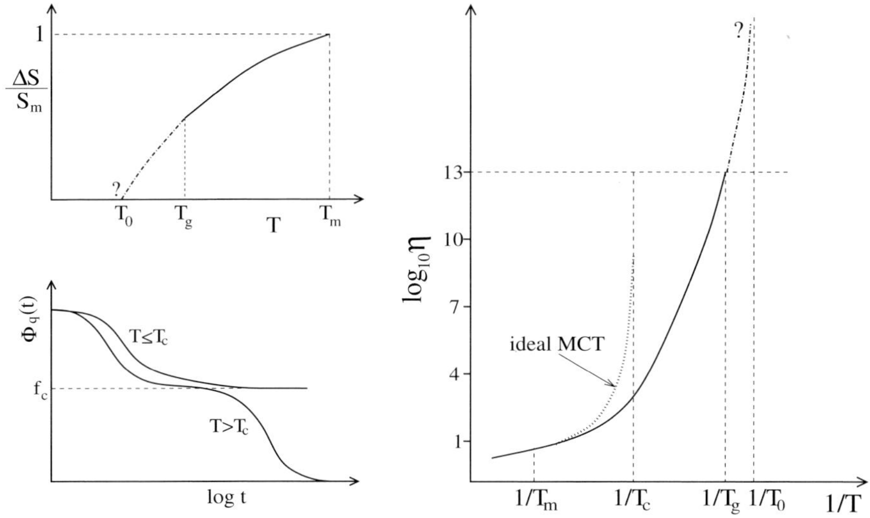

Fig. 7.2. Right figure: Schematic plot of the viscosity $\eta(T)$ of a fluid (note $\eta(T) \propto \tau$) vs. inverse temperature $1/T$. The location of the melting temperature (T_m), the critical temperature of mode coupling theory (T_c) [3], the glass transition temperature (T_g) and the Vogel-Fulcher-Kauzmann temperature [1,15] (T_0) are shown on the abscissa. The glass transition temperature T_g is defined empirically requiring [1] $\eta(T = T_g) = 10^{13}$ Poise. Two complementary concepts to explain the glass transition are indicated by the schematic plots on the left: The lower figure shows the time correlation function $\Phi_q(t)$ for density fluctuations at wave-vector q which according to idealized mode coupling theory shows at a temperature T_c a nonzero "non-ergodicity parameter" f_c [3]. For T somewhat larger than T_c, $\Phi_q(t)$ exhibits a plateau and the "lifetime" τ of this plateau (as well as η) diverge as one approaches T_c [3]. The upper figure shows the entropy difference $\Delta S(T) = S_{\text{fluid}} - S_{\text{crystal}}$, with $S_m \equiv \Delta S(T_m)$. The linear extrapolation of ΔS for $T < T_g$ defines the Kauzmann temperature T_0 via $\Delta S(T = T_0) = 0$ [15]. Adapted from Binder [16].

As is well known, it is already a problem to characterize the static structure of a glass: the structure of an amorphous material is not regular like a crystalline solid, but shows only short range order similar to a liquid. However, the latter flows, while the amorphous solid is rigid! In fact, if one makes a scattering experiment, it is hard to distinguish from the structure whether one has a fluid above the glass transition temperature T_g or a solid below T_g (Fig. 7.1) [13,14]. If one approaches T_g, the structural relaxation time τ which is related to the viscosity $\eta(T)$, for instance - increases smoothly by up to 15 decades, as shown schematically in Fig. 7.2, without any accompanying significant structural change detectable by scattering experiments (Fig. 7.1). This increase of $\eta(T)$ is often fitted to the Vogel-Fulcher relation [1]

$$\eta(T) \propto \exp[E_{VF}/(T - T_{VF})] \,, \tag{7.1}$$

where E_{VF} is an effective activation barrier. From this functional form it is clear that $\eta(T)$ is predicted to diverge at the Vogel-Fulcher temperature T_{VF}, which is lower than T_g, of course, if one invokes the empirical definition of T_g via $\eta(T = T_g) = 10^{13}$ Poise [1]. However, it is questionable whether the temperature dependence of η as given by (7.1) really holds.

Another very common relation often used to describe various relaxation functions of glassforming fluids is the Kohlrausch-Williams-Watts function, also called stretched exponential, [1],

$$\varphi(t) \propto \exp[-(t/\tau)^{\beta}] \, . \tag{7.2}$$

This relation involves an exponent $\beta \leq 1$, whose precise physical significance is somewhat obscure. Again it is unknown under which circumstances (if any) (7.2) is exact, and in which it is just a convenient fitting formula to represent data.

Often it is claimed that the glass transition is a purely kinetic phenomenon, and if one would be able to wait long enough (which could mean times like the age of the universe, however!) one could see that glass is not really a solid but still a fluid that flows. However, this idea is not generally accepted, since there are some indications that there may be an underlying quasi-equilibrium phase transition between metastable phases, namely from the supercooled metastable fluid to an (also metastable) ideal glass phase (the stable phase for temperatures lower than the melting temperature T_m is of course the crystal). Such an indication is Kauzmann's entropy paradox [15]: By studying the difference in entropy between liquid and crystal one finds that near T_g the difference $\Delta S(T) = S_{\text{fluid}} - S_{\text{crystal}}$ has decreased to about 1/3 of its value S_m at the melting/crystallization temperature T_m. If this trend is extrapolated (linearly in T) to even lower temperatures, $\Delta S(T)$ would become negative below the Kauzmann temperature T_0 (which is usually quite close to the Vogel-Fulcher-temperature T_{VF}), see Fig. 7.2. It would indeed be paradox if the entropy of the supercooled fluid (with its disordered structure) were less than the entropy of the ordered solid! One possibility to bypass the problem is to assume that this "entropy catastrophe" is avoided by a phase transition at T_0 (or at some temperature in between T_0 and T_g). In fact, for the glass transition of polymer melts Gibbs and Di Marzio [17] proposed an approximate theory that shows such a vanishing of the entropy at T_0, and subsequently Adam and Gibbs [18] suggested arguments to show that (7.1) holds with $T_{VF} = T_0$. However, although these concepts enjoy some popularity, all these arguments are based on very crude and hardly justifiable assumptions and approximations, and hence they are not accepted by many researchers. For instance, the mode coupling theory of the glass transition (MCT)[3] claims that there is indeed an underlying transition but this is *not* a phase transition in the sense of thermodynamics but rather a "dynamical transition" from an ergodic to a nonergodic behavior. This transition should occur at a critical temperature T_c and can be seen in the form of the time dependence of the correlation function of density fluctuations or its Fourier transform $\Phi_q(T)$, see Fig. 7.2. Above T_c, this correlator decays to zero, but as T_c is approached a plateau develops whose "lifetime" gets larger and larger until it diverges, in the ideal case: the system gets "stuck", the decay of $\Phi_q(t)$ stops at the "nonergodicity parameter" f_c, an order parameter for the glass transition that appears discontinuously at T_c. The physical idea behind this theory is the "cage picture": the motion of any atom in a dense fluid is constrained by its neighbors, which form a cage around it. At low enough temperatures the escape out of the cage gets blocked. MCT predicts that close to this dynamical transition τ and $\eta(T)$ show a power-law divergence as one approaches T_c,

$$\tau \propto \eta(T) \propto (T - T_c)^{-\gamma} . \qquad (7.3)$$

In reality this dependence is, however, observed only in a limited temperature interval. The way out of this dilemma is the argument that one must not neglect (as "idealized" mode coupling theory does [3]) thermally activated processes, so-called "hopping processes", by which atoms supposedly can escape from their cage when T is less than T_c. The theory then claims [19] that a simple Arrhenius behavior results in this region, $\log \tau \propto 1/T$ for $T < T_c$, and in the vicinity of T_c the power-law divergence of (7.3) is rounded off to a smooth crossover from the power law to the Arrhenius divergence. Thus this theory does not involve any phase transition, there is just a smooth crossover from one type of dynamical behavior to another one near T_c, and T_g means that relaxation times have grown so large that the system falls out of equilibrium.

In real systems this crossover seems to occur at a viscosity somewhere between 10 and 10^3 Poise, i.e. a time window that can be explored with molecular dynamics (MD) simulations. Hence such simulations are able to investigate the beginning of the approach to the critical temperature T_c which MCT describes [3] and hence are very useful to check the validity of this theory. However, the following 10 decades of the viscosity between T_c and T_g are out of reach for MD simulations so far. Unfortunately this is precisely the region that one needs to explore, for a definite distinction between the theories!

Thus although straightforward atomistic MD methods [20,21] clearly face a dilemma, we shall nevertheless describe how far one can push this type of approach, choosing SiO_2 as an example (Sect. 7.2). A method to extend the range of times and accessible temperatures somewhat is the concept of "parallel tempering" [22–26], and this approach and its problems will be presented in Sect. 7.3. For the sake of contrast, Sect. 7.4 will then describe the 10-state Potts glass model. Although this model is only an abstract caricature for a real glass, it has the merit that quite a few results are known analytically and that Monte Carlo simulations are possible at T_c and even at lower temperatures, if one considers only systems of a few hundred Potts spins. The disappointing aspect is, Sect. 7.4, that even in this very idealized case one learns relatively little about the glass transition of the infinite system, since one has to fight against dramatic finite size effects [27–30]!

Then we shall describe briefly (Sect. 7.5) a coarse-grained model of short polymer chains [14,31–37]. This beadspring model is quite successful in reproducing a number of experimental results qualitatively, as already exemplified in Fig. 7.1. The cooling rates that one can reach are about 3 orders of magnitude smaller than for SiO_2. Thus the model is very useful for testing mode coupling theory [36,37]. However, also for this system there is actually only a dim hope that one can get distinctly below the critical temperature T_c! With respect to that it may be better to work with the so-called bond fluctuation model on the lattice [6,38–47] - a system which allows to equilibrate melts at low temperatures with artificial moves [44–47]. By simulations of this model also the configurational entropy and its temperature dependence can be extracted [44,45] and thus it can be shown that the entropy theory of Gibbs and Di Marzio [17] is rather inaccurate and misleading (Sect. 7.6). Finally, an interesting variant (Sect. 7.7) of the bond fluctuation model will be considered. Here one uses effective potentials that are constructed such that a real material is mimicked, e.g. polyethylene [48]. This trick allows that part of the problem of bridging the time scales is taken care of by a "time rescaling factor" [48], a special translation factor between the physical time and the Monte Carlo time. Sect. 7.8 then will summarize some of the conclusions emerging from all this work.

7.2 Towards the Simulation of Real Glassy Materials: The Case of SiO_2

Molten SiO_2 is a prototype of a network glassformer. Furthermore it is a system that is well suited for molecular dynamics simulations since a very well-tested pair potential based on quantum-chemical calculation has been developed [49]. By a suitable combination of long-range Coulomb interactions and short range forces, chosen in the form

$$V_{ij}(r) = \frac{q_i q_j e^2}{r} + A_{ij} \exp(-B_{ij} r) - C_{ij}/r^6 \qquad \text{with } i,j \in \{\text{Si,O}\}, \qquad (7.4)$$

the effective interaction between the ions can be described reliably. Here e is the charge of an electron, $q_{\text{O}} = -1.2$, $q_{\text{Si}} = 2.4$, and the values of the parameters A_{ij}, B_{ij} and C_{ij} can be found in Ref. [49]. This potential is able to describe the formation of covalent bonds without the explicit assumption of three-body forces, whose calculation would be very time consuming. Due to the long range of the electrostatic interactions, Ewald summation techniques have to be used, while the short range part of the potential can be cut off at a suitable radius r_c. It turns out that $r_c = 5.5$Å yields good results [50]. The MD time step, however, must be chosen relatively small, namely $\delta t = 1.6$fs.

Note that the presence of the long range Coulombic interactions make the calculation of the forces still a quite CPU-intensive task. Furthermore one has to average the results over several independent runs in order to improve the statistics.

In a first set of simulations, the method used to cool the sample was very similar to the procedure used in glass factories where the temperature is reduced linearly with time t, starting from some initial temperature T_i such that $T(t) = T_i - \gamma t$, where γ is the cooling rate [50]. The main difference between the simulation and the cooling of the real material are the actual numbers used here: The initial temperature that had to be chosen in the simulation was very high, $T_i = 7000$K, and cooling rates were between $\gamma = 10^{15}$K/s and $\gamma = 10^{12}$K/s [50]. In contrast the glass factory uses typical initial temperatures of 1600K and cooling rates of 1K/s or even less, so the simulation is at least 12 orders of magnitude off! Despite these extremely high cooling rates - which are inevitable due to the heavy computational burden - the generated structures are qualitatively reasonable. In particular one obtains random tetrahedral networks in which almost all Si atom sit in the center of a tetrahedron and most of the O atoms sit at the corners.

Earlier investigators (for a review see [7,8,50]) were so bold to claim that such glass structures are identical to those occurring in nature, denying any significant dependence on cooling rate. However, as we have shown [50], such a claim is foolish since one sees a pronounced dependence on cooling rate in many quantities, including the structure. As a typical example we show in Fig. 7.3 how the distribution of the length n of rings depends on γ[50]. (See the figure caption for a definition of this length.) It is seen that over the range of γ that is accessible there is a significant increase of $P(n = 6)$ and a significant decrease of $P(n = 3)$ and $P(n = 4)$, while $P(n = 5)$, $P(n = 7)$ and $P(n = 8)$ almost stay constant. Clearly an extrapolation of such data to the physically relevant cooling rate $\gamma = 1$K/s is very difficult, and perhaps not yet possible: Perhaps $P(n = 3)$ and $P(n = 4)$ are already practically equal to zero for $\gamma = 1$K/s - we don't really know. Even simple quantities, such as the density of the glass at low temperatures, are hard to predict reliably. (This problem is also complicated by the fact that molten SiO_2 has at relatively high temperatures a density anomaly where the thermal expansion coefficient changes sign.)

A particular dramatic failure with extrapolations to lower values of γ was encountered in an attempt to determine the cooling rate dependence of the glass transition temperature $T_g(\gamma)$. As is done in many experimental studies, one can fit a smooth function of temperature to the liquid branch of the enthalpy (where the melt has not yet fallen out of equilibrium) and another smooth function to the glass branch of the enthalpy, and estimate $T_g(\gamma)$ from the temperature where these two branches intersect. Fig. 7.4 shows a plot of $T_g(\gamma)$ versus γ - note the logarithmic scale for γ! - for the simulation of SiO_2. One sees, first of all, that there is a very strong dependence of $T_g(\gamma)$

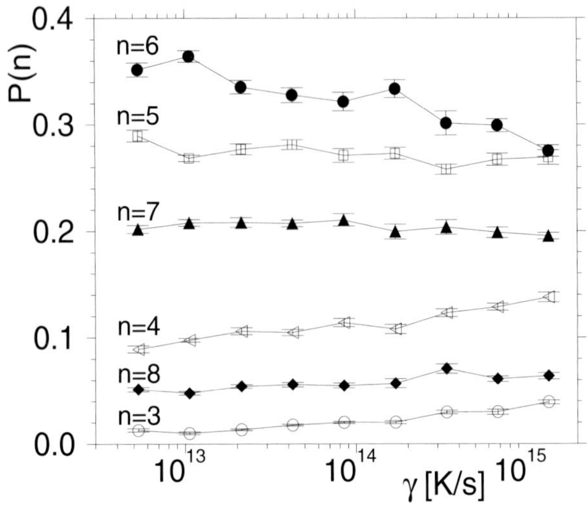

Fig. 7.3. Cooling rate dependence of the probability $P(n)$ that in the network structure of SiO_2 a ring of size n is present. A ring is defined as the shortest connection of consecutive Si–O elements that form a closed loop and n is the number of these segments. In this simulation we used 668 oxygen and 334 Si-atoms, and cooled the sample at constant pressure $p = 0$ in an NpT simulation, cooling from the initial temperature $T_i = 7000$K to the final temperature $T = 0$ K. An average over 10 independent runs was performed, allowing to estimate the statistical errors given in the figure. From Vollmayr *et al.* [50].

on γ, with $T_g(\gamma) \approx 4000$K for $\gamma = 10^{15}$K/s, while $T_g(\gamma)$ has decreased down to about $T_g(\gamma) \approx 2900$K for $\gamma = 10^{13}$K/s. In this range of cooling rates, the dependence of $T_g(\gamma)$ is not linear in $\log(\gamma)$. Nonlinear variations of $T_g(\gamma)$ that are qualitatively similar to those of Fig. 7.4 have been reported in the experimental literature, too [51], and are typically described by

$$T_g(\gamma) = T_{VF} - B/[\log(\gamma A)] \tag{7.5}$$

This dependence can be justified by assuming that the fluid falls out of its (metastable) equilibrium when the time constant of the cooling, γ^{-1}, equals the structural relaxation time $\tau(T)$ at $T = T_g(\gamma)$, and by using the Vogel-Fulcher law from (7.1) for $\tau(T)$, $\tau(T) = A\exp[B/(T - T_{VF})]$. Obviously, (7.5) does provide a very good fit to the data of the SiO_2 simulation, but the resulting $T_{VF} = 2525$K is rather unreasonable: Remember that the *experimental* glass transition temperature is $T_g \approx 1450$K, the melting temperature of crystalline SiO_2 is around 2000K, and T_{VF} should be significantly lower than T_m and even somewhat below T_g, cf. Fig. 7.2. We emphasize here that the failure of the simulation to predict $T_g(\gamma = 1\text{K/s})$ is not primarily due to the inaccuracy of the pair potential since, as will be explained in detail below, a different analysis of SiO_2 simulation data yields much more reasonable results. The failure implied by Fig. 7.4 simply comes from the fact that the

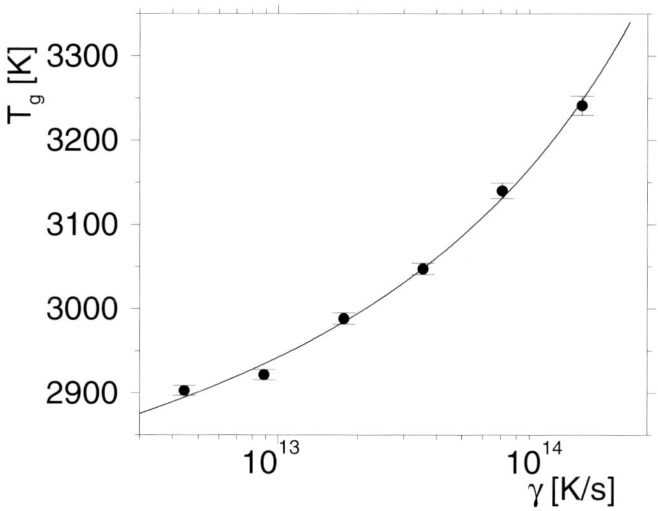

Fig. 7.4. Effective glass transition temperature $T_g(\gamma)$ plotted vs. the cooling rate γ, for molecular dynamics simulations of SiO_2 using the BKS potential [49] and estimating $T_g(\gamma)$ from intersection points of fit functions to the enthalpy, as described in the text. All data are based on averages over 10 statistically independent runs. The curve is a fit to the function given in (7.5) of the text, resulting in $T_{VF} = 2525K$. From Vollmayr *et al.* [50].

10-1000 picosecond timescale that is basically probed here is too many orders of magnitude off from the time scale relevant for the glass transition and that therefore an extrapolation of the results becomes a insecure undertaking.

A better way to study amorphous silica is to fix density at a reasonable value, for instance the experimental value, and equilibrate the system at a temperature which is as low as possible. Present day simulations can propagate a system of around 8000 ions over a time span of around 20ns which allows for a full equilibration at $T = 2750$K [52]. Longer time are accessible for smaller systems. However, it was found that if one has fewer than $O(10^3)$ ions the results are plagued with finite size effects [53]. Simulating a large system over this time scale are on the forefront of what is feasible today, and require the use of multi-processor super computers such as CRAY-T3E, making use of a parallelization of the force calculation [52–54].

This well-equilibrated melt can then be used as a starting condition for a cooling run at constant density. The advantage of this procedure is that a state at $T = 2750K$ at the correct density is much closer in local structure to the real glass, than the structures generated by the procedures described above, and hence the spurious effects of the by far too rapid quench are much less pronounced. This conclusion is corroborated by a comparison of the simulated structure factor with experiment [55], see Fig. 7.5. Given the fact that the comparison in Fig. 7.5 does not involve any adjustable parameter whatsoever, the agreement between simulation and experiment is quite remarkable, and this reiterates our above conclusion that the potential used {Eq. (7.4)} is accurate enough, and should not be blamed for discrepancies as discussed in connection with Fig. 7.4.

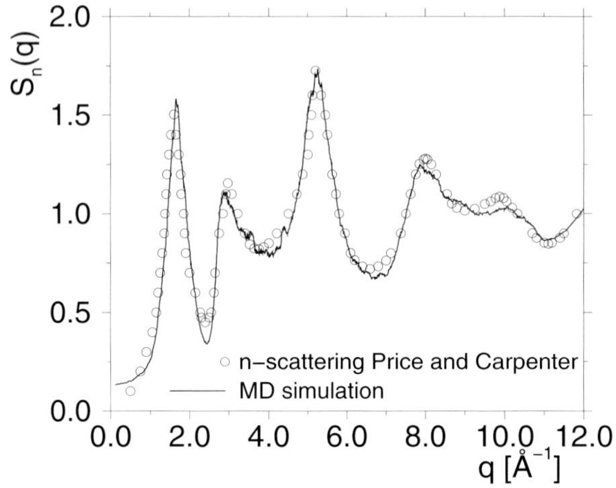

Fig. 7.5. Static neutron structure factor of SiO_2 at room temperature ($T = 300K$) plotted versus wave-vector q. The full curve is the molecular dynamics simulation of Ref. [52], using the experimental neutron scattering lengths for Si and O atoms, while the symbols are the neutron scattering data of Ref. [55]. From Horbach and Kob [52].

For the temperatures at which one can equilibrate the system, i.e. here 2750K and higher, it is also possible to determine the self-diffusion constants of Si and O atoms from the simulation. This is done by calculating the mean square displacements $\langle|\boldsymbol{r}_i(t) - \boldsymbol{r}_i(0)|^2\rangle = \Delta r^2_\alpha(t)$ of the particles of type $\alpha \in$ {Si,O}, and apply the Einstein relation $\Delta_\alpha(t) = 6D_\alpha t$ in the regime of late times where the dependence of $\Delta_\alpha(t)$ on t is in fact linear [52–54]}. The result is shown in Fig. 7.6, where also the respective experimental data [56,57] are included. As one can see from Fig. 7.6, one needs to cover 16 decades, from $10^{-4}\text{cm}^2/\text{s}$ to $10^{-20}\text{cm}^2/\text{s}$, to cover the full range including simulation results and experiments, but the simulation results alone are actually restricted to the first four decades of this range only. The straight lines fitted on this Arrhenius plot to the experiment as well as to the simulation show that in this case a bold extrapolation actually is rather successful - but of course there is no guarantee that this will work similarly well in other cases.

A very interesting aspect of the temperature dependence of the diffusion constants is that there are strong deviations from Arrhenius behavior at very high temperatures. It turns out that this region is rather well described by a power law, $D \propto (T - T_c)^\gamma$, as it is implied by mode coupling theory, see (7.3)

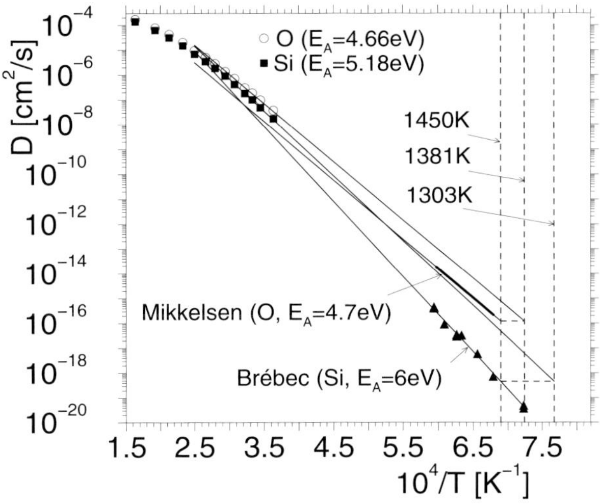

Fig. 7.6. Plot of the self-diffusion constant D of silicon atoms (Si) and oxygen atoms (O) in molten SiO_2 as a function of inverse temperature. The symbols in the upper left part are the results from molecular dynamics simulations and the data in the lower right part stems from experiments [56,57]. The thin straight lines show simple Arrhenius behavior $\{D \propto \exp(-E_A/k_BT)\}$ with various choices of the activation energy E_A, as indicated in the figure. The vertical broken lines indicate the experimental glass transition temperature, $T_g = 1450\text{K}$, as well as values for T_g that one obtains if one extrapolates the data from the simulations to low temperatures and then estimates T_g from the experimental value of the O diffusion constant $\{D_{\mathrm{O}}(T = T_g^{\mathrm{sim}}) = 10^{-16}\mathrm{cm}^2/\mathrm{s} \Rightarrow T_g^{\mathrm{sim}} = 1381\text{K}$ or the Si diffusion constant, respectively $\{D_{\mathrm{Si}}(T = T_g^{\mathrm{sim}}) = 5 \cdot 10^{-19}\mathrm{cm}^2/\mathrm{s} \Rightarrow T_g^{\mathrm{sim}} = 1303\text{K}$. From Horbach and Kob [52].

with $D \propto \tau^{-1}$. In fact, this conclusion is strongly corroborated by a detailed analysis of the intermediate scattering function $\phi_q(t)$ for wave-vector q and various other quantities [52]. This finding is somewhat surprising, however, since $T_c \approx 3330K$ [52], i.e. far above the melting temperature of crystalline SiO_2! Thus it is no surprise that experimental results had not given hint that mode coupling theory also describes a "strong" glassformer such as SiO_2 (where τ and $\eta(T)$ follow a simple Arrhenius behavior over a wide range of temperature). Nevertheless, this discovery that a critical temperature exists also for SiO_2 is of great interest, because it suggests that the differences of the relaxation dynamics between different glassforming fluids are of a quantitative nature only, while qualitatively the behavior is always the same.

7.3 Parallel Tempering

One of the major reasons for the slowing down of the dynamics of (atomistic) glass forming systems is that at low temperatures each atom is trapped in a cage formed by its surrounding neighbors. On the other hand the atom itself is part of a cage that trap the neighboring atoms. With decreasing temperature each of these cages becomes stiffer and stiffer and finally each atom can perform only a rattling motion, i.e. the system has become a fluid that doesn't flow anymore, i.e. a glass. The basic idea of the parallel tempering method is to help the particles to escape their local cage by supplying them with sufficient kinetic energy to overcome the local barrier. Originally proposed for spin models [22,23], the method has been found to be also useful for off-lattice systems. A recent review on the method can be found in Refs. [24,25]. In the following we discuss briefly how the method is implemented in practice.

If we denote the Hamiltonian of the system as $H = K(\mathbf{p}) + E(\mathbf{q})$, where K and E are the kinetic and potential energy, respectively, and $\mathbf{p} = (p_1, p_2, \ldots, p_N)$ and $\mathbf{q} = (q_1, q_2, \ldots, q_N)$ are the momenta and coordinates of the particles, we construct a new Hamiltonian $\mathcal{H}$ as follows:

Make M independent copies of the Hamiltonian H: $H_i = K(\mathbf{p}_i) + E(\mathbf{q}_i)$. Here the $\mathbf{p_i}$ and $\mathbf{q_i}$ are the momenta and coordinates belonging to the i–th subsystem. $\mathcal{H}$ is then defined as

$$\mathcal{H}(\mathbf{p}_1, \ldots, \mathbf{p}_M, \mathbf{q}_1, \ldots \mathbf{q}_M) = \sum_{i=1}^{M} H_i(\mathbf{p}_i, \mathbf{q}_i) = \sum_{i=1}^{M} K(\mathbf{p}_i) + \Lambda_i E(\mathbf{q}_i). \quad (7.6)$$

The $1 = \Lambda_1 > \Lambda_2 > \ldots \Lambda_M$ are constants which we will use later. We now make a molecular dynamics simulation of the Hamiltonian $\mathcal{H}$ at a constant temperature $T = \beta_0^{-1}$. After a certain time interval Δt_{PT} we attempt to exchange the two configurations m and n belonging to two neighboring systems (i.e. $m = n \pm 1$). Whether or not the swap of these two configurations is accepted depends on a Metropolis criterion with a acceptance probability

$$w_{m,n} = \begin{cases} 1, & \Delta_{m,n} \leq 0 \\ \exp(-\Delta_{m,n}), & \Delta_{m,n} > 0, \end{cases} \quad (7.7)$$

where $\Delta_{m,n} = \beta_0(\Lambda_n - \Lambda_m)(E(\mathbf{q}_m) - E(\mathbf{q}_n))$. Since the normal molecular dynamics simulation as well as the Monte Carlo procedure on time scale Δt_{PT} fulfill the condition of detailed balance, the whole algorithm does so also, i.e. after a sufficiently long time the system composed by the subsystem will converge to a Boltzmann distribution. Note that in the systems with a small value of Λ the interaction between the particles is weakened (see (7.6)). Therefore it can be expected that the particles in these systems move faster than those in systems with a large value of Λ. Another way to see this is to say that each system is simulated at a different temperature and that periodically the temperature of the system is increased or decreased (hence the name of the algorithm). This walk in temperature space should thus allow the system to overcome the local barriers formed by the above mentioned cages and thus to propagate faster in configuration space.

Note that this algorithm has a substantial number of parameters, all of which influence its efficiency considerably. In order that the acceptance probabilities of (7.7) are reasonably high, the coupling constants Λ_i should not be too different. On the other hand one wants that Λ_M is as small as possible since this will lead to a fast propagation of the system at this temperature. Therefore one is forced to choose a relatively large value of M. This in turn is, however, not good for the overall performance of the algorithm since in order to be ergodic each configuration has to make a random walk in Λ−space, and the time to do this increases like M^2. Last not least there is the exchange time Δt_{PT} which should not be too small since then the system just swaps back and forth configurations that are very similar. On the other hand Δt_{PT} should also not be too large, since one needs these type of moves in order to explore the Λ−space quickly. The optimal choice of these parameters is currently not known and still the focus of research [58]. A further problem is to find out after which time the system $\mathcal{H}$ has really equilibrated. It seems that to guarantee this it is not sufficient that every subsystem has visited every point in Λ−space [58,59]. A good random walk should look like the one shown in Fig. 7.7. Furthermore we point out that it might be possible that a suboptimal choice of these parameters might make the whole algorithm rather inefficient [59].

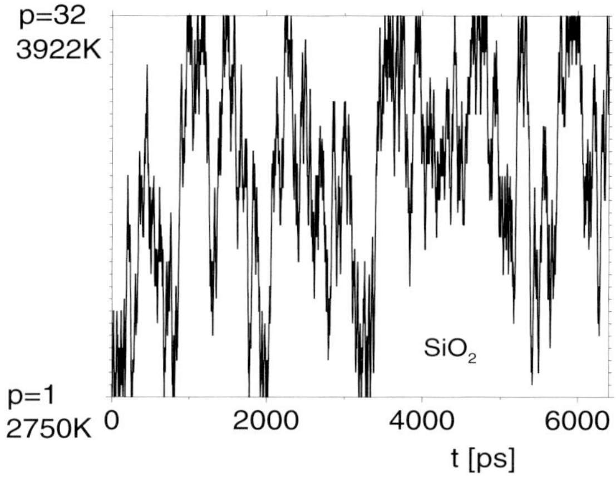

Fig. 7.7. Time dependence of the coupling constant for a parallel tempering simulation of liquid SiO_2. The number of particles was 336 and the number of subsystems was 32. Note that the shown subsystem visits all the different coupling constants several times, thus giving evidence that the overall system has indeed reached equilibrium. From Kob *et al.* [26].

If the above mentioned parameters of the algorithm are chosen well, the parallel tempering method can indeed speed up the equilibration of the system considerably. This is demonstrated in Fig. 7.8 where we show the mean squared displacement of the silicon atoms in SiO_2 as a function of time. From the figure we see that at the lowest temperatures the mean square displacement increases by about a factor of 100 faster than the corresponding curve obtained from the conventional molecular dynamics simulation. From the figure it becomes also clear that the parallel tempering slows *down* the dynamics of the system at high temperatures. This is due to the fact that these systems are coupled to the ones at the low temperatures and hence cannot propagate as fast anymore.

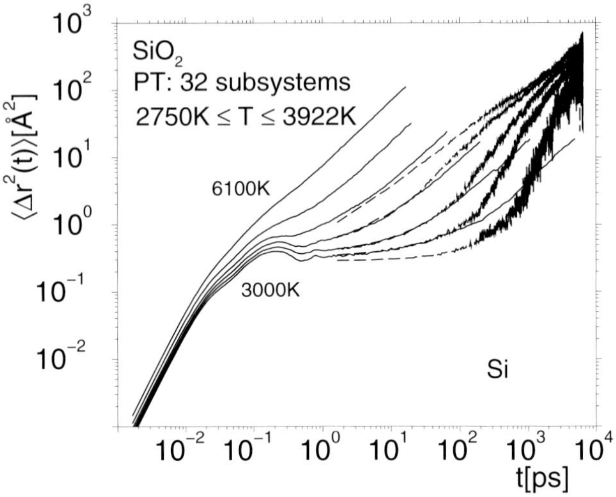

Fig. 7.8. Time dependence of the mean squared displacement of Si in SiO_2 at different temperatures. The dashed lines are from parallel tempering runs and correspond to temperatures 3922K, 3585K, 3235K, 3019K, and 2750K (top to bottom). The solid lines are from conventional molecular dynamics runs and correspond to temperatures 6100K, 4700K, 4000K, 3580K, 3250K, and 3000K (top to bottom). From Kob *et al.* [26].

Before we conclude this section we mention that the parallel tempering algorithm has been found to be also very efficient for the equilibration of the Potts glass discussed in the next section. Thus, although the algorithm might have some problems for certain systems or values of parameters, there are models where it seems to work very well.

7.4 An Abstract Model for Static and Dynamic Glass Transitions: The 10-State Mean Field Potts Glass

In this section we are concerned with a model for which it is known exactly that there is a *dynamical* (ergodic to nonergodic) transition at a temperature T_D and a second, *static*, transition at a lower temperature $T_0 < T_D$, where a static glass order parameter q appears discontinuously: the infinite range p-state Potts glass with $p > 4$ [60–66]. In this model, one has Potts "spin" variables σ_i which can take one out of p discrete values which we simply label from 1 to p, $\sigma_i \in \{1, 2, \ldots, p\}$, where i labels the "sites", $i = 1, 2, \ldots, N$. An energy pJ_{ij} is gained if two spins σ_i, σ_j are in the same state,

$$\mathcal{H} = -\sum_{i<j} J_{ij}(p\delta_{\sigma_i \sigma_j} - 1) . \tag{7.8}$$

Every spin interacts with every other spin via an interaction J_{ij} which is Gaussian distributed, i.e.

$$P(J_{ij}) = \left[\sqrt{2\pi}(\Delta J)\right]^{-1} \exp\{-(J_{ij} - J_0)^2/[2(\Delta J)^2]\} . \tag{7.9}$$

Here the mean J_0 and the width ΔJ are normalized such that

$$J_0 \equiv [J_{ij}]_{av} = \widetilde{J}_0/(N-1), \quad (\Delta J)^2 \equiv [J_{ij}^2]_{av} - [J_{ij}]_{av}^2 = \Delta\widetilde{J}/(N-1) , \tag{7.10}$$

a choice that ensures a sensible thermodynamic limit. We fix the temperature scale by choosing $\Delta\widetilde{J} \equiv 1$, and set the mean of the distribution "antiferromagnetic", $\widetilde{J} = 3 - p$, in order to avoid any tendency towards ferromagnetic order. (Note that for $p = 2$ this model would reduce to the standard Ising mean field spin glass (Sherrington-Kirkpatrick model) [67], but we shall be concerned with $p = 10$ here.) This model, which due to the choice (7.9) exhibits quenched random disorder already in the high temperature phase above the glass transition, can be solved exactly in the thermodynamic limit [60–66]. One finds (Fig. 7.9) that slightly above T_D the dynamic auto-correlation function of the spins exhibits a two-step decay, in that a plateau develops whose life-time diverges at T_D. It is important to note that this behavior is described *exactly* by mode coupling equations of the same type as they occur for the structural glass transition [3]! This shows that this rather abstract model might be more similar to a real structural glass than one would expect at a first glance. At a lower temperature T_0, a static glass transition occurs [60–66], where a static order parameter appears discontinuously. Interestingly the static response function does not diverge at T_0, i.e. the glass susceptibility is still finite here. The entropy does not have a jump at T_0, but shows only a kink. Thus there is no latent heat associated with this transition! A Kauzmann temperature T_K, where the (extrapolated) entropy of the high temperature phase would vanish, also exists, but in this case clearly $T_K < T_0$ and T_K does not have a physical meaning.

It is of course interesting to know if computer simulations can identify the static and dynamic glass transition in a model for which one knows from the exact solutions [60–66] that all these glass transitions do indeed exist. Surprisingly, the answer to this question is "no" since very strong finite size effects are present. In particular it is even hard to see that the above mentioned plateau in the autocorrelation function develops as one approaches the temperature T_D of the dynamical transition (Fig. 7.10) [28]. This is demonstrated in Fig. 7.10 where we show the autocorrelation function of the Potts spins as a function of Monte Carlo time, for $160 \leq N \leq 1280$. Note that

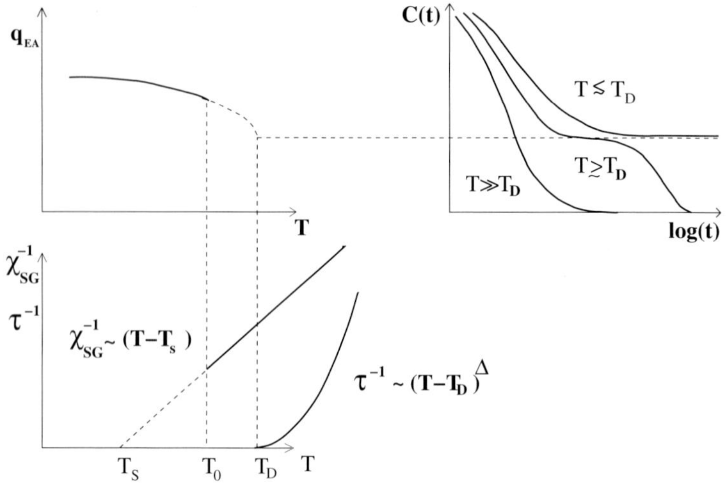

Fig. 7.9. Mean-field predictions for the p-state Potts glass with $p > 4$. The spin glass order parameter, $q_{\rm EA}$, is nonzero only for $T < T_0$ and jumps to zero discontinuously at $T = T_0$. The spin glass susceptibility $\chi_{\rm SG}$ follows a Curie-Weiss-type relation with an apparent divergence at $T_{\rm S} < T_0$. The relaxation time τ diverges already at the dynamical transition temperature T_D. This divergence is due to the occurrence of a long-lived plateau of height $q_{\rm EA}$ in the time-dependent spin autocorrelation function $C(t)$. The discontinuous transition of the order parameter, however, is not accompanied by a latent heat. Therefore, there is no jump in the entropy at T_0, but only a kink occurs. The extrapolation of the high-temperature branch of the entropy would vanish at a "Kauzmann temperature" $T_K = [(1/4)(p-1)/\ln p]^{1/2} T_{\rm S} \approx 0.988 T_{\rm S}$. From Brangian *et al.* [30].

this range is of the same order of magnitude as the particle numbers used for simulations of the structural glass transition, using models such as the binary Lennard-Jones fluid [12] or similar models. No evidence for strong finite size effects was ever found for the latter models if N was larger than ≈ 1000 [68]. Thus, *a priori* it is not at all obvious that system sizes of the order 10^3 are completely insufficient to characterize the dynamics of a system in the thermodynamic limit. However, from Fig. 7.10 we must conclude that for the present system this is indeed the case, at least for temperatures close to the dynamical transition temperature T_D. This is in contrast with the behavior at a high temperature, e.g. $T = 1.8$. From the figure we recognize that at this temperature there are hardly any finite size effects and that the data have nicely converged to the thermodynamic limit even for modest system sizes.

Brangian *et al.* [27–30] defined a relaxation time τ from the time t that it takes the autocorrelation function to decay to the value $C(t = \tau) = 0.2$ (broken straight line in Fig. 7.10). This time is plotted logarithmically versus $1/T$ in Fig. 7.11, so an Arrhenius behavior would be a straight line on this plot. One can see rather clearly a crossover from a power law divergence (that

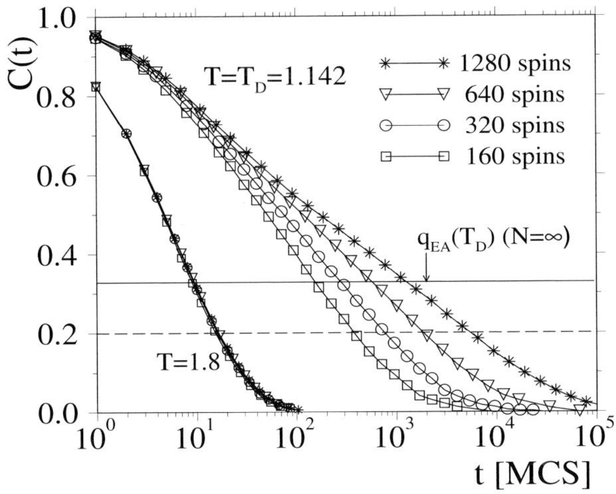

Fig. 7.10. Time dependence of the autocorrelation function $C(t)$ of the spins in the 10-state mean field Potts glass. $C(t)$ is normalized such that $C(t = 0) = 1$ and $C(t \to \infty) = 0$ for $T > T_D$. Time is measured in units of Monte Carlo steps per spin [MCS]. Two temperatures are shown, $T = 1.8$ and $T = T_D = 1.142$ [66], for several values of N. The solid horizontal line indicates the theoretical value of the Edwards-Anderson order parameter $q_{\mathrm{EA}}(T_D) \equiv C(t \to \infty)$ at $T \to T_D^-$ for $N \to \infty$[66]. The horizontal dashed line shows the value used to define the relaxation time τ, $C(t \equiv \tau) = 0.2$. From Brangian *et al.* [28].

would emerge fully in the limit $N \to \infty$ for $T > T_D$) to the Arrhenius law at low T. This behavior qualitatively resembles the behavior expected for structural glasses where the different valleys in the rugged energy landscape for $T < T_0$ are separated by finite (free) energy barriers. In contrast to this one knows that in the Potts glass in the limit $N \to \infty$ these barriers are truly infinite if $T < T_D$, and hence the dynamics is strictly nonergodic.

Similar finite size effects affect also the behavior of static properties [27–30]. One might wonder whether it is possible to use these finite size effects to apply standard finite size scaling analyses to extract reliable information on the location of the static transition temperature from the simulations. Unfortunately the answer is "no": As Fig. 7.12 shows, the standard method [69] of locating a static transition from the intersection point of the order parameter cumulant gives rather misleading results here since the curve seem(!) to intersect at a wrong temperature. Thus one must conclude that there is need to better understand finite size effects for such unconventional glass transitions as sketched in Fig. 7.9, before one can study them reliably with simulations.

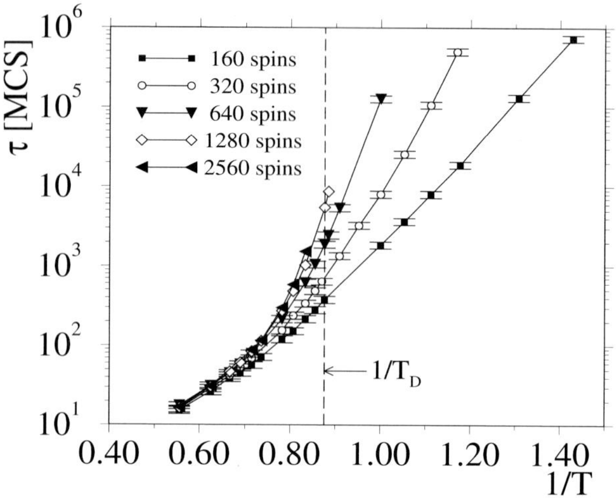

Fig. 7.11. Arrhenius plot of the relaxation time τ of the 10-state mean field Potts glass model for different system sizes. Error bars of τ are mostly due to sample-to-sample fluctuations. The vertical dashed line is the location of T_D where, for $N \to \infty$, the relaxation time is predicted to diverge with a power-law. From Brangian *et al.* [28].

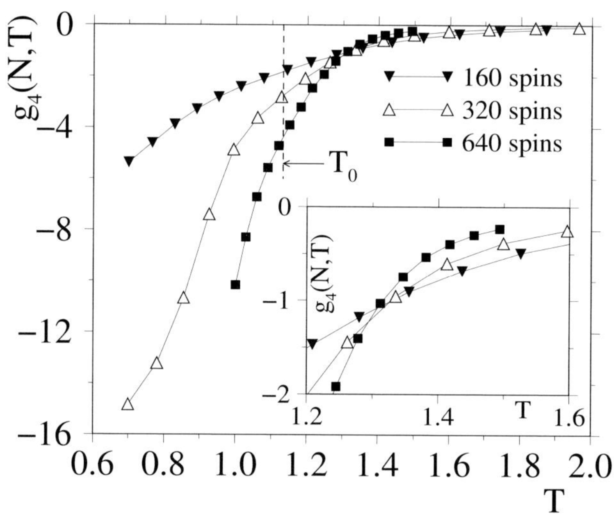

Fig. 7.12. Temperature dependence of the order parameter cumulant $g_4(N,T) = \frac{(p-1)^2}{2}\left\{1 + \frac{2}{(p-1)^2} - \frac{[\langle q^4 \rangle]_{av}}{[\langle q^2 \rangle]_{av}}\right\}$ for the 10-state mean field Potts glass for three choices of N ($N = 160$, 320 and 640). (Here q is the order parameter.) The vertical straight line shows the location of the static transition temperature T_0 as predicted by the exact solution [66]. The inset is an enlargement of the region where the three curves intersect. From Brangian *et al.* [28].

7.5 The Bead-Spring Model: A Coarse-Grained Model for the Study of the Glass Transition of Polymer Melts

We now draw attention to a model which is intermediate between the abstract model as considered in the previous section and the chemically realistic model of silica melts discussed in Sect. 7.2. This intermediate model is a coarse-grained model of glassforming polymer melts. Short polymer chains are described by a bead-spring model, with a chain length of $N = 10$. The (effective) monomers interact with each other via a truncated and shifted Lennard-Jones potential,

$$U_{LJ}(r) = 4\varepsilon[(\sigma/r)^{12} - (\sigma/r)^{6}] + C\,, \quad r \le r_c = 2.2^{1/6}\sigma \tag{7.11}$$

while $U_{LJ}(r) = 0$ if $r > r_c$. The constant C is chosen such that U_{LJ} is continuous at $r = r_c$.

The spring potential present between two neighboring beads is given by

$$U_{\mathrm{FENE}}(l) = -(k/2)R_0^2 \log[1 - (l/R_0)^2] \tag{7.12}$$

with the following values of the constants [31]:

$$\varepsilon = 1\,, \sigma = 1\,, k = 30\,, R_0 = 1.5\,. \tag{7.13}$$

This choice for the parameters creates frustration in the model: the minimum of the bond potential along the chain occurs at a position $l_{\mathrm{min}} \approx 0.97$ that is incompatible with the minimum position $r_{\mathrm{min}} \approx 1.13$ of the Lennard-Jones potential, as far as the formation of simple crystal structures is concerned. This conflict between these two length scales prevents crystallization very efficiently, and the resulting structure of the melt and the corresponding glass resembles corresponding experimental data very nicely, as has already been demonstrated in Fig. 7.1.

If one carries out "slow" cooling experiments one finds that the volume per monomer shows at a temperature $T_g \approx 0.41$ a kink [32]. This signals that the system has changed from the liquid branch to the glass branch and hence has fallen out of equilibrium. Qualitatively, the data looks again very similar to that of corresponding experiments [70]. However, if one compares experiment and simulation more quantitatively, one notes again a big disparity in the cooling rates: In the simulation the temperature was reduced by $\Delta T = 0.02$ every 500000 MD time steps, each time step being $\delta t = 0.002\tau_{MD}$ with $\tau_{MD} = \sigma(m/\varepsilon)^{1/2}$, m being the effective mass of the monomeric units. If one estimates that τ_{MD} corresponds roughly to 10^{-11}s, and that $T = 1$ corresponds to 500K, one arrives at a cooling rate of $\Delta T/\Delta t \approx 10^9$K/s. While this estimate is three orders of magnitude smaller than the corresponding cooling rate for the silica melts [50], it is still many orders of magnitude larger than the corresponding experimental cooling rates. Hence also in this case there is a huge gap between the cooling rates accessible in simulations and those used in real experiments.

This model yields also qualitatively very reasonable results for the relaxation dynamics: The self-diffusion constant can be fitted well by the Vogel-Fulcher law given by (7.1), with $T_{VF} \approx 0.34$, below the kink temperature $T_g \approx 0.41$. The mode coupling critical temperature is located at $T_c \approx 0.45$, above the kink temperature, and the ratios T_c/T_g and T_c/T_{VF} are quite reasonable. Although in the simulation only 1200 monomers were used, a nice plateau is found in the intermediate incoherent scattering function $\phi_q^s(t)$, see Fig. 7.13. Hence one can conclude that no strong finite size effects are present for this model.

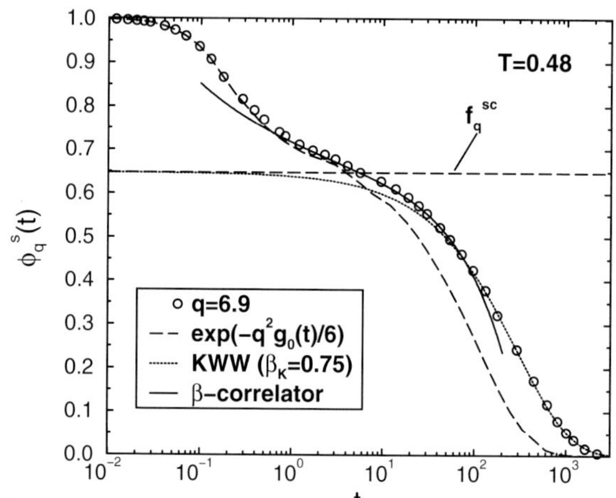

Fig. 7.13. Comparison of the incoherent intermediate scattering function $\phi_q^s(t)$ for the bead-spring model at $T = 0.48$ and $q \approx 6.9$ [$\approx$ maximum of $S(q)$, cf. Fig. 7.1] with various approximations: a Gaussian approximation (dashed line), $\phi_q^S = \exp[-q^2 g_0(t)/6]$, where $g_0(t)$ is the mean square displacement of the monomers. The mode coupling fit for the regime of the so-called "β−relaxation" (solid-line) and a fit with the Kohlrausch function {Eq. (7.2), dotted line} also are included. The non-ergodicity parameter f is indicated as a horizontal dashed line. From Baschnagel *et al.* [14].

Also the Rouse modes [71] which describe the mesoscopic Brownian motion of the polymer chains on length scales that are between monomer-monomer distances and the coil size, are found to relax over almost two decades in $T - T_c$ with relaxation times that show the mode coupling power law [33], see Fig. 7.14. Only very close to T_c, for $T \leq 0.46$, can one see small indications that the singularity at T_c is in fact rounded off. This model has allowed many very impressive tests [35,37] of mode coupling theory, similar to an often studied binary Lennard-Jones mixture [12,72]. But similar to the

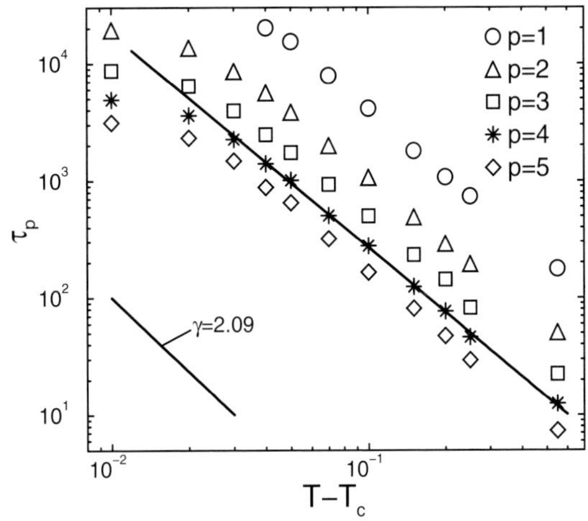

Fig. 7.14. Variation of the relaxation time τ_p of the Rouse modes with the mode index p for the bead spring model plotted vs. $T - T_c$, showing also a power law fit for $p = 4(\gamma_p = 1.83 \pm 0.02)$. Within the error bars, this slope provides a reasonable fit for all p shown. From Baschnagel *et al.* [33].

case in the latter model, it has so far turned out impossible to study temperatures for $T < T_c$ in thermal equilibrium. And none of these models - neither the model for SiO_2, nor the binary Lennard-Jones model [12] nor the present beadspring model - could provide any clarification about the validity of the entropy theory [17].

7.6 The Bond Fluctuation Model Approach to Glassforming Polymer Melts

The bond fluctuation model [6,38–47] is an even more abstract model of polymers than the bead-spring model discussed in the previous section, since it forces the chains to "live" on a simple cubic lattice, and all motions on scales smaller than a lattice constant are completely suppressed. In this model a polymer is represented again as a chain of effective monomers connected by effective bonds, but now each effective monomer is described by an elementary cube on the lattice that blocks all 8 sites at the corners of the cube from further occupation (Fig. 7.15). The length of the effective bonds is allowed to vary from 2 to $\sqrt{10}$ lattice constants (taken as length unit in this section). The only nonbonded interaction is the one of excluded volume. The dynamics of the random conformational changes of the real polymer is represented in a crude way by attempted hops of randomly chosen monomers in randomly chosen lattice directions. If about one half of all lattice sites are occupied, the

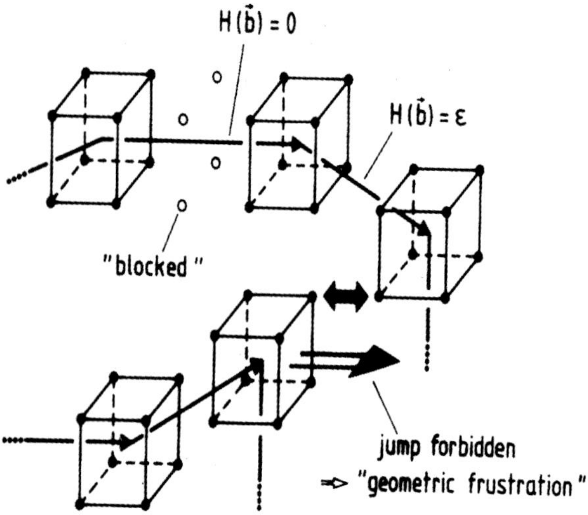

Fig. 7.15. Sketch of a possible configuration of monomers belonging to two different chains in the bond fluctuation model of a polymer melt. For one monomer of the lower chain, an attempted move is indicated; this jump is forbidden, however, since it violates the excluded volume constraint. Also the choice of a two-state energy function is indicated, namely $\mathcal{H}(\boldsymbol{b}) = 0$ if the bondvector $\boldsymbol{b}$ equals $\boldsymbol{b}_{\min} = (0, 0, \pm 3a)$ or a permutation thereof (a is the lattice spacing, chosen as unit of length in the following), and $\mathcal{H}(\boldsymbol{b}) = \varepsilon = 1$ else. Note that if a bond takes a ground state bond $\boldsymbol{b}_{\min}$ it blocks automatically 4 sites (the 4 sites are highlighted by empty circles). From Baschnagel *et al.* [39].

system behaves like a dense melt, and even short chains with chain length $N = 10$ show already typical polymer-properties, e.g. the scaling of the radius of gyration with $\sqrt{N}$, etc.

Since real polymers show with decreasing temperature an increase of the persistence length and hence of the chain radius, it is natural to model this effect by an effective potential $U(l)$ for the length of the effective bonds, energetically favoring long bonds. If one chooses as a minimum of this potential $U(l_{\min} = 3) = 0$ while $U(l) = \varepsilon = 1$ for all other bond lengths l, one also incorporates "geometric frustration" (Fig. 7.15) into the model: Each bond that reaches its ground state wastes the four lattice sites in between the adjoining effective monomers, which are completely blocked for further occupation. From the point of view of packing as many effective monomers as possible in a dense melt on the lattice, the bonds that waste lattice sites are very unfavorable. Hence configurational entropy favors short bonds that do not waste any other lattice sites for further occupation. Thus a conflict between entropy and energy is created, which is responsible for the glass transition observed in the Monte Carlo simulations of this model.

This model has the big technical advantage that it can be equilibrated even at relatively low temperatures by the so-called "slithering snake algorithm". In this type of Monte Carlo moves one randomly attempts to remove a bond from one chain end and attach it to the other chain end in a randomly chosen orientation [44]. Although this algorithm does not correspond to any physically realistic dynamics of polymers it is a perfectly admissible Monte Carlo move for studying equilibrium properties. Using this algorithm, thermal equilibrium can be established at rather low temperatures, such as $T = 0.16$, where after 10^7 steps with the conventional "random hopping" algorithm the autocorrelation of the end-to-end vector of the chains still has not decayed below 90% of its starting value [46]. If we wish to study dynamical properties of this model, we first perform a run with this slithering snake algorithm, to obtain initial states that are characteristic for thermal equilibrium. Subsequently we can start a run with the normal random hopping moves of the effective monomers, which thus yields a physically reasonable description of the dynamics [46]. If one estimates that one Monte Carlo Step per monomer corresponds to about 10^{-12} seconds in real time, a run of 10^7 steps would reach a physical time of 10^{-5} seconds, which is several orders of magnitude longer than the typical time scales accessible with molecular dynamics. Using this algorithm it was hence possible to make a very nice test of mode coupling theory [42,43], resulting in $T_c \approx 0.15$ while [46] $T_{VF} = 0.125 \pm 0.005$. However, the investigation of the relaxation dynamics in the regime $T_{VF} < T \leq T_c$ seems to be very difficult also in the framework of this lattice model, and in fact has not yet been attempted.

Using the bond fluctuation model it was also determined how the glass transition temperature T_g depends on the length of the chain [41] and the results are compatible with the law

$$T_g(\infty) - T_g(N) \propto 1/N\,. \tag{7.14}$$

Such a dependence has also been found experimentally [73], and is one of the most notable predictions of the entropy theory of Gibbs and Di Marzio [17]. Therefore many experimentalists believe that this theory is correct. However, this conclusion is premature, as a study of the configurational entropy for the present lattice model shows (Fig. 7.16). While the entropy does indeed decrease rather strongly with increasing value of inverse temperature, starting out from an "athermal melt" (corresponding to infinite temperature), this decrease becomes slower when one approaches the vicinity of T_c, and the simulation data do not show that the entropy vanishes, although they also cannot rule it out that this happens at a T far below T_c. However, if one works out the Gibbs-Di Marzio theory [17] explicitly for the present lattice model (all the input parameters of the theory [17] can also be extracted from the simulation, so there are no adjustable parameters whatsoever in this comparison!), one sees that the theory underestimates the actual entropy considerably at all temperatures. In particular this failure is responsible for the

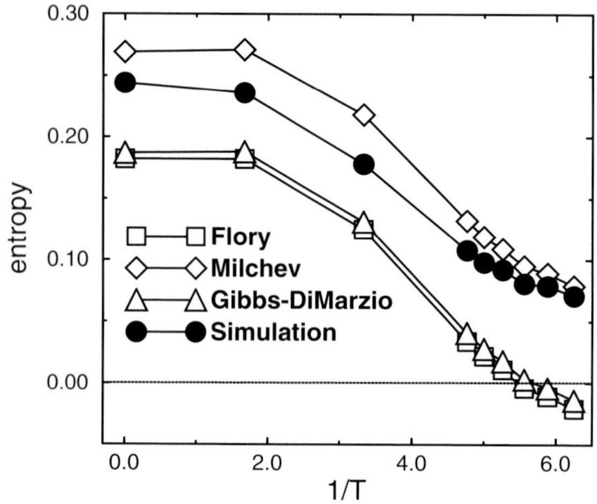

Fig. 7.16. Comparison of the temperature dependence of the entropy per lattice site as obtained from the simulation of the bond fluctuation model (open circles) with the theoretical predictions of Gibbs and Di Marzio [17], Flory [74] and Milchev [75]. Note that the estimates for T_c and T_{VF} are $T_c \approx 0.15$ and $T_{VF} \approx 0.125$. Therefore the vanishing of the entropy at $T \approx 0.18$ is an artifact due to inaccurate approximations involved in the calculation of $S(T)$ via the entropy theory [17]. From Wolfgardt *et al.* [45].

vanishing of the entropy at $T_K \approx 0.18$, which obviously is a spurious result, since this temperature is even higher than T_c, well in the melt regime where the polymer system is a liquid and not a glass. In fact, a slightly different approximation due to Milchev [75] renders the entropy nonnegative at all temperatures, but deviates now a bit from the simulation data in the other direction. Thus, these investigations show that although (7.14) does indeed hold it does not imply anything about the validity of the Kauzmann "entropy catastrophe".

7.7 Can One Map Coarse-Grained Models onto Atomistically Realistic Ones?

From the above comments it is clear that in simulations of simplified coarse-grained models the range of times one can span is much larger than the one for chemically realistic models that include atomistic detail (microseconds rather than nanoseconds). On the other hand, the simplified models may elucidate general concepts but they fail to make quantitative predictions on the properties of particular materials. Thus the question arises whether one can somehow combine the advantages of both approaches.

An idea to do this is to make the coarse-graining process in a more systematic way and to construct coarse-grained models that "remember" from which atomistic system they come from. For a polymer chain, coarse-graining along the backbone of the chain may mean that if we label the covalent bond consecutively $(1, 2, 3, 4, 5, 6, \dots)$ the bonds $1, 2, 3$ form the effective bond I, the bonds $4, 5, 6$ form the effective bond II, etc [76]. The potentials on the atomistic scale (e.g. potentials controlling the lengths of covalent bonds, the angles between them, the torsional angles, etc.) have then to be translated into suitable effective potentials for the length l of the effective bonds and the angle Θ between them. The simplest choice would be to assume potentials of the form

$$U_{\text{eff}}(l) = \frac{1}{2} u_0 (l - l_0)^2 \;, V_{\text{eff}}(\Theta) = \frac{1}{2} v_0 (\cos \Theta - \cos \Theta_0)^2 \quad . \tag{7.15}$$

In the past potentials of this type have indeed be extracted from the probability distributions $P(l) \propto \exp[-U_{\text{eff}}(l)/k_B T]$, $P(\Theta) \propto \exp[-V_{\text{eff}}(\Theta)/k_B T]$ observed in the simulations of single chains (where long range interactions need to be truncated, however) [76,77]. Of course, the effective parameters u_0, l_0, v_0, Θ_0 are somewhat temperature dependent, and in principle one should deduce them from simulations of atomistically described melts containing many chains, rather than from single-chain simulations [77]. The practical implementation of how one constructs best the effective potentials that mimic one particular material is still an active topic of research [48,77].

A further important aspect is the question to what extent the dynamics with such a coarse grained system reflects the dynamics of a real chain. Here one needs to focus on the slowest local process, which are hops of small groups of monomers to a new conformation, such that a barrier of the torsional potential is crossed. Without such moves involving barrier crossing no conformational changes can occur. In a typical case, e. g. for polyethylene at $T = 500$K, the time scale for such hops is about two orders of magnitude larger than the vibration times of bond lengths and bond angles. Only because of this separation of time scales one can hope that a coarse-grained model can describe the essential features of the slow dynamics in the polymer melt at all, if the time units are properly rescaled. As shown by Tries *et al.* [48], the knowledge of the torsional potentials allows, using a an approach that resembles transition state theory, to construct a "time rescaling factor", that gives the translation of the time unit of the Monte Carlo simulations (attempted Monte Carlo steps per monomer) into physical time units (Fig. 7.17). One sees that for polyethylene 1 Monte Carlo step corresponds to 0.1 to 10ps, in the temperature region of interest. At high temperatures, namely for $T = 509$K, the accuracy of the coarse-grained model of $C_{100}H_{202}$ was tested by running a molecular dynamics simulation of a united atom model for about a nanosecond (which is of the order of the Rouse relaxation time at this temperature) for comparison [48]. It is found that the agreement between both approaches is almost quantitative. The advantage of the Monte

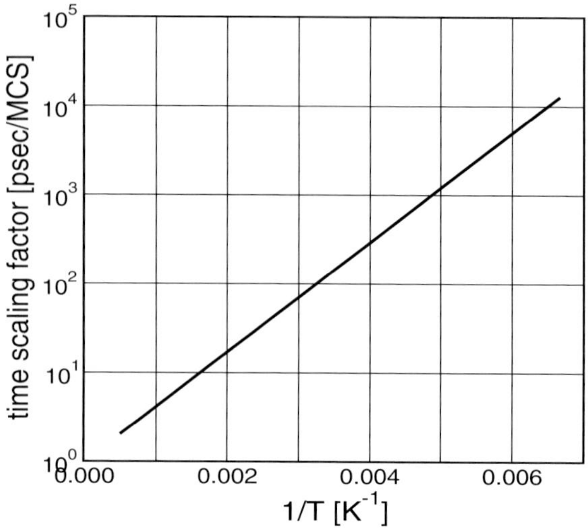

Fig. 7.17. Temperature dependence of the time scaling factor converting the time unit of the Monte Carlo simulation into femtoseconds, for the case of polyethylene. The straight line shows that a simple Arrhenius law is a good approximation. Adapted from Ref. [48].

Carlo simulation of the coarse-grained model is, however, that one can easily study a supercooled melt also at $T = 250$K, a temperature which is basically inaccessible to the molecular dynamics approach.

If one compares the results of the coarse-grained model to experimental data, e. g. for the viscosity and its temperature dependence, the agreement is encouragingly good but not perfect [48]. One aspect which is clearly missing in the coarse-grained model is the description of attractive intermolecular forces. Thus, while this approach of mapping atomistic models to coarse-grained ones clearly has a great potential, there are still nontrivial problems that need to be solved.

7.8 Concluding Remarks

In this brief review, the "state of the art" of computer simulations of glassy systems was summarized. The main problem in this field is the problem of bridging time scales - a supercooled fluid close to the glass transitions exhibits a nontrivial dynamic behavior that extends from very fast processes (in the picosecond time scale range) to very slow processes (with relaxation times of the order of hours). Atomistic molecular dynamics simulations of chemically realistic models (as exemplified here for the case of molten SiO_2) can treat only a very small part of this broad range of time scales, and also special techniques such as the parallel tempering method can add only one or two decades to this range but not more. (Note also that there are still some unsolved technical problems with this method [59]). While such atomistic simulations are nevertheless useful, in particular since they complement the time range directly accessible to experiment, and give a very detailed insight into the interplay between structure and dynamics in supercooled fluids, they clearly cannot answer questions on the nature of relaxation processes for temperatures close to (the experimental) T_g, and the possible existence for an underlying static phase transition (from a metastable supercooled fluid to a metastable ideal glass) at a temperature $T_K < T_g$. Also molecular dynamics studies of coarse-grained models for melts of short, unentangled polymer chains suffer from similar problems, although the effective cooling rates in these models are about a factor of 10^3 smaller than in the model for silica, and one can access relaxation times that are almost in the microsecond range. These models are very useful as a testbed for the mode coupling description of the glass transition in fragile glassformers, however. Furthermore they have also allowed to gain very useful insight on the relaxation between the local motions responsible for the glass transition (cage effect etc.) and the more mesoscopic Brownian motion of the polymer chains (as described by the "Rouse modes", for instance).

A slightly more abstract model of the same systems, the bond fluctuation model of glassforming polymer melts, corroborates these conclusions, although due to its discrete nature it is somewhat less suitable to describe the local structure of packing effective monomers in a polymer melt or their motion on small scales (confined in a cage). However, this model has the merit that it allows to compute the temperature dependence of the configurational entropy $S(T)$ and thus to test the correctness of theories like the one of Gibbs and Di Marzio. While it is found that the entropy $S(T)$ decreases significantly if the polymer melt approaches the glass transition, there is clear evidence that the theory of Gibbs and Di Marzio is quantitatively very unreliable since it underestimates $S(T)$ significantly at all temperatures, and the "entropy catastrophe" that it predicts is clearly an artifact of inaccurate approximations.

Finally, studies of an even more abstract model were discussed, the 10-state Potts glass with mean field infinite range interactions. This model has the advantage that it is known exactly that it has a dynamical (ergodic to nonergodic) transition at T_D as well as a static transition at a (slightly) lower temperature T_0, at which a glass order parameter appears discontinuously and the entropy shows a kink. The conceptional disadvantage of this model, however, is that it has a built-in quenched random disorder (via its random exchange couplings) at all temperatures, unlike systems that undergo a structural glass transition, which have no quenched disorder in the high temperature phase (the supercooled fluid for $T > T_g$). Monte Carlo studies of this model, intended to serve as a general testbed for systems with both a dynamical and a static glass transition, show that unexpectedly large finite size effects occur, which are poorly understood. Thus even for this "simple" model much more work is necessary.

While the anticipated progress in computer hardware and algorithmic improvements will allow to extend the time ranges accessible in all these simulations somewhat, there is not real hope that one can bridge the desired 15 (or more) decades in time in this way. More promising in principle is the approach of providing an explicit mapping between atomistic models (which cover the fast processes) and coarse-grained models (which describe the somewhat slower processes, in the 10ps to 1μs range), so that one effectively considers the same model system but with different approaches on different time scales. Of course, this idea is difficult to work out consistently in practice, and only modest first steps towards its realization have been taken. Much more work in this direction is certainly very desirable in the future.

Acknowledgements

We are particularly grateful to C. Bennemann, C. Brangian, J. Horbach, T. Stühn, K. Vollmayr, and M. Wolfgardt for their valuable collaboration on parts of the research described here, and acknowledge financial support from the Deutsche Forschungsgemeinschaft (DFG/SFB 262), the Bundesministerium für Bildung und Forschung (BMBF grant No 03N6015) and SCHOTT Glas. We thank the NIC Jülich and the HLRS Stuttgart for generous allocations of computer time.

References

1. J. Jäckle: Rep. Progr. Phys. **49**, 171 (1986)
2. K. L. Ngai (Ed.): *Proceedings of the First International Meeting on Relaxations in Complex Systems*, J. Non-Cryst. Solids **131-133** (1991)
3. W. Götze, J. Sjögren: Rep. Progr. Phys. **55**, 214 (1992)
4. K. L. Ngai (Ed.): *Proceedings of the Second International Discussion Meeting on Relaxations in Complex Systems*, J. Non-Cryst. Solids **172-174** (1994)
5. K. L. Ngai (Ed.): *Proceedings of the Third International Discussion Meeting on Relaxations in Complex Systems*, Non-Cryst. Solids **235-237** (1998)
6. W. Paul, J. Baschnagel: "Monte Carlo simulation of the glass transition of polymers". In: *Monte Carlo and Molecular Dynamics Simulations in Polymer Science* ed. by K. Binder (Oxford University Press, New York 1995) pp. 307-355.
7. K. Binder, W. Kob: "How can computer simulations contribute to understand the static structure of glasses". In: *Analysis of the Composition and Structure of Glass and Glass Ceramics* ed. by H. Bach and D. Krause (Springer, Berlin 1999) pp. 255-267
8. W. Kob, K. Binder: "How can computer simulations contribute to the understanding of the dynamics of glasses and glass melts?" In: *Analysis of the Composition and Structure of Glass and Glass Ceramics* ed. H. Bach and D. Krause (Springer, Berlin 1999) pp. 344-356.

9. K. Binder: Computer Phys. Commun. **121-122**, 168 (1999)
10. K. Binder: J. Non-Cryst. Solids **274**, 332 (2000)
11. K. Binder, J. Baschnagel, W. Paul: Progr. Polymer Sci. (in press)
12. W. Kob: J. Phys.: Cond. Matter **11**, R85 (1999)
13. A. Arbe, D. Richter, J. Colmenero, B. Farago: Phys. Rev. E**54**, 3853 (1996)
14. J. Baschnagel, C. Bennemann, W. Paul, K. Binder: J. Phys.: Condens. Matter **12**, 6365 (2000)
15. W. Kauzmann: Chem. Rev. **43**, 219 (1948)
16. K. Binder: Ber. Bunsenges. Phys. Chem. **100**, 1381 (1996)
17. J. H. Gibbs, E. A. Di Marzio: J. Chem. Phys. **28**, 373 (1958)
18. G. Adam, J. H. Gibbs: J. Chem. Phys. **43**, 739 (1965)
19. M. Fuchs, W. Götze, S. Hildebrand, A. Latz: J. Phys.: Condens. Matter **4**, 7709 (1992)
20. M. P. Allen, D. J. Tildesley: *Computer Simulation of Liquids* (Clarendon Press, Oxford, 1987)
21. K. Binder, G. Ciccotti (eds.): *Monte Carlo and Molecular Dynamics of Condensed Matter Systems* (Societa Italiana di Fisica, Bologna, 1996)
22. E. Marinari and G. Parisi, Europhys. Lett. **19**, 451 (1992)
23. K. Hukushima, K. Nemoto: J. Phys. Soc. Jpn. **65**, 1604 (1996)
24. U. H. E. Hansmann and Y. Okamoto, Annual Rev. of Comp. Phys. Vol. VI, (World Scientific, Singapore, 1999) p. 179
25. Y. Iba, Int. J. Mod. Phys. C **12** 623 (2001)
26. W. Kob, C. Brangian, T. Stühn, R. Yamamoto: "Equilibrating Glassy Systems with Parallel Tempering" In: *Computer Simulation Studies in Condensed Matter Physics XIII* ed. by D. P. Landau, S. P. Lewis, H. P. Schüttler (Springer, Berlin, 2001) pp. 153-166.
27. C. Brangian, W. Kob, K. Binder: Europhys. Lett. **53**, 756 (2001)
28. C. Brangian, W. Kob, K. Binder: J. Phys. A: Math. Gen. (in press)
29. C. Brangian, W. Kob, K. Binder: "Static and dynamics glass transitions in the 10-state Potts glass: What can Monte Carlo simulations contribute?" In: *Proceedings of the STATPHYS 21 Satellite meeting: Computational Statistical Mechanics - Challenges for the* 21st *Century* ed. by D. P. Landau (in press)
30. C. Brangian, W. Kob, K. Binder: "Monte Carlo study of the dynamic and static glass transition in the 10-state Potts glass: A scenario for the structural glass transition?" In: H. E. Stanley et al, eds. *Proceedings of the NATO ARW on New kinds of phase transitions in disordered materials, Wolga River, Russia, May 24-28, 2001* (Kluwer Acad. Publ., in press)
31. C. Bennemann, W. Paul, K. Binder, B. Dünweg: Phys. Rev. E**57**, 843, (1998)
32. K. Binder, C. Bennemann, J. Baschnagel, W. Paul: "Anomalous diffusion of polymers in supercooled melts near the glass transition". In: R. Kutner, A. Pekalski, and K. Sznajd-Weron, eds. *Anomalous Diffusion: From Basics to Applications* (Springer, Berlin, 1999) pp. 124-139.
33. C. Bennemann, J. Baschnagel, W. Paul, K. Binder: Comput. Theoret. Polymer Sci. **9**, 217 (1999)
34. C. Bennemann, W. Paul, J. Baschnagel, K. Binder: J. Phys.: Condens. Matter **11**, 2179 (1999)
35. C. Bennemann, C. Donati, J. Baschnagel, S. C. Glotzer: Nature **399**, 246 (1999)
36. C. Bennemann, J. Baschnagel, W. Paul: Eur. Phys. J. B**10**, 323 (1999)
37. M. Aichele, J. Baschnagel: Eur. Phys. J. E**5**, 229, 245 (2001)

38. H.-P. Wittmann, K. Kremer, K. Binder: J. Chem. Phys. **96**; 6291 (1992)
39. J. Baschnagel, K. Binder, H.-P. Wittmann: J. Phys.: Condens. Matter **5** 1597 (1993)
40. J. Baschnagel, K. Binder: Physica A**204**, 47 (1994)
41. B. Lobe, J. Baschnagel: J. Chem. Phys. **101**, 1616 (1994)
42. J. Baschnagel: Phys. Rev. **49**, 135 (1994)
43. J. Baschnagel, M. Fuchs: J. Phys.: Condens. Matter **7**, 6761 (1995)
44. M. Wolfgardt, J. Baschnagel, K. Binder: J. Phys. (Paris) II **5**, 1035 (1995)
45. M. Wolfgardt, J. Baschnagel, W. Paul, K. Binder: Phys. Rev. E**54**, 1535 (1996)
46. K. Okun, M. Wolfgardt, J. Baschnagel, K. Binder: Macromolecules **30**, 3075 (1987)
47. K. Binder, J. Baschnagel, S. Böhmer, W. Paul: Phil. Mag. B**77**, 591 (1998)
48. V. Tries, W. Paul, J. Baschnagel, K. Binder: J. Chem. Phys. **106**, 738 (1997)
49. B. W. H. van Beest, G. J. Kramer, R. A. van Santen: Phys. Rev. Lett. **64**, 1995 (1990)
50. K. Vollmayr, W. Kob, K. Binder: Phys. Rev. B**54**, 15808 (1996)
51. R. Brüning, K. Samwer: Phys. Rev. B**46**, 11318 (1992)
52. J. Horbach, W. Kob: Phys. Rev. B**60**, 3169 (1999)
53. J. Horbach, W. Kob, K. Binder, A. C. Angell: Phys. Rev. E**54**, R5897 (1996)
54. J. Horbach, W. Kob, K. Binder: Phil. Mag. B**77**, 297 (1998)
55. D. L. Price, J. M. Carpenter: J. Non-Cryst. Solids **92**, 153 (1987)
56. J. C. Mikkelsen: Appl. Phys. Lett. **45**: 1187 (1984)
57. G. Brebec, R. Seguin, C. Sella, J. Bevenot, J. C. Martin: Acta Metall. **28**, 327 (1980)
58. T. Stühn, Diploma Thesis, University of Mainz, 2000
59. C. De Michele and F. Sciortino, preprint cond-mat/0112056.
60. D. Elderfield, D. Sherrington: J. Phys. C: Solid State Phys. **16**, L491, L971, L1169 (1983)
61. D. J. Gross, I. Kanter, H. Sompolinsky: Phys. Rev. Lett **55**, 304 (1985)
62. T. R. Kirkpatrick, P. G. Wolynes: Phys. Rev. B**36**, 8552 (1987)
63. T. R. Kirkpatrick, D. Thirumalai: Phys. Rev. B**37**, 5342 (1988)
64. D. Thirumalai, T. R. Kirkpatrick: Phys. Rev. B**38**, 4881 (1988)
65. G. Cwilich, T. R. Kirkpatrick: J. Phys. A: Math. Gen **22**, 4971 (1989)
66. E. De Santis, G. Parisi, F. Ritort: J. Phys. A: Math. Gen. **28**, 3025 (1995)
67. K. Binder, A. P. Young: Rev. Mod. Phys. **58**, 801 (1986)
68. K. Kim and R. Yamamoto, Phys. Rev. E **61**, R41 (2000).
69. K. Binder: Rep. Progr. Phys. **50**, 783 (1987)
70. G. B. Mc Kenna: 'Glassformation and glassy behavior'. In: *Comprehensive Polymer Science, Vol. 2* ed. by C. Booth and C. Proce. Pergamon Press, Oxford 1986, pp. 311-362.
71. M. Doi, S. F. Edwards: *The Theory of Polymer Dynamics* (Clarendon, Oxford, 1986)
72. M. Nauroth and W. Kob, Phys. Rev. E **55**, 657 (1997).
73. G. Pezzin, F. Zilio-Grandi, P. Sanmartin: Eur. Polymer J. **6**, 1053 (1970)
74. P. J. Flory: Proc. Roy. Soc. London A**234**, 60 (1956)
75. A. Milchev: C. R. Acad. Bulg. Sci. **36**, 1415 (1983)
76. J. Baschnagel, K. Binder, W. Paul, M. Laso, U. W. Suter, I. Batoulis, W. Jilge, T. Bürger: J. Chem. Phys. **95**, 6014 (1991)
77. W. Tschöp, K. Kremer, I. Batoulis, T. Bürger, O. Hahn: Acta Polymer **49**, 61, 75 (1998)

Part VI

Lattice Models

8 Monte Carlo Methods for Bridging the Timescale Gap

Nigel Wilding[1] and David P. Landau[2]

[1] Department of Physics, University of Bath, Bath BA2 7AY. U.K.
[2] Center for Simulational Physics, The University of Georgia, Athens, Georgia 30602, U.S.A.

Abstract. We identify the origin, and elucidate the character of the extended time-scales that plague computer simulation studies of first and second order phase transitions. A brief survey is provided of a number of new and existing techniques that attempt to circumvent these problems. Attention is then focused on two novel methods with which we have particular experience: "Wang-Landau sampling" and Phase Switch Monte Carlo. Detailed case studies are made of the application of the Wang-Landau approach to calculate the density of states of the 2D Ising model and the Edwards-Anderson spin glass. The principles and operation of Phase Switch Monte Carlo are described and its utility in tackling 'difficult' first order phase transitions is illustrated via a case study of hard-sphere freezing. We conclude with a brief overview of promising new methods for the improvement of deterministic, spin dynamics simulations.

8.1 General Introduction

During the past half century there has been intense study of phase transitions in an extremely broad range of materials. From the theoretical perspective this has led to the development of relatively simple models that seek to capture the essential qualitative features of real systems. To obtain the phase behaviour of such models, a wide variety of analytical techniques have been developed. More recently, these pen and paper approaches have been supplemented by computer simulations.

The fundamentals of determining phase behaviour have long been understood–statistical mechanics tells us that any equilibrium thermodynamic properties of interest can be determined once the partition function is known. However, the partition function is defined as a sum over *all* microstates of the system. This immediately leads to difficulties because the number of microstates is huge for all but the very smallest systems. As a consequence, exact enumeration of the partition function generally becomes impossible. Only for a handful of special models is it possible to solve the problem exactly (at least for some quantities), and such models consequently serve as invaluable testing grounds for more generally applicable analytical and computational techniques.

In view of the difficulties in solving model system exactly, one often resorts to stochastic sampling schemes such as Monte Carlo (MC) simulation in order to gain information on the thermodynamic properties of model systems. MC methods have proven themselves quite powerful and flexible in the study of phase transitions in various areas of statistical physics [1]. Nevertheless, serious problems can arise. Specifically, in the neighborhood of a first order phase transition one encounters metastable states and hysteresis effects that lead to extended sampling times and systematic errors in the estimation of phase boundary loci. Similarly, at a second order phase transition, critical slowing down renders it difficult to attain the large system sizes necessary for the accurate determination of critical point parameters.

In this article we shall review in some detail the sources of difficulty with traditional MC simulations of phase transition and shall describe a few of the methods that have been developed to circumvent these difficulties. We shall also briefly describe recent advances in simulation methods for the study of deterministic time behaviour. We wish to emphasize at the outset that our treatment is not meant to be exhaustive, and will only consider a few of the many approaches that have been developed. We have surely devoted disproportionately much space to our own contributions; this is mainly because we can explain them best.

8.2 Problems and Challenges

8.2.1 Introduction to Metropolis Importance Sampling

MC simulation methods have been employed for over half a century and are, in many respects, quite mature. The prototype approach, importance sampling, was introduced by Metropolis *et al* [2] in 1953 and has been employed extensively for a wide range of simulation studies. Owing to its simplicity and ease of implementation it is still in widespread use today.

To illustrate the operation of the Metropolis algorithm let us conside the simple Ising model–a lattice based two-state spin model. The Ising model Hamiltonian is

$$E = -J \sum_{<i,j>} \sigma_i \sigma_j , \tag{8.1}$$

where J is the coupling constant, $\sigma = \pm 1$, and the sum extends over all nearest neighbours.. There are several variations of the Metropolis approach, but basically a single spin is chosen and flipped with a probability p depending on the associated energy change ΔE:

$$p = \min\left(1, \exp(-\Delta E/k_B T)\right) \tag{8.2}$$

The resulting spin configurations provide good visual information about the nature of the clusters that develop as a phase transition is approached. This is illustrated in Fig. 8.1, which shows a typical configuration of the 2d Ising model near its second order phase transition (critical point).

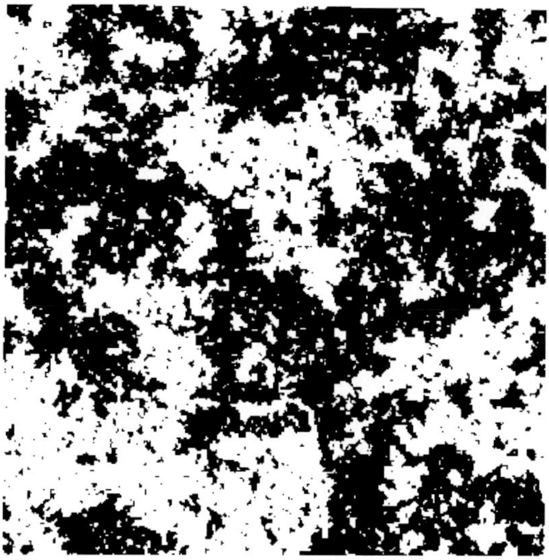

Fig. 8.1. A typical configuration near the critical temperature of the two dimensional Ising model of 1024×1024 spins. White areas correspond to up spins and black areas to down spins.

The sequence of states produced by the Metropolis algorithm constitutes a random walk through the phase space of the model system. It is so designed as to yield configurations with the correct equilibrium (Boltzmann) probabilities. Consequently, estimates for ensemble averages of equilibrium observables (such as the magnetisation or internal energy) can be obtained simply by accumulating their values over many independent configurations.

The Metropolis method is, of course, not limited to lattice based spin systems and is also straightforward to apply to systems with continuous degrees of freedom. A simple example is a fluid of hard particles, such as that considered in Sect. 8.3.2.4. However, when systems with continuous symmetry are considered, the matter of the dynamical evolution of the system arises. It is important to appreciate that the MC simulation dynamics of, for example, a fluid, does not necessarily bear any relation to the time development that would occur in a physical system. This is a general feature of most MC algorithms, the utility of which is therefore limited in cases where one is not interested in obtaining accurate dynamical information. Fortunately, this disadvantage is potentially compensated in a variety of ways. Specifically, freed from the strictures of realistic dynamics, the simulator is at liberty to construct any number of imaginative sampling schemes, which go beyond the importance sampling embodied by the Metropolis algorithm, and which allow the system to explore phase space much more efficiently than would be possible with realistic dynamics. This aspect of MC simulation will be a recurring theme of the methods we describe below.

8.2.2 Origin of Time-Scale Problems

The Metropolis method outlined above generally works well in single phase regions of phase diagram. However it can become very inefficient in the vicinity of a phase transition. The problems encountered (and the methods that lead to their resolution) depend qualitatively on whether the transition is second order or first order in character. Here we briefly summarize the principal issues associated with each case.

On approaching a second order phase transition (critical point), the correlation length ξ of order parameter fluctuations diverges like $\xi \sim (T-T_c)^{-\nu}$, where T is the temperature, T_c is the critical temperature and $\nu > 0$ a critical exponent. The magnitude of the correlation length is a measure of the degree of correlated configurational structure (essentially the cluster diameter). That this is very large near the critical temperature is confirmed in Fig. 8.1. The problem for MC simulation is one of autocorrelation time: since the clusters are very large, it takes many steps of a local (single spin flip) algorithm to produce a new statistically independent configuration, which in turn hinders accumulation of unbiased statistics. This 'critical slowing down' can best be quantified in terms of a characteristic relaxation time τ. As $T \to T_c$ this relaxation time τ diverges as

$$\tau = (T - T_c)^{-\nu z} , \tag{8.3}$$

where z is known as the dynamical critical exponent. Since in a simulation, the correlation length ξ is bounded by the linear dimension of the system L, one finds that as $T \to T_c$, τ diverges as L^z. Consequently we are limited in how close we can approach T_c, or in what size of system we can simulate, because the correlations begin to dominate and the amount of time required to produce uncorrelated configurations becomes excessive (c.f. Fig. 8.2). It is also important to note that different observables have correlation times that differ by a multiplicative factor so their correlation times may differ greatly even though the dynamic critical exponent z is the same.

Turning now to first order phase transitions, the problems encountered in MC simulations generally stem from barriers to sampling which hinder the exploration of the coexisting phases. In order to obtain accurate estimates for the location of a first order phase boundary, it is necessary for the simulation to pass back and forth many times between the coexisting phases; doing so permits an accurate estimate of their relative statistical weights. However, (and as we describe below), a simulation launched near a first order coexistence point will not readily explore both coexisting phases. Instead it will tend to remain in the phase within which it was initiated–this is the phenomenon of metastability. Transitions between the coexisting pure phases are suppressed due to the intrinsically low probability (large free energy) of the interfacial states through which the system must pass in traversing the inter-phase route. The result is extended sampling times which grow rapidly with system size. For systems with "complex" order, e.g. spin glasses, the

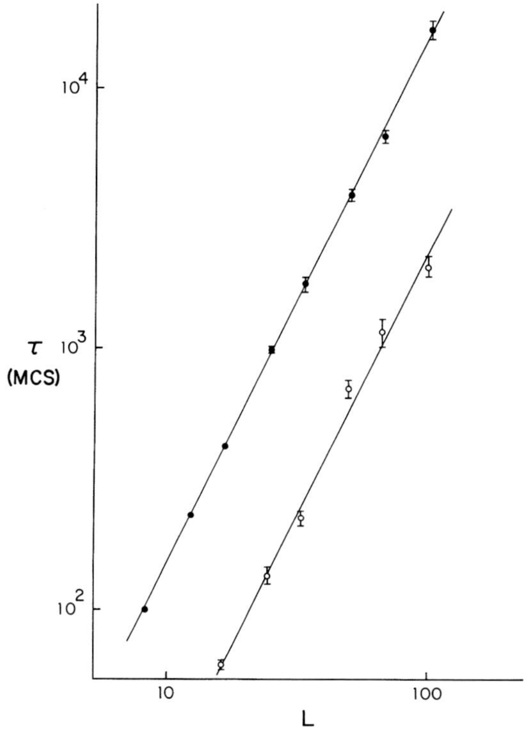

Fig. 8.2. Divergence of the correlation times with increasing lattice size for the magnetization and energy for the three dimensional Ising model at T_c. From Wansleben and Landau [3].

problem is accentuated due to the presence of many competing statistically important regions of phase space, each separated by high barriers from the others.

8.2.3 Traditional Computational Solutions

In some cases it is possible to optimize the Metropolis method using clever algorithms and computer coding schemes. For example, in the case of lattice spin models, instead of flipping a single spin at each MC step, one can flip many spins. One such approach, suitable for vector or massively parallel computers, is "checkerboard updating" which decomposes the lattice into inter-penetrating sub-lattices that are considered alternately. Further speed-ups can be realized using "multispin coding" in which several spins are packed into a single computer word and are operated upon simultaneously. Unfortunately, such methods do not generalize well to off-lattice models such as fluids. Moreover, their utility remains limited even for lattice-based spins models because they do not always allow increases in system sizes of the magnitude needed to overcome the growth of timescales outlined above. A few special purpose Ising model computers have been constructed, and although these have provided enhanced performance they cannot solve the intrinsic limitations of the Metropolis method.

8.3 Some "Recent" Developments

8.3.1 Second Order Transitions

Many of the MC simulation methods commonly deployed for the study of phase transitions and critical phenomena, have been described in detail elsewhere [1]. Although the workhorse for many years was the Metropolis algorithm, new, efficient algorithms have allowed simulation to achieve the resolution which is needed to accurately locate and characterize phase transitions. For the examination of second order transitions in lattice-based spin models, cluster-flip algorithms, beginning with the seminal work of Swendsen and Wang [4], have been used to reduce critical slowing down near second order transitions. We consider these in some detail below.

Another invaluable technique in the context of both second and first order transitions, is histogram extrapolation [15]. The method (which is outlined in Sect. 8.3.2.3) can be applied in the analysis of simulation data to increase the amount of information that can be gleaned from it. However, its applicability is limited in large systems by the statistical quality of the "wings" of the histogram. This latter effect is quite important in systems with competing interactions for which short range order effects might occur over very broad temperature ranges, or even give rise to frustration that produces a very complicated energy landscape and limits the efficiency of other methods.

8.3.1.1 Cluster Flipping

Successive configurations generated by a MC simulation of a spin model can be more rapidly decorrelated if each trial update involves more than one spin flip at each trial update. The question of how to do that in an intelligent way eluded researchers for many years until a little known theorem in theoretical physics was used to design new methods that flip correlated clusters of spins. The first steps were taken by Kasteleyn and Fortuin [5] who showed that it was possible to map a ferromagnetic Potts model onto a corresponding percolation model. In the percolation problem states are produced by throwing down particles, or bonds, in an uncorrelated fashion; hence there is no critical slowing down. The Fortuin-Kasteleyn transformation thus permits a problem with slow critical relaxation to be mapped into one where such effects are largely absent. The Swendsen-Wang approach [4] replaces each pair of interacting Potts spins on the lattice by a bond on an equivalent lattice with probability p where

$$p = 1 - \exp(K) \tag{8.4}$$

and $K = -J/(k_BT)$. All clusters of sites which are produced by a connected network of bonds are identified, and then each cluster is randomly assigned a new spin value, using a random number, i.e. each site in a cluster must have the same new spin value.

Since the probability of placing a bond between pairs of sites depends on temperature, the resultant cluster distributions will vary dramatically with temperature. Near a critical point a rich array of clusters is produced and each resultant configuration differs substantially from its predecessor. The dynamic critical exponent z is reduced from a value of just over 2 for Metropolis single-site spin flipping to a value of about 0 (actual log) in 2-dim. and 0.5 in 3-dim. [6]. The overall performance of the algorithm also depends strongly on the complexity of the code which is usually much greater than for single spin-flip methods. Hence, for small lattices the Swendsen-Wang technique may actually be slower in real time!, but for sufficiently large lattices it will eventually become more efficient.

One obvious shortcoming of the Swendsen-Wang approach is that significant effort is expended in dealing with small clusters as well as large ones. These small clusters do not contribute to the critical slowing down, so their consideration does not accelerate the algorithm. In order to partially eliminate this constraint, Wolff [7] proposed an alternative algorithm based on the Fortuin-Kasteleyn theorem in which single clusters are grown and flipped sequentially; the resultant performance generally exceeds that of the Swendsen-Wang method. The algorithm begins with the (random) choice of a single site. Bonds are then drawn to all nearest neighbors which are in the same state with the same probability as for Swendsen-Wang sampling. One then moves to all sites in turn which have been connected to the initial site and places bonds between them and any of their nearest neighbors which are in the same state with probability p. The process continues until no new bonds are formed and the entire cluster of connected sites is then flipped. Another initial site is chosen and the process is then repeated. The Wolff dynamics has a smaller prefactor and smaller dynamic exponent than does the Swendsen-Wang method. Of course the measurement of MC time is more complicated since a different number of spins is altered by each cluster flip. The generally accepted method of converting to MCS/site is to normalize the number of cluster flips by the mean fraction of sites flipped at each step.

8.3.1.2 The N-fold Way and Extensions

At this point we briefly review a fairly old algorithm that has found new utility through the development of powerful extensions. At very low temperatures the flipping probability for the Metropolis method becomes quite small and virtually nothing happens for a long time. In order to avoid this wasteful procedure Bortz et al [8] introduced an event driven algorithm (the "N-fold Way") in which a flip occurs at each step of the algorithm and then the lifetime of the preceding state is calculated.

First we observe that there are only a small number of possible local environments which a spin can possibly have and consequently a limited number of different flipping probabilities. All spins are collected into lists, in which each member has the identical energetic local environment. For an Ising square lattice there are a total of 10 possible combinations of a spin and "interacting environment", i.e. every spin in the system can belong to one of only 10 classes. (For other interactions the number of classes may differ, but in all cases will be some modest size integer N. Hence the name N-fold way.) The total probability of any spin of class l flipping in a given step is

$$p_l = n_l \exp\left(-\Delta E_l / k_B T\right) \tag{8.5}$$

where n_l is the number of spins which are in class l. The integrated probability Q_N of "some" event occurring in a given step is simply the sum of the probabilities for all N classes. A random number generated to determine the class from which the next spin to be overturned will come, and once the class has been chosen, another random number must be chosen to pick a spin from among those in the class. Finally, a 3rd random number will be used to determine how much time has elapsed before this event has taken place. Each time a spin is flipped, it changes class so it must then be removed from the list belonging to its original class and added to the new list corresponding to its new class. In addition, all of its (interacting) near neighbors change class. Efficient bookkeeping obviously is important. Treating the flipping event as a stochastic process, we can generate a random number ζ between 0 and 1 and find that the "lifetime" of the state before flipping occurs is

$$\Delta t = -(\tau_o / Q_N) \ln \zeta \tag{8.6}$$

where τ_o is the time needed to carry out a spin flip. The thermodynamic averages of properties of the system are then calculated by taking the lifetime weighted average over the different states which are generated. The N-fold way is rather complicated to implement, but at low temperatures the net gain in speed can be many orders of magnitude.

A recent generalization of the N-fold way algorithm ("absorbing Markov chains", or MCAMC [9]), was shown to offer substantial advantage for the study of magnetization switching in nanoscale ferromagnets and related problems. At low temperatures a strongly magnetized ferromagnet will not immediately reverse when an oppositely directed magnetic field is applied because there is a large nucleation barrier to the formation of a cluster of overturned spins. In a Metropolis simulation very long times are then needed to see the magnetization reversal. The MCAMC approach extends the N-fold way algorithm to allow the simultaneous flipping of more than one spin to facilitate formation of a nucleation cluster. The "level" of the method determines how many spins may be overturned in a single step. The level-1 MCAMC is essentially the N-fold way [9] and is best used for an initial state in which all spins are up. A random number ζ is picked and used to determine the lifetime of the state. A spin is then randomly chosen and overturned. Level-2 MCAMC offers a decided advantage in the case that the nucleation cluster size is at least two, since it avoids the tendency to merely overturn those spins that have just been flipped. The level-2 MCMAC begins with a fully magnetized state and overturns 2 spins. Then a transient submatrix T is defined to describe the single timestep transition probabilities, i.e. for overturning one spin to reach a transient (intermediate) state, and the recurrent submatrix $\mathbf{R}$ which gives the transition probabilities from the transient to the absorbing (final) states. Again a random number ζ is chosen and the lifetime of the state is determined by $\nu^{T^m} < \zeta < \nu^{T^{m-1}} e$ where ν is the vector describing the initial state and e is the vector with all elements equal to one. Another random number is then generated to decide which spins will actually flip. Following generation of this "initial cluster", the N-fold way may then be used to continue. This method may be systematically extended to higher order when the size of the nucleation cluster is larger so that the process of overturning a cluster is "seeded".

8.3.1.3 "Wang–Landau" Sampling

We shall now describe a new, efficient MC algorithm that offers substantial advantages over existing approaches and takes a different approach to sampling for statistical systems [10]. In contrast to "traditional" Monte Carlo methods that generate canonical distributions at a given temperature $g(E) \times e^{-E/K_B T}$, this method estimates the density of states $g(E)$ accurately via a random walk which produces a "flat" histogram in energy space. Of course, multiple random walks, each restricted to a different range of energy, may be performed to further improve the efficiency. The resultant pieces of $g(E)$ can be joined together and used to produce canonical averages for thermodynamic quantities at essentially any temperature.

The algorithm relies on the observation that if a random walk in energy space is performed with a probability proportional to the reciprocal of the density of states $\frac{1}{g(E)}$, then a flat histogram is generated for the energy distribution. This is done by modifying the estimated density of states systematically to produce a "flat" histogram over the allowed range of energy and simultaneously making the density of states converge to the correct value. Some initial estimate is made for the density of states, e.g. $g(E) = 1$. The random walk in energy space proceeds by flipping spins randomly; if E_1 and E_2 are energies before and after a spin is flipped, the transition probability of a spin flip is

$$p(E_1 \rightarrow E_2) = \min(\frac{g(E_1)}{g(E_2)}, 1). \qquad (8.7)$$

Each time an energy level E is visited, $g(E)$ is updated by multiplying the existing value by a modification factor $f > 1$, i.e. $g(E) \rightarrow g(E) * f$, starting with a large enough value that $g(E)$ grows quickly, e.g. $f_0 = e^1 \simeq 2.71828....$ The random walk continues until the accumulated histogram $H(E)$ is "flat", the modification factor is then reduced using some simple recipe, e.g. $f_1 = \sqrt{f_0}$, the histogram is reset to $H(E) = 0$, and a new random walk is begun. This process continues for n iterations, until f_n is smaller than some predefined final value (such as $f_{\mathrm{final}} = \exp(10^{-8}) \simeq 1.00000001$). Typically, the phrase "flat histogram" means that the histogram $H(E)$ for all possible E is not less than $\sim 80\%$ of the average value $\langle H(E)\rangle$. Since $g(E)$ is modified every time the state is visited, only a relative density of states is produced at the end of the simulation, and the results must be normalized. This can be done by using the condition that the number of ground states for the Ising model is 2 to re-scale the density of states; and if multiple walks are performed within different energy ranges, they must be matched up at the boundaries in energy.

During the early stages of iteration the algorithm does not satisfy detailed balance, since $g(E)$ is modified continuously; however, after many iterations the modification factor approaches 1 and detailed balance is recovered to high precision. Then, from (8.7):

$$\frac{1}{g(E_1)} p(E_1 \to E_2) = \frac{1}{g(E_2)} p(E_2 \to E_1) \tag{8.8}$$

where $\frac{1}{g(E_1)}$ is the probability at the energy level E_1 and $p(E_1 \to E_2)$ is the transition probability from E_1 to E_2 for the random walk. Consequently, the detailed balance condition is satisfied to within an accuracy proportional to $\ln(f)$.

The convergence and accuracy of this algorithm may be tested for a system with a second order transition, the $L \times L$ Ising square lattice with nearest neighbor coupling [11,15,12]. In Fig. 8.3, final results for the densities of states for several finite lattice sizes are compared with exact results [13]. The quality of the data can best be assessed by looking at the relative error $\varepsilon(\log(g(E)))$defined as $\varepsilon(X) \equiv |(X_{\mathrm{sim}} - X_{\mathrm{exact}})/X_{\mathrm{exact}}|$ for a general quantity X, and as the inset shows the agreement is excellent. With this algorithm $g(E)$ can be estimated efficiently even for large systems; moreover, the free energy and the entropy are also accessible, unlike in conventional MC simulations. The free energy is given by

$$F(T) = -k_{\mathrm{B}}T \ln(Z) = -k_{\mathrm{B}}T \ln(\sum_E g(E) e^{-\beta E}). \tag{8.9}$$

A comparison between the simulational data and the exact free energy for the Ising square lattice [14] for $L = 256$ is shown in Fig. 8.4. Here, too, the agreement is excellent!

8.3.2 First Order Transitions

It is common when discussing the character of first order phase transitions to do so in terms of the mathematical properties of thermodynamic response

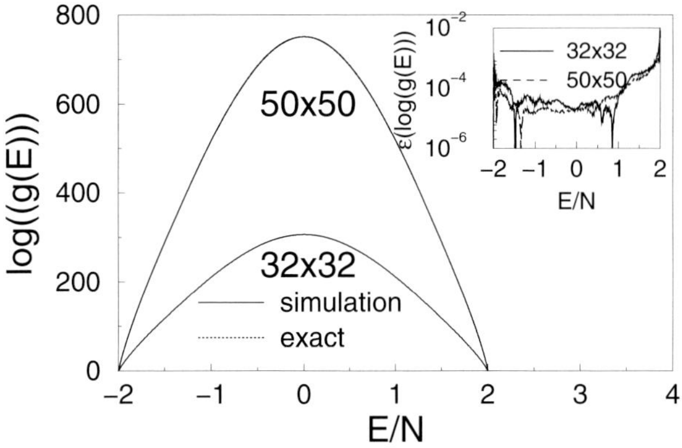

Fig. 8.3. Density of states for two different size Ising square lattices.

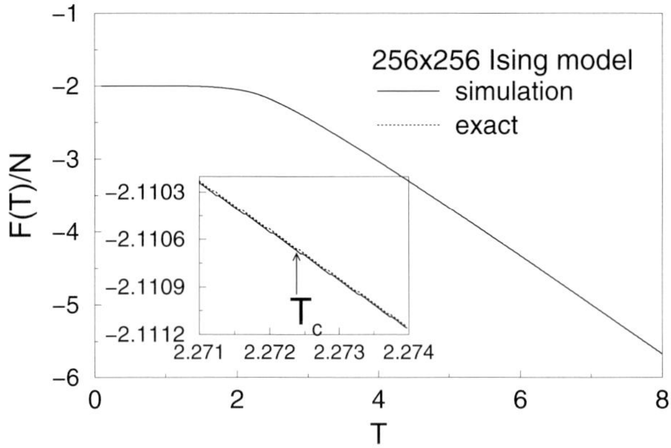

Fig. 8.4. free energy for the two dimensional Ising model.

functions such as the heat capacity or the order parameter as the transition is traversed [16]. An alternative view, one that is more natural in the context of Monte Carlo simulation, considers instead the probabilities associated with the sets of microstates (configurations) identifiable as belonging to each of the pure phases that coexist on a first order phase boundary.

8.3.2.1 Free Energy Comparison: The Statistical Mechanics Perspective

The situation we shall consider is illustrated schematically in Fig. 8.5 which depicts the configuration space of some system that can exist in either of two phases A and B. The sets of pure phase configurations are shown as disjoint "islands". The region between the pure phase states corresponds to

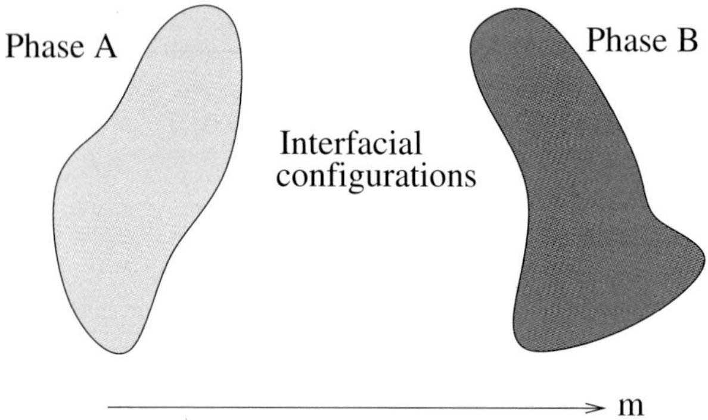

Fig. 8.5. Schematic representation of configuration space for a system that can exist in two phases.

interfacial configurations in which a portion of the system is found in phase A, separated by an interface from the remainder which is in phase B.

Typically the location of the system in configuration space is specified in terms of a fluctuating order parameter m which serves to distinguish whether the system is in phase A, phase B or somewhere in between. For example, in a simple ferromagnet such as the Ising model, m is the magnetization, while for a fluid undergoing a liquid-gas phase transition it is usually taken to be the fluid density.

The key to locating a first order phase transition by simulation is the ability to measure the free energy difference between the two coexisting phases. Unfortunately, a single simulation cannot deliver the absolute free energy of a given phase. However, subject to certain provisos, it can measure the free energy *difference* between two phases. To see this, consider the canonical distribution of microstates, which is given by

$$p(\{\sigma\}) = \frac{e^{-\beta H(\{\sigma\})}}{Z}, \tag{8.10}$$

where $\{\sigma\}$ labels a microstate (i.e. a configuration), $H(\{\sigma\})$ is the system Hamiltonian, β is the inverse temperature and Z is the partition function. Suppose, now, that we sum over all microstates identifiable as belonging to a certain phase γ, then

$$\begin{aligned} p_\gamma &= \sum_{\{\sigma\}\in\gamma} p(\{\sigma\}) \\ &= \frac{1}{Z}\sum_{\{\sigma\}\in\gamma} e^{-\beta H(\{\sigma\})} \\ &\equiv \frac{Z_\gamma}{Z}, \end{aligned} \tag{8.11}$$

where the last step serves to define the *configurational weight* Z_γ of the phase γ. The relative probabilities of two phases A and B is then

$$\mathcal{R}_{AB} \equiv \frac{p_A}{p_B} = \frac{Z_A}{Z_B} \qquad (8.12)$$

$$\equiv \frac{e^{-\beta F_A}}{e^{-\beta F_B}}$$

where F_γ denotes the free energy of phase γ. It follows that the free energy difference between two phases A and B is simply proportional to the logarithm of the ratio of their a-priori probabilities:

$$F_A - F_B = -\frac{1}{\beta}\ln \mathcal{R}_{AB}, \qquad (8.13)$$

Clearly this equation implies that precisely at coexistence ($F_A = F_B$) the system will be found with equal probability in each of the two phases.

In order to exploit (8.13) a simulation procedure is needed that will measure the requisite probability ratio. To achieve this the simulator must contrive a scheme by which both phases are visited many times in the course of single simulation run. Monitoring the relative frequency with which the simulation is found in each phase provides a direct estimate of $\mathcal{R}_{AB}$. Practically this is can be done by appeal to the measured form of the order parameter distribution function $p(m)$. In a simulation this distribution is normally accumulated in the form of a histogram. Formally it is given by

$$p(m) = \frac{1}{Z}\sum_{\{\sigma\}} \delta(m - m(\{\sigma\}))e^{-\beta H\{\sigma\}}\,, \qquad (8.14)$$

where the sum extends over all microstates of the system.

At a first order phase transition, the form of $p(m)$ is strongly double peaked in character, as shown schematically in Fig. 8.6. States having m values close to the peak positions of $p(m)$ correspond to pure phase configurations, while those in the trough between the peaks correspond to mixed-phase (interfacial) configurations. Compared to pure phase states, interfacial configurations have an intrinsically small probability on account of their high surface tension. Accordingly such states are visited only rarely in the course of a simulation at coexistence.

In principle, measurement of the ratio of the integrated weights (areas) under the two peaks provide a direct estimate of $\mathcal{R}_{AB}$, whence the location of coexistence can be estimated. Unfortunately, the strongly double peaked character of $p(m)$ complicates accurate measurement of its form. The low probability of interfacial states renders spontaneous transitions between the two coexisting phases very infrequent and results in greatly extended correlation times. This in turn hinders the accumulation of independent statistics

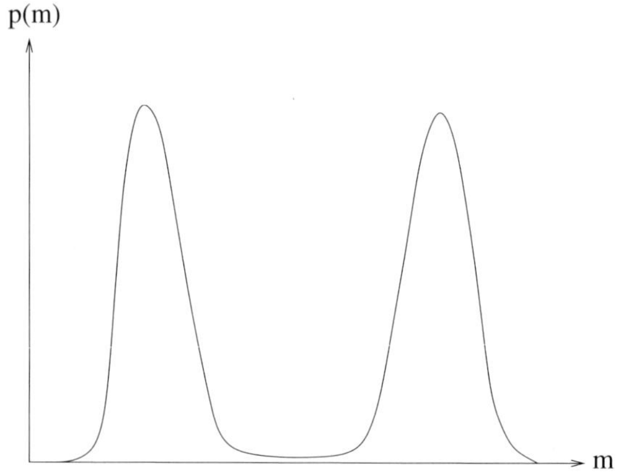

Fig. 8.6. Schematic of the form of the order parameter distribution function at a first order phase boundary.

on the relative peak weights and estimates for coexistence parameters. In the next subsection we discuss one effective method of bridging this time scale gap, although others exist, see e.g. ref. [17,18].

8.3.2.2 Multicanonical Monte Carlo

The basic idea underpinning Multicanonical Monte Carlo [19] is to preweight the evolution of the MC Markov chain so as to preferentially sample the interfacial configurations of intrinsically low probability. So doing allows the simulation to pass with ease from one pure phase to the other, thereby increasing the frequency of transitions and hence the statistical quality of the estimate for $\mathcal{R}_{AB}$. The effects of the imposed bias on this estimate (as well as those for other observables) can be corrected straightforwardly. The situation is illustrated schematically in Fig 8.7.

The bias in a multicanonical simulation is imposed via the Hamiltonian. Instead of simulating with the bare Hamiltonian of the system of interest we define an "effective Hamiltonian" given by

$$\tilde{H}(\{\sigma\}) = H(\{\sigma\}) + \eta(m) , \tag{8.15}$$

where $\eta(m)$ is a preweighting function, which must (as described below) be prescribed in advance. Simulating with this effective Hamiltonian, we measure the preweighted order parameter probability distribution function $\tilde{p}(m)$, given by

$$\tilde{p}(m) = \frac{1}{\tilde{Z}} \sum_{\{\sigma\}} \delta(m - m(\{\sigma\})) e^{-\beta \tilde{H}(\{\sigma\})} . \tag{8.16}$$

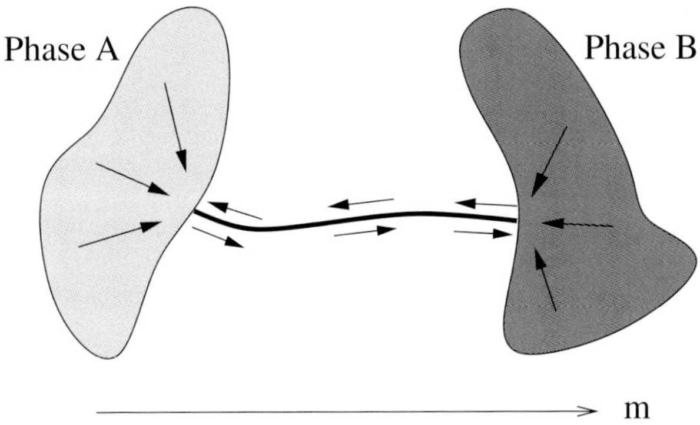

Fig. 8.7. Schematic illustration of the multicanonical method. By appropriate preweighting of the MC Markov chain, the probability of sampling interfacial configurations is enhanced.

The true (Boltzmann distributed) weight function $p(m)$ is recovered by unfolding the effects of the imposed weights:

$$p(m) = e^{\eta(m)} \tilde{p}(m) . \tag{8.17}$$

It transpires that there exists a choice of $\eta(m)$ for which $\tilde{p}(m)$ is constant in the region of m between the two peaks of $p(m)$. Under these circumstances, the system performs a one-dimensional random walk over the entire domain of m, thereby permitting efficient accumulation of statistics for $\tilde{p}(m)$. The effects of the bias are subsequently unfolded from $\tilde{p}(m)$ (via (8.17)) to obtain the desired unbiased function $p(m)$. Figure 8.8 shows an example for the subcritical Ising model.

Inspection of (8.14), (8.16), and (8.17) reveals that to realize a flat distribution, $\tilde{p}(m) =$ constant, requires that the preweighting function be prescribed as $\eta(m) = \ln p(m)$. But $p(m)$ is of course just the function we are trying to find. Thus it is not possible in general to immediately implement an optimal multicanonical preweighting. Instead it is necessary to construct a suitable weight function from scratch via an iterative procedure. A number of strategies for achieving this have been proposed and tested in the literature. We shall not review them here, but refer the interested reader to the recent review of Berg [23].

8.3.2.3 Tracking Phase Boundaries: Histogram Extrapolation

Use of the multicanonical method permits efficient determination of a point on a first order phase boundary. However, the task of tracing out a whole coexistence line in a (possibly) multidimensional space can nevertheless be a laborious one due to the need to determine many coexistence points. Moreover, at first sight it would seem necessary that at each state point studied a suitable preweighting function be determined from scratch. Fortunately, as

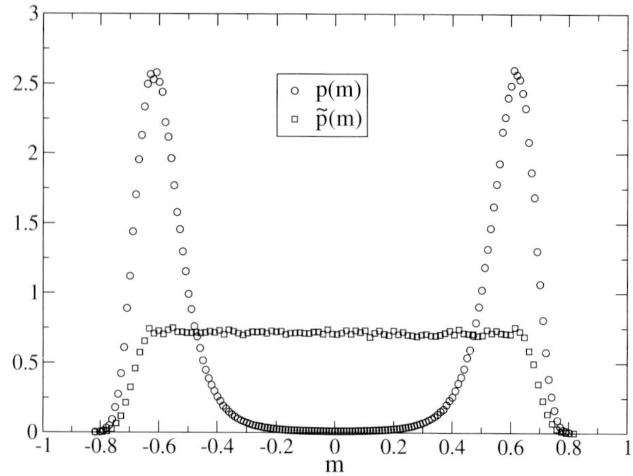

Fig. 8.8. Results from a multicanonical simulation of the $L = 10$, 3D-Ising model at $T = 0.95T_c$ in zero field. Shown is the preweighted form of the order parameter distribution function $\tilde{p}(m)$ and the corresponding form of $p(m)$.

we shall now show, use of histogram extrapolation greatly simplifies these tasks.

Histogram extrapolation rests on the observation that histograms of observables accumulated at one set of model parameters (e.g. β and a field h conjugate to the order parameter) can be reweighted to provide estimates of histograms for other values of these parameters. Consider the *joint* probability distribution of energy and order parameter at the particular parameter values $\beta = \beta_0$ and $h = h_0$. Formally $p(m, E|\beta_0, h_0)$ is given by

$$p(m, E|\beta_0, h_0) = \frac{1}{Z_0} \sum_{\{\sigma\}} \delta(E - E(\{\sigma_i\}))\delta(m - m(\{\sigma_i\}))\, e^{-\beta_0 H_0} , \quad (8.18)$$

where $H_0(\{\sigma\}, N) \equiv E(\{\mathbf{r}\}) + h_0 m$. It is easy to show that an estimate for the form of $p(m, E)$ at the parameter values $\beta = \beta_1, h = h_1$ can be obtained from the measured $p(m, E|\beta_0, h_0)$ by the simple reweighting [15] :

$$p(m, E|\beta_1, h_1) = \frac{Z_1}{Z_0} e^{-(\beta_1 H_1 - \beta_0 H_0)} p(m, E|\beta_0, h_0) , \quad (8.19)$$

where the ratio Z_1/Z_0 is an unimportant constant that is effectively absorbed into the normalization. If desired, this joint distribution can then be integrated to yield the order parameter probability density function at β_1, h_1:

$$p(m|\beta_1, h_1) = \int dE\, p(m, E|\beta_1, h_1) . \quad (8.20)$$

In principle, a single simulation at one state point in the phase diagram suffices to obtain information for all other state points. Unfortunately the reality is less auspicious. Owing to finite sampling time, it is not possible in practice to extrapolate a single histogram obtained at β_0, h_0 to arbitrary values of β_1, h_1. Instead, the parameters to which we extrapolate must be fairly close to those at which the simulation was actually performed or the procedure loses accuracy. One way of dealing with this problem is to perform a sequence of separate simulations at strategic intervals across the range of model parameters of interest. The role of histogram reweighting is then to interpolate to the regions of parameter space between the simulation points. We note in passing that it is possible to combine (in a self-consistent fashion) the results of a number of different simulations at different model parameters and perform histogram extrapolation on the *aggregate* data. For a description of this more sophisticated procedure, we refer the reader to Refs. [15] and [20].

Histogram extrapolation can be combined with the multicanonical ensemble method to deliver a powerful method for tracking a phase boundary through a multidimensional parameter space. The procedure comprises two stages. Firstly it is bootstrapped by finding the form of $p(m)$ at some arbitrary near-coexistence state point. This is achieved via a multicanonical simulation conducted at some guessed values of the coexistence parameters. Generally speaking, however, unless this guess is very good, the peaks in the resulting form of $p(m)$ will have unequal areas. The guess is therefore subsequently refined by tuning the model parameters (within the extrapolation scheme) to give equal areas under the two peaks.

In a similar way, the histogram for $p(m)$ can be extrapolated to obtain estimate of other state points further along the phase boundary, provided of course that they lie within the range of reliable extrapolation. Additionally, for each phase boundary point considered, the extrapolation delivers an estimate of the associated form of $p(m)$. The latter serves as a suitable multicanonical preweighting function for a fresh simulation conducted at a new coexistence point, the results of which can themselves be used to extrapolate even further along the phase boundary. Clearly by iterating this procedure one traces out the full coexistence line without ever having to determine a new preweighting function from scratch. Further details of this approach can be found in refs.[21,22].

8.3.2.4 Phase Switch Monte Carlo

Use of multicanonical methods to surmount the free energy barrier between coexisting phases has proved itself an effective and efficient means of studying many varieties of first order phase transition. Systems to which the method has been applied span a broad range from simple lattice spin models to complex fluids [23]. Unfortunately, the approach is not effective in all situations; a case in point is the freezing of a simple fluid.

The particular difficulties presented by the freezing transition stem from the distinctive symmetries of the coexisting fluid (F) and crystalline solid

(CS) phases. In a multicanonical simulation the natural inter-phase route traverses mixed phase (interfacial) states. Such states are kinetically problematic in the F-CS transition because the crystal that forms from the fluid is (one finds) often extensively defect-ridden. These defects do not normally anneal out on accessible simulation time-scales and can cause the system to become trapped in states from which it cannot escape.

In view of this problem, computational studies for the freezing transition have, to date, relied primarily on indirect approaches, specifically Thermodynamic Integration [24,20]. Here, instead of linking the two phases directly, the free energy of each phase is computed separately for states of a range of densities, using integration techniques, which connect their thermodynamic properties with those of known reference states. The two branches of the free energy are then matched to determine the coexistence parameters. Thermodynamic integration can be computationally laborious because of the need to perform many simulations at different values of the model parameters defining the integration path. Additionally the integration path may encounter singularities in the form of other first order phase transitions.

In view of these difficulties, a new MC simulation approach to first order transition has recently been proposed [25,26]. The method, known as Phase Switch Monte Carlo (PSMC), was originally developed as a means of computing free energy differences between distinct crystalline structures, where interfacial states are computationally problematic. It has recently been extended to allow the freezing transition to be tackled. In outline, the new method [26] samples the disjoint configuration spaces of two coexisting phases within a single simulation. At its heart is a global coordinate transformation or "phase switch" (implemented as a MC move) which directly maps one pure phase onto the other. Biased sampling methods are employed to enhance the probability of certain "gateway" states in each phase from which the switch can be successfully launched (cf. Fig. 8.9). The method permits direct determination of equilibrium coexistence-point parameters and prescribes statistical uncertainties transparently.

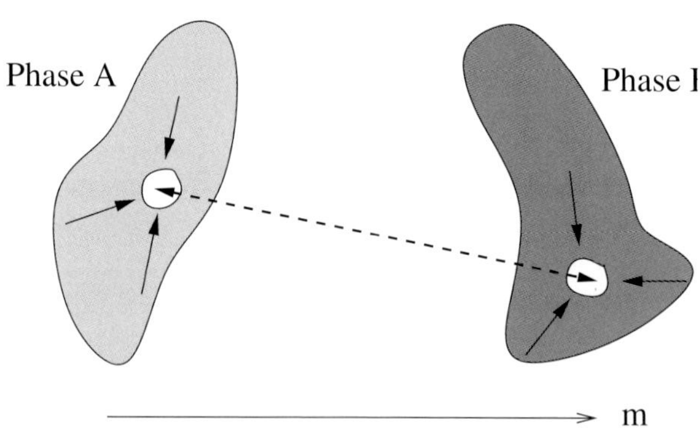

Fig. 8.9. Schematic illustration of phase switch method.

To illustrate the method we consider the freezing transition of hard spheres simulated within a constant-NpT simulation ensemble [20] with periodic boundary conditions. The configurational weight of a phase may be written as

$$\mathcal{Z}_\gamma(N,p) = \int_0^\infty dV e^{-pV} Z_\gamma(N,V) \tag{8.21}$$

where N is the particle number, V the system volume and p the reduced pressure [27]. γ (CS or F) labels the phase, while

$$Z_\gamma(N,V) = \frac{1}{N!}\prod_{i=1}^{N}\int_{V,\gamma} d\boldsymbol{r}_i e^{-E(\{\boldsymbol{r}\})} \; . \tag{8.22}$$

Here E is the hard sphere configurational energy [27], while the factor of $(N!)^{-1}$ corrects for indistinguishability. The γ-label on the integral stands for some *configurational constraint* that picks out configurations $\{\boldsymbol{r}\}$ that 'belong' to phase γ. In a MC simulation, this constraint is formulated implicitly as follows. Let $\boldsymbol{R}_1^\gamma \ldots \boldsymbol{R}_N^\gamma \equiv \{\boldsymbol{R}\}^\gamma$ denote some *representative configuration* of phase γ; we shall refer to $\{\boldsymbol{R}\}^\gamma$ as the reference configuration. Then the constraint picks out those configurations which can be reached from $\{\boldsymbol{R}\}^\gamma$ on the simulation time-scale. This time-scale is presumed to be sufficiently long to allow exploration of one phase, but still short compared to spontaneous inter-phase traverses. Such a situation is realized if the freezing transition is sufficiently strongly first order.

Let us now designate the reference sites $\{\boldsymbol{R}\}^\gamma$ as the origins of the particle coordinates via some arbitrary association between the N particles and the N reference sites. The set of particle positions can then be written as

$$\boldsymbol{r_i}^\gamma = \boldsymbol{R_i}^\gamma + \boldsymbol{u}_i \; ,$$

which serves to define the set of displacement vectors $\{\boldsymbol{u}\}$ (independent of the phase label γ) linking each particle i to its associated reference site $\boldsymbol{R}_i$. The configurational energy follows accordingly as

$$E^\gamma(\{\boldsymbol{u}\}) \equiv E(\{\boldsymbol{R}^\gamma + \boldsymbol{u}\}) \; .$$

We now examine the implications of the above formulation for the configurational weights of the F and CS phases. In the case of the F-phase *all* contributing configurations are reachable from any one and may simply write

$$Z_F(N,V) = \frac{1}{N!}\prod_{i=1}^{N}\int_{V,\{\boldsymbol{R}\}^F} d\boldsymbol{u}_i e^{-E^F(\{\boldsymbol{u}\})} \tag{8.23}$$

where $\{\boldsymbol{R}\}^F$ is some specific but arbitrary fluid configuration, which can be selected at random in the course of MC exploration of the fluid phase.

In the case of the CS phase we choose $\{\boldsymbol{R}\}^{CS}$ to be the sites of a FCC lattice (the stable crystalline phase of hard spheres [25]) having the appropriate scale. Here though, it should be recognized that in contrast to the F phase, the MC simulation does not sample the complete CS configuration space. This actually comprises a number of distinct mutually inaccessible *fragments* corresponding essentially to the different permutations of particles amongst lattice sites. In the absence of self diffusion, a Monte Carlo simulation of the CS phase will visit (and thus count) only the states within the fragment in which it is initiated. By symmetry each fragment should contribute equally to the configurational weight. Hence the total configurational weight of the CS phase is given by multiplying the contribution of one fragment by the number of fragments. The number of such fragments is the number of distinct permutations of N distinguishable particles amongst N fixed lattice sites in a periodic system. This number is not $N!$ but $(N-1)!$ since certain permutations are reachable from others via a global translation (permitted via the boundary conditions) [26]. Thus we have

$$Z_{CS}(N,V) = \frac{(N-1)!}{N!}\prod_{i=1}^{N}\int_{V,\{\boldsymbol{R}\}^{CS}} d\boldsymbol{u}_i e^{-E^{CS}(\{\boldsymbol{u}\})}$$

$$= \frac{1}{N}\prod_{i=1}^{N}\int_{V,\{\boldsymbol{R}\}^{CS}} d\boldsymbol{u}_i e^{-E^{CS}(\{\boldsymbol{u}\})}\,, \tag{8.24}$$

The relative free energies of the two phases is given via (8.13). Combining this equation with (8.21), (8.23) and (8.24) we can write

$$\mathcal{R}_{\mathrm{F,CS}} = \frac{P(F|N,p)}{P(CS|N,p)} = \frac{\mathcal{Z}_F(N,p)}{\mathcal{Z}_{CS}(N,p)}$$

$$= \frac{\int_0^\infty dVe^{-pV}\prod_{i=1}^{N}\int_{V,\{\boldsymbol{R}\}^F} d\boldsymbol{u}_i e^{-E^F(\{\boldsymbol{u}\})}}{(N-1)!\int_0^\infty dVe^{-pV}\prod_{i=1}^{N}\int_{V,\{\boldsymbol{R}\}^{CS}} d\boldsymbol{u}_i e^{-E^{CS}(\{\boldsymbol{u}\})}} \tag{8.25}$$

To determine $\mathcal{R}_{F,CS}$ (Eq. 8.25), we require a MC procedure which visits both solid and fluid regions of configuration space in a single run. The key to achieving this is the observation that, by construction, the system may be transformed between the CS and F reference states simply by switching the representative vectors ($\boldsymbol{R}_i^F \rightleftharpoons \boldsymbol{R}_i^{CS}\ \forall i$). Hence, by continuity, *any* CS (F) configuration 'sufficiently close' to the reference one will also transform to a F (CS) state under this operation. This phase switch can itself be implemented as a MC step, so that the phase label γ becomes a stochastic variable.

However, the set of configurations for which the MC switch will be *accepted* will in general constitute only a small fraction of the respective configuration spaces. To ensure effective two-phase sampling the MC procedure must be multicanonically biased to enhance appropriately the probabilities with which these 'gateway' regions are visited. To that end we define an order parameter designed to allow the system to be drawn into the gateway regions:

$$M = M_\gamma(\{\boldsymbol{u}\}) = \sum_i \{O_i[1 - \theta(u_i - u_c)] + T_i\theta(u_i - u_c)\} \tag{8.26}$$

Here θ is the step function. $T_i \equiv \alpha u_i$ measures the length of a notional tether connecting site i to its associated particle. O_i measures the overlap (between particle i and its neighbors) which *would* be created by a phase switch. The parameter α controls the relative importance of T_i and O_i; u_c controls the tether-length domain in which each contributes.

The equilibrium states of both phases are characterized by large M values. The 'overlap' term contributes in both phases: swapping the $\{\boldsymbol{R}\}$ vectors will (in general) produce a configuration of the 'other' phase in which spheres overlap. The 'tether' term contributes only in the F-phase (as we describe below) where particles may drift arbitrarily far from the sites with which they are nominally associated; the tethers provide the means to 'pull' the fluid towards the reference sites. The gateway states are those for which $M = 0$ i.e. for which a phase switch can be implemented without incurring hard sphere overlaps.

The entire region of configuration space relevant to the problem can be sampled in the extended ensemble defined by

$$\tilde{\mathcal{Z}}(N, p, \{\eta\}) \equiv \sum_\gamma \int_0^\infty dV \prod_i^N \int_\gamma d\boldsymbol{u}_i e^{-H^\gamma(\{\boldsymbol{u}\}, V)} \tag{8.27}$$

where

$$H^\gamma(\{\boldsymbol{u}\}, V) = E^\gamma(\{\boldsymbol{u}\}) + pV + \eta_\gamma(M) - \delta_{\gamma,CS} \ln{(N-1)!}$$

while $\{\eta\}$ represents a set of weights associated with each of the set of M-macrostates associated with each phase. These weights are customized to sample the full range of M-space and to enhance the probabilities of the $M = 0$ gateway states [28].

Simulations in this ensemble allows one to measure the joint probability distribution $P(M, V, \gamma|N, p, \{\eta\})$. From this one can unfold the bias due to the weights to infer the true equilibrium distribution $P(M, V, \gamma|N, p)$. The desired free energy difference between the two phases (Eq. 8.25) follows by integrating over the contributions associated with each γ to give the *a priori* probabilities of the respective phases. Additionally, histogram extrapolation techniques (cf. Sect. 8.3.2.3) can be employed to determine the value of $\mathcal{R}_{\mathrm{F,CS}}$ at neighboring pressures, thereby permitting a very precise determination of the coexistence pressure.

We now turn to the MC procedure required for exploration of the space spanned by the configuration variables $\{\boldsymbol{u}\}$, V and γ. It comprises four types of configuration update, each of which is accepted with a probability defined by a Metropolis rule [20] reflecting the associated change in the effective

Hamiltonian H. The first two –particle position updates and volume updates (implemented as dilations)– are effected in standard ways [20]. The third –like the first two– also preserves the phase label; but it is novel. In this process, we choose two sites at random (i and j say) and identify the corresponding displacement vectors $\boldsymbol{u}_i$ and $\boldsymbol{u}_j$. The candidate configuration is defined by the replacements

$$\boldsymbol{u}_i \to \boldsymbol{u}_i' \equiv \boldsymbol{u}_j + \boldsymbol{R}_j - \boldsymbol{R}_i \quad \text{and} \quad \boldsymbol{u}_j \to \boldsymbol{u}_j' \equiv \boldsymbol{u}_i + \boldsymbol{R}_i - \boldsymbol{R}_j$$

This process can be thought of as an *association* update: the particle initially associated with ('tethered to') site j is subsequently associated with site i (and *vice versa*). It changes the *representation* of the configuration (the coordinates $\{\boldsymbol{u}\}$); but it leaves the physical configuration invariant. It is only *required* in the fluid phase where particles diffuse far from the sites with which they are initially associated and the associated tethers become large. Association updates allow the tethers to respond efficiently to the influence of the tether contribution to $\{\eta\}$. Finally, the 'phase switch' entails replacing one set of representative vectors, $\{\boldsymbol{R}\}^\gamma$ say, by the other, $\{\boldsymbol{R}\}^{\gamma'}$, with the volumes scaled appropriately and the displacement coordinates $\{\boldsymbol{u}\}$ held fixed. In the switch, the volume is scaled by $\alpha_v \equiv \bar{V}^{\gamma'}/\bar{V}^\gamma$ where $\bar{V}^\gamma$ is the equilibrium volume of phase γ.

Before the simulation is performed, values must be assigned to the parameters u_c and α appearing in the definition of the order parameter (8.26). There is some license in making this choice. We set $u_c = D/2$ (the hard sphere radius) which ensures that the overlap contribution to M 'switches on' whenever a particle is closer to its reference site than to any other reference site. Such a choice avoids the possibility that the phase switch might lead to a glassy configuration. The sole constraint on the assignment of the parameter α is that it be set sufficiently large to ensure a smooth hand-over between the tether contribution and overlap contribution to M. Satisfactory results were found using the value $\alpha = 1.7$. As an efficiency measure we chose to keep one particle *fixed* at its representative site in each phase; this suppresses the global translation mode in the CS phase and eliminates the need for association updates in this phase.

Simulations were performed using systems of $N = 32$, 108 and 256 particles. Suitable weights were obtained by iterative means using some of the techniques described in ref. [28]. In Fig. 8.10 we show a typical portion of the evolution of the preweighted order parameter M as a function of MC time. For clarity of representation, states in the F phase are denoted by positive values of M, while negative values correspond to CS phase states. We note that the range of M values sampled in the CS phase is quite small because particles are localized near their reference sites by the suppression of the global translation mode. By contrast, much larger values of M are explored in the CS phase because the particles can drift far from the reference site to which they are associated. Nevertheless the whole range can be spanned

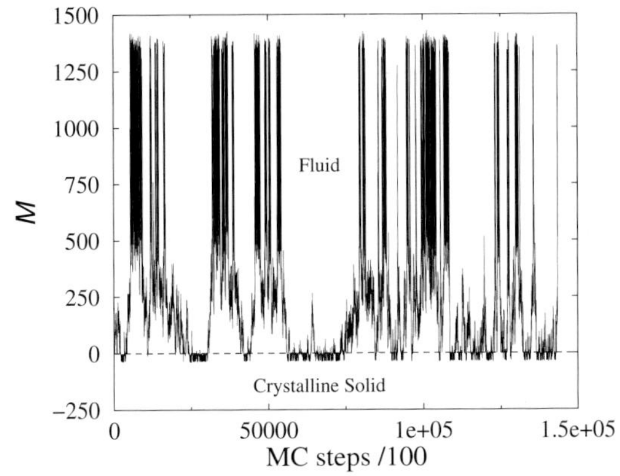

Fig. 8.10. The MC time evolution of the order parameter M for the $N = 256$ system. Phase switches occur between $M = 0$ states. For further details, see text.

relatively quickly by virtue of the highly efficient associations updates which permit large-scale changes in tether lengths.

The density distribution $p(\rho)$ was obtained from the measured distribution $p(M, V, \gamma|N, p, \{\eta\})$ by marginalising with respect to the volume V and unfolding the effect of the weights [22]. The results for the $N = 256$ system in the vicinity of the coexistence pressure are shown in Fig. 8.11. The distributions derive from histogram reweighting of simulation data obtained at $p = 11.18$. Coexistence, identified by the equality of the area under each peak, occurs for $p = 11.23(3)$.

Figure 8.12 shows the coexistence pressure for our three system sizes plotted as a function of the scaling variable $1/N$. The associated uncertainties $\sigma[p]$ are given by $\sigma[p] = \sigma[\mathcal{R}]/(N \mid \Delta v \mid)$ where $\Delta v = [\bar{V}_F - \bar{V}_{CS}]/N$ and $\sigma[\mathcal{R}]$ is the uncertainty in the measured ratio of the peak-weights, whose statistics depends on the frequency of the inter-phase switch. The three points are consistent with the presumed scaling form and the extrapolated prediction ($p = 11.49(9)$) is, within error, in accord with [24] and [29] (see Fig. 8.12 inset).

8.3.2.5 First Order Transitions and Wang–Landau Sampling

A simple model system that exhibits first order transitions and serves as an ideal testing ground for diverse algorithms is the 2-dimensional $Q - state$ Potts model on $L \times L$ square lattice with nearest-neighbor interactions and periodic boundary conditions. The Hamiltonian for this model can be written as:

$$H = - \sum_{<ij>} \delta(q_i, q_j) \tag{8.28}$$

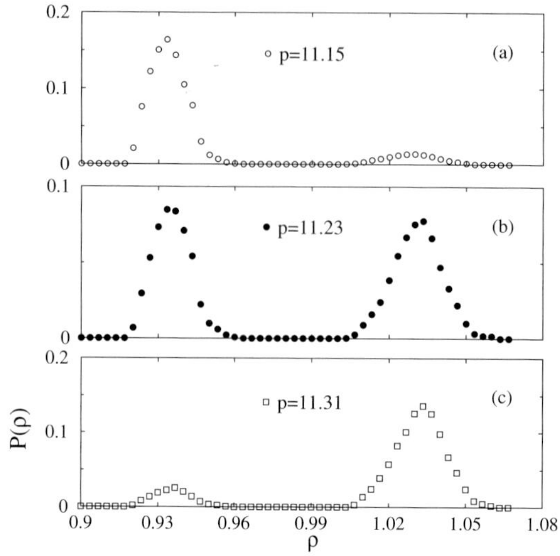

Fig. 8.11. The distribution of the density of the system of $N = 256$ particles at pressures (a) just below, (b) at and (c) just above coexistence for this N. The mean single phase density averages are $\rho_F = 0.934(3)$ and $\rho_{CS} = 1.031(4)$ in accord with the coexistence parameters reported in [29].

and $q = 1, 2, ...Q$. For $Q = 10$ the transition is strongly first order and long time scales for tunneling between coexisting states pose severe problems for standard methods. Wang-Landau sampling can be performed as for the Ising model but integers are chosen randomly between $[1 : Q]$ for possible new Potts spin values. The maximum density of states generated in this way for $L = 200$ is very close to 10^{40000}!

A canonical distribution $P(E, T)$ can then be determined from

$$P(E,T) = g(E)e^{-E/k_BT} \tag{8.29}$$

From the simulational result for the density of states $g(E)$, we can calculate the canonical distribution and in Fig. 8.13, we show the resultant double peaked canonical distribution [30], at the transition temperature T_c . Note that the peaks of the distributions are normalized to 1 in this figure. The valley between two peaks is quite deep, e.g. is 7×10^{-5} for $L = 100$. The latent heat for this temperature driven first-order phase transition can be estimated from the energy difference between the double peaks.

Because of the double peak structure at a first-order phase transition, conventional Monte Carlo simulations are not efficient since an extremely long time is required for the system to travel from one peak to the other in energy space. With the algorithm proposed in this paper, all possible energy levels are visited with equal probability, so it overcomes the tunneling barrier between the coexisting phases in the conventional Monte Carlo simulations.

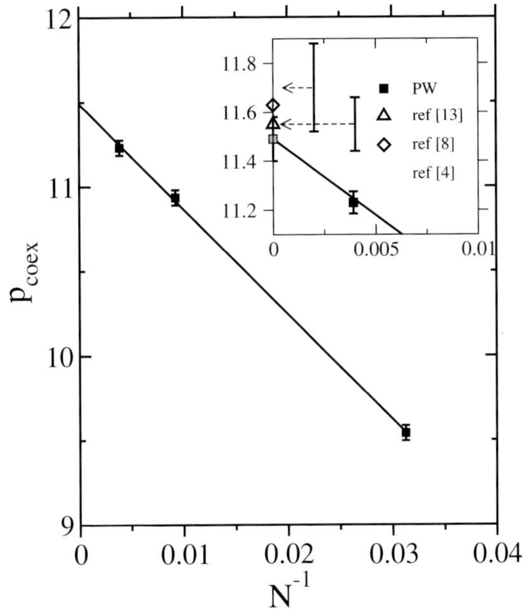

Fig. 8.12. The coexistence pressure for systems of different N using (8.25) both with (•) and without (○) the $1/N$ prefactor in the CS configurational weight. The solid line is a fit to the former; the dashed line is lower by $\ln N/[N\Delta v]$. The inset compares our extrapolated value with the results of others, with error bars shifted for clarity.

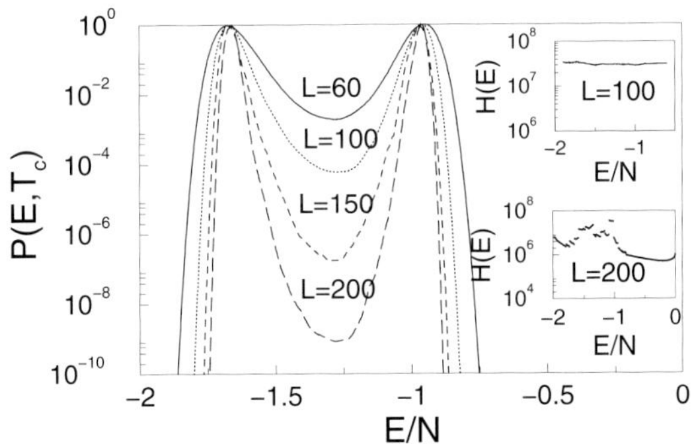

Fig. 8.13. Probability distribution for the $Q = 10$ Potts model at the finite lattice "transition temperature".

The histograms for $L = 60$, 80 and 100 are shown in the inset of the Fig. 8.13 which are very flat. The histogram in the figure is the overall histogram defined by the total number of visits on each energy level for the random walk.

With this algorithm, if the system is not larger than 100×100, the random walk on important energy regions (such as that which includes the two peaks of the canonical distribution at T_c) can be carried out with a single processor and will give an accurate density of states within about 10^7 visits per energy level. However, for a larger system, we can use a parallelized algorithm by performing random walks in different energy regions, each using a different processor.

The histograms from individual random walks are shown in the inset of Fig. 8.13 both for 100×100 and 200×200 lattices. In this case, we only require that the histogram of the random walk in the corresponding energy segment is sufficiently flat without regard to the relative flatness over the entire energy range. In Fig. 8.13, the results for large lattices show clear double peaks for the canonical distributions at temperatures $T_c(L) = 0.70127$ for $L = 150$ and $T_c(L) = 0.701243$ for $L = 200$. Considering the valley which we find for $L = 200$ is as deep as 9×10^{-10}, we can understand why it is impossible for conventional Monte Carlo algorithms to overcome the tunneling barrier with available computational resources.

This approach allows the estimation of the transition temperature, and the transition temperature extrapolated from the simulational data is consistent with the exact solution for the infinite system. Results can also be compared with existing numerical data such as estimates of transition temperatures and double peak locations obtained with the multicanonical simulational method by Berg and Neuhaus [19] and the Multibondic cluster algorithm by Janke and Kappler [31]. Wang-Landau sampling allowed calculation of the density of states up to 200×200 within 10^7 visits per energy level to obtain a good estimate of the transition temperature and locations of the double peaks. Using the multicanonical method and a finite scaling guess for the density of states, Berg *et. al.* only obtained results for lattices as large as 100×100 [19], and multibondic cluster algorithm data [31] were not given for systems larger than 50×50.

8.3.3 Systems with Complex Order

There are many systems in statistical physics that do not have simple groundstates and conventional phase transitions, e.g. spin glasses [32] for which the interactions between the magnetic moments produce frustration because of structural disorder. One of the simplest theoretical models for such systems is the Edwards-Anderson model [33] (EA model). For such systems analytical methods provide limited information and, because of the rough energy landscape of such disordered systems, the relaxation times of the conventional Monte Carlo simulations are very long. Thus, simulations can be performed only on rather small systems, and many properties of spin glasses are still left unclarified [35–42].

Using a random walk in energy space, we can estimate the ground state energy and $g(E)$ very easily. For spin glass systems it is also important to determine the order parameter which can be defined by [33]

$$q^{\mathrm{EA}}(T) \equiv \lim_{t\to\infty} \lim_{N\to\infty} q(T,t), \quad q(T,t) \equiv \langle \sum_{i=1}^{N} S_i(0) S_i(t)/N \rangle. \tag{8.30}$$

Here, $N = L^3$ is the total number of the spins in the system, L is the linear size of the system, $q(T,t)$ is the auto-correlation function, which depends on the temperature T and the evolution time t, and $q(T,0) = 1$. When $t \to \infty$, $q(T,t)$ becomes the order parameter of the spin glass. This parameter takes the following values

$$q^{\mathrm{EA}}(T) \begin{cases} = 1 \text{ if } T = 0 \\ = 0 \text{ if } T \geq T_{\mathrm{g}} \\ \neq 0 \text{ if } 0 < T < T_{\mathrm{g}} \end{cases}, \tag{8.31}$$

Note that the value at $T = 0$ may differ from 1 if the ground state is highly degenerate. It is more efficient to perform a random walk in a single system than two replicas so an order parameter can be defined as

$$q \equiv \langle \sum_{i=1}^{N} S_i^0 S_i / N \rangle. \tag{8.32}$$

where $\{S_i^0\}$ is one ground state spin configuration and $\{S_i\}$ is any configuration during the random walk. The behavior of q is essentially the same as the order-parameter defined by the Edwards and Anderson [33]. (It is not quite the order-parameter defined by Edwards and Anderson, but was used in the early simulations [43,44].)

After a bond configuration is generated a one-dimensional random walk in energy space is generated to find a spin configuration for the ground states. Since the order-parameter is not directly related to the energy, to get a good estimate of this quantity a two-dimensional random walk is required to obtain the density of states $G(E,q)$. (In this way barriers in parameter space, or configuration space, may be overcome using the same rule for the 2D random walk as for a 1D random walk in energy space.) From the density of states $G(E,q)$, all quantities may then be determined.

The probability distribution can be determined as for the Ising and Potts models, and we show this in Fig. 8.14. At low temperature there are over 30 orders of magnitude difference in the probability for neighboring values of the order parameter. Standard methods could not possibly access all states with such large differences in probability.

The energy landscape is then given by:

$$U(q,T) = \frac{\sum\limits_{E,q} E G(E,q) e^{-\beta E}}{\sum\limits_{E,q} G(E,q) e^{-\beta E}} \tag{8.33}$$

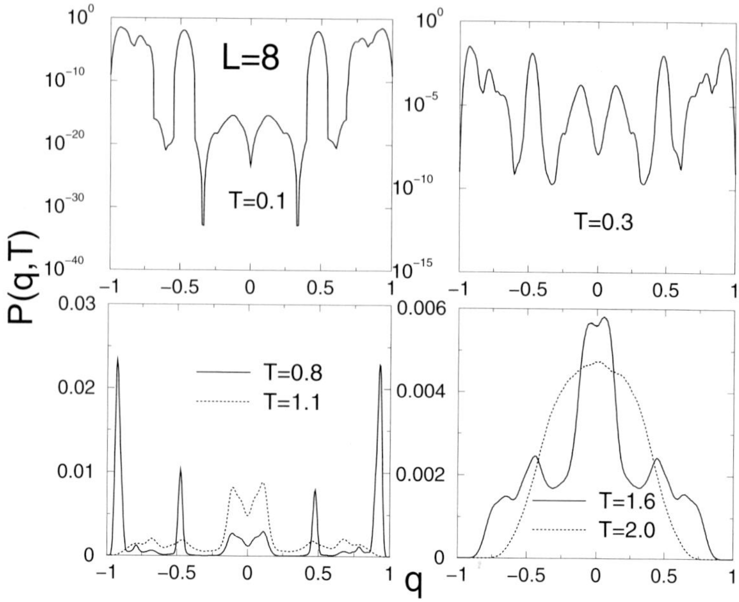

Fig. 8.14. Probability distributions for the Edwards-Anderson spin glass in three dimensions.

This landscape is very rough at low temperatures and is not accessible by standard Monte Carlo methods. (Even two decades after the model was proposed, there is still some doubt as to the existence of a phase transition in the 3d Ising spin glass [45]. Their simulational data can be described equally well by a finite-temperature transition or by a $T = 0$ singularity of an unusual type. Kawashima and Young's recent simulations could not rule out the possibility of $T_{\mathrm{g}} = 0$ [36]. As the temperature is reduced, the canonical distribution and energy landscape become increasingly rough and it is almost impossible for conventional Monte Carlo methods to overcome the barrier between local and global minima. It is possible to heat the system to increase the possibility of escape from local minima by simulated annealing and simulated tempering [46] and parallel tempering methods [47,48], but it is still very difficult to perform equilibrium simulations at low temperatures. Crossings of the fourth order cumulant as a function of temperature at around $T = 1.2$ suggest that there is a transition at this temperature, a conclusion that is consistent with the finite size behavior of the order parameter.

8.3.4 "Dynamic" Behavior: Spin Dynamics with Decompositions of Exponential Operators

Many magnetic systems have true dynamics that cannot be investigated by stochastic simulations. Instead spin dynamics simulation methods can be used to integrate the coupled equations of motion that govern the time dependent behavior. As an example we consider a model Hamiltonian for continuous degrees of freedom represented by a three-component spin $\mathbf{S}_k$ with fixed length $|\mathbf{S}_k|=1$ for each lattice site k

$$\mathcal{H} = -J \sum_{<k,l>} (S_k^x S_l^x + S_k^y S_l^y + \lambda S_k^z S_l^z) - D \sum_k (S_k^z)^2 , \tag{8.34}$$

where J is the exchange integral, $< k,l >$ denotes a nearest-neighbor pair of spins $\mathbf{S}_k$, λ is an exchange anisotropy parameter, and D determines the strength of a single-site or crystal field anisotropy. For $\lambda = 1$ and $D = 0$ (8.34) represents the classical isotropic Heisenberg ferromagnet or the corresponding antiferromagnet for $J > 0$ or $J < 0$, respectively. Of course, realistic descriptions of specific magnetic materials may require additional interactions, but we restrict ourselves to this simple Hamiltonian to describe the essential features of modern spin dynamics methods. We shall see that the excitations may occur at quite low frequencies and critical slowing down appears near second order phase transitions when systems are examined in this manner. Long time scales thus become problematic even though their origins are quite different than for the problems considered earlier.

The *dynamic* properties of these spin systems are given by the solution to the equations of motion [49]

$$\frac{d}{dt}\mathbf{S}_k = \frac{\partial \mathcal{H}_{tot}}{\partial \mathbf{S}_k} \times \mathbf{S}_k \tag{8.35}$$

that must be integrated numerically. Initial conditions are *equilibrium* configurations generated by Monte-Carlo simulations. The dynamic structure factor $S(\mathbf{q},\omega)$ can be measured by inelastic neutron scattering and is the space-time Fourier transform of the spin-spin correlation function

$$\mathcal{G}^{\alpha,\beta}(\mathbf{r}_k - \mathbf{r}_l, t-t') \equiv \langle S_k^\alpha(t) S_l^\beta(t')\rangle , \tag{8.36}$$

where $\alpha,\beta = x,y,z$ denote the spin component, $\mathbf{r}_k$ and $\mathbf{r}_l$ are lattice vectors, and the average $\langle \dots \rangle$ must be taken over a large number of independent initial *equilibrium* configurations.

The most time consuming part of a spin dynamics simulation is the numerical integration of (8.35). Consequently, the largest possible time step is desirable; however, standard methods restrict the size of the time step if the *conservation laws* of the dynamics are to be retained. Clearly $|\mathbf{S}_k|$ for each lattice site k and the total energy are to be conserved, and specific symmetries of the Hamiltonian impose additional conservation laws. Conservation of spin length and energy is important because the condition $|\mathbf{S}_k|=1$ is a major part of the definition of the model and the energy of a configuration determines its statistical weight.

Predictor-corrector methods provide a very general tool for the numerical integration of initial value problems like (8.35) with an equilibrium configuration as the initial value. Schematically (8.35) can be written as $\dot{y} = f(y)$

with the initial condition $y(0) = y_0$. The predictor step is given, e.g., by the explicit Adams-Bashforth four-step method [50]

$$y(t+\delta t) = y(t) + \frac{\delta t}{24}\left[55f(y(t)) - 59f(y(t-\delta t)) + 37f(y(t-2\delta t)) - 9f(y(t-3\delta t))\right] \tag{8.37}$$

which has a local truncation error of the order $(\delta t)^5$. The corrector step is one iteration of some implicit method, e.g. an Adams-Moulton three-step method [50]

$$y(t+\delta t) = y(t) + \frac{\delta t}{24}\left[9f(y(t+\delta t)) + 19f(y(t)) - 5f(y(t-\delta t)) + f(y(t-2\delta t))\right] \tag{8.38}$$

which also has a local truncation error of order $(\delta t)^5$.

For $D = 0$, $\lambda = 0$ the predictor-corrector method conserves the magnetization to within machine accuracy, and its implementation is independent of the special structure of the right-hand side of the equation of motion. Other conservation laws are only obeyed to within the accuracy set by the truncation error of the method. In practice, this limits the time step to typically $\delta t = 0.01/J$ [50].

Recently, a new method has been developed that relies on the decomposition of exponential operators. The motion due to (8.35) may be visualized as a Larmor precession of the spin $\mathbf{S}$ around an effective axis Ω which is itself time dependent. For the simple case $D = 0$, but arbitrary values of λ, the evaluation of the right-hand side of (8.35) is best performed by decomposing the lattice into two sub-lattices such that a spin on one sub-lattice performs a Larmor precession in a local field Ω of neighbor spins which are *all* located on the other sub-lattice. The basic idea of the algorithm is to perform a *rotation* of a spin about its local field Ω by an angle $\alpha = \mid \Omega \mid \delta t$, rather to integrate (8.35) by some standard method. This procedure conserves the spin length to within machine accuracy. Exploiting a sub-lattice decomposition of (8.35) also ensures *energy* conservation. If the two sub-lattices are denoted by $\mathcal{A}$ and $\mathcal{B}$, respectively, (8.35) has the form

$$\frac{d}{dt}\mathbf{S}_{k\in\mathcal{A}} = \Omega_{\mathcal{B}}\{\mathbf{S}\} \times \mathbf{S}_{k\in\mathcal{A}} \quad , \quad \frac{d}{dt}\mathbf{S}_{k\in\mathcal{B}} = \Omega_{\mathcal{A}}\{\mathbf{S}\} \times \mathbf{S}_{k\in\mathcal{B}}, \tag{8.39}$$

where $\Omega_{\mathcal{A}}\{\mathbf{S}\}$ and $\Omega_{\mathcal{B}}\{\mathbf{S}\}$ denote the local fields produced by the spins on sub-lattice $\mathcal{A}$ and $\mathcal{B}$, respectively. Either of the equations in (8.39) reduces to a *linear* system of differential equations if the spins on the other sub-lattice are kept fixed. The spins $\mathbf{S}_{k\in\mathcal{A}}$ are rotated for fixed values of $\mathbf{S}_{k\in\mathcal{B}}$ and vice versa. The energy is *exactly* conserved during this alternating update scheme. Note, that each sub-lattice rotation is performed with the *actual* values of the spins on the other sub-lattice, so that only one copy of the spin configuration is kept in memory at any time. The magnetization will not be conserved during the rotation operations; moreover, the two alternating rotation operations do not commute so a closer examination of the sub-lattice decomposition of the spin rotation is required.

We again represent a full configuration by a vector y which is now decomposed into two "sub-lattices" $y_{\mathcal{A}}$ and $y_{\mathcal{B}}$, i.e. $y = (y_{\mathcal{A}}, y_{\mathcal{B}})$. The cross products in (8.39) can expressed by matrices A and B which are the infinitesimal generators of the rotation of the spin configuration $y_{\mathcal{A}}$ on sub-lattice $\mathcal{A}$ at fixed $y_{\mathcal{B}}$ and of the spin configuration $y_{\mathcal{B}}$ on sub-lattice $\mathcal{B}$ at fixed $y_{\mathcal{A}}$, respectively. The update of the configuration y from time t to $t + \delta t$ is then determined by an exponential (matrix) operator i.e.

$$y(t + \delta t) = e^{(A+B)\delta t} y(t). \tag{8.40}$$

Although the exponential operator in (8.40) rotates each spin of the configuration, it has no simple explicit form because the rotation axis for each spin depends on the configuration itself. However, the operators $e^{A\delta t}$ and $e^{B\delta t}$ which rotate $y_{\mathcal{A}}$ at fixed $y_{\mathcal{B}}$ and $y_{\mathcal{B}}$ at fixed $y_{\mathcal{A}}$, respectively, *do* have explicit forms. For example, for $\lambda = 1$ and $D = 0$

$$\Omega_{\mathcal{A}}[\{\mathbf{S}\}] = -J \sum_{l=NN(k)} \mathbf{S}_l \equiv \Omega_k, \tag{8.41}$$

where the sum is over nearest neighbors of k (which belong to $y_{\mathcal{B}}$). Of course, (8.41) can be generalized for $\lambda \neq 1$.

The alternating update scheme for the integration of (8.39), i.e., (8.35) now amounts to the replacement $e^{(A+B)\delta t} \to e^{A\delta t} e^{B\delta t}$ in (8.40), which is only correct up to terms of the order $(\delta t)^2$ [51]. The magnetization will therefore only be conserved up to terms of the order δt (global truncation error), which is insufficient for practical purposes. The remedy is to use higher order Suzuki-Trotter decompositions of the exponential operator in (8.40) to increase the local truncation error of the algorithm and thus improve the magnetization conservation. The simplest possible improvement is given by the second order decomposition [51]

$$e^{(A+B)\delta t} = e^{A\delta t/2} e^{B\delta t} e^{A\delta t/2} + \mathcal{O}(\delta t^3), \tag{8.42}$$

which will be used for comparison with the predictor-corrector method. Note, that (8.42) is equivalent to the midpoint integration method applied to (8.39). One possible fourth order decomposition [51] is

$$e^{(A+B)\delta t} = \prod_{i=1}^{5} e^{p_i A\delta t/2} e^{p_i B\delta t} e^{p_i A\delta t/2} + \mathcal{O}(\delta t^5) \tag{8.43}$$

with the parameters

$$p_1 = p_2 = p_4 = p_5 \equiv p = 1/(4 - 4^{1/3}) \quad \text{and} \quad p_3 = 1 - 4p. \tag{8.44}$$

Equations (8.42) and (8.43) show the analogy to symplectic integrators obtained from the Liouville operator formalism for Molecular Dynamics simulations. [50] The second order decomposition given by (8.42) is just equivalent to the velocity Verlet algorithm for Molecular Dynamics simulations which is also equivalent to the leapfrog algorithm.

The extra computational effort needed for the decomposition method can be compensated for by using larger time steps. The evaluation of the trigonometric functions that appear in the expressions $S_k(t+\delta t)$ can also be readily approximated to reduce cpu time needed. In contrast to the predictor - corrector method, conservation of magnetization is only observed within the truncation error of the decomposition method, because according to (8.42) and (8.43) the sublattices $\mathcal{A}$ and $\mathcal{B}$ are no longer strictly equivalent.

This method can also be used to study the case $D \neq 0$ in (8.34). For a spin in sublattice $\mathcal{A}$ the equation of motion is

$$\frac{d}{dt}\mathbf{S}_{k\in\mathcal{A}} = \Omega_{\mathcal{B}}[\{\mathbf{S}\}] \times \mathbf{S}_{k\in\mathcal{A}} - 2DS^z_{k\in\mathcal{A}}\mathbf{e}_z \times \mathbf{S}_{k\in\mathcal{A}}, \tag{8.45}$$

where $\mathbf{e}_z$ is the unit vector in the z-direction, and spins in sublattice $\mathcal{B}$ obey an equation of the same form. Now the equation of motion for each individual spin on each sublattice is *nonlinear*. For the sublattice decomposition of the spin rotation in the isotropic case discussed above the requirement for energy conservation in the presence of a single site anisotropy is

$$\Omega_k \cdot \mathbf{S}_k(t+\delta t) - D\left[S^z_k(t+\delta t)\right]^2 = \Omega_k \cdot \mathbf{S}_k(t) - D\left[S^z_k(t)\right]^2 \tag{8.46}$$

for $k \in \mathcal{A}$ and $k \in \mathcal{B}$, where Ω_k is given by (8.41). In order to perform a rotation operation an effective rotation axis must be identified. This can be done by rewriting (8.46) in the form $\widetilde{\Omega}_k \cdot (\mathbf{S}_k(t+\delta t) - \mathbf{S}_k(t)) = 0$, where $\widetilde{\Omega}_k$ is given by

$$\widetilde{\Omega}_k = \Omega_k - D\left(0, 0, S^z_k(t) + S^z_k(t+\delta t)\right), \tag{8.47}$$

i.e., in order to perform the rotation S^z_k at the future time $t+\delta t$ must be known in *advance*. This problem can only be solved iteratively and the speed of the integration algorithm is correspondingly reduced.

An evaluation of the different integration methods depends upon the accuracy within which the conservation laws are fulfilled. For the isotropic case, $D=0$, the implementation of (8.42) and (8.43) is straightforward. The energy $e(t)$ for the predictor-corrector method increases linearly with time whereas the decomposition methods both yield a constant value for $e(t)$ that fluctuate but increase quite slowly with time. The predictor-corrector method conserves $m(t)$ *exactly*, whereas the second order decomposition produces fluctuations of $m(t)$ on all time scales, even for times greater than $2000/J$. Somewhat remarkably, the fourth order decomposition with $\delta t = 0.2/J$ gives a substantially better magnetization conservation than the second order decomposition with $\delta t = 0.04/J$. The net increase in speed is eightfold for both decomposition methods, but if the overall quality of the magnetization conservation is also taken into account, there is a clear advantage for the fourth order decomposition according to (8.43) for the isotropic case $D = 0$.

In the strongly anisotropic case, e.g. $D = J$, the predictor-corrector method can be applied as before, but the decomposition scheme needs modification because the spin rotation axis depends on the spin value S_k^z at the future time $t + \delta t$ (see (8.47)). This gives rise to a self consistency problem which can be solved iteratively, where the quality of the energy conservation depends on the number of iterations performed.

If priority is placed on overall energy conservation *and* speed, the second order decomposition has some advantages over the predictor-corrector method. If priority is put on energy conservation alone, the fourth order decomposition shows the best performance, but it is only slightly faster than the predictor-corrector method.

To summarize, the advantages of the predictor-corrector method are its versatility and its ability to conserve the magnetization exactly. The decomposition method is less general than the predictor-corrector method. for example, crystal field anisotropies leave the performance of the predictor-corrector method almost unaffected, whereas the decomposition method is reduced in speed. The greatest advantage of the decomposition method is its ability to handle large time steps and conserve spin length exactly. In the absence of anisotropies it also conserves the energy exactly and it maintains reversibility. For anisotropic Hamiltonians energy conservation and reversibility can be obtained to a high accuracy using iterative schemes but exact magnetization conservation is lost. The time steps typically used are much bigger than for the predictor-corrector method so that much longer timescales can be accessed.

8.4 Summary and Outlook

In this article we have seen that Monte Carlo studies of phase transitions using Metropolis sampling and standard Boltzmann importance sampling are often inadequate for the study of phase transitions. Severe difficulties are encountered in the form of a huge disparity between typically accessible simulation timescales and those on which the phenomena of interest occur.

The manner in which this time scale "gap" is manifest differs qualitatively between continuous and first order transitions. In the continuous case, the divergent correlation length causes critical slowing down, resulting in extended correlation times for simple local update algorithms. In the case of some spin models, this problem can be largely alleviated via the use of collective coordinate ("cluster") updating schemes. These focus the computational effort on relaxing the large scale configurational structure, the temporal persistence of which constitutes the bottleneck for phase space evolution.

In the context of first order phase transitions and frustrated systems such as spin glasses, the timescale gap is manifest in the form of long-lived metastable states which hinder the sampling of all the regions of phase space important for measurements of free energies. We have seen that in terms of the structure of the probability distribution of some order parameter, one finds two or more sharp maxima separated by deep minima. At first order transitions, the maxima are associated with pure phase states while the minima are associated with mixed-phase (interfacial) configurations. Owing to the high free energy of interfacial states that separates the pure phase regions, the frequency of transitions between the coexisting pure phases is suppressed. A number of approaches for dealing with this problem were discussed. It was seen that in some models the use of non-Boltzmann sampling such as Multicanonical Monte Carlo or "Wang-Landau" sampling can be used to bridge the probability valley between the maxima, or even calculate the entire density of states over some region of interest. In a number of cases, however (specifically those in which one or more of the phases is crystalline), traversing interfacial states is not kinetically tolerable because the system gets trapped in defect ridden configurations. In such cases, we detailed how one can employ the methods of Phase Switch MC to perform a global coordinate transformation that maps pure phase states of one phase onto those of another.

As regards the general outlook for bridging timescale gaps in problems in volving phase transitions, MC (in some form or another) looks likely to remain the method of choice for a great many problems. Its strength derives from the flexibility it offers in allowing the simulator to engineer large scale configurational changes in a single step, unhindered by the inherently slow physical dynamics. It can also be used in combination with biased sampling to visit regions of phase space of intrinsically low probability, thereby facilitating exploitation of short cuts through phase space. In the coming years such methods are likely to be further refined, allowing progress on a number of previously intractable problems. One area in which the present authors would like to see progress is the extension of cluster algorithms (currently only applicable to spin systems) to allow more efficient simulation of continuous phase transitions in off-lattice models such as fluids. Doubtless too, it would be of interest to combine the new methods for studying phase transition (such as PSMC) with techniques allowing first principles calculation of electronic structure, thereby facilitating accurate study of phase transitions in models of real substances.

Acknowledgement

We wish to acknowledge fruitful collaborations with A. Bunker, A.D. Bruce, M. Krech, S.-H. Tsai, and F. Wang on various aspects of the work reported here. DPL acknowledges support by the National Science Foundation under Grant No. DMR-9727714.

References

1. D. P. Landau and K. Binder, *A Guide to Monte Carlo Methods in Statistical Physics*, (Cambridge U. Press, Cambridge, 2000).
2. N. Metropolis, A. W. Rosenbluth, M. N. Rosenbluth, A. M. Teller, and E. Teller, J. Chem. Phys. **21**, 1087 (1953).
3. S. Wansleben and D. P. Landau, Phys. Rev. B **43**, 6006 (1991).
4. R. H. Swendsen and J.-S. Wang, Phys. Rev. Lett. **58**, 86 (1987).
5. P. W. Kasteleyn and C. Fortuin, J. Phys. Soc. Japan Suppl. **26s**, 11 (1969); C. Fortuin and P. W. Kasteleyn, Physica **57**, 536 (1972).
6. J.-S. Wang, Physica A **164**, 240 (1990).
7. U. Wolff, Phys. Rev. Lett. **62**, 361 (1989).
8. A. B. Bortz, M. H. Kalos and J. L. Lebowitz, J. Comput. Phys. **17**, 10 (1975).
9. M. A. Novotny, Phys. Rev. Lett. **74**, 1 (1995); M. A. Novotny, Computers in Physics **9**, 46 (1995)
10. F. Wang and D.P. Landau, Phys. Rev. Lett. **86**, 2050 (2001); F. Wang and D. P. Landau, Phys. Rev. E **64**, 056101-1 (2001).
11. D. P. Landau, Phys Rev. B **13**, 2997 (1976).
12. J.-S. Wang, T. K. Tay and R. H. Swendsen, Phys. Rev. Lett. **82**, 476 (1999).
13. P. D. Beale, Phys. Rev. Lett. **76**, 78 (1996).
14. A. E. Ferdinand and M. E. Fisher, Phys. Rev. **185**, 832 (1969).
15. A. M. Ferrenberg and R. H. Swendsen, Phys. Rev. Lett. **61**, 2635-2638 (1988); *ibid.* **63**, 1195-1198 (1989).
16. M. Plischke and B. Bergersen, *Equilibrium Statistical Physics*, (World Scientific, Singapore, 1994).
17. A.Z. Panagiotopoulos, Mol. Phys. **61**, 813 (1987).
18. A.Z. Panagiotopoulos, J. Phys. Condens. Matter **12** R25 (2000);
19. B. A. Berg and T. Neuhaus, Phys. Rev. Lett. **68**, 9 (1992), Phys. Lett. B **267**, 249 (1991), J. Lee, Phys. Rev. Lett. **71** , 211 (1993). B. A. Berg J. Stat. Phys. **82**, 323 (1996), B. A. Berg, Nucl. Phys. B **63**, 982 (1998).
20. D. Frenkel and B. Smit, *Understanding Molecular Simulation* (Academic, New York, 1996).
21. N.B. Wilding. Phys. Rev. E **52**, 602 (1995).
22. N.B. Wilding, Am. J. Phys. **69**, 1147 (2001).
23. B.A.Berg, cond-mat/0110521.
24. W.G. Hoover and F.H. Ree, J. Chem. Phys, **49**, 3609 (1968).
25. A.D. Bruce, N.B. Wilding and G.J. Ackland, Phys. Rev. Lett. **79**, 3002 (1997); A.D. Bruce, A.N. Jackson, G.J. Ackland and N.B. Wilding, Phys. Rev. E **61**, 906 (2000).
26. N.B Wilding and A.D. Bruce, Phys. Rev. Lett. **85**, 5138 (2000).
27. p, V and E are measured in units of k_BT/D^3 and D^3 and k_BT respectively, where D is the hard sphere diameter.
28. G.R. Smith and A.D. Bruce, Phys. Rev. E **53**, 6530 (1996).
29. R.J. Speedy, J. Phys: Condens. Matter, **9**, 8591 (1997).
30. M.S.S. Challa, D.P. Landau and K. Binder Phys. Rev. B **34**, 1841 (1986).
31. W. Janke and S. Kappler, Phys. Rev. Lett. **74**, 212 (1995).
32. K. Binder and A. P. Young, Rev. Mod. Phys. **58**, 801 (1986).
33. S.F. Edwards and P.W. Anderson, J Phys. F. Metal Phys. **5** 965 (1975).

34. A. R. Lima, P. M. C. de Oliveira and T. J. P. Penna, J. Stat. Phys. **99**, 691 (2000).
35. R.N. Bhatt and A.P. Young, Phys. Rev. Lett. **54** 924 (1985).
36. N. Kawashima and A.P. Young, Phys. Rev. B **53** R484 (1996).
37. M. Palassini and A. P. Young, Phys. Rev. Lett. **82** 5128 (1999).
38. M. Palassini and A. P. Young, Phys. Rev. Lett. **83** 5126 (1999).
39. M. Palassini and A. P. Young, Phys. Rev. Lett. **85** 3017 (2000).
40. E. Marinari, Phys. Rev. Lett. **82** 434 (1999).
41. M. A. Moore, H. Bokil and B. Drossel, Phys. Rev. Lett. **81** 4252 (1999).
42. J. Houdayer and O. C. Martin, Phys. Rev. Lett. **82** 4934 (1999).
43. I. Morgenstern and K. Binder, Phys. Rev. **B22**, 288 (1980).
44. K. Binder in *Fundamental Problems in Statistical Mechanics V*, edited by E. G. D. Cohen, (North-Holland, 1980).
45. E. Marinari, G. Parisi and F. Ritort, J. Phys. A **27** 2687 (1994).
46. E. Marinari and G. Parisi, Europhys. Lett. **19**, 451 (1992).
47. K. Hukushima and K. Nemoto, J. Phys. Soc. Japan **65**, 1604 (1996).
48. K. Hukushima, in *Computer Simulation Studies in Condensed Matter Physics* XIII, eds. D. P. Landau, S. P. Lewis and H.-B. Schüttler (Springer, Berlin, Heidelberg, 2000).
49. D. P. Landau and M. Krech, J. Phys.: Condens. Matter **11**, R179 (1999).
50. M. Krech, A. Bunker, and D. P. Landau, Comp. Phys. Commun. **111**, 1 (1999).
51. M. Suzuki and K. Umeno in: *Computer Simulation Studies in Condensed Matter Physics VI*, eds. D. P. Landau, K. K. Mon, and H.-B. Schüttler (Springer, Berlin, 1993).

9 Go-with-the-Flow Lattice Boltzmann Methods for Tracer Dynamics

Christopher P. Lowe[1] and Sauro Succi[2]

[1] Department of Chemical Engineering, University of Amsterdam, Nieuwe Achtergracht 166, 1018 WV Amsterdam, The Netherlands
[2] Istituto di Applicazioni Calcolo, viale Policlinico 137, 00161 - Roma, Italy

Abstract. Because of its underlying basis in kinetic theory, we discuss the advantages of using the lattice-Boltzmann equation (LBE) as a component in multi-scale simulations. As an example of upward coupling, we examine how the simple problem of the dynamics of tracers can be studied within the LBE framework. We describe how, by utilizing the kinetic view of the model, very efficient techniques can be developed to study this problem. For the specific example of hydrodynamic dispersion (the extra spreading of the tracer due to fluid flow) we apply the methods to a problem where there is an analytic value to compare with, namely flow in a tube, and a more complex system, namely a close packed cubic array of spheres. For the former we show that the method accurately reproduces known results, even with a crude representation of the tube. For the latter we show that the dispersion coefficient asymptotes on a time-scale determined by molecular diffusion. This behaviour is not what is expected for a random medium. Our results thus illustrate that a periodic system is too crude an approximation, as far as dispersion in random media is concerned. Nonetheless, our values for the dispersion coefficient agree nicely with experimental results on periodic systems.

9.1 Introduction

The dynamics of multicomponent/multiphase flows plays a major role in many real life-applications, ranging from engineering, material science, environmental and life sciences. It also represents a standing challenge for mathematical and numerical methods, due the concurrent and simultaneous interaction of multiple space-time scales. Many numerical methods are available to deal with these complexities including Eulerian and Lagrangian grid methods, as well as a variety of particle methods. However, it is increasingly evident that only a multi-scale approach can handle the type of complexities encountered in modern, interdisciplinary, high-tech applications. By multi-scale we mean that different physical processes might be handled with different techniques, each best suited to deal with the most relevant scales associated with the given process. For instance, fluid flow could be handled by a grid method, while diffusion and dispersion phenomena could be handled by, say, a Lagrangian passive tracer technique. Many options are available, and the key to success is to develop flexible and robust interfaces orchestrating the different techniques into seamless computational procedures.

Very advanced multi-scale (or multiphysics) applications have been developed in the context of brittle fracture dynamics, involving a three-level procedure based upon the use of Finite Elements in the region far from the crack, a molecular dynamics treatment of atomic motion near the crack and a quantum mechanical description of bonding in the crack tip's immediate neighborhood [1]. Such type of complex multiphysics applications require ingenious and careful interfaces in the so-called "hand-shaking" regions where boundary information has to be exchanged.

An appealing alternative is to explore methods which can host multiple physical levels, say atomistic, kinetic and fluid, within a single mathematical framework, the obvious advantage being of doing away with the aforementioned hand-shaking interfaces [2].

The most natural conceptual framework to develop this program is provided by kinetic theory, which formulates three-dimensional atomistic N-body dynamics in terms of fluid-motion in a $6N$-dimensional phase space. This approach is traditionally doomed by the infamous curse of dimensionality. Even at the lowest level, $N = 1$, of one-body kinetic equations, such as Vlasov or Boltzmann, we are left with six-dimensional fluids, beyond the capabilities of any conventional grid method. Until recently, the only way out of this dimensional curse was offered by stochastic particle methods, such as Bird's Direct Simulation Monte Carlo [4]. The common tenet behind this being that grid discretization of a six dimensional space involves of the order of N_g^6 degrees of freedom, N_g being the number of discrete points per linear dimension.

In the last decade however, this conclusion has been successfully challenged, with the realization that in many instances the 6-dimensional phase space does *not* need to be sampled (or discretized) uniformly. In particular, velocity (momentum) degrees of freedom prove amazingly nice to us, in that discretization over a handful of discrete speeds (between ten and twenty) is sufficient to capture the complexities of most hydrodynamic phenomena. The resulting discrete kinetic equation, known as the Lattice Boltzmann equation, has indeed met with remarkable success for the numerical simulation of a variety of complex hydrodynamic phenomena, ranging from multiphase flows in grossly irregular geometries, up to fully developed turbulent flows of industrial interest [5,6]. The critical reader might object that there is little surprise that ten or twenty discrete speeds can capture hydrodynamics, since the latter is known to be described by a few low-order kinetic moments, such as the fluid density, speed and temperature.

There are at least two good reasons why LBE survives this plausible criticism. First, the mathematical structure of LBE makes it much simpler to simulate on a serial computer, and more efficiently on parallel ones, as compared to the non-linear partial differential equations of fluid dynamics. Second, and more importantly for multiscale applications, judicious extensions of the plain LBE scheme, including generalized equilibria and/or effective potentials, may take LBE in the mesoscopic territory [7] where continuum equations are generally ill-posed, and often just not known altogether!

Summarizing, LBE is easily coupled "upwards" to macroscopic grid methods since smooth kinetic information leaves on a discrete grid of macroscopic size. It is also easily coupled "downwards" with atomistic methods whenever the underlying microscopic physics can be lumped into generalized equilibria and/or effective potentials. Either way, no "handshaking" regions are required, which makes LBE-based multiscale applications particularly appealing.

In this work we shall be concerned with a specific instance of "upward" coupling, namely how to combine LBE-based fluid flow with the dynamics of passive/active tracers. This problem has numerous applications, combustion and polymer flows being just two relevant ones.

9.2 LBE Schemes with Tracer Dynamics

As of today, there are at least four classes of methods coupling LBE with tracer dynamics:

- Extra-dimensional methods
- Grid-grid methods
- Hybrid grid-particle methods
- Go-with-the-flow (GWF) kinetic methods

9.2.1 Extra-dimensional Methods

Extra-dimensional methods, the earliest ones, are based on the simple observation that a LBE flow in 4 dimensions, x, y, z, w with the fourth dimension w quenched, is formally equivalent to a 3-dimensional flow with a passive scalar. Here, quenching means that any dependence on w is erased by collapsing the fourth dimension into a single layer of cells [3]. This method enjoyed some popularity in the early days of Lattice Boltzmann research, when no three-dimensional LBE lattice ensuring correct hydrodynamics was known, so that three-dimensional hydrodynamics was to be simulated in four dimensions anyway. As of today, extra-dimensional methods have lost most of their appeal since LBE flows can be simulated in three-dimensions without any problem.

9.2.2 Hybrid Grid–Grid

These are pragmatic methods in which the tracer equation is handled by any of the well known grid methods, such as finite differences, volumes or elements. In particular, LBE has been coupled to traditional Patankar's finite volumes for the study of catalytic flows [9]. This option has the merit of simplicity, but may introduce a relatively high amount of numerical diffusion. As a result, it is practical for relatively low Reynolds flows, but it cannot be recommended for applications where a stringent control on numerical viscosity is a must (say, pollutants in turbulent flows, thin flame propagation).

9.2.3 Hybrid Grid–Particle

Since the tracer obeys a convection-diffusion (Fokker-Planck) equation, it can clearly be handled with on or off-lattice stochastic particle methods. These are generally computationally more expensive than grid methods, but provide a potential advantage whenever sharp features in the tracer distribution need to be tracked by the simulation. Resolving such features with grid methods is generally quite expensive (static grids) or cumbersome (dynamic grids). Off-lattice particles are attractive because they provide a natural form of adaptivity.

9.2.4 Go-with-the-Flow Kinetic Methods

By "go-with-the-flow" (GWF) methods we imply on-lattice schemes in which the populations describing the tracer are moved with transition rates which depend on the local flow conditions. These methods are particularly appealing because, 1) they are tightly coupled to LBE (in particular, they live on the same grid), 2) they can minimize numerical diffusion without taxing computer storage.

To date, we are aware of two implementations, by Lowe and Frenkel [10], and Succi et al. [11]. In the former, the transition rates at each given lattice site are computed by using the actual values of the discrete distribution at that site. In the latter, the same rates are computed based on the local flow fields, plus additional correction terms which permit one to virtually cancel numerical viscosity (actually, to push it to fourth-order in grid spacing). The resulting scheme is therefore suitable to high-Reynolds applications, and it is currently used also for dynamic turbulence models [12]. Since details of this scheme have been published elsewhere, in the following we focus on the Lowe-Frenkel implementation.

9.3 Hydrodynamic Dispersion

The simplest example of a problem that can be tackled with a "go with the flow" approach is that of hydrodynamic dispersion. Although there are more complex applications of the same kind of approach [13,14], this problem is useful for illustrative purposes. It emphasises that a lot can be achieved by considering the kinetic - particle level - basis of the LBE, rather than regarding it simply as a discretization of the Navier-Stokes equations.

The phenomenon of hydrodynamic dispersion has been known at least since Slichter [15]. It is the additional spreading of a passive component dispersed in a solvent flowing through a porous medium. This extra spreading

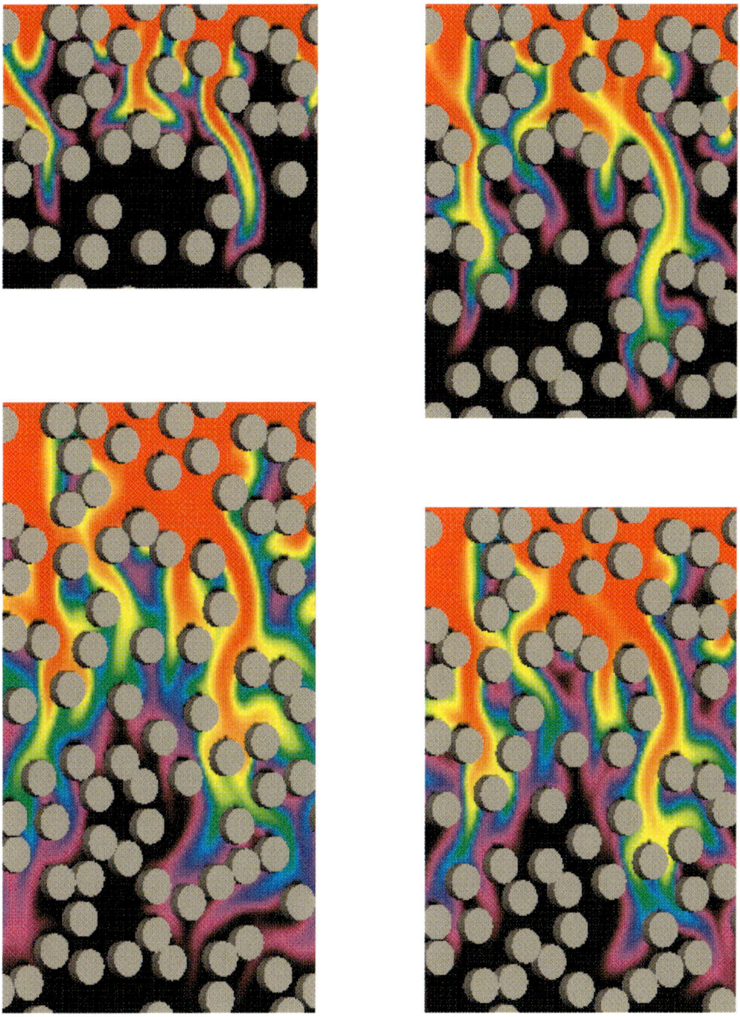

Fig. 9.1. Hydrodynamic dispersion. A fluid (not shown) is flowing, from top to bottom, between randomly placed disks. A constant uniform concentration of tracer is introduced at the top and is transported through the system. The pictures correspond to progressive increments of time, starting from top left and proceeding clockwise. For this simulation molecular diffusion is small. In the absence of hydrodynamic dispersion one would simply expect a narrow front moving at a uniform velocity through the system. Note how the tracer is spread by the fluctuations of the flow velocity about the mean.

is due to fluctuations in the local flow velocity about the mean (see Fig. 9.1). Thus, the diffusion coefficient (calculated in a co-moving frame of reference) for a flowing fluid, D, can substantially exceed the molecular diffusion coefficient, D_0. The latter characterizes transport in the stationary fluid. Following convention, we will refer to the former as the dispersion coefficient. While conceptually simple, this is by no means a trivial or irrelevant problem. It is of practical importance, in that it is closely related to the spreading of pollutants in groundwater. It is also of fundamental interest in that, as we shall see, it depends very strongly on the characteristics of the flow velocity field [16].

9.3.1 The Moment Propagation Method

A useful starting point for studying scalar transport within the framework of the the lattice Boltzmann method, is to return to the lattice gas models from which the model originally derives. One of the successful applications of lattice gases was in studying transport phenomena in fluids. Techniques such as the moment propagation method [17], developed within this framework, allowed many subtle aspects of theoretical predictions for transport processes in fluids to be tested to an unprecedented degree of accuracy [18,19]. The idea behind the method was to define the probability that a particle, located at a lattice site $\mathbf{r}$, is travelling with a particular (discrete) velocity $\mathbf{c}_i$ after a collision at time t, as $n_i(\mathbf{r},t)/\rho(\mathbf{r},t)$. Here $n_i(\mathbf{r},t)$ is the boolean variable defining whether there is or is not such a particle with the required velocity, and $\rho(\mathbf{r},t)$ the total number of particles located at site $\mathbf{r}$. Given this definition, the moment propagation method provided a means, from some initial distribution, of simultaneously summing the probabilities of all trajectories ending at a given site at a given time. Furthermore, the method enabled one not just to calculate probabilities, but to calculate functions that can be written in the form of weighted averages of some initial and current property of the trajectories. The usefulness of this will become apparent when we discuss calculating the velocity autocorrelation function (VACF) using the method. If one can calculate these functions for all possible trajectories in one go, it is clear that the statistical error decreases with increasing time, simply because more trajectories are sampled. For correlation functions, which generally start from some initial value (defined by an equilibrium property) then decay to zero, this is very useful. It allows one to study this decay for longer periods, without the signal disappearing into the noise.

If we now turn to the lattice-Boltzmann model, then in the same spirit we can treat tagged particle motion by considering that the probability a particle located at a given site has some post-collisional velocity, c_i, is

$$p_i(\mathbf{r},t) = \frac{f_i(\mathbf{r},t)}{\rho(\mathbf{r},t)} \tag{9.1}$$

where $f_i(\mathbf{r},t)$ is the single particle distribution function for discrete velocity i. This is the central quantity one computes for the lattice Boltzmann equation [6], and quantities such as the density, $\rho(\mathbf{r},t)$, and momentum density,

$\mathbf{j}(\mathbf{r}, t)$, are simply moments of this distribution

$$\rho(\mathbf{r}, t) = \sum_{i=1}^{b} f_i(\mathbf{r}, t)$$

$$\mathbf{j} = \sum_{i=1}^{b} f_i(\mathbf{r}, t)\mathbf{c}_i = \rho(\mathbf{r}, t)\mathbf{u}(\mathbf{r}, t) \tag{9.2}$$

Here b is the number of discrete velocities and $\mathbf{u}(\mathbf{r}, t)$ the flow velocity.

Let us now suppose we wish to calculate one component (taken as $\alpha\beta$) of the VACF, $C_{\alpha\beta}(t)$. This is defined as

$$C_{\alpha\beta}(t) =< c_{i\alpha}(0)c_{i\beta}(t) > \tag{9.3}$$

where the average is taken over all particles and all trajectories. Suppose further that we know the quantity $P_\alpha(\mathbf{r}, t)$, which we define as the sum of the probabilities of all trajectories ending at $\mathbf{r}$ at time t, weighted by the α component of the initial velocity of the trajectory. Following our definition above (equation (9.1)), we can write down the time evolution of $P_\alpha(\mathbf{r}, t + 1)$

$$P_\alpha(\mathbf{r}, t+1) = \sum_{i \neq i_b} \frac{P_\alpha(\mathbf{r} - \mathbf{c_i}, t) f_i(\mathbf{r} - \mathbf{c}_i, t)}{\rho(\mathbf{r} - \mathbf{c}_i, t)} + \sum_{i = i_b} \frac{P_\alpha(\mathbf{r}, t) f_i(\mathbf{r}, t)}{\rho(\mathbf{r}, t)} \tag{9.4}$$

where we have defined a time step of unity. The first summation is for links (connections between sites) that do not intersect a solid boundary, the second ($i = i_b$) for links that do. It is implicit here that the "bounce back at half time-step" rule is being used to apply stick boundary conditions [23]. After the site collisions (integer times), the contribution to the VACF arising from all trajectories ending at $\mathbf{r}$, $C^r_{\alpha\beta}(\mathbf{r}, t)$, is the product of $P_\alpha(\mathbf{r}, t)$ with the probability that a particle currently has a given β component of velocity,

$$C^r_{\alpha\beta}(\mathbf{r}, t) = P_\alpha(\mathbf{r}, t) \sum_{i, i_b} \frac{f_i(\mathbf{r}, t)}{\rho(\mathbf{r}, t)} c_{i\beta} = P_\alpha(\mathbf{r}, t) u_\beta(\mathbf{r}, t) \tag{9.5}$$

the summation being in fact equal to the β component of the local flow velocity, $u_\beta(\mathbf{r}, t)$. The full VACF can the be obtained simply by summing the contribution from all nodes and dividing by the the number of particles in the system $N = \sum_{\mathbf{r}, i} f_i = N\rho_0$,

$$C_{\alpha\beta}(t) = \frac{1}{N} \sum_{\mathbf{r}} C^r_{\alpha\beta}(\mathbf{r}, t) \tag{9.6}$$

Care needs to be taken if solid boundaries are present because from half integer times onwards (i.e. after the boundary collisions have changed the velocities), (9.5) becomes

$$C^r_{\alpha\beta}(\mathbf{r}, t+1/2) = P_\alpha(\mathbf{r}, t)\left(\sum_{i\neq i_b} \frac{f_i(\mathbf{r}, t)}{\rho(\mathbf{r}, t)} c_{i\beta} - \sum_{i=i_b} \frac{f_i(\mathbf{r}, t)}{\rho(\mathbf{r}, t)} c_{i\beta}\right)$$

$$= P_\alpha(\mathbf{r}, t) u_\beta^{1/2}(\mathbf{r}, t) \tag{9.7}$$

As they stand, these equations are no use because we do not know $P_\alpha(\mathbf{r}, t)$. However, we can calculate it explicitly for some initial time. Taking our time origin just after the node collisions (+), we have

$$P^+_\alpha(\mathbf{r}, 1) = \sum_{i\neq i_b} f_i(\mathbf{r} - \mathbf{c}_i, 0) c_{i\alpha} - \sum_{i=i_b} f_i(\mathbf{r}, 0) c_{i\alpha} \tag{9.8}$$

and we can explicitly calculate

$$C^+_{\alpha\beta}(0) = \frac{1}{N} \sum_{\mathbf{r}, i} c_{i\alpha} c_{i\beta} \tag{9.9}$$

$$C^+_{\alpha\beta}(1/2) = C^+_{\alpha\beta}(0) - \frac{2}{N} \sum_{\mathbf{r}, i=i_b} c_{i\alpha} c_{i\beta}$$

Taking our time origin just after the boundary collisions (−) on the other hand, gives the same value got $C_{\alpha\beta}(0)$, but an explicit value of P_α for half a time-step

$$P^-_\alpha(\mathbf{r}, 1/2) = \sum_{i\neq i_b} f_i(\mathbf{r} - \mathbf{c}_i, 0) c_{i\alpha} + \sum_{i=i_b} f_i(\mathbf{r}, 0) c_{i\alpha} \tag{9.10}$$

That is, because the velocities of particles on boundary links are reversed they contribute to the VACF in the opposite sense compared to the (+) case. We are now in a position to calculate the full VACF, averaged over all possible trajectories of the tagged particles. For the (+) time origin we use (9.9) to calculate $C^+(0)$ and $C^+(1/2)$, (9.8) then specifies P_α at $t = 1$, which can be propagated in time according to (9.4). Once P_α is known, the VACF, $C^+(t)$ can then be calculated at integer and half integer times using (9.5) and (9.7) respectively. Similarly, for the (−) time origin, P_α is specified explicitly from (9.10), propagated in time according to (9.4), then the VACF, $C^-(t)$ calculated at integer time from (9.7) and at half integer times from (9.5). The true VACF is then obtained by averaging over the two time origins

$$C_{\alpha\beta}(t) = \frac{1}{2}\left(C^+_{\alpha\beta}(t) + C^-_{\alpha\beta}(t)\right) \tag{9.11}$$

The dispersion coefficient, $D_{\alpha\beta}$, is related to the time integral of the VACF via the Green-Kubo relation

$$D_{\alpha\beta} = \int_0^\infty C_{\alpha\beta}(t) - C_{\alpha\beta}(\infty) dt \tag{9.12}$$

The second term is required because if there is flow there is nett momentum in the system. The VACF, as normally defined, will not decay to zero, but to some constant value (this will be discussed in more detail later). This simply reflects the fact that the mean square displacement will increase quadratically in time, because of the nett drift, rather than linearly, as expected for a diffusive process. This problem is addressed by simply transforming into a co-moving frame of reference. To do so, in (9.12) we simply subtract off the infinite time value of the normal VACF, $C_\alpha(\infty)$. Allowing for the discrete time-steps used in the lattice-Boltzmann model, the integral becomes a summation, so we arrive at the following expression for the dispersion coefficient

$$D_{\alpha\beta} = \frac{1}{4}\left(C_{\alpha\beta}(0) - C_{\alpha\beta}(\infty)\right) + \frac{1}{4}\sum_{i=1}^{\infty} C^{+}_{\alpha\beta}(i/2) + C^{-}_{\alpha\beta}(i/2) - C_{\alpha\beta}(\infty) \quad (9.13)$$

It should be noted that although this procedure, of calculating the VACF at both integer and half integer times and averaging over the two time origins, may seem unnecessarily involved, it is important. To see this, let us consider the motion of tracer particles in a system where there is no flow ($u_\alpha(\mathbf{r}) = 0$), but there are solid obstacles present. The presence of the obstacles will hinder the motion of the tracer, reducing the diffusion coefficient, D, relative to its value in the absence of the obstacles. Thus, the time integral of the VACF must be reduced by the presence of the obstacles, D_0. However, examining the moment propagation equations above, since $u_\alpha(\mathbf{r}) = 0$, (9.5) for the VACF at integer times will always give a VACF of zero whether the objects are present or not. This contradicts our statement that the obstacle should suppress the diffusion coefficient because their presence does not change this contribution to the VACF. On the other hand, for links adjacent to the surface of the obstacle $u_\alpha^{1/2}(\mathbf{r})$ is not generally zero so a contribution arises to the VACF at the surface of the obstacles at half integer times. It is this contribution that gives a negative region in the VACF and the corresponding reduction in the diffusion coefficient. Similarly, the value we specify for P_α will always be zero for the $(-)$ time origin, but non zero for the $(+)$ time origin. Thus the $(-)$ time origin makes no contribution, reducing the true VACF by a factor of 2 (equation (9.11)). This factor is nonetheless necessary to obtain the correct value for the diffusion coefficient. As a quantitative example, we consider a random packing of spheres, the spheres occupying 58% of the available space. Following the procedure outlined, taking all contributions to the VACF into account, we obtain a value $D/D_0 = 0.55$. This is in good agreement with experimental values for random packings of beads [10].

This in essence is how we can use the moment propagation method within the framework of the lattice-Boltzmann model. There are however a few additional considerations worthy of note which we will now consider.

9.3.2 Galilean Invariance

Defining "local molecular diffusion" as being characterized by the behaviour at a single lattice site over one time-step, the summation over i in (9.9) is related to the corresponding element of the momentum flux tensor $\mathbf{\Pi}$, so

$$C_{\alpha\beta}(0) = \frac{\Pi_{\alpha\beta}}{\rho(\mathbf{r})} \tag{9.14}$$

From (9.9), we also see that, in the absence of boundary links (or in the "thermodynamic limit" where the proportion of boundary links becomes negligible) we also have $C_{\alpha\beta}(0) = C^{+}_{\alpha\beta}(1/2)$. If the fluid is at rest then we also have $C_{\alpha\beta}(t > 1) = 0$. It follows from (9.12) that the local diffusion coefficient is

$$D_{\alpha\beta} = \frac{\Pi_{\alpha\beta}(\mathbf{r})}{2\rho(\mathbf{r})} \tag{9.15}$$

Limiting our attentions to the diagonal terms ($\alpha\alpha$), the equilibrium momentum flux tensor is [23]

$$\Pi_{\alpha\alpha} = c_s^2\rho(\mathbf{r}) + \rho(\mathbf{r})u_\alpha(\mathbf{r})u_\alpha(\mathbf{r}) \tag{9.16}$$

where c_s is the speed of sound. In our model $c_s^2 = 1/2$, yielding

$$D_{\alpha\alpha} = D_0 = 1/4 \tag{9.17}$$

If, on the other hand, the fluid is in uniform motion with a velocity u_α, we have $C_{\alpha\alpha}(t \geq 1) = {u_\alpha}^2$. Substituting this in (9.13) we again obtain $D_{\alpha\alpha} = 1/4$. The model is Galilean invariant. If the equilibrium distribution does not contain the $\mathbf{uu}$ term then we find $D = 1/4 - {u_\alpha}^2$. The system is not Galilean invariant, the local dispersion coefficient depends on the local flow velocity. It is usual practice to simulate low Reynolds number (Re) flows by solving the "creeping flow" equations, valid for $Re << 1$. To do so, one omits the $\mathbf{uu}$ term in the equilibrium momentum flux tensor (at the macroscopic level it is this term which gives rise to the non-linear convective term in the Navier-Stokes equations). This procedure simplifies the collision operator. However, we see here that, when studying dispersion in flowing fluids, this approximation introduces a velocity dependent local diffusion coefficient. It is only valid if $D_\alpha >> u_\alpha^2$.

9.3.3 Varying the Peclet Number

The value of the molecular diffusion coefficient $D_0 = 1/4$, implies that the system has a low natural Peclet number. The Peclet number Pe quantifies the relative displacements of a tracer particle due to convection and molecular diffusion. In terms of a characteristic length scale $\bar{l}$, velocity $\bar{u}$, and diffusion coefficient $\bar{D}$, it is defined as $Pe = \bar{u}\bar{l}/\bar{D}$. For high values of Pe convective transport dominates. For low values molecular diffusion dominates. In the lattice Boltzmann model the velocity fields must satisfy the unsurprising condition that the flow remain significantly sub-sonic, that is $u(\mathbf{r}) << c_s$, where c_s is this the speed of sound. There are tight stability limits on c_s and it must take a value close to $c_s \sim 1/\sqrt{2}$. If we define a grid Peclet number Pe^G, for which $\bar{D} = D_0$, $\bar{u} =< u(\mathbf{r}) >$ and $\bar{l}$ is the lattice spacing, then, using the equation for the molecular diffusion coefficient (equation (9.17)) this implies $Pe^G << 2$. That is, the system has a low "natural" Peclet number. This problem can be eased somewhat if we decouple the tracer from the lattice-Boltzmann fluid by noting that, for any state of the lattice-Boltzmann equation, we can hypothetically convert some of the tracer particles into rest particles. We then have a modified tracer distribution $f_i'(\mathbf{r}) = f_i(\mathbf{r}) - \rho(\mathbf{r})\Delta/\rho_0$ and a rest particle distribution $f^0(\mathbf{r}) = b\Delta\rho(\mathbf{r})/\rho_0$, where b is the number of discreet velocities. Thus, $b\Delta/\rho_0$ is the probability of a particle being at rest after the action of the collision operator. This procedure leaves the important hydrodynamic quantities (local density, flow velocity and pressure gradient) unchanged. It does, however, reduce the pressure term in the tracer momentum flux tensor to $\rho(\mathbf{r})(c_s^2 - b\Delta/2\rho_0)$. Considering one lattice site and substituting the modified momentum flux tensor into (9.5) we find that, for our model (for which $b = 18$),

$$D_0 = \frac{1}{4} - 6\frac{\Delta}{\rho_0} \tag{9.18}$$

The propagation equation, equation (9.4), becomes

$$P_\alpha(\mathbf{r}, t+1) = \sum_{i \neq i_b} \frac{P_\alpha(\mathbf{r} - \mathbf{c}_i, t) f_i'(\mathbf{r} - \mathbf{c}_i, t)}{\rho(\mathbf{r} - \mathbf{c}_i, t)} + \sum_{i = i_b} \frac{P_\alpha(\mathbf{r}, t) f_i'(\mathbf{r}, t)}{\rho(\mathbf{r}, t)} + \frac{f_i^{\,0}(\mathbf{r})}{\rho(\mathbf{r})} P_\alpha(\mathbf{r}, t) \tag{9.19}$$

whilst all the other equations can simply be written with $f_i'(\mathbf{r})$ replacing $f_i(\mathbf{r})$ (rest particles make no contribution to the VACF). By using this procedure we remove the restriction outlined above, making it possible to access higher Peclet numbers.

9.3.4 The VACF at Infinite Time

In order to calculate dispersion coefficient (equation (9.3)) we need to either know, or subtract off automatically, the contribution from the infinite time value of the VACF. This can be done quite simply as follows. In the limit of long times the distribution of the quantity we propagate, $P_\alpha(\mathbf{r}, t)$, will tend to its equilibrium distribution. That is, a distribution independent of t. If we write the distribution in the form $P_\alpha(\mathbf{r}, \infty) = A\rho(\mathbf{r})$ and substitute this for $P_\alpha(\mathbf{r})$ in the propagation equation (equation (9.9)) we see that this distribution is in fact invariant under propagation. This is, therefore, the form of the equilibrium distribution. The value of the constant A is fixed by the fact that the sum of $P_\alpha(\mathbf{r},t)$ over all sites in the system is conserved. Thus,

$$P_\alpha(\mathbf{r},\infty) = \frac{\rho(\mathbf{r})}{\rho_0} \langle P_\alpha(\mathbf{r},1) \rangle \tag{9.20}$$

Having calculated $P_\alpha(\mathbf{r},\infty)$ it is possible to perform the entire calculation in center of mass coordinates by changing the initial conditions to

$$\bar{P}_\alpha(\mathbf{r},1/2) = P_\alpha(\mathbf{r},1/2) - P_\alpha(\mathbf{r},\infty) \tag{9.21}$$

then propagating $\bar{P}_\alpha(\mathbf{r},t)$ rather than $P_\alpha(\mathbf{r},t)$. The second term is, by definition, invariant under the propagation operation. Both the components of the VACF will now decay to zero and the $C_{\alpha\beta}(\infty)$ terms in (9.3) will automatically be included using this modified initial condition. This approach is computationally useful because any round-off errors are applied equally to both terms and automatically cancel. It is therefore the method we prefer to employ.

9.3.5 Generalization

There are two notable respects in which the method outlined above can be generalized. Firstly, it can be used to calculate any function that can be written in the form of the average of the product of an initial probability, a trajectory probability and a final probability. For example, using a simple scalar quantity instead of $\mathbf{c_i}$ to weight the trajectories we simply solve the convection diffusion equation for the initial scalar distribution (see Fig. 9.1. More interestingly, the self dynamic structure factor (SDSF) $F(k,t)$ can also be calculated in this way, for a given value of the wave-vector k . As with the VACF, the result will automatically be averaged over all possible starting points and all possible trajectories in the system, giving an extremely high statistical accuracy [20,21]. The SDSF is the Fourier transform of the distribution function for the time dependent displacement of tagged particles $(\mathbf{r}_i(t))$

$$F(k,t) = \langle \exp\left(-i\mathbf{k} \cdot (\mathbf{r}_i(0) - \mathbf{r}_i(t)\right) \rangle \tag{9.22}$$

As such, it gives more detailed information about the actual form of the distribution. The VACF only provides information about the second moment. In principle, by making use of a cumulant expansion, the SDSF can be used to extract information on higher order moments.

The second generalization is to problems in which the flow of the fluid is time-dependent [22], not in a steady state. The procedure simply involves retaining the dynamic LBE distributions, $f_i(\mathbf{r},t)$. One could envisage that a combination of these two generalizations would be very useful for studying dispersion in time dependent turbulent flows, although, to our knowledge, this has not been attempted.

9.4 Applications of the Model

9.4.1 Dispersion in a Tube

The classic example of non-equilibrium tracer motion is dispersion in a fluid flowing through a tube. This problem was first treated experimentally and theoretically by Taylor [24]. For this system, where the velocity field in known (so long as the Reynolds number is small), the dispersion coefficient along the tube axis can be calculated analytically

$$\frac{D}{D_0} = 1 + \frac{\mathrm{Pe}^2}{48} \tag{9.23}$$

where the Peclet number Pe is defined $\mathrm{Pe} = R < u > /D_0$, with R the tube radius. Taylor's approach was subsequently generalized by Aris [25] and this process is now known as Taylor-Aris dispersion. To simulate Taylor-Aris dispersion our procedure is as follows. We map a cylinder onto the lattice and run the lattice-Boltzmann simulation under the influence of a uniform external force density $\mathbf{F}$, directed along the axis of the tube, until a steady state is reached. More efficiently, the iterative procedure described by Verberg and Ladd [26] could be used to reach the steady state. Using the method described above, we then apply the moment propagation equations to calculate the VACF for tagged particles in the system. By using periodic boundary conditions in the plane perpendicular to the tube axis, the problem reduces to two dimensions. Only one lattice spacing being required for the tube cross-section. Note that this is only possible because we are calculating the VACF, a quantity depending only on the flow velocity. Since the flow velocity does not vary along the tube axis, the VACF is unaffected by the periodic boundary conditions. If the calculation is performed by actually following the spacial and temporal evolution of the tracer this is not the case and fairly extensive three dimensional simulations are required [27]. A degree of approximation is inevitably introduced by mapping the cylinder cross-section onto a lattice so we calculate an effective radius, R^*, for the model cylinder by calculating the the average flow velocity and then using the exact result derived from the Navier-Stokes equations to define $< u >= F{R^*}^2/8\eta$, where η is the fluid viscosity. The values of R^* approach the nominal radius R as we increase the system size.

If we examine the expression from which we calculate the dispersion coefficient (equation (9.13)), we can make the following observation: the first two terms are independent of the flow velocity, only terms in the summation depend on the flow field. These terms are thus hydrodynamic, in the sense that in the absence of any flow they are zero. This leads us to define a "flow" contribution to the VACF, $C_f(t)$,

$$C_f(t) = C(t+1) - C(\infty) \tag{9.24}$$

and a corresponding time dependent hydrodynamic dispersion coefficient, $D_f(t)$,

$$D_f(t) = \frac{1}{2}\sum_{0}^{t} C_f(t) \tag{9.25}$$

corresponding to the summation of terms in (9.13). In the limit $t \to \infty$, we therefore have

$$D_f(\infty) = D - D_0 \tag{9.26}$$

That is, $D_f(t)$ characterizes the dispersion due to flow, over and above simple molecular diffusion. According to the Taylor-Aris result this is simply,

$$D_f(\infty) = \frac{\mathrm{Pe}^2}{48} \tag{9.27}$$

In Fig. 9.2 we have plotted the dimensionless flow component of the VACF, $C_f(t)/<u>^2$ as a function of the dimensionless time tD_0/R^{*2}. Normalizing the function by $<u>^2$ makes the initial value independent of the Peclet number (so long as the Reynolds number is small). The dimensionless time is the time divided by the time it takes a particle to diffuse, by molecular diffusion alone, a distance the order R^{*2}. Because of the symmetry of the flow in the tube, the only way a particle can "forget" its original velocity is by

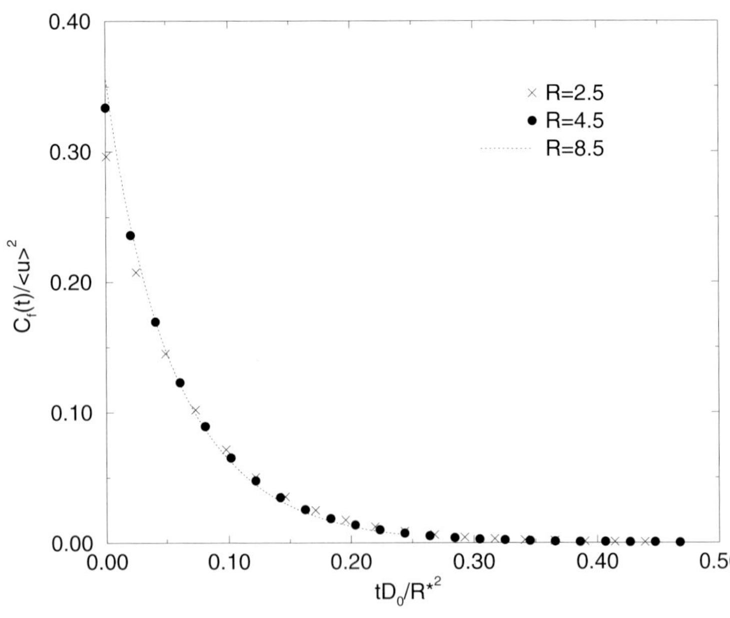

Fig. 9.2. Tube flow. The dimensionless flow component of the velocity autocorrelation function as a function of the dimensionless diffusive time. The data correspond to tubes with different nominal radii R, given in lattice units.

diffusing in a direction transverse to the flow direction. Therefore, molecular diffusion always determines the decay of the VACF. The data plotted in Fig. 9.2 are for various nominal tube radii and for a fixed Reynolds number $R_e = 0.01$. This means that the average flow velocity and molecular diffusion coefficient are not the same for the three sets of data. Nonetheless, we see that, plotted in these dimensionless terms, the data rapidly collapse onto a single curve, independent of the tube radius. The decay is in fact a simple exponential. It should also be noted that this diffusion controlled decay of the VACF, implying $D \sim 1/D_0$ for a fixed flow velocity, always leads to Taylor-Aris-like behaviour in the sense that $D/D_0 \sim Pe^2$.

If we integrate the dimensionless VACF then, in these units, the Taylor-Aris result corresponds the integral, at long times, approaching a value 1/48 [28]. In Fig. 9.3 we have plotted the time integral of the dimensionless VACF, multiplied by 48. If we recover the Taylor-Aris result the plot should approach unity. As we see, for the tube with nominal radius $R = 2.5$ lattice spacings (which is very small) there is an error of a few percent but as we increase the resolution of the simulation, by increasing R, the exact result is recovered. The results also show that the dispersion coefficient reaches a constant on a time-scale $t \sim R^2/3D_0$. On shorter time-scales the dispersion coefficient cannot be regarded as a constant and a simple diffusive picture for the transport process is inadequate.

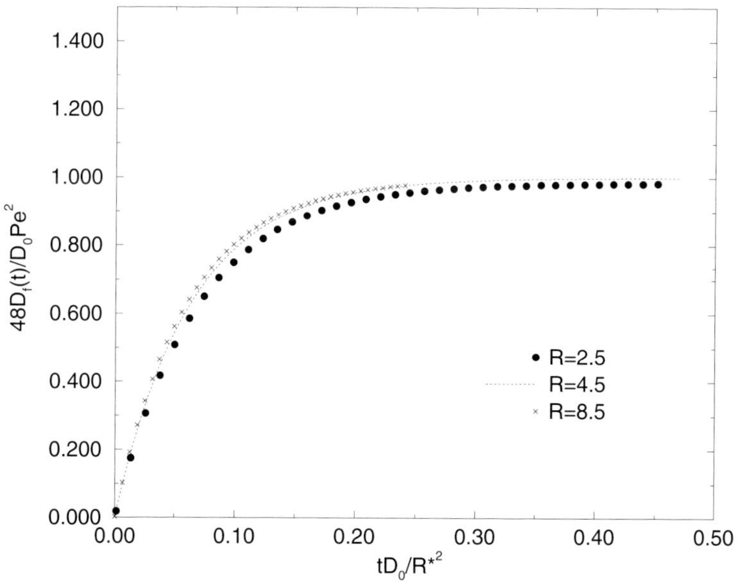

Fig. 9.3. Tube flow. The dimensionless time dependent flow diffusion coefficient as a function of the dimensionless diffusive time. The data correspond to tubes with different nominal radii R, given in lattice units.

9.4.2 Dispersion in Cubic Periodic Arrays

A more complex system we can consider is that of a close packed cubic array of spheres. This system, while still possessing a high degree of symmetry, is notably more complex in that the speed and direction of the streamlines are no longer constant. There is a probability that a particle's velocity becomes decorrelated from its initial velocity by convection along, as well as diffusion between, streamlines. Thus, one might imagine, it mimics more closely the system of most practical interest, a random medium. Indeed, extensive studies on periodic systems have been made with the aim of understanding dispersion in random media [29].

To calculate the VACF for this system we again make use of the symmetry by simulating only one cubic cell and applying periodic boundary conditions. Again this is an advantage of working at the velocity, rather than the positional level. The latter would require the explicit simulation of a large number of cells. The simulations were repeated using larger cells (and therefore spheres with a larger nominal radius, R) until the results, to within a percent, became independent of nominal radius. We will not focus here on the low Peclet number behaviour, but rather on the high Peclet number asymptotic behaviour. In Fig. 9.4 we have made the same dimensionless plot we used to analyse the Taylor-Aris results. The dimensionless flow component of the VACF is plotted in terms of the dimensionless diffusive time. The data shown are for Peclet numbers $8, 16, 32, 64$ and 128. Clearly, the data do not all collapse onto the same curve, as they do for Taylor-Aris dispersion. In particular, at short times, the decay depends on the Peclet number and at high enough Peclet number oscillations in the rate of decay can clearly be seen. These correspond to the tracer particles repeatedly traversing the cubic cells and experiencing, because of the symmetry, a partial recorrelation of their velocities. What is must notable from the figure though, is that the curves do approach one unique curve with increasing Peclet number. That is, at high Peclet numbers the decay of the VACF is dictated by molecular diffusion. Because of the symmetry, it is clear that the particle velocities do not become decorrelated by convection and Taylor-Aris type behaviour ($D/D_0 \sim Pe^2$) should be recovered.

To check this, we have calculated the dispersion coefficient at high Peclet numbers. The values we obtained are shown in Fig. 9.5, where we also compare with the experimental measurements of Gunn and Price [30]. Firstly, the agreement with the experimental values is good (there is a substantial statistical error associated with the experimental values). If we allow for the fact that, in the experiments, increasing the Peclet number also involves increasing the Reynolds number there is a small correction to the simulation results that does not influence the agreement significantly. It is clear from the data that we are approaching an asymptotic scaling $D/D_0 \sim Pe^2$. That is, despite the increased complexity of this system, we recover simple Taylor-Aris type dispersion. Brenner has argued on theoretical ground that this should

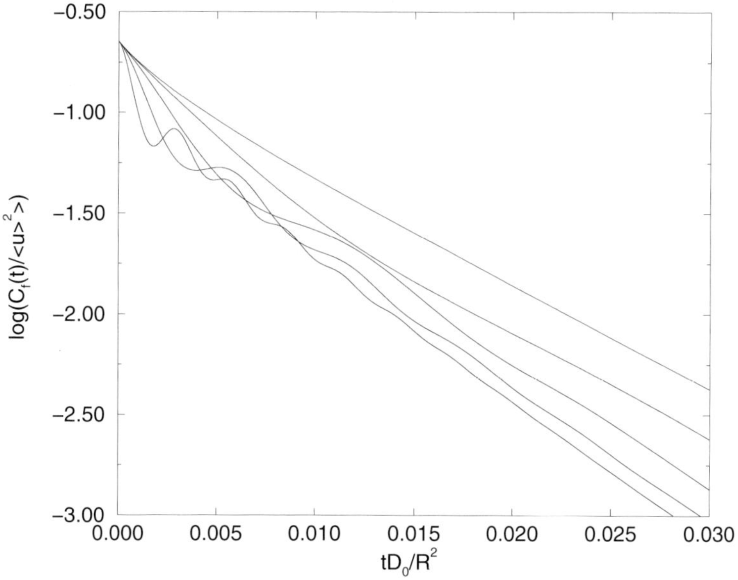

Fig. 9.4. Periodic close packed cubic array. The dimensionless flow component of the velocity autocorrelation function as a function of the dimensionless diffusive time. The lines correspond, from bottom to top, with Peclet numbers 128, 64, 32, 16 and 8.

be the case [31]. It should be noted that this is not the behaviour one expects for a truly random media, where velocity correlations can decay purely by convection [10]. This calls into question the practice of using spacially periodic media as models for dispersion in random porous media and indicates that, even where one has a random model medium, care must be exercised in excluding the effect of boundary conditions.

9.5 Conclusions

We hope to have convinced the reader that while with hindsight the lattice-Boltzmann equation can be regarded, at its simplest, as a finite difference scheme for solving the Navier-Stokes equations, there is added value in maintaining its historical roots. These are firmly embedded in kinetic theory. To this end we have pointed out the conceptual and practical values of using this concept within the framework of multiscale computations. We have furthermore given a specific illustration of the advantages of maintaining the kinetic perspective to study tracer transport in flowing fluids. Hopefully the new millennium will yield still more applications in computational science of the equation that, over a hundred years ago, Boltzmann himself "despaired" of solving,

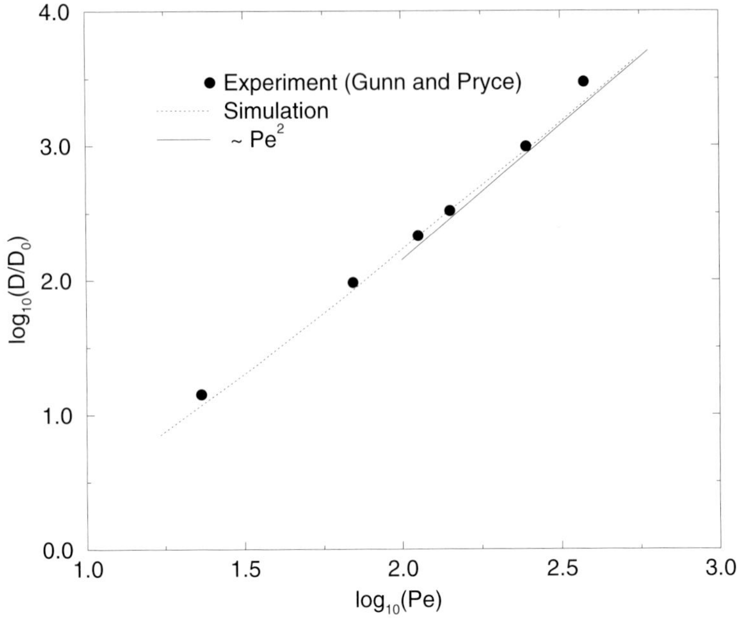

Fig. 9.5. Close packed cubic array. The dispersion coefficient at high Peclet numbers.

Acknowledgements

Work performed under NATO Grant PST.CLG.976357. SS acknowledges a scholarship from the Physics Department at Harvard University. C.P. Lowe would like to acknowledge the contribution of Daan Frenkel.

References

1. F. Abraham, J. Broughton, N. Bernstein, E. Kaxiras, Comp. in Phys., **12**, 538 (1998).
2. S. Succi, O. Filippova, G. Smith, E. Kaxiras, Comput. in Science and Engineering, **3**, 26(2001).
3. D. d'Humieres, P. Lallemand, U. Frisch, Europhys. Lett., **4**, 291 (1986).
4. G. A. Bird, *Molecular Gas Dynamics*, Oxford University Press (London 1976).
5. G. Mc Namara, G. Zanetti, Phys. Rev. Lett., **61**, 2332 (1988), F. Higuera, S. Succi, R. Benzi, Europhys. Lett., **9**, 345 (1989), R. Benzi, S. Succi and M. Vergassola, Phys. Rep., **222**, 145 (1992).
6. S. Succi, *The Lattice Boltzmann Equation*, Oxford Univ. Press, Oxford, 2001.
7. G. Gonnella, E. Orlandini and J.M. Yeomans, Phys. Rev. **E58**, 480 (1998).
8. Y. Qian, D.d'Humieres, P. Lallemand, Europhys. Lett., **17**, 149 (1989).
9. M. Presti, G. Bella, S. Succi, Society Automotive Engineers Paper 972957., 1, 1997.

10. C.P. Lowe and D. Frenkel, Phys. Rev. Lett. **77**, 4552 (1996).
11. S. Succi, G. Bella, H. Chen, K. Molvig, C. Teixeira, J. Comp. Phys., **152**, 493 (1999).
12. C. Teixeira, Int. J. Mod. Phys. C, **9**, 1281 (1998).
13. P.B. Warren, Int. J. Mod. Phys. **C8**, 889 (1997).
14. J. Horbach and D. Frenkel, Phys. Rev. **E64**, 061507 (2001).
15. C.S. Slichter, *Water Supply Papers* **14R** (U.S. Geological Survey, 1905).
16. D.L. Koch and J.F. Brady, J. Fluid Mech. **154**, 399 (1985).
17. D. Frenkel and M.H. Ernst, Phys. Rev. Lett **63**, 2165 (1989).
18. M.A. van der Hoef and D. Frenkel, Physica **D 47**, 191 (1990).
19. M.A. van der Hoef and D. Frenkel, Phys. Rev. Lett. **66**, 3459 (1991).
20. C.P. Lowe, D. Frenkel and M.H.J. Hagen, J. Stat. Phys. **87**, 1229 (1997).
21. M.H.J. Hagen, D. Frenkel and C.P.Lowe, J. Chem. Phys. **109**, 275 (1998).
22. C.P.Lowe and D.Frenkel, Physica **A 220**, 251 (1995).
23. A.J.C. Ladd, J. Fluid Mech. **271**, 285 (1994).
24. G. I. Taylor, *Proc. Roy. Soc., Ser A* **219**, 186 (1953).
25. R. Aris, *Proc. Roy. Soc., Ser A* **253**, 67 (1956).
26. R. Verberg and A.J.C. Ladd, Phys. Rev. **E60**, 3366 (1999).
27. R.M.H. Merks, A.G. Hoekstra and P.M.A. Sloot in R.L. Lagendijk; J.W.J. Heijnsdijk; A.D. Pimentel and M.H.F. Wilkinson, editors, Proceedings of the 7th annual conference of the Advanced School for Computing and Imaging, pp. 137-141. ASCI, (2001).
28. M.A. Paine, R.G. Carbonell and S. Whitaker, Chem. Eng. Sci. **38**, 1781 (1983).
29. A. Eidsath, R.G. Carbonell, R.G. Whitaker and L.R. Herrman, Chem. Eng. Sci. **38**, 1803 (1983).
30. D.J. Gunn and C. Pryce, Trans. Inst. Chem. Engrs **47**, T341 (1969).
31. H. Brenner, Trans. R. Soc. Lond. **A297**, 81 (1980).

Part VII

Multiscale Modelling in Materials Science

10 Atomistic Simulations of Solid Friction

Martin H. Müser

Institut für Physik, Johannes Gutenberg-Universität, Mainz, Germany

Abstract. Friction between two solid bodies in relative sliding motion takes place on a large spectrum of length and time scales: From the nanometer/second scale in an atomic force microscope up to the extremely macroscopic scales of tectonic motion. Despite our familiarity with the effects of friction, fundamental questions remain unanswered. The atomistic origins of well-established phenomenological friction laws are controversial. Many explanations, seemingly well-established, have recently been called into question by new experimental results. Computer simulations have also revealed flaws in previous theoretical approaches and led to new insights into the atomistic processes responsible for friction. In this chapter, selected computer simulation studies of friction and their implications will be discussed. Emphasis will be given to the question what one can learn from a friction simulation and how it is possible to avoid effects that merely arise due to poor models. Moreover, it will be outlined how it is possible to gain insight into tribological processes that take place on macroscopic time scales with the help of atomistic computer simulations, which are typically constrained to the nanosecond regime.

10.1 Introduction

The possibility of developing an atomistic theory of the friction between solid bodies has increased significantly over the last decade. New experimental techniques have made it possible to study well-defined mechanical single-asperity contacts, typically a few nanometers broad in an atomic force microscope [1,2] (AFM) experiment and a few micrometers broad in a surface force apparatus (SFA) experiment [3,4]. The physical laws observed in such nano and microscale contacts often deviate qualitatively from those observed in macroscopic systems. For instance the friction-load dependence in a single asperity contact [5,6] usually deviates strongly from the linear relationship that is almost commonly observed in macroscopic multi-asperity contacts [7]. Another example for scale effects is the onset of oscillatory depletion forces between approaching surfaces when the confinement of a lubricant is reduced to a few nanometers and the frequently observed concurrent dramatic increase in shear forces that oppose relative lateral sliding of the two solids in contact [8,9]. It is obvious that the fundamental understanding of tribology can only be achieved by a combination of experimental, theoretical, and computational efforts. This is not only an interesting, scientific endeavor, but improving the understanding and ultimately controlling tribology (science of friction, lubrication, and wear) has been and will remain useful for the development of new technologies. One example is the design of novel micro mechanical devices which one may expect to be facilitated through a better atomistic theory of friction. From a computational point of view, even the small scale is not trivial to model realistically, because the chemically detailed, atomistic simulation of an AFM tip scratching on a surface requires the simultaneous use of techniques that are usually employed to address different regimes in space and time. Yet, despite these difficulties, atomistic simulations will not only yield insight into nanotribology but moreover become increasingly important in explaining macrotribological phenomena, among other reasons because they can provide constitutive equations for the use in finite element methods.

Another reason for the importance of simulations in tribological contexts with respect to purely analytical approaches is that there is no principle like minimization of free energy that determines the steady state of non-equilibrium systems. But even if there was, simulations would be needed to address the complex systems of interest, just as in many equilibrium problems.

This chapter is meant as a help to conduct atomistic simulations of friction between solids in a meaningful and efficient way, rather than to give a broad overview of the field. A far more comprehensive review on results of friction simulations was given recently [10]. Here, technical issues will be emphasized. In particular, I will try to point out the traps (of which there are many) that can significantly depreciate the scientific relevance of friction simulations. Their results depend to a large degree on the boundary conditions, the choice of the initial geometry, and the way in which the system is driven. This statement might sound trivial, but the idealized framework or artificial features encountered in simulations often make it difficult to compare the results of the simulations to experiments. Hence it is necessary to know *prior* to the simulation which interactions of the system are relevant and need to be included into the model and which features of the model are irrelevant. In the present context, I call a feature irrelevant if it does not change the tribological behavior in a qualitative way. The relevant features of a model may include the degree of correlation in the surface corrugations between the two sliding objects, the degree and chemical nature of surface contamination, the dimensionalities of the interface and that of the slider, the surface roughness on mesoscopic scales, the elasticity of the solids in contact, thermal and quantum fluctuations of the atomic motion, as well as the age of the contacts, to name a few. Naturally, the relevance of one feature is most always intertwined with that of another feature. Moreover the chemical composition of the interface, which is reflected in the choice of the force field, can also play an important role. In order to understand specific behavior, it may be insufficient to analyze models that merely employ simple two-body potentials.

A typical contact between two real solid bodies is such that prior to contact the bodies are three-dimensional, their two-dimensional surfaces are locally curved, contaminated with dirt or intentionally lubricated, and the surface corrugations of the bodies are uncorrelated. The solids can yield plastically and long-range elastic deformations are possible, which however, are strongly suppressed by long-range elastic interactions. In simulations one often encounters the opposite situation. The surfaces are already flat before contact takes place, there is no contamination, and by some magical force, the corrugation profiles of both bodies are not only identical, but also perfectly aligned. The atoms are sometimes pinned to ideal lattice sites, disabling long-range elastic deformations, or alternatively, they are coupled as a two-dimensional harmonic sheet, thus neglecting long-range elastic interactions. Whether these discrepancies of the real world and the simulation is relevant depends on the problem under investigation. In many cases, however, the tribological properties are qualitatively altered by the simplifications.

Not only the geometry and the interactions of the model are important, but in addition, the results of tribological simulations often depend sensitively on the way in which the system is driven [11]. Driving the slider in an unrealistic way poses the second class of potential traps. One can get qualitatively different behavior if one assumes constant sliding velocity or if the slider is pulled with a weak spring. Furthermore the equivalence of different ensembles that are valid for large systems in equilibrium thermodynamics (constant separation vs. constant load, constant temperature vs. constant energy) usually breaks down in non-equilibrium situations.

The reminder of this chapter will start with the presentation of a simple but rather generic case study emphasizing the role of boundary conditions and other details. In Sect. 10.2, a simple model for friction will be introduced, originally proposed by Prandtl but commonly referred to as Tomlinson model. The important lesson to be learned is that we need to identify relevant mechanical instabilities in order to understand solid friction. A discussion of dry friction will be given in Sect. 10.3, including the analysis of friction forces as a function of disorder and dimensionality of interface and slider. Selected studies of simulations incorporating lubricants will be presented in Sect. 10.4. Sect. 10.5 focuses on technical issues, in particular on how to model the confining walls and the way in which they are driven. Sect. 10.6 will contain the conclusions.

10.1.1 The Relevance of Details: A Simple Case Study

Let us have a look at the relevance of 'details' in a simple model system. Consider two rigid, impenetrable, identical walls separated by a boundary lubricant, like a quarter layer of simple, non-reactive spherical molecules. A schematic one-dimensional representation is given in Fig. 10.1. The lower wall, which will be called the substrate, is supposed to be fixed, while the upper wall is pulled via a spring of varying stiffness k at a small, fixed velocity v. For large k, one expects smooth sliding of the upper wall similar to fixed sliding velocity, while at small k one expects stick-slip motion, which may best be characterized as a jerky motion of the top wall: The slider is stuck for a long time and then suddenly pops forward before it is stuck again. Stick-slip motion occurs when mechanical stress builds up sufficiently slowly in a contact. After a certain threshold force is reached, namely the static friction force, the two solids start sliding and the stored elastic energy is quickly released as kinetic energy, ultimately leading to generation of heat and/or plastic deformation. From a macroscopic description, unstable, stick-slip trajectories are obtained if the friction force between slider and substrate decreases with increasing velocities. Stick-slip motion is observed from nanoscale junctions up to tectonic plates.

Fig. 10.1. Schematic representation of boundary lubrication. The slider is moved with respect to a fixed substrate at velocity v via a spring of stiffness k. The two solids do not interact directly but only via the boundary lubricant (full circles).

The idealized situation of Fig. 10.1 can be analyzed in terms of a numerical simulation. This is done by first defining a suitable Hamiltonian for the system. One can then integrate the resulting equations of motion numerically. The friction force can be calculated by averaging the force in the driving spring or alternatively from the force that is exerted from the boundary lubricant and the substrate on the upper wall. Once steady state is reached, the two methods must give the same expectation value since otherwise the top wall would be accelerated. See Sect. 10.5 for a more thorough description of technical issues. The results of such a molecular dynamics simulation for a two-dimensional interface are shown in Fig. 10.2. Both surfaces are identical, namely triangular lattices with identical lattice constants. In one simulation, the surfaces are oriented perfectly, in another simulation, one surface is rotated by 90°, which makes the two surfaces essentially incommensurate. Two crystals are called incommensurate if they do not share a common periodicity. Of course, in a computer simulation two solids cannot be perfectly incommensurate due to the finite system size. But if the smallest common periodicity coincides with the linear length of the simulation cell, one can usually call the surfaces incommensurate for all practical purposes.

On the left-hand side of Fig. 10.2, one can see that the friction forces depend sensitively on the relative orientation of the two surfaces, even though

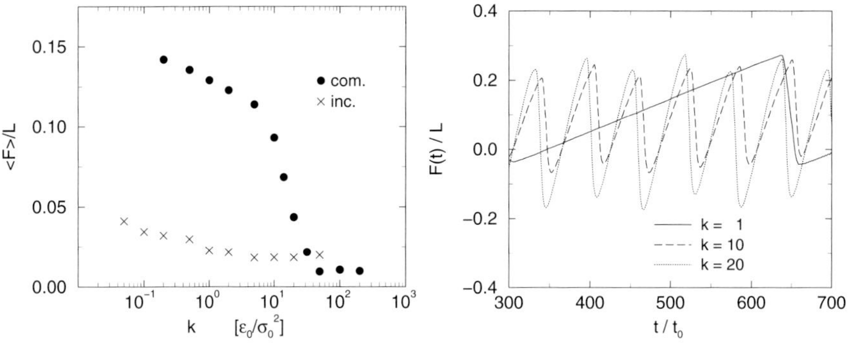

Fig. 10.2. Left: Average force per load $\langle F \rangle / L$ acting on a block pulled at small velocity with spring constant k. Block and substrate are separated by a boundary lubricant. Commensurate (com) and incommensurate (inc) orientations between the identical, confining walls are considered. Right: Instantaneous force acting on the spring of the commensurate system. From Ref. [12].

they are not in direct contact. In particular, the commensurate system shows large friction forces in the stick-slip regime at small pulling spring constants k and small forces in the smooth-sliding regime at large k. In the commensurate case, the instantaneous force behaves in a very periodic way. In particular, at those values of k, where the friction forces start to decrease dramatically, one can observe spikes of two different heights in the force spikes, see the right-hand side of Fig. 10.2. The different spikes can be related to hcp and fcc like configurations of the boundary lubricant (ABA and ABC type layering for lower wall, lubricant, upper wall). However, one needs to be aware that insights into such detailed mechanisms are only useful, if we try to understand experiments in which the effort has been made to orient two surfaces with respect to one another.

Unlike the perfectly aligned walls, the incommensurate system does not show the dramatic drop in $\langle F \rangle$ as it crosses over from stick slip to smooth sliding. Hence, in the present study, we may identify the crossover regime for the commensurate surfaces as an artifact of the commensurability. Moreover the spikes in the instantaneous friction force (not shown here) are rather erratic [12]. To some degree, the resulting trajectory of the incommensurate case is even reminiscent of earthquake dynamics as discussed in Ref. [12]. This can be seen as unexpected, because the microscopic origins of friction between tectonic plates certainly involve much more complicated processes than those considered in the simple molecular dynamics (MD) study of a confined boundary lubricant. So even if we reproduce the effects observed in nature or experiment satisfactorily, it does not necessarily mean that we have identified the relevant atomistic processes leading to friction. The case study also reveals that the friction force derived from the simulations depends strongly on the driving device, namely the harder the spring the smaller the friction force. This effect is particularly strong for the commensurate surfaces.

10.2 Solid Friction Versus Stokes Friction

The classical laws of friction go back to Coulomb and Amontons [13]. Static friction F_s, the force needed to initiate relative sliding between two solid bodies, is proportional to the load L (first law) and independent of the (apparent) area of contact (second law), thus

$$F_s = \mu_s L. \tag{10.1}$$

μ_s is called the static friction coefficient that (to a good approximation) only depends on the chemo-physical properties of the interface and the two solids in contact. The third friction law says that the kinetic friction force F_k, the force needed to keep two solids in relative sliding motion, is (rather) independent of the sliding velocity v. F_k also satisfies the first two laws and in most cases does not vanish in the limit small of sliding velocities v. Of course, in true thermodynamic equilibrium, or alternatively in the mathematical limit $v \to 0$, one would expect kinetic friction to vanish. For most practical and experimental applications, however, one is far away from this limit. The situation is similar to that of the static shear modulus C_{44} of window glass at room temperature. While the true equilibrium value of C_{44} is commonly believed to be zero, the experimentally measured result is in the order of a few dozen Gigapascals.

The solid friction laws are different from those for Stokes or viscous friction which are valid for the motion of a (Brownian) particle in a fluid or a gas. In the case of Stokes friction, there is a linear relationship between driving force and average velocity v, provided that F is sufficiently small. The proportionality coefficient, which is related to the viscosity, can be calculated (at least in principle) from equilibrium statistical mechanics in terms of linear response theory [14]. The viscous force turns out to be the natural consequence of the interaction of one particle with many other particles. The (linear) response of one particle or excitation to an external force can be described as if the other particles were acting like a heat bath composed by a friction term linear in velocity plus random forces. This concept cannot only be applied to the famous Brownian motion, but it can be extended to many different cases such as the damping of phonons or other elementary excitations in solids. Note that the fluctuation-dissipation theorem also predicts the response to a small time-dependent force.

If the coupling of a degree of freedom to a heat bath is a condition for Stokes friction or viscous damping, what is the requirement for F_s and F_k to be different from zero? If two solids interlock geometrically, there will of course be a finite F_s, see Sect. 10.3.1. However, finite F_s does not imply that $F_k(v \to 0)$ remains finite in the limit $v \to 0$. In particular if we disregard the internal degrees of freedom of the two solids, there is no way to dissipate energy and $F_k = 0$ simply owing to energy conservation. In a first approximation, one may reflect the internal degrees of freedom in terms of a viscous force, but then F_k would still vanish linearly with v in contradiction to most experimental observations.

In 1928, Prandtl suggested that elastic instabilities change the picture qualitatively [15]. Usually this insight is attributed to Tomlinson, who published similar ideas in 1929 [16]. In his model, Prandtl describes the substrate as completely rigid. The slider is moved at constant velocity v with respect to the substrate. The slider's surface atom feel a force from the substrate that is periodic with the substrate's lattice constant b. Furthermore the surface atoms are supposed to be coupled harmonically to their lattice sites by a spring constant k and there is dissipation linear in the atom's velocity v_n. If we assume that the microscopic origin of the dissipation is the consequence of the interaction of atom n with a heat bath, i.e. the substrate's phonons, then its equation of motion reads:

$$m_n \ddot{x}_n + m\gamma \dot{x}_n = -k(x_n - x_n^0) + f_0 \sin(2\pi x_n / b), \tag{10.2}$$

where we assume that the equilibrium position x_n^0 of atom n in the slider moves at constant velocity, for example $x_n^0 = v\,t$.

Let us analyze the motion of atom n qualitatively for large values of k and small values of k. If k is larger than $2\pi f_0/b$, then it is easy to show that each atom has only one well-defined mechanical equilibrium site, irrespective of the value of x_n^0. When the upper solid is moved at a constant, small v, each atom will always be close to its unique equilibrium position. This equilibrium position moves with a velocity that is in the order of v. Hence the dissipated friction force is of the order of $m\gamma v$ and consequently F_k vanishes linearly with v as v tends to zero. The situation becomes different for $k < 2\pi f_0/b$. Atoms with more than one stable equilibrium position will now pop from one stable position to another one when the slider is moved laterally. For small pulling velocities v, such a process occurs when an atom does not have a mechanically stable position at time $t + \delta t$ in the vicinity of the old stable mechanical position near which it was located at time t. Such a situation is discussed in Fig. 10.3 in terms of the time-dependent potential energy $V(x)$ associated with the conservative forces, namely

$$V(x) = \frac{1}{2\pi} f_0 b \cos(2\pi x / b) + \frac{1}{2} k (x_n - x_n^0)^2 \tag{10.3}$$

with $\dot{x}_0 = v_0 > 0$. In the 'popping' processes (indicated by the thick solid line in Fig. 10.3), the velocities $\dot{x}_n$ will exceed v by orders of magnitudes for $v \to 0$. At small v, the dynamics along most of the sinusoidal line is rather independent of the precise value of v_0 and the dissipated energy $\int dx \gamma m v$ has a well-defined positive limit $F_k(v_0 = 0)$. Hence in the absence of thermal fluctuations that have been disregarded in this discussion, F_k remains finite even in the limit of infinitely small v.

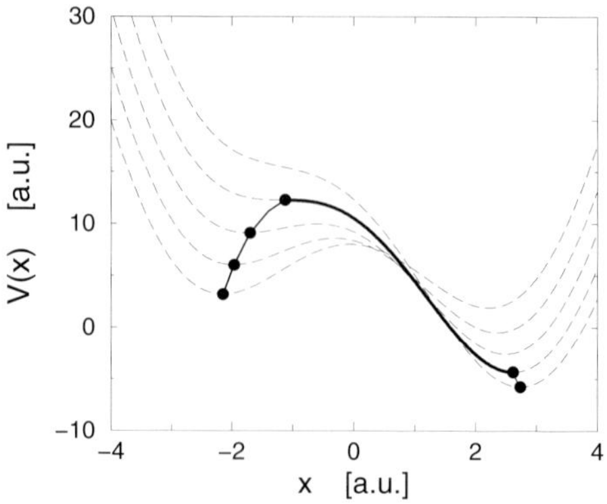

Fig. 10.3. Schematic representation of the time evolution of the potential energy in the Prandtl model (dashed lines), see (10.3). All curves are equidistant in time, separated by a time interval Δt. The circles denote mechanically stable positions and the solid line connects mechanically stable points, indicating the motion of an overdamped point particle.

Based on his model, Prandtl formulated the condition for finite F_k in the limit of small sliding velocities: *If the (elastic) coupling of the mass points is chosen such that at every instance of time a fraction of the mass points possesses several stable equilibrium positions, then the system shows hysteresis....* In the context of friction, hysteresis translates to finite static friction or to a finite kinetic friction that does not vanish in the limit of small sliding velocities. The word 'elastic' in Prandtl's statement is put into parenthesis, in order to make the statement more general, for example, applicable to the case of boundary lubrication. From this analysis, we see that solid friction arises from instabilities. Hence an important lesson to be learned from computer simulations is what these instabilities on a microscopic level are and how these instabilities affect the tribological behavior of a junction. A separate issue is how the heat generated in the pops is transported away from the interface.

Incorporating thermal fluctuations changes the picture qualitatively. In the strict limit $v \to 0$ when all atoms have sufficiently much time to find their true thermal equilibrium for all relative wall positions, friction will vanish linearly with v according to linear response theory. However, in a large regime of small, but finite v, Prandtl predicted that owing to thermal fluctuations, the friction force in his model should only have small, logarithmic corrections in the order of $\ln v$. Prandtl's hypothesis was rederived many times and an analysis of experimental AFM data seemed to confirm that picture [17]. However, a more refined analysis of the thermal fluctuations in the Prandtl-Tomlinson model making use of the theory of fluctuation induced spinodal decompositions [18] suggests corrections in the order of $(k_B T \ln v)^{2/3}$, which indeed fits the experimental results distinctly better than the simple $\ln v$ corrections.

10.3 Dry Friction

The term dry friction obtains a novel meaning in computer simulation, because one can easily prepare absolutely non-contaminated surfaces. Experimentalists often use a less strict definition and mean to express that no lubricant has been added intentionally.

10.3.1 Rigid Walls and Geometric Interlocking

Early theories of friction were based on the purely geometric argument that friction is caused by interlocking of impenetrable and rigid surface asperities [19,20]. The idea (Fig. 10.4) is that the top solid must be lifted up a typical slope $\tan\alpha$ determined by roughness on the bottom surface. If there is no microscopic friction between the surfaces, then the minimum force to initiate sliding is

$$F_s = \mu_s L \tag{10.4}$$

with $\mu_s = \tan\alpha$. This result satisfies Amontons's laws with a constant coefficient of friction $\mu_s \equiv F_s/L = \tan\alpha$. In 1737, Bélidor obtained a typical experimental value of $\mu_s \approx 0.35$ by modeling rough surfaces as spherical asperities arranged to form commensurate crystalline walls [20]. However, asperities on real surfaces do not match as well as envisioned in these models or sketched in Fig. 10.4. On average, for every asperity or atom going up a ramp, there is another going down. One concludes that the mean friction between rigid surfaces vanishes unless they happen to have the same periodicity and alignment.

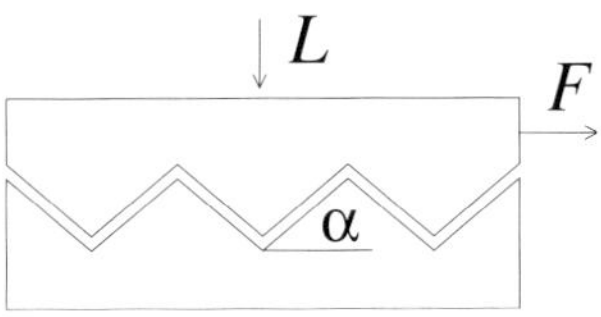

Fig. 10.4. Sketch of two surfaces with interlocking asperities. The top surface experiences a normal load L and a lateral force F, which attempts to pull the top surface up the slope $\tan\alpha$. The bottom wall is fixed.

A cancellation between 'up' and 'down' on an atomic scale happens particularly easily between two flat, incommensurate surfaces. There have been a significant number of computer simulations showing that the *wearless* static friction becomes extremely small, except in the artificial case of identical and perfectly aligned walls. Atomistic computer simulations of iron on iron [21], a blunt pyramidal diamond tip on a metal surface [22], a blunt pyramidal copper tip on incommensurate copper substrates [23], interlayer sliding in multiwalled carbon nanotubes [24], and Lennard Jonesium on Lennard Jonesium [25] all confirm the simple prediction that lateral forces cancel out to a significant degree. These simulations have in common that the solids were treated truly three-dimensionally and that the atoms (interacting via microscopic interaction potentials) were allowed to relax thus making long-range elastic deformations possible. Significant lateral forces were observed only in combination with strongly irreversible processes such as plastic deformation [22], the production of wear [23,26], material mixing, or cold welding [25], but *not* due to geometric interlocking.

When two rigid solids in contact are disordered, the cancellation of lateral forces is less systematic than for an incommensurate contact between rigid bodies. For flat solids with atomic scale roughness only, one can expect that the lateral forces F_s grow proportional to the square root of the intimate area of contact A for a given, constant normal pressure σ_n. Since the load L is given by $A\sigma_n$, the ratio F_s/L vanishes with $1/\sqrt{A}$ as A goes to infinity [27]. But geometric interlocking only explains static friction. Kinetic friction can still vanish, because the energy required to lift up the slider in Fig. 10.4 to the top of the hill can be regained in principle by moving it downhill in a controlled way. As discussed in Sect. 10.2, the situation can change dramatically in principle if the elastic interactions within the bulk are sufficiently weak to cause elastic or other instabilities.

10.3.2 Elastic Deformations: Role of Disorder and Dimensions

Although we are concerned with computer simulations of friction rather than with theoretical arguments, it is instructive to analyze the interplay of disorder, dimensionality, and elastic deformations. To do this, let us consider a d_{obj}-dimensional elastic solid, in which neighbored atoms are coupled via simple springs. We may safely assume the free elastic solid to be mechanically stable, meaning that the tensor of the elastic constants is positive definite. The dimension of the interface between the slider and disordered substrate is denoted by d_{int}.

In such a situation, there will be a competition between the random substrate-slider interactions and the elastic coupling within the solids. An important question to ask is how the interactions change when we change the scale of the system, for example, how strong are the random and the elastic interactions on a scale $2L$ if we know their respective strengths on a scale L. Here L gives the linear dimension of the solids in all directions, that is to say parallel and normal to the interface. As discussed above, the random forces between substrate and slider will scale with the square root of the interface's size, hence the random forces scale with $L^{d_{\text{int}}/2}$. The elastic forces on the other hand scale[1] with $L^{d_{\text{obj}}-2}$. In the thermodynamic limit

[1] A linear chain can be more easily compressed if we replace one spring by two springs coupled in series. In two dimensions, springs are not only coupled in series but also in parallel, so that the elastic coupling remains invariant to a 'block transformation'. Each additional dimension strengthens the effect of 'parallel' coupling.

$L \to \infty$, the effect of disorder will always dominate the elastic interactions or vice-versa, *unless*

$$L^{d_{\rm int}/2} \propto L^{d_{\rm obj}-2}. \tag{10.5}$$

For $\lim_{L\to\infty} L^{d_{\rm int}/2}/L^{d_{\rm obj}-2} >> 1$, the random interactions will dominate and hence pinning via elastic instabilities cannot be avoided. This disorder-induced elastic pinning is then similar to that of compliant, ordered systems as discussed above within the Prandtl-Tomlinson model. For $\lim_{L\to\infty} L^{d_{\rm int}/2}/L^{d_{\rm obj}-2} << 1$, the long-range elastic forces dominate the long-range random forces. The slider's motion can only be opposed by elastic instabilities if the elastic coupling is sufficiently weak at finite L in order to make local pinning possible, again akin of the case $\lambda > 1$ in the Prandtl-Tomlinson model.

The so-called *marginal* situation, in which both contributions scale with the same exponent, $d_{\rm int}/2 = d_{\rm obj} - 2$, occurs in the important case of 3-dimensional solid bodies with 2-dimensional surfaces. In the marginal situation, the friction force can stay finite, however, one may expect the friction force per unit area and hence the friction coefficient to be exponentially small [28]. The *marginal* dimension in the case of $d_{\rm obj} = d_{\rm int}$ (adsorbed monoatomic layers, charge density waves, etc.) is $d_{\rm mar} = 4$ [29]. But even in dimensions smaller than the marginal dimension, friction forces may turn out to be small. One example is an experimental quartz crystal microbalance study [30] of solid and liquid krypton films on disordered gold surfaces ($d_{\rm obj} = d_{\rm int} = 2$), for which the pinning forces were undetectably small.

10.3.3 Extreme Conditions and Non-elastic Deformations

In many cases, the atomistic topology of chemical bonds changes when two solids come into intimate contact and are start to slide. As a consequence, the surface will be altered dramatically when the two solids are removed from one another after the sliding process. A description in terms of *elastic* deformations is not applicable any longer in such a situation. There can be many reasons for non-elastic deformations: (i) plastic flow during contact formation [26,31] or other thermodynamically driven cold welding [25], (ii) plastic deformations due to large normal pressures [22], (iii) large sliding velocities [32], and (iv) sliding induced generation of dislocations [23], to name a few.

The processes that occur in these strongly non-equilibrium situations may become rather complex. This makes it even more important to set up the simulations and to analyze the processes in a meaningful way. For example, it is important to choose the initial conditions such that the bulks have room to yield and that the generated debris does not necessarily remain within the contact. A simulation methodology which allows debris to be transported away from the interface is shown in Fig. 10.6 in the context of sliding of lubricated surfaces.

One particularly nice study of strongly irreversible processes consists of large scale molecular dynamics simulations of the indentation and scraping of metal by Belak and Stowers [22]. Their simulations show that tribological properties are strongly affected by wear or the generation of debris, which in turn may again depend strongly on such 'details' as the system's and the interface's dimensionality. Belak and Stowers considered a blunted carbon tip that was first indented into a copper surface and then pulled over the surface. Since diamond is a hard material, the tip was treated as a rigid unit. Interactions within the metal were modeled with an embedded atom potential and Lennard-Jones potentials were used between Si and Cu atoms.

In the two-dimensional (2D) simulation, indentation was performed at a constant velocity of about 1 m/s. The contact followed Hertzian behavior up to a load $L \approx 2.7$ nN and an indentation of about 3.5 Cu layers. The surface then yielded on one side of the tip, through the creation of a single dislocation edge on one of the easy slip planes. The load needed to continue indenting decreased slightly until an indentation of about five layers. Then the load began to rise again as stress built up on the side that had not yet yielded. After an indentation of about six layers, this side yielded, and further indentation could be achieved without a considerable increase in load. The hardness, defined as the ratio of load to contact length (area), slightly decreased with increasing load once plastic deformation had occurred.

After indentation was completed, the carbon tip was slid parallel to the original Cu surface. The work to scrape off material was determined as a function of the tip radius. A power law dependence was found at small tip radii that did not correspond to experimental findings for micro-scraping. However, at large tip radii, the power law approached the experimental value. Belak and Stowers found that this change in power law was due to a change in the mechanism of plastic deformation from intragranular to intergranular plastic deformation.

In the three-dimensional (3D) simulations, the substrate contained as many as 36 layers or 72,576 atoms. Hence long-range elastic deformations were included. The surface yielded plastically after an indentation of only 1.5 layers, through the creation of a small dislocation loop. The accompanying release of load was much bigger than in 2D. Further indentation to about 6.5 layers produced several of these loading-unloading events. When the tip was pulled out of the substrate, both elastic and plastic recovery was observed. Surprisingly, the plastic deformation in the 3D studies was confined to a region within a few lattice spacings of the tip, while dislocations spread several hundred lattice spacings in the 2D simulations. Belak and Stowers concluded that dislocations were not a very efficient mechanism for accommodating strain at the nanometer length scale in 3D.

When the tip was slid laterally at v = 100m/s during indentation, the friction or 'cutting' force fluctuated around zero as long as the substrate did not yield (Fig. 10.5). This nearly frictionless sliding can be attributed to the

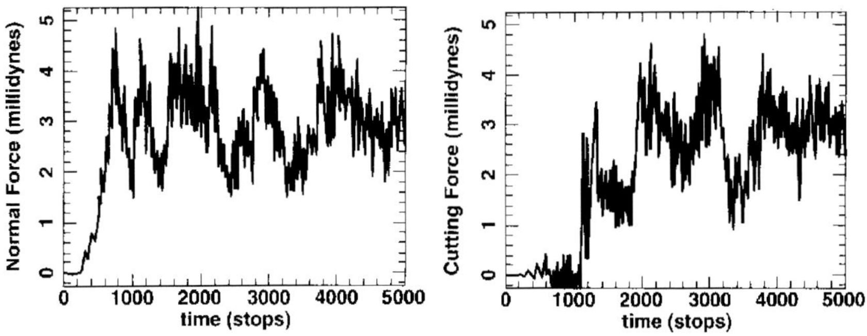

Fig. 10.5. Normal (left) and lateral (right) force on a three dimensional, pyramidal diamond tip on a copper surface as a function of time. No plastic flow was reported up to 1,000 time steps. The indentation stopped at about 5 layers after 2,000 time steps. From Ref. [22].

fact that the surfaces were incommensurate and the adhesive force was too small to induce locking. Once plastic deformation occurred, the cutting force increased dramatically. Fig. 10.5 shows that the lateral and normal forces are comparable, implying a friction coefficient of about one. This large value was expected for cutting by a conical asperity with small adhesive forces [37].

10.4 Lubrication

Any atom or molecule that is bonded weakly to a surface can be considered to perform as a lubricant as long as they do not get squeezed out of the microscopic points of contact. Examples are not only synthetic and mineral oils, but also molecules from the atmosphere physisorbed between the surfaces such as hydrocarbon chains or sometimes even simple nitrogen molecules. Lubricants hinder two surfaces to come into intimate contact. Their presence inhibits or at least reduces the generation of wear and debris. For example, an appropriate additive in a lubricant reacts with a fresh metal surface to form protective surface films. In many cases, a lubricant film glassifies in a point of large normal stress, i.e., in a microscopic point of contact. When the junction breaks, the contact breaks within the lubricant and not within one of the contacting asperities.

The traditional view is that lubricants do not only reduce wear but generally reduce friction between two solids. This is certainly true for most macrotribological processes. However, as we have seen in many examples discussed above, we do not expect any significant solid friction in a microscopic contact as long as no plastic deformation or other strongly irreversible processes occur. In the absence of such processes, the presence of a few adsorbed lubricant particles can increase solid friction, as it is able to accommodate the surface corrugation of both walls simultaneously and thus lock the confining walls together.

The overwhelming majority of experimental and technological sliding systems incorporates lubricants. It is important to include the effect of these lubricants in simulations if one wants to compare simulations to the bulk of tribological experiments. It is then desirable to mimic the experimental, mechanical and thermodynamic boundary conditions such as temperature T, pressure p, chemical potentials μ_i, etc. as closely as possible. One should of course keep in mind the Gibbs-Duhem relation, which states that it is not possible to specify all intensive thermodynamic variables (such as T, p, the various μ_i) independently from one another.

In a molecular dynamics (MD) or in a Monte Carlo (MC) simulation it is rather simple to keep T and p constant. It is more difficult to keep μ constant, because this requires the use of grand canonical moves. These moves typically equilibrate rather slowly and are likely to interfere seriously with the dynamics in MD or the pseudo-dynamics in MC simulations[2]. Since most tribological effects are non-equilibrium effects and therefore intimately linked to the dynamics of the system rather than to their thermodynamics, it is important not to alter the dynamics in an artificial way.

In some cases, it is nevertheless desirable to change the number of atoms in the contact. It was suggested to include this possibility with the help of reservoirs [33] as shown on the left-hand side of Fig. 10.6. While the total particle number N is kept fixed in such a simulation, the number of atoms in the contact can change and one may refer to such a simulation as pseudo grand canonical. The externally applied pressure (tensor) and temperature can then be imposed like in an equilibrium simulation, for instance in such a way that the lubricant remains fluid in the reservoir.

The simulation of the reservoir necessitates simulating many bulk-liquid molecules outside the real contact, producing a certain computational overhead. It was suggested to replace the need for a reservoir by fixing T and the parallel pressure $p_{\parallel}$ to bulk values [34]. It was furthermore suggested to control $p_{\parallel}$ by adjusting the normal separation D between the solids and to leave the contact area A constant. The strategy would be to repeat simulations with different values of N, all under the same fixed A, $p_{\parallel}$, and T. One of the results would be the depletion force, which is the average normal pressure $\langle p_{\perp} \rangle$ times A, as a function of N similar to that shown on the left-hand side of Fig. 10.6. Note that the differentiation between $p_{\parallel}$ and $p_{\perp}$ implies that the lubricant does not correspond to an isotropic fluid any longer. Indeed, the oscillations in the depletion forces are accompanied by strong layering in the lubricant. In order to understand the concurrent dramatic increase in the resistance to sliding observed in experiments [8,9], one also has to inves-

[2] If the MC simulation consists of small local moves only, then the generated trajectories correspond to overdamped dynamics that may provide valuable information on solid friction. However, there is no general principle for determining the appropriate probability distribution of steps in a non-equilibrium MC simulation.

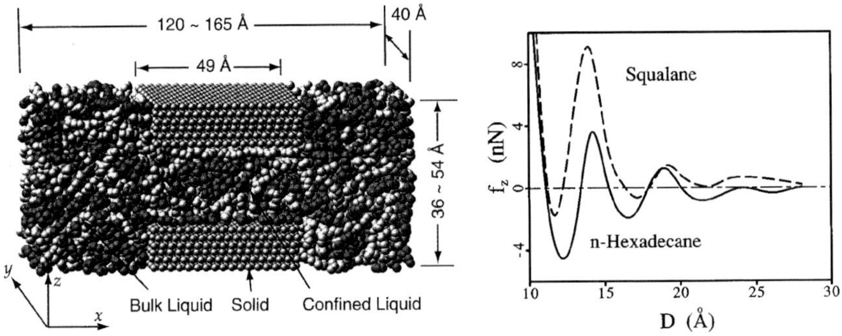

Fig. 10.6. Left: Geometry of a pseudo grand canonical simulation box. Small, regularly arranged spheres represent gold atoms from the solid bulks, while larger circles are associated with lubricant molecules. The lubricant can go back and forth between contact and reservoir. Right: Normal force as a function of separation for two different lubricants as a function of surface separation. From Ref. [33].

tigate the effect of confinement on the viscosity [35] and the way in which the (atomic-scale) corrugations imprint into the lubricant [36].

10.4.1 Boundary Lubrication

Boundary lubrication refers to a situation in which most of the lubricant has been squeezed out of the contact and the remaining lubricant glassifies. It usually occurs in a mechanical contact under high load and low speed conditions. In the extreme limit, only a few atoms remain in the contact. But even a few atoms alter dramatically the friction between two surfaces.

Since in most cases lubricants are only weakly bonded to surfaces, the most commonly used form for the lubricant-lubricant and lubricant-wall interactions are Lennard Jones (LJ) potentials. LJ potentials reflect the important effect that atoms attract when separated by a sufficiently large distance, but repel and behave like hard disks in the presence of a large external pressure.

It has been shown in a series of molecular dynamics simulations [27,38–40] that the presence of a small contamination layer even as thin as a submonolayer leads to static and kinetic friction between two incommensurate surfaces. Within the model calculations the two walls would have had zero static friction if no weakly bonded molecules had been introduced into the interface. As argued above, the weakly bonded atoms are able to accommodate the surface roughness of both confining walls simultaneously, which makes the walls lock together. Of course, in order to obtain not only static friction but also non-zero kinetic friction, it is necessary for the atoms to exhibit mechanical instabilities, see Sect. 10.2. Whether or not the atoms experience instabilities may depend on the special properties of the system such as dimensionality and symmetry of the confining walls and their relative orientation, the details of the lubricant wall interactions such as the sign of the first higher harmonic, and further details. Fig. 10.2 suggests that there are more instabilities present for incommensurate surfaces than for commensurate surfaces in the smooth sliding regime where the pulling spring is hard, while in the stick-slip regime (small values of k) the upper wall in the commensurate case experiences instabilities as a whole owing to geometric interlocking.

The microscopic friction-load relation for boundary-lubricated flat surfaces turns out to be similar to that of Amontons's macroscopic friction law $F = \mu L$, or a simple generalization thereof, namely

$$\tau_k = \tau_0 + \alpha p_\perp , \tag{10.6}$$

where $\tau_k = F_k/A$ is the (kinetic) shear pressure, τ_0 an adhesive offset, and α a constant that can be associated with the friction coefficient μ if the externally imposed normal pressure $p_\perp = L/A$ is distinctly larger than τ_0. A nearly linear relation between τ_s and p can be observed over a wide range of pressures, which is shown exemplarily on the left-hand side of Fig. 10.7 for the case of smooth sliding.

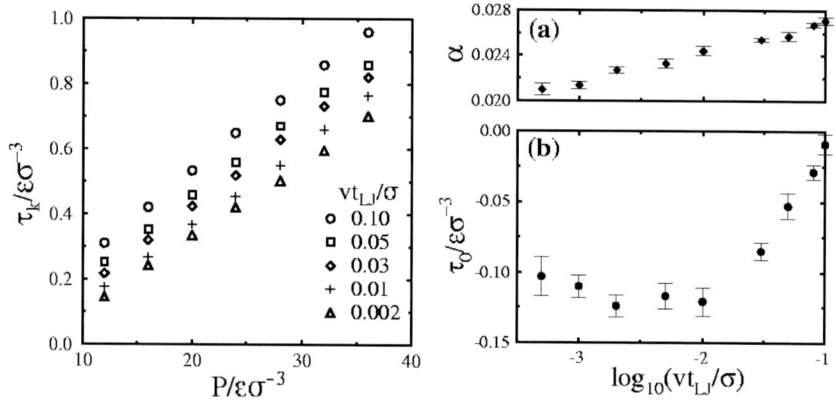

Fig. 10.7. Left: Kinetic shear stress τ_k as a function of normal pressure for different velocities. The data of data like that on the left was fitted to (10.6). Right: The resulting fit parameters α (top) τ_0 (bottom). From Ref. [40]

It is interesting to analyze the velocity dependence of the coefficients τ_0 and α, which is shown on the right-hand side of Fig. 10.7. In agreement with Coulomb's observation, the kinetic friction force F_k is barely sensitive to the sliding velocity v: The parameters α and τ_0 only vary logarithmically with v. The overall decrease of τ_k with v can be associated with thermal activation and diffusion of the lubricant atoms out of their metastable traps [40], confirming previous interpretations of simple rate-state models of friction [41].

10.4.2 Hydrodynamic Lubrication and Its Breakdown

When the contact geometry and the operating conditions are such that the load is fully supported by a fluid film, the surfaces are completely separated. This is generally referred to as the hydrodynamic lubrication and can be well described in most cases by Reynold's equation. When the surfaces are easily deformable, as in rolling contact bearings or human and animal joints, the equations of elasticity and the pressure dependence of lubricant dependence must also be included in the solution of the problem [42]. The hydrodynamic or elastohydrodynamic continuum theory begins to break down as atomic structure becomes important.

A general assumption of continuum theories is the stick condition of the lubricant near a solid wall, that is to say, the tangential velocity of the fluid at the fluid wall interface is set equal to that of the wall. When two surfaces come closer and the confinement is increased, slip can occur at the interface. It is then convenient to introduce a slip length $\mathcal{S}$ into elastohydrodynamic continuum calculations. $\mathcal{S}$ represents the distance into the wall at which the velocity gradient would extrapolate to zero. The calculation of slip length from velocity profiles has some ambiguity. The least ambiguous resolution to the definition of slip length may be given in Ref. [43]. Additional effects due to increasing confinement have been discussed at the beginning of this Sect. 10.4. A more thorough discussion of the literature on atomistic simulations in the hydroelastic lubrication regime is given in Ref. [10]. Some technical issues relevant to the thermostatization of lubricants will also be given below in Sect. 10.5.5.

10.5 Setting Up a Tribological Simulation

10.5.1 The Essential Ingredients

When designing a computer simulation of tribological phenomena, one needs to model (i) the physo-chemical properties of the two materials in contact and the lubricant or the atmosphere involved, (ii) the initial conditions/geometry and (iii) in addition to what needs to be specified in an equilibrium simulation, the driving device. Let us assume that we do have a model for the interactions between the atoms involved in the simulation. We then need to set up Newton's equations of motion for the various degrees of freedom and integrate these equations of motion just like in a computer simulation of an equilibrium system [14,44,45].

The relevant degrees of freedom are: The center of mass of the confining top and bottom solid $\mathbf{R}_{t,b}$, the coordinates $\mathbf{r}$ of those atoms which are coupled directly to the confining solids (typically the atoms in the outermost layers), and the coordinates $\mathbf{x}$ of all other atoms, which may include lubricant atoms and additional wall atoms that do not belong to the outermost layers.

A quite general form to treat the driving device is to keep the substrate fixed ($\mathbf{R}_b$ = const) and to couple the center-of-mass coordinate $\mathbf{R}_t$ of the upper solid to an externally driven harmonic spring as shown schematically in Fig. 10.1. The equation of motion for the top wall then reads:

$$M_t \ddot{\mathbf{R}}_t = K\left(\mathbf{R}_t^0 - \mathbf{R}_t\right) + \mathbf{F}_t\left(\mathbf{R}_t, \{\mathbf{r}\}\right). \tag{10.7}$$

Here M_t denotes the inertia of top solid, K is the stiffness of the spring that couples the solid to the driving device $\mathbf{R}_t^0$. In most cases, K is a diagonal matrix, with three independent components that reflect the normal coupling, and the coupling parallel to the two interfacial directions. Two popular choices for $\mathbf{R}_t^0$ are

$$\mathbf{R}_t^0 = R_{t,y}^0 \mathbf{e}_y + R_{t,z}^0 \mathbf{e}_z + \begin{cases} v_x \mathbf{e}_x\, t & \text{`tribological' driving} \\ \Delta R_{t,x}^0 \mathbf{e}_x \cos(\omega t) & \text{`rheological' driving} \end{cases} \tag{10.8}$$

with $\mathbf{e}_\alpha$ a vector of dimension unity pointing in α direction. Here, v_x corresponds to the average sliding velocity of the upper wall if the spring (see Fig. 10.1) is pulled at constant velocity, and $\Delta R_{t,x}^0$ and ω are the amplitude and frequency of the driving device if the response of the system to oscillations is probed. Despite the simplicity of equations (10.7) and (10.8), it is important to realize the implications that various choices can have. This issue is discussed further below in Sect. 10.5.4.

We still need to specify the coupling $\mathbf{F}_t\left(\mathbf{R}_t, \{\mathbf{r}\}\right)$ between the coordinates of the top solid $\mathbf{R}_t$ and the atoms belonging to the uppermost layer. The discussion is of course equivalent for the bottom wall, even though one usually does not move its position $\mathbf{R}_b$. It is convenient to define equilibrium positions $\mathbf{r}_n^0$ of atom n in the uppermost layer relative to the top solid under the condition that no other atoms are present in the simulation. The equilibrium positions $\mathbf{r}_n^0$ will then always have the same relative displacement $\Delta\mathbf{r}_n^0$ with respect to the top solid, thus

$$\mathbf{r}_n^0(t) = \mathbf{R}_t(t) + \Delta\mathbf{r}_n^0, \tag{10.9}$$

where periodic boundary conditions are usually employed only in directions parallel to the interface. The real position of atom n will then couple to its equilibrium position. To lowest order this coupling can be considered harmonic and the force that the *ideal* lattice point exerts on an atom of the outermost layer can be written as

$$\mathbf{f}_{t,n} = -k\left[\mathbf{r}_n(t) - \mathbf{r}_n^0(t)\right]. \tag{10.10}$$

k is again a matrix in the most general case and the actual choice of k is in principle highly non-trivial. To a large extent, k reflects the elastic properties of the top solid, see the discussion of the choice for k in Sect. 10.5.6. Since $\mathbf{f}_{t,n}$ is the force that the equilibrium site and hence the top solid exerts on atom n, we can finally write down the expression for the last term on the right-hand side of (10.7):

$$\mathbf{F}_t\left(\mathbf{R}_t, \{\mathbf{r}\}\right) = -\sum_n \mathbf{f}_{t,n}. \tag{10.11}$$

As a next step, one has to specify the interactions between atoms within the uppermost layer. A reasonable choice is to do this is again within the harmonic approximation, thus

$$V_{nn'} = \frac{1}{2}\sum_{n<n'}\sum_{\alpha,\beta} k_{nn'}^{\alpha\beta}\left(r_n^\alpha - r_{n'}^\alpha\right)\left(r_n^\beta - r_{n'}^\beta\right), \tag{10.12}$$

with r_n^α being the α's component of the vector $\mathbf{r}_n$. Note that the force constants $k_{nn'}^{\alpha\beta}$ do not only approximate the direct interactions between the atoms in the uppermost layer, but they are effective quantities which may also contain information of the elastic and geometric properties of the top solid. A more detailed discussion is given in Sect. 10.5.6.

Lastly, the particles in the outermost layer must also be coupled to the remaining atoms in the simulation cell, for instance to lubricant atoms or those wall atoms that are not coupled to specific lattice sites. If the forces between wall atoms with coordinate $\mathbf{r}_n$ and remaining atoms with coordinates $\mathbf{x}_k$ are given by $\mathbf{f}_{kn}(\mathbf{r}_k, \mathbf{x}_n)$, then the net force $\mathbf{f}_n$ on atom n in an outermost layer is given by

$$\mathbf{f}_n = \mathbf{f}_{t,n} + \sum_{n'\neq n}\sum_{\alpha,\beta} \mathbf{e}^\alpha k_{nn'}^{\alpha\beta}\left(r_n^\beta - r_{n'}^\beta\right) + \sum_k \mathbf{f}_{nk}(\mathbf{r}_n, \mathbf{x}_k), \tag{10.13}$$

where the sum in the last term on the right-hand-side of (10.13) includes all atoms that do not belong to the uppermost layer. All other interactions, i.e., those in between lubricant atoms or between slider and substrate atoms, should be treated just like in equilibrium simulations of materials [44].

10.5.2 Physo-chemical Properties

The physo-chemical properties of the interface are reflected through the choice of the atomic model potentials. One needs to specify the *intra*bulk and the *inter*bulk atomic interaction potentials as well as the interaction of a lubricant or another adsorbed atom with all other atoms. *Intra*bulk properties are often modeled as entirely elastic, i.e., atoms belonging to one solid are connected to each other and/or to their ideal lattice site via simple harmonic springs. Such simplifying modeling will be sufficient if plastic deformation, cold welding, etc. do not play a significant role in the processes of interest.

Many studies use simple Lennard Jones (LJ) potentials for the interbulk interactions and the interactions between lubricant and walls. LJ interactions are given by

$$V(r_{ij}) = 4\epsilon_{ij}\left[\left(\frac{\sigma_{ij}}{r_{ij}}\right)^{12} - \left(\frac{\sigma_{ij}}{r_{ij}}\right)^{6}\right], \tag{10.14}$$

where ϵ_{ij} and σ_{ij} have units of energy and length, respectively. Typically, the interactions are truncated at a distance $r_{ij}^{(c)}$ and shifted in such a way that $V(r_{ij})$ is continuous at $r_{ij}^{(c)}$. There is a lot one can learn from systems modeled in this way, since LJ potentials are sufficient to describe the generic feature that two atoms attract each other when they are separated by a long distance, while they repel upon close approach. Moreover, there is a lot of flexibility in the choice of the parameters ϵ_{ij}, σ_{ij}, and also the radius $r_{ij}^{(c)}$. The effect of adhesive interactions can be switched on or switched off depending on the choice of $r_{ij}^{(c)}$. A typical choice for $r_{ij}^{(c)}$ for which adhesive effects are eliminated is the cutoff in the minimum of $V(r_{ij})$, hence $r_{ij}^{(c)} = 2^{1/6}\,\sigma_{ij}$.

The use of LJ potentials is also widespread in coarse grained models of polymers. Simulations of friction between polymer bearing surfaces, i.e., simulations of shear forces between polymer brushes, are done in terms of bead-spring models [46–48] In bead-spring models, all interactions between polymer segments consist of LJ interactions and additional non-linear potentials keep the segments bound together [49]. Note that the effect of a good or a bad solvent can be included into such a coarse grained model via the choice of cutoff radius $r^{(c)}$ and temperature T, making it possible to access distinctly larger length and time scales than if all atoms had been included explicitly into the simulation.

Often, one tries to understand the friction between two specific solids. In such a case, LJ potentials are usually not sufficient and more realistic descriptions are required just like in an equilibrium simulation. Many materials show complicated surface reconstruction after cleavage, which often alters surface properties and hence friction dramatically. This concerns metals, whose surface cannot be described accurately in terms of simple two-body potentials. A popular choice for the simulation of metals are so-called embedded atom potentials. Usually, layered materials and carbon nanotubes cannot be modeled with simple two-body potentials either and more realistic description are required, see Refs. [44,50] and references therein for a more detailed overview of empirical many-body potentials.

10.5.3 Initial Geometry

The frictional properties of a slider-substrate system do not only depend on the chemical nature of the two solids, the lubricant, and the thermodynamic conditions such as temperature, normal pressure, etc., but also on the way in which the system is initially set up. Most simulations of friction take place between two flat surfaces, as shown on the left-hand side of Fig. 10.8, while most well-controlled experimental single asperity contacts employ curved tips, either an AFM tip scratching over a smooth substrate or two crossed mica cylinders in SFA experiments. A geometry with similar contact mechanics as those occurring in AFM or SFA experiments is shown on the right-hand side of Fig. 10.8.

Fig. 10.8. Schematic representation of initial geometries. Left: Flat surfaces. Middle: Blunt tip on substrate. Right: Curved tip on substrate.

There are many processes that can differ qualitatively as a function of the initial geometry, because of the differences in the distribution of the normal stress in the contact. One obvious example is the squeeze-out dynamics of a fluid lubricant when the asperity approaches the substrate [52]. But also the dry friction depends strongly on the geometry. For flat, disordered surfaces, there is a well-defined friction coefficient μ_s, which however, decreases with the area of contact A. This is shown on the left-hand side of Fig. 10.9. In the case of a dry, curved tip, F_s depends only linearly on L if tip and substrate are commensurate, which is shown on the right-hand side of Fig. 10.9. The presence of a few contaminating or lubricating atoms changes these dependencies again.

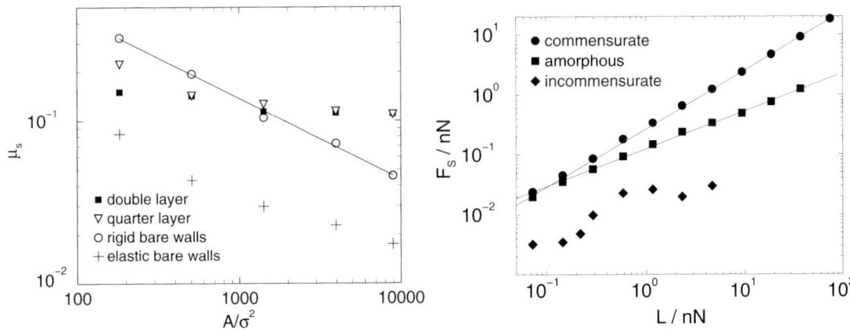

Fig. 10.9. Left: Static friction coefficient $\mu_s = F_s/L$ as a function of interface size A for flat, amorphous walls with different degrees of contamination. From Ref. [27] Right: F_s as a function of load L for a curved tip (commensurate, amorphous, and incommensurate) on a crystalline substrate. Right: From Ref. [51].

Only the quartz crystal microbalance (QCM) [53], which enables one to measure the viscous friction between an adsorbate layer and a smooth crystalline surface does not incorporate curved surfaces. However the simple scaling arguments presented in Sect. 10.3.2 show that the processes and the potential instabilities occurring in QCM experiments are likely to be different from those occurring between two three-dimensional solids.

Before setting up the simulation it is advisable to consider whether surface curvature plays an important role in the process of interest. However, one needs to be aware that the computationally feasible radii of curvature R_c are distinctly smaller than experimental R_c. Employing curved surfaces also reduces the real contact and thus produces computational overhead outside of the contact. Moreover it would be advisable to do simulations for different R_c and to test whether scaling arguments to extrapolate to larger R_c can be employed.

10.5.4 Driving Device

There are many ways in which the solids can be coupled to a driving device. A rather generic way is to couple the upper wall to a spring that is moved according to a well-described procedure, i.e. rheologically or tribologically. The scenario is visualized in Fig. 10.1 and described in more detail in between (10.7) and (10.11). The use of one simple spring K is a simplistic way to model not only the experimental driving device, but to some degree also the elasticity of the slider.

There are two important limits for the value of K. One limit is to use an infinitely weak spring $K \to 0$ and to set the (hypothetical) position of the driving device a distance ΔZ away from the actual position of the top wall such that $K\Delta Z = L$. This mode is equivalent to impose a constant force and is frequently used to drive the system in the direction normal to the interface, in which case L corresponds to the load. It was shown that small normal spring constants K result in smaller friction forces than if the system was driven with large K, at a given average load L [54]. The reason is that the slider has more possibilities to cross energy-barriers if the spring is weak and hence the slider chooses the path of minimum resistance. Of course, in terms of implementing the condition normal load, one would replace the first term of the right-hand side of (10.7) simply with the externally imposed load L. The other limit is $K \to \infty$, in which case the position of the top wall $\mathbf{R}_t$ is identical with that of the driving device $\mathbf{R}_t^0$. For tribological driving, this typically implies a constant sliding velocity mode for the lateral motion and in case of zero normal velocity a constant separation constraint. The effects of the lateral spring constant on the average friction has already been discussed in the simple case study of Sect. 10.1.1.

The driving device can have different modes in different directions. A natural choice would be to apply a constant load mode in the normal z-direction, while the slider is pulled with a weak spring parallel to the x-axis, and coupled to another non-moving spring in y direction.

10.5.5 Thermostating

Most experiments are conducted in such a way that, far away from the sliding surfaces, constant temperature is maintained. The heat produced in the sliding process is conducted away from the interface via phonons and in the case of metals heat dissipation also occurs via electrons. One can mimic the effect of heat conduction in a simulation by employing thermostats or heat baths similar to those used in equilibrium simulations. However, there are a few additional difficulties in a non-equilibrium simulation involving sliding surfaces.

(i) There is no well-defined frame of reference. (ii) The driving device does work on the system. The final thermal energy or temperature and thus the system's properties will depend on the rate with which energy is pulled out of the system. In equilibrium simulations, static properties including thermal energy do not depend at all on the strength of that coupling and dynamical properties should depend only little on the coupling to the heat bath.

These three difficulties can be easily overcome if certain rules are respected. (i) Only the outermost layers should be thermostated and thermostatization should take place within the frame of reference defined by the motion of $\mathbf{R}_t$. Alternatively, dissipative particle dynamics (DPD) [55] may be used to thermostat locally in the center-of-mass system of two neighbored wall atoms. DPD ensures that the Gibbs distribution is recovered as the stationary solution to the Fokker-Planck equation. As compared to regular Langevin-type thermostats, DPD is more difficult to implement and slightly more CPU time intensive. However, it has the distinct advantage to act only on a local scale. In some cases, one may also want to use a DPD thermostats within the lubricant, for example to mimic the effect of collisions with solvent atoms that are not simulated explicitly, but only taken into account via effective interactions. If implemented correctly, Navier-Stokes coefficients are recovered in the hydrodynamic limit [56]. (ii) In extreme conditions, the effects of heating cannot be neglected any longer. One possibility is to define the thermostat's coupling strength such that the heat flow into the thermostat corresponds approximately to that which one would have if the system was made infinitely large normal to the interface (size consistency).

10.5.6 Methods to Treat the Wall's Elasticity

From a computational point of view, we do not want to spend most of the CPU time with the simulation of the bulk in order to reproduce the proper elastic behavior. However, the discussion above shows that a proper description of elastic effects may be crucial. If we only couple the atoms to the equilibrium sites, we suppress elastic deformations and as a consequence unrealistic pressure profiles in the contact may be obtained. On the other hand, we do not want to neglect the effect of long-range elastic interactions by simply connecting the surface atoms with effective springs. This would favor elastic long-range deformations in an unrealistic way and thus make possible elastic instabilities.

It is possible to reproduce the bulk's proper static, elastic response by a suitable choice of the coefficients k in (10.7) and $k_{nn'}^{\alpha\beta}$ in (10.12), because all harmonic modes can be integrated out in principle. The treatment can even be generalized to a dynamic response in terms of Green's functions [57]. Such sophisticated treatments, however, are very demanding in several aspects, and simplifying procedures are pursued in practice. The easiest approach is to treat the bulk's elasticity in a mean-field approximation, in which case one only needs to define the coupling strength k of a wall atom to its ideal lattice site. From the scaling arguments in Sect. 10.3.2, one may withdraw that k would only have a well-defined value in the limit of the slider's linear system size L, if the spatial dimension $d_{\rm obj}$ of the slider is larger than two. In $d = 2$, the proper mean-field choice for k would be $k \propto \ln L$.

A good compromise is to use elastic coupling to the ideal lattice site and next neighbor coupling. Such a model is not only an interesting generalization of the Prandtl-Tomlinson model for the analysis of dry friction [58], but also useful for computer simulations involving lubricants, because the lubricant gets squeezed less easily through the confining walls. For flat surfaces, one may use simple harmonic springs. If one of the two surfaces is curved as shown in Fig. 10.10, then the harmonic springs have to be replaced with anharmonic interactions: Within the radius of contact r_c, the normal deflection $\delta z(r)$ of the tip atoms with respect to an uncompressed tip is proportional to r^2, where r represents the lateral distance from the center of the tip. In order to reproduce the proper Hertzian contact profile $p_\perp \propto (r - r_c)^{3/2}$ for a tip pressed against a hard but non-adhesive substrate, the normal restoring force $f_\perp$ to the lattice site must be chosen according to [51]

$$f_\perp(\delta z) = \sqrt{\delta z / R_{\rm c}} A_a / K, \tag{10.15}$$

where $R_{\rm c}$ is the tip's radius of curvature and A_a the surface area per atom represented in the surface. Note that this procedure does not impose Hertzian contact mechanics, but the final pressure distribution depends among other things on the adhesive interaction between tip and substrate.

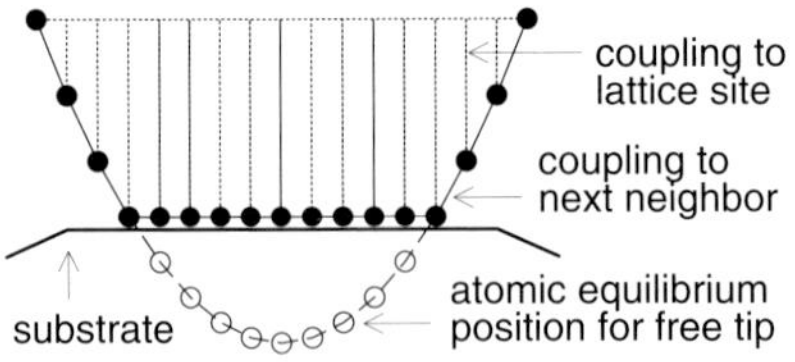

Fig. 10.10. Schematic representation of elastic interactions within a soft tip pressed on a hard surface. Both couplings can have one component normal to the interface and two transversal components.

10.5.7 Calculation of the Friction Force

There are various ways to calculate the kinetic friction force. Using the notation introduced in Sect. 10.5.1, one can measure the force that the external driving device exerts on the wall atoms, i.e., $K(R^0_{t,x} - R_{t,0})$ with x the sliding direction, or alternatively the total force that the top-wall atoms exert on their equilibrium positions, i.e. the projection of $\sum_n \mathbf{f}_{t,n}$, see (10.10), onto the sliding direction. We may also sum up the force between top-wall atoms and all other atoms to obtain the friction force. The different ways should be identical since otherwise the top wall would be accelerated. The argument for the equivalence of the different ways to calculate the friction force breaks down when the surface experiences a time-dependent external force as is the case in the rheological driving mode. This issue has been discussed in detail in Ref. [11] within the Prandtl-Tomlinson model.

In the absence of thermal fluctuations, static friction F_s is defined as the maximum external shear force necessary to invoke lateral sliding. In order to calculate F_s, one can ramp up an external force arbitrarily slowly $F_{\text{ext}}(t) = \dot{F}_{\text{ext}} t$ and identify the time t_{dep} when the system depins and starts sliding. F_s can then be associated with $F_{\text{ext}}(t_{\text{dep}})$. At finite temperature, the precise value of F_s will depend on the rate $\dot{F}_{\text{ext}}$. Near the depinning threshold, thermal fluctuations will assist the system to overcome the barrier and small corrections to F_s, approximately in the order of $\ln \dot{F}_{\text{ext}}$, will presumably apply, similar to the kinetic friction in the Prandtl-Tomlinson model or boundary lubrication [40].

10.5.8 Interpretation of Time Scales and Velocities

The interpretation of the length and time scales in a computer simulation becomes important when comparison is made to experiment. The length scale is defined in a computer simulation through the size of an atom with a typical radius of about 1.5 Å. The definition of the time scale is much less clear cut. According to the international system of units, a second is defined as the duration of $9.192\,10^9$ and a few periods of a particular vibration of the isotope ^{133}Cs. However, the physics of this vibration is completely irrelevant in a tribological context and a different definition of time scales will be more appropriate. This definition will depend on the particular problem. In the following two exemplary discussions, it will be assumed implicitly that there is a time-scale separation between the vibronic motion of the atoms and all processes requiring thermal activation.

Let us first consider the case of dry friction, i.e., an AFM tip in contact with a substrate. Equation (10.2) is then frequently used to model phenomenologically the results of a real experiment or of a computer simulation. If we carry out a simulation of a specific experiment and use realistic model potentials, then we will withdraw a similar value for the tip-substrate interaction strength f_0 from the simulation as from experiment. We might also be successful in reproducing the rate γ with which energy is dissipated and the (effective) coupling strength k between tip and driving device. However, in almost all cases, we will not assign the experimental device/tip mass m_{exp} to our top wall, but a mass m_{sim} that is many orders of magnitude smaller. This means that the relevant frequencies defined either via $\sqrt{k/m}$ or via $\sqrt{f_0/m}$ are also many orders of magnitude smaller in experiment than in the simulations. Thus, in order to compare to experiment our velocities need to be scaled by a factor $\sqrt{m_{\mathrm{exp}}/m_{\mathrm{sim}}}$ and comparing absolute velocities would be meaningless.

Let us next consider the case of two boundary lubricated surfaces under shear. The film confined between two surfaces is interpreted as a glassy state, which is able to accommodate the atomistic surface corrugation of both walls simultaneously. After a typical relaxation time τ, a lubricant atom will undergo thermally activated motion, which eventually leads to lateral creep motion of the slider. In correspondence to the concept of the Deborah number D, which is defined as time of relaxation τ divided by the time of observation Δt, one may expect the friction force to depend on the ratio of τ and the time Δt necessary for the slider to move laterally a typical atomic length scale $\Delta l = \Delta t/v$. A distinct advantage of computer simulation is that one can alter the Deborah number at will either by varying τ via the normal pressure or via a change in sliding velocity v.

10.6 Conclusions

This contribution is meant as a guideline for setting up a meaningful simulation of frictional processes. To do this successfully, it is helpful to have both a good understanding of the underlying theoretical approaches and the knowledge of recent advances in numerical algorithms.

Understanding the underlying theoretical ideas is important in order to prevent the simulations to be meaningless when we want to compare our results to experiments. After all, the goal is to interprete and to understand experiments. Since it is unfeasible to treat all relevant degrees of freedom in a simulation, we are forced to make a model before we carry out the simulation. In my opinion, there are two classes of mistakes that one can make, one class is due to reductional modeling while the other class is due to fictitious modeling. Reductional modeling means that we have not captured all aspects of the system, but what we simulate is at least part of the fuller picture. Fictitious modeling includes a (friction) mechanism that is not relevant in experiment but arises as a consequence of our model. For instance if the walls are described as elastic media although plastic deformation can occur and be a relevant contribution to the net friction, then we see a reduced part of the full picture. An example for fictitious modeling would be a simulation in which long-range elastic interactions are neglected and elastic instabilities occur, although they might only play a minor role in the real system and not contribute to friction. The following remaining compilation of recommendations summarize many aspects of this chapter.

The cardinal and frequently encountered mistake is the use of commensurate surfaces. Even if there is 'stuff' between the surfaces, there will always remain some finite (artificial) resistance to the initiation of sliding. The reason is that commensurability breaks translational invariance in a very specific way. It is certainly true that the effect often becomes negligibly small if the distance between the walls is sufficiently large. It has yet been shown that not only fluids but even a gas can pin two commensurate solids even if they do not interact directly with each other [39]. Similar comments apply to the simulation of friction between walls that have dense Langmuir-Blodgett type layers grafted onto them. The atomistic processes occurring between commensurate layers in simulations will be strikingly different from those between non-matching surfaces in experiment.

The presence of a few adsorbed atoms on surfaces most always also alters the tribological behavior qualitatively. This concerns in particular reactive and reconstructing surfaces that have much higher friction in UHV than in ambient conditions. In the other extreme, when two (flat) surfaces in contact are believed not to deform irreversibly at all, then sliding in UHV will mainly be opposed by a drag force. In this case, an adsorbed layer will be responsible for a dramatic increase in solid friction.

Other issues discussed in this chapter include the effect of wall curvature and initial geometry, the effects of long-range elasticity, the importance of properly implementing the driving device (artifacts due to constant separation and constant, sliding velocity constraints), and thermostats that allow to transport heat away from the interface in a non-equilibrium situation. Concerning all these aspects, one should of course attempt to mimic the experimental situation as closely as computationally feasible. In some cases, however, it is advantageous not to mimic experiment. For instance, if the mass of the slider is small in the simulation, the gap between the macroscopic processes and the microscopic motion is reduced. This makes it sometimes possible to simulate processes on rather small time scales that occur only on macroscopic time scales in experiment.

Multi-scale techniques that have been used in simulations of fracture like in Ref. [59] will certainly prove valuable in friction simulations as well. These techniques combine ab-initio, atomistic, and coarse-grained modeling within one simulation. In particular, the simulation of an AFM tip substrate interaction seems to be a well-suited problem: The intimate contact can be modeled in terms of ab-initio, the area further outside with an atomistic description, and the proper contact mechanics can be guaranteed with continuum methods for the areas even further away from the intimate contact.

Acknowledgments

The author thanks K. Binder, C. Brangian, M. O. Robbins, and M. Urbakh for useful discussions. This work was supported through the Israeli-German D.I.P.-Project No 352-101 and the Materialwissenschaftliche Forschungszentrum (MWFZ) Rheinland-Pfalz.

References

1. G. Binning, C. F. Quate, and C. Gerber, Phys. Rev. Lett. **56**, 930 (1986).
2. C. M. Mate, G. M. McClelland, R. Erlandsson, and S. Chiang, Phys. Rev. Lett. **59**, 1942 (1987).
3. J. N. Israelachvili, P. M. McGuiggan, and A. M. Homola Science, **240**, 189 (1988).
4. J. N. Israelachvili, Surf. Sci. Rpt., **14**, 109 (1992).
5. U.D. Schwarz, O. Zwörner, P. Köster, and R. Wiesendanger, Phys. Rev. B **56**, 6987 (1997); *in ibit.* 6997.
6. A. Berman, C. Drummond, and J. N. Israelachvili, Tribol. Lett. **4**, 95 (1998)
7. F. P. Bowden and D. Tabor, *The Friction and Lubrication of Solids* (Clarendon Press, Oxford, 1958).
8. J. Klein and E. Kumacheva, Science **269**, 816 (1995); J. Chem. Phys. **108**, 6996.
9. A. L. Demirel and S. Granick, Phys. Rev. Lett. **77**, 2261 (1996).
10. M. O. Robbins and M. H. Müser, in *Modern Tribology Handbook I*, pp. 717-765, Ed. B. Bhushan (CRC Press, Boca Raton, FL, 2001); cond-mat/0001056.
11. V. Zaloj, M. Urbakh, and J. Klafter, Phys. Rev. Lett. **81**, 1227 (1998); J. Chem. Phys. **110**, 1263 (1999).
12. M. H. Müser, in *Fundamentals of Tribology and Bridging the Gap between Macro- and Micro/Nanoscales*, pp. 235-240, Ed. B. Bhushan (Kluwer Academic Publishers, Netherlands, 2001); cond-mat/0012100.
13. B.N.J. Persson, *Sliding Friction: Physical Principles and Applications* (Springer, Berlin, 1998).
14. D. Frenkel and B. Smit, *Understanding Molecular Simulation: From Algorithms to Applications*, (Academic Press, San Diego, 1996).
15. L. Prandtl, ZS. f. angew. Math. u. Mech., **8**, 85 (1928).
16. G. A. Tomlinson, Phil. Mag. Series, **7**, 905 (1929).
17. E. Gnecco et al., Phys. Rev. Lett. **84**, 1172 (2000).
18. Y. Sang, M. Dubé and M. Grant, Phys. Rev. Lett. **87**, 174301 (2001).
19. F. P. Bowden and D. Tabor, *The Friction and Lubrication of Solids*, (Clarendon Press, Oxford, 1986).
20. D. Dowson, *History of Tribology* (Longman Inc., New York, 1979) pp. 153-67.
21. M. Hirano and K. Shinjo, Phys. Rev. B **41**, 11837 (1990).
22. J. Belak and I. F. Stowers, in *Fundamentals of Friction*, ed. I. L. Singer and H. M. Pollock (Springer-Verlag, Berlin, 1990).
23. M. R. Sørensen, K. W. Jacobsen, and P. Stoltze, Phys Rev. B **53**, 2101 (1996).
24. A. N. Kolmogorov and V. H. Crespi, Phys. Rev. Lett. **85**, 4727 (2000).
25. M. H. Müser, Tribol. Lett. **10**, 15 (2001); cond-mat/0004494.
26. A. Buldum and S. Ciraci, Phys. Rev. B **55**, 12892 (1997).

27. M. H. Müser, L. Wenning, and M. O. Robbins, Phys. Rev. Lett. **86**, 1295 (2001).
28. B. N. J. Persson and E. Tosatti, Solid State Comm. **109**, 739 (1999).
29. D. S. Fisher, Phys. Rev B, **31**, 1396 (1985).
30. C. Mak and J. Krim, Faraday Disc. **107**, 389 (1997).
31. U. Landman and W. D. Luedtke, J. Vac. Sci. Technol. B, **9**, 414 (1991).
32. J. E. Hammerberg, B. L. Holian, J. Röder, A. R. Bishop, and J. J. Zhou, Physica D, **123**, 330 (1998).
33. J. Gao, W. D. Luedtke, and U. Landman, J. Chem. Phys. **106**, 4309 (1997).
34. J.-C. Wang and K. A. Fichthorn, J. Chem. Phys. **112**, 8252 (2000).
35. I. Bitsanis, S. A. Somers, H. T. Davis, and M. Tirrell J. Chem. Phys. **93**, 3427 (1990).
36. P. A. Thompson and M. O. Robbins, Phys. Rev. A **41**, 6830 (1990).
37. N. P. Suh, *Tribophysics*, (Prentice-Hall, Englewood Cliffs, 1986).
38. G. He, M. H. Müser, and M. O. Robbins, Science **284**, 1650 (1999).
39. M. H. Müser and M. O. Robbins, Phys. Rev. B **61**, 2335 (2000).
40. G. He and M. O. Robbins, Tribol. Lett. **10**, 7 (2001).
41. J. H. Dieterich, J. Geophys. Res. **81**, 2169 (1979).
42. A. Z. Szeri, in *Modern Tribology Handbook I*, pp. 383-453, Ed. B. Bhushan (CRC Press, Boca Raton, FL, 2001).
43. L. Bocquet and J.-L. Barrat, Phys. Rev. E **49**, 3079 (1994).
44. K. Ohno, K. Esfarjani, and Y. Kawazoe, *Computational Materials Science*, (Springer, Berlin, 1999).
45. J. M. Thijssen, *Computational Physics* (Cambridge University Press, Cambridge, 1999).
46. For reviews, see e.g., G.S. Grest, in *Advances in Polymer Science 138*, p. 149, ed. S. Granick (Springer, Berlin, 1999); G. S. Grest and M. Murat, in *Monte Carlo and Molecular Dynamics Simulations in Polymer Science*, p. 476, ed. K. Binder (Oxford University Press, New York, 1995).
47. P. S. Doyle, E. S. G. Shaqfeh, and A. P. Gast, Macromolecules **31**, 5474 (1998).
48. T. Kreer, M. H. Müser, K. Binder, and J. Klein, Langmuir **17**, 7804 (2001).
49. K. Kremer and G. S. Grest, J. Chem. Phys. **92**, 5057 (1990).
50. S. Erkoc, Phys. Rep. **278**, 79 (1997).
51. L. Wenning and M. H. Müser, Europhys. Lett. **54**, 693 (2001).
52. B. N. J. Persson, J. Chem. Phys. **113**, 5477 (2000).
53. J. Krim and A. Widom, Phys. Rev. B **38**, 12184 (1988).
54. V. Zaloj, M. Urbakh and J. Klafter, Phys. Rev. Lett. **82**, 4823 (1999).
55. P. Espanol and P. Warren, Europhys. Lett. **30**, 191 (1995).
56. M. Ripoll, M. H. Ernst, and P. Espanol, J. Chem. Phys. **115**, 7271 (2001).
57. A list of references for treating elastic solids in terms of Green's function can be found at: http://www.ctcms.nist.gov/gf/references.php3.
58. M. Weiss and F.-J. Elmer, Phys. Rev. B, **53**, 7539 (1996).
59. F. F. Abraham, J. Q. Broughton, N. Bernstein, and E. Kaxiras, Europhys. Lett. **44**, 783 (1998); J. Q. Broughton, F. F. Abraham, N. Bernstein, and E. Kaxiras, Phys. Rev. B **60** 2391 (1999).

Part VIII

Methodological Developments in MD and MC

11 Bridging the Time Scale Gap with Transition Path Sampling

Christoph Dellago[1] and David Chandler[2]

[1] Department of Chemistry, University of Rochester, Rochester, New York 14627, USA
[2] Department of Chemistry, University of California, Berkeley, California 94720, USA

Abstract. Transition path sampling is a methodology which overcomes both the long timescale problem and the lack of prior knowledge about transition mechanisms. Here we briefly review the basic principles of transition path sampling, illustrate its application using autoionization in liquid water, and emphasize the capabilities and limitations of the methodology.

11.1 Why Transition Path Sampling Is Needed

Many interesting processes in nature are characterized by the presence of different relevant time scales. In a chemical reaction, for instance, the reaction time can be many orders of magnitudes longer than the molecular vibration period usually measured in units of femtoseconds [1]. Such a separation of time scales creates serious problems for the computer simulation: on one hand the resolution in time needs to be fine enough to capture the properties of fast motions (such as molecular oscillations) and on the other hand the simulation must be extended to times longer then the longest relevant time scale in order to observe the events of interest (such as chemical reactions). This is the notorious time scale gap problem addressed in this conference. It is not only a problem in chemical physics. For example some comets exhibit rapid transitions between heliocentric orbits inside and outside the orbit of Jupiter [2]. While the transition, during which the comet transiently orbits Jupiter for a few periods, is swift, many revolutions of the comet around the sun can occur between transition.

Often, widely separated time scales are caused by energy (or free energy) barriers preventing the system from quickly visiting a representative sample of pertinent configurations. In the past, many efficient computer simulation techniques [3] such as umbrella sampling [4], the multiple histogram method [5], and, most recently, the Laio-Parrinello approach [6], to mention just a few, have been developed to overcome such barriers and sample the free energy surface for a specified control parameter (or order parameter). While biasing schemes of this sort can be used to determine structural quantities such as chemical potentials and equilibrium constants, they are of limited use if one wishes to study the dynamics or kinetics of rare transitions. In this case, detailed knowledge of the underlying mechanism in terms of a reaction coordinate is necessary to apply standard techniques such as the reactive flux method [7–9].

For numerous problems of current interest, however, reaction coordinates are not known, and in applying the techniques mentioned above one has to rely on guessing the relevant, possibly collective, variables. If the selected degrees of freedom do not capture the essence of the mechanisms under study, such biasing schemes are bound to fail. Fortunately, there are diagnostic tools, such as committor distributions (to be described later in the article), which can be used to detect if the relevant degrees of freedom have been identified correctly or not [10,11]. But what can one do when such failure has become apparent?

To answer this question, it is important to distinguish whether the system under consideration is simple or complex. The former is where dynamics is dominated by energetic (as opposed to entropic) barriers, and the topography of the energy landscape is not excessively complicated. In this case, insight can be obtained by locating minima (stable states) and saddle points (transition states) on the potential energy surface. This basic idea has been successfully implemented in various methods including eigen vector following [12,13], the nudged elastic band method [14], and hyperdynamics [15]. In complex (i.e., non-simple) systems, however, the potential surface can exhibit a huge number of distinguishing features, such as local minima, maxima and saddle points. Explicitly enumerating all these features is impractical. Further, dynamical bottlenecks need not coincide with these features in any straightforward way. This fact is illustrated by the "golf course" landscape, in which an entropic barrier hinders the system from finding its energetically most favorable configuration [16]. The failure of methods relying on searching the potential energy surface for stationary points can be illustrated even more drastically in entropy driven phase transitions, such as the freezing of hard spheres experimentally observed in colloidal suspensions [17,18]. In such processes all configurations accessible to the system are isoenergetic and any transition is purely driven by entropic imbalances. No potential energy minima or saddle points exist, but there still are stable states separated by a free energy barrier and rare transitions between these stable states can occur. So what then can we do for a complex system where we don't know the mechanism and cannot find it by searching for specific points on the potential energy surface?

Transition path sampling offers an answer by considering trajectories instead of single configurations [19–21]. This change in perspective permits the generation of rare transitions between stable states without prior knowledge of mechanisms or reaction coordinates. Rather than requiring such prior information as an input, transition path sampling can help in finding it. In the following section we briefly review the essential ideas of transition path sampling. More detailed descriptions of the method including several illustrative examples are given in two recently published review articles [10,11]. As with any newly developed technique, there are misconceptions about what transition path sampling can and cannot do. As such, we follow our summary of the basic principles of transition path sampling with a discussion of its capabilities and its limitations, with emphasis on how these limitations might be surmounted in the future.

11.2 How Transition Path Sampling Works

11.2.1 Probabilities of Trajectories

The transition path sampling method is based on a statistical mechanics of trajectories in which every trajectory $x(t)$ of length t is assigned a statistical weight $P[x(t)]$. The set of all pathways $x(t)$ consistent with the path probability $P[x(t)]$ is called the transition path ensemble. Here x is a possibly high dimensional vector including all variables necessary to specify the state of the system under study. For instance, in a molecular system x may consist of coordinates and momenta of all particles. The trajectory $x(t)$ is a sequence of such states, in which x_0 denotes the first state on the trajectory and x_t denotes the last one. For practical reasons $x(t)$ is represented by a or chain of states, but in principle $x(t)$ can be thought of as a general trajectory, which can be continuous or discrete depending on the process one intends to study.

The form of the probability functional $P[x(t)]$ specifying the weight of a given path $x(t)$ in the transition path ensemble depends on dynamical rules governing the time evolution of the system. Let's for simplicity assume that the system evolves according to some set of equations of motion (for instance Newton's equations of motion) and that the dynamics has the Markov property, i.e. the state x of the system at time t completely determines the probability to find the system in a certain other state x' a short time later. Then, the probability of a particular trajectory to be observed is:

$$P[x(t)] = \varrho(x_0) \prod_{i=0}^{L-1} p(x_{i\Delta t} \rightarrow x_{(i+1)\Delta t}) \,, \tag{11.1}$$

where we have imagined that the pathway is represented by an ordered chain of L states and $p(x_{i\Delta t} \rightarrow x_{(i+1)\Delta t})$ is the (properly normalized) conditional probability to observe the system in state $x_{(i+1)\Delta t}$ at time $t + \Delta t$ provided it was in state $x_{i\Delta t}$ a time Δt earlier. In the case of Newton's equations of motion (and of any other deterministic set of equations of motion) the path probability $P[x(t)]$ consists of a product of Dirac delta functions describing deterministic trajectories flowing from the respective initial conditions. If the system under study is more conveniently described by a stochastic equation of motion (such as Langevin's), the short time transition probability $p(x_{i\Delta t} \rightarrow x_{(i+1)\Delta t})$ is not singular and more than one trajectory is allowed to emanate from the same initial condition x_0. In any case, the path functional describes trajectories generated with a particular dynamical rule. Therefore, members of the transition path ensemble are physical trajectories (as opposed, for instance, to artificial minimum energy pathways) that can be used to study kinetics and dynamics.

In (11.1), $\varrho(x_0)$ denotes the distribution of starting points. This distribution of initial conditions might be the canonical distribution if one is studying a system in thermal equilibrium. In other situations, $\varrho(x_0)$ may represent a specific non-equilibrium distribution of initial conditions generated in a particular experiment.

11.2.2 Defining the Transition Path Ensemble

Let us now restrict the transition path ensemble to trajectories starting in a certain region A in phase (or configuration) space and ending in a different region B:

$$P[x(t)] = h_A(x_0)\varrho(x_0)\left[\prod_{i=0}^{L-1} p(x_{i\Delta t} \to x_{(i+1)\Delta t})\right] h_B(x_t)/Q_{AB} \ , \qquad (11.2)$$

where Q_{AB} normalizes the path ensemble. In the above equation, $h_A(x)$ (or $h_B(x)$) is a function which is one if its argument is located in A (or B) and vanishes otherwise. Such a restriction is useful, when one is interested in rare transitions between regions A and B, where A and B may characterize reactants and products of a chemical reaction, different phases of condensed material, different conformations of a biomolecule, or different kinds of orbits followed by a comet. Careful definition of regions A and B is a crucial step in a path sampling study of rare transitions between long lived states [22]. On one hand, A and B should be large enough to accommodate equilibrium fluctuations when the system resides in one of these stable states. On the other hand, it is important to avoid overlap of regions A and B with the basin of attraction of the other stable state. If an overlap exists due to non-discriminating definitions of states A and B, large portions of the transition path ensemble might reside entirely in one of the stable states without displaying a real transition. Often, proper definition of the initial and final regions is far from trivial requiring several trial and error attempts [1,10].

Introducing the characteristic functions $h_A(x)$ and $h_B(x)$ is not the only way to restrict the transition path ensemble. In general, we may introduce any particular functional $F[x(t)]$ that specifies a subset of trajectories flowing from x_0:

$$P[x(t)] = F[x(t)]\,\varrho(x_0)\left[\prod_{i=0}^{L-1} p(x_{i\Delta t} \to x_{(i+1)\Delta t})\right] /Q_F \ , \qquad (11.3)$$

where Q_F again normalizes the distribution. Such path ensembles with a restriction different from $h_A(x_0)h_B(x_t)$ are appropriate when probing non-equilibrium properties, as has been illustrated in studying the relaxation of energy gaps in solvation dynamics [23]. In general, a restricting functional $F[x(t)]$ could be used to introduce a bias that facilitates sampling of any sort of rare dynamical structures that might be of interest. An example could be the space-time bubbles in glass formers [24].

11.2.3 Sampling the Transition Path Ensemble

The equations above provide abstract definitions of ensembles of pathways in terms of the statistical weight of dynamical trajectories. In order to extract useful information from these ensembles, such as mechanisms and rate constants, one must sample the transition path ensemble and harvest trajectories according to their probability. This can be done efficiently by using a Monte Carlo (MC) algorithm [19]. In such a procedure, one starts from a trajectory with non-zero weight and then uses this trajectory to generate a new one. The new trajectory is then accepted or rejected according to a detailed balance criterion which ensures correct sampling of the ensemble. Repeating this procedure over and over again, one carries out a random walk through the space of all trajectories visiting points in this space according to their weight in the transition path ensemble. Such a Monte Carlo procedure is analogous to the Monte Carlo simulation of, say, a liquid, in which one explores the potential energy surface by carrying out a biased random walk through configuration space. A transition path sampling simulation proceeds in an analogous way with the difference that the sampled objects are dynamical trajectories instead of configurations. It is, however, important to realize that while pathways are sampled with a Monte Carlo (MC) procedure, the trajectories themselves are physical trajectories as, for instance, obtained in a molecular dynamics (MD) simulation. In this case, transition path sampling is an MC sampling of MD trajectories.

The efficiency of a transition path sampling simulation depends on the specific recipe used to generate a new trajectory from an old one. We have found that the "shooting and shifting" algorithm provides the highest efficiency of all techniques applied so far [20]. In addition of being highly efficient, this algorithm is the only one so far applicable for the sampling of deterministic trajectories [25], which have singular short time transition probabilities, such as they are generated in most classical and *ab initio* molecular dynamics simulations [1,26]. (Stochastic trajectories, which have continuous transition probabilities, can be sampled with numerous different algorithms similar to those developed to sample path integrals [27].) In the "shooting" part of the algorithms one generates a new deterministic trajectory from an old one by first randomly selecting a point along the old trajectory. Then one slightly modifies the momenta belonging to this point by adding a small perturbation δp to the momenta p. (Some care must be exercised when doing that in order to maintain symmetry between forward and backward step [11]). Starting from this perturbed state one "shoots off" new trajectory segments of appropriate length both forward an backward in time. Finally, the newly created trajectory is accepted of rejected depending on how its statistical weight compares to the weight of the old trajectory. Due to the chaoticity of the equations of motion, the new trajectory quickly diverges from the old one leading to rapid progress in the random walk through path space and, hence, to efficient sampling. In the "shifting" part of the algorithm one merely shifts the starting point of the trajectory in time by integration of the equations of motion forward or backward in time. Of course, such an operation does not create a completely new trajectory. Rather, a shifting move just selects a slightly translated part of the same trajectory. Nevertheless, shifting is useful when collecting path averages and should be added regularly to shooting moves.

Modifications of this "shooting and shifting" algorithm can also be applied to sample stochastic trajectories, as obtained from solutions of the Langevin equation, and also to sample Monte Carlo "trajectories". In principle, "shooting and shifting" can be applied whenever a dynamical rule to generate trajectories is available. In all cases, deterministic and stochastic, the efficiency of the "shooting and shifting" algorithm stems from the fact that the propagation rule of the underlying dynamics is used to generate new trajectories from old ones. As a consequence, much cancellation occurs in the expressions for the acceptance probability [11]. Trajectories generated in this way have a high acceptance probability even if they are very different (of course, this is true only provided they are consistent with the imposed selection criterion $F[x(t)]$). Using the algorithms described above one can efficiently explore pathspace and collected large numbers of pathways on which the rare, but important event of interest occurs. In the next section we discuss how these pathways can be used to obtain knowledge of mechanisms and kinetics.

11.3 What Transition Path Sampling Can Do

As discussed earlier, the transition path sampling methodology was developed to study rare, but important events occurring in complex systems for which prior mechanistic knowledge is unavailable. Transition path sampling overcomes both problems, the rare event problem *and* the lack of a proper reaction coordinate, by formulating a statistical mechanics of trajectories in which the weight functional for trajectories incorporates the required occurrence of the rare event of interest as an additional factor. Due to this additional factor, reactivity is maintained throughout the path sampling simulation and no computing time is wasted generating trajectories along which the rare event does not occur. In this sense, transition path sampling does to dynamics what umbrella sampling does to statics. In this section we discuss these capabilities of transition path sampling using a study of autoionization in liquid water as an example [1].

In particular, we consider a volume of liquid water and imagine following the motion of a particular water molecule. Typically, it will take about

10 hours before this molecule will dissociate and form a pair of solvated hydronium and hydroxide ions. The time scale for this process has been determined experimentally by generating excess concentrations of the two ionic species and monitoring their recombination as a functions of time [28,29]. The detailed mechanism of this process, however, is difficult to probe both experimentally and in simulations and has been unknown until recently [1].

11.3.1 The Rare Event Problem

From a computational point of view, one of the main problems stems from the fact that during the dissociation process, covalent bonds are broken and new ones are formed. This problem has been solved and today simulations of chemically reactive systems consisting of hundreds of atoms are routinely carried out using the Car-Parrinello molecular dynamics (CPMD) method based on density functional theory [30]. But since the longest time scales accessible to *ab initio* molecular dynamics simulations of this kind are of the order of dozens of picoseconds (a 1ps trajectory of 32 water molecules can be carried out in about one day on a small PC cluster), the rare event problem still persists. In fact, assuming the speed of computers keeps growing at the current pace doubling every 18 month, one has to wait until the year 2080 to be able to observe one dissociation event per CPU-day in a sample of 32 water molecules. Thus, autoionization in liquid water cannot be studied with straightforward computer simulation in the foreseeable future.

Naively, one might simply select the interionic distance as a reaction coordinate. While such a choice can clearly distinguish between the intact water molecule and the separated ion pair (provided the ions are far enough apart), it fails to capture the essential physics of the dissociation process. Therefore it can be used in calculations of equilibrium constants (again, this is true only for a sufficiently large system), but it cannot be employed to produce trajectories, in which the dissociation event actually takes place.[1]

11.3.2 Solving the Rare Event Problem with Transition Path Sampling

Since transition path sampling does not require definition of a reaction coordinate and is not adversely affected by the separations of time scales present in the system, it can be used to study autoionization in liquid water. When

[1] In their pioneering *ab initio* computation of the equilibrium constant for autoionization in liquid water Trout and Parrinello used the distance between particular hydrogen and oxygen nuclei as a reaction coordinate [31,32]. This reaction coordinate permits the calculation of the potential of mean force required to separate the ions in the initial stages of the dissociation. But because the separation of the ions proceeds through a Grotthuss mechanism, this choice of reaction coordinate is neither capable of a full characterization of the dissociation event nor able to distinguish between the intact water molecule and the charge dissociated state.

doing that, the first step consists in defining an order parameter capable of distinguishing the intact water molecule from the dissociated ion pair. As mentioned above, the interionic distance can serve as such an order parameter provided one selects a sufficiently large distance to characterize the ion pair. In a small system, such as the 32 water molecule sample studied in [1], however, the interionic distance can not discriminate between the two stable states. Through trial and error we have found that the length of the shortest hydrogen bond wire between the ions can accomplish that. Hydrogen bond wires are known to provided effective routes for proton transfer in aqueous systems and the absence of such a wire can stabilize a pair of hydroxide and hydronium ions even in a system as small as 32 water molecules. Using the presence (or absence) of a hydrogen bond connection as a criterion to define initial region A and final region B dozens of reactive trajectories have been generated with the shooting and shifting algorithm described above [1]. Since the dissociation occurs rapidly when it takes place, sampled trajectories can be as short as 300 fs.

Transition path sampling maintains the reactivity of these trajectories throughout the simulation. Therefore no CPU time is wasted computing the time evolution of the system between reactive events. It is this concentration on the reactive event which makes transition path sampling an efficient way to overcome the time scale gap discussed in the Introduction. The basic principle of a transition path sampling simulation is essentially the same as the one of a umbrella sampling Monte Carlo simulation. Just like in an umbrella sampling simulation a biasing potential confines the sampled configurations to a certain region in configuration space, transition path sampling confines the sampling to a restricted region in trajectory space (in the case of the autoionization problem this is the set of trajectories along which dissociation occurs).

The CPMD trajectories collected with transition path sampling indicate that the dissociation of a water molecule occurs in two basic steps. First a rare solvent electric field fluctuation destabilizes an OH bond of a particular water molecule. The proton initially belonging to this now broken bond then transfers to a neighboring water molecule along a hydrogen bond. As the electric field generated by the solvent fluctuation keeps acting on the ion pair, both proton and hydroxide can separate further along hydrogen bonds in a Grotthuss like fashion [33,34]. At this point the two ions are still connected by a wire of hydrogen bonds along which the ions can recombine when the electric field subsides. Indeed, most of the ions pairs formed in this way are of a transient nature and eventually recombine. In a few rare cases, however, the hydrogen bond wire connecting the nascent ions breaks preventing rapid recombination and trapping the ions in a charge separated metastable state. The lifetime of this metastable state is long enough to allow the ions to separate completely and diffuse to large distances. Thus, a successful dissociation requires two events: first, a rare solvent electric field fluctuation must destabilize and OH bond and drive the two ions apart along a hydrogen bond wire and, second, the hydrogen bond wire must break interrupting a route for rapid recombination of the ions.

11.3.3 Interpreting the Ensemble of Harvested Paths

While some information about the transition mechanism can be gleaned by mere inspection of the harvested trajectories, usually more stringent analysis techniques are necessary to obtain the sought information. We found that one useful way to do that consists in locating transition states along the harvested trajectories [10,11,35]. Here, a transition state is defined in a statistical sense as a configuration which is has the same likelihood to relax into either one of the longlived stable states A and B. For a given configuration q relaxation probabilities, or committors, p_A and p_B can be computed by initiating many short trajectories at q with random momenta and observing wether they relax into A or into B. The name "committor" indicates that p_A and p_B describe how committed a particular configuration q is to the two states A and B. If $p_A = p_B \approx 0.5$ the configuration q is committed to neither A or B and can be regarded as a transition state. Now, along a certain reactive trajectory connecting A and B there exists at least on transition state defined in this statistical sense. Locating transition states along all pathways harvested in a transition path sampling simulation yields the "transition state ensemble". Analysis of this ensemble is usually far simpler than analysis of sets of entire pathways and can produce valuable information about reaction mechanism.

We determined the transition state for autoionization for a single, but typical trajectory. We found that for this particular trajectory the transition state coincided with the breaking of a hydrogen bond in the hydrogen bond chain connecting the nascent ions. Trajectories initiated from configurations with intact hydrogen bond chain occurring before the transition state predominantly lead to recombination of the ions. Trajectories initiated from configurations with a broken hydrogen bond chain, on the other hand, typically relaxed into the metastable charge separated state. This behavior demonstrates that the breaking of the hydrogen bond wire connecting the charges is indeed a crucial event in the autoionization process stabilizing the ions in their charge separated state from which they can then diffuse to large distances and separate completely. In another study, analysis of the transition state ensemble for ionic dissociation in aqueous solution has revealed the importance of specific solvent degrees of freedom for the dissociation mechanism [36].

Another analysis method not directly related to transition path sampling, but is useful in conjunction with it, consists in calculating distributions of committors (p_A or p_B) for certain subsets of configurations [11,10,36].

11.3.4 Rate Constants

Since transition path sampling is based on harvesting physical dynamical trajectories it can be used to calculate kinetic rate constants [37]. All information on the kinetics of transitions between longlived stable states A and B is contained in the time correlation function

$$C(t) = \frac{\langle h_A(x_0) h_B(x_t) \rangle}{\langle h_A \rangle}, \tag{11.4}$$

where $\langle \cdots \rangle$ denotes a thermal average. $C(t)$ is the conditional probability to observe the system in state B at time t given it was in state A at time 0 and it provides a link between the microscopic rate equations and the microscopic dynamics of the system. If a separation of time scale exists, i.e. the reaction time $\tau_{\rm rxn}$ is much longer than the time $\tau_{\rm mol}$ characteristic for molecular motions, $C(t)$ grows linearly in the time regime $\tau_{\rm mol} \leq t \ll \tau_{\rm rxn}$ and the reaction rate constant k_{AB} for the reaction from A to be B can be inferred from the slope of $C(t)$ in this regime [9].

Within the transition path sampling formalism, the correlation functions $C(t)$ can be calculated by exploiting an isomorphism between time correlations functions and free energies [10,11]. Physically, this isomorphism can be understood in terms of volumes in configuration and trajectory space. Free energy differences in ordinary statistical mechanics essentially measure volumes in configuration space populated by two different systems. Similarly, the time correlation function $C(t)$ measures how many of all trajectories emanating from region A reach region B within a certain time t. These two sets of trajectories fill different volumes in path space and the time correlation function $C(t)$ corresponds to the ratio of these two volumes.

Using this isomorphism one can compute $C(t)$ by determining the reversible work required to confine the endpoints of pathways of length t originating in region A to region B. For this purpose one can employ mature techniques developed for the estimation of free energy differences, such as umbrella sampling [4] or parallel tempering [38,39] to mention just two popular and powerful methods. In contrast to other approaches, transition path sampling does not require any prior knowledge about the transition mechanism in order to calculate rate constants. Therefore, transition path sampling is the only viable route to study the kinetics of some complex system, for which such knowledge is unavailable.

11.4 What Transition Path Sampling Cannot Do (Yet)

11.4.1 One and Two Point Boundary Problems

In this section we discuss some important limitations of the transition path sampling methodology. One such limitation is that transition path sampling cannot by applied to find the final state B if it is not known in advance. In contrast to other methods discussed in this meeting (for instance, the contributions by Tuckerman, Jónsson and Röthlisberger), transition path sampling requires definition of both regions A and B. In other words, one must know where the system is coming from and where it is going to. Thus, transition path sampling solves the two point boundary problem, but it cannot be applied to accelerate the search for the final state B given that the system starts in A. Once the final state B has been found with other methods, bias-free dynamical pathways connecting A and B can be found and studied with transition path sampling.

11.4.2 Chains of States with Long Time Steps

In principle, transition path sampling could be used to sample long-time step pathways as discussed by Elber at this meeting. In this stochastic path approach trajectories with time steps of up to 500 ps are determined by minimization of an action functional [40,41]. The functional is obtained by considering the distribution of errors one makes when integrating Newton's equation of motion with a finite time step. During the minimization of the action the path end points are held at fixed positions and it is this constraint that filters out all pathways which run into an instability (straightforward MD simulations "explode" with large time step). Thus, shooting and shifting moves, which rely on the propagation rules of molecular dynamics, cannot be used to sample such ensembles of long time step trajectories

The long time step stochastic path method is a kind of transition path sampling, similar to that suggested by Pratt [42]. But the precise form of this algorithm applies an unphysical statistical weight for the pathways. Namely, the single time step transition probability does not conserve the equilibrium distribution. According to the long time step action functional, the dynamics is that of a system with random forces, but without friction. Hence, the fluctuation-dissipation theorem is violated and the system must heat up.

Now, one might hope of fixing the problem by adding a frictional term to the action. Unfortunately, this fix by itself does not help much, because integration algorithms for the Langevin equation conserve the canonical distribution only in the small time step limit. To conserve the equilibrium distribution with large time steps, the transition probability for those steps must obey detailed balance. One might think, therefore, that applying something like Metropolis MC trajectories with large displacements should do the trick. This thought may be part of the answer, but in practice, there is more to be considered. If using atomic coordinates in a dense system, like the proteins considered by Elber, large displacements will lead to overwhelming rejection of trial moves.

Any systematic derivation of a large time step algorithm must ultimately employ some type of coarse graining in time. High frequency components of the motion must be integrated out. In so doing, however, it is natural to imagine that space as well as time should be coarse grained. In that way, reasonable large time step transition probabilities have a chance of being less than negligible. It is not obvious, of course, how to proceed in picking useful coarse grained variables. These choices could be highly system dependent. But for the purposes of this discussion, let us suppose that coarse graining is done with a spatial grid that tiles space. The state of a grid cell will be characterized according to the average configuration (and perhaps motion) of atoms that fall within that cell over a coarse graining time Δt. A matrix of transition probabilities can then be either estimated on physical grounds, or perhaps calculated by transition path sampling of the atomic motions in a cell and neighboring cells over that time period Δt. With a matrix of transition probabilities assembled, long time steps method might then be applied with reasonable efficiency. Future research will show if these ideas have any merit.

11.4.3 Pattern Recognition

Finally, we mention the problem of pattern recognition that exists for pathways whether coarse grained or not. As discussed in Sect. 11.3, insight into the transition mechanism can be obtained by determining the transition state ensemble. But while transition path sampling provides the algorithms to generate such an ensemble, its analysis is still based on a trial and error procedure: one first guesses which parameters are relevant for the transition and than verifies if one's guess is correct by computing appropriate distribution in the transition state ensemble. A systematic or automated way to identify important variables would be highly valuable. Solution of this pattern recognition problem would also aid characterization of stable states, which is far from trivial in many complex systems.

Acknowledgments

We thank the conference chairmen and in particular Peter Nielaba for organizing an inspiring meeting and inviting us to present our work. Our research on this topic was supported in its nascent stages by the US National Science Foundation, and subsequently by the Department of Energy Basic Energy Sciences Program.

References

1. P. L. Geissler, C. Dellago, D. Chandler, J. Hutter, M. Parrinello: *Science* **291**, 2121 (2001)
2. C. Jaffé, S. D. Ross, M. W. Lo, J. Marsden, D. Farrelly, T. Uzer: *Phys. Rev. Lett.* **89**, 011101 (2002)
3. D. Frenkel, B. Smit: *Understanding Molecular Simulation*, 2nd. edn. (Academic Press, San Diego 2002)
4. G. M. Torrie, J. P. Valleau: *J. Comp. Phys.* **23**, 187 (1977)
5. A. M. Ferrenberg, R. H. Swendsen: *Phys. Rev. Lett.* **63**, 1195(1989)

6. A. Laio, M. Parrinello: communicated to *Proc. Natl. Acad. Sci. USA* (2002)
7. J. B. Anderson: *J. Chem. Phys.* **58**, 4684 (1973)
8. C. H. Bennett: 'Molecular Dynamics and Transition State Theory: the Simulation of Infrequent Events'. In *Algorithms for Chemical Computations.* ed. by R. E. Christoffersen (Amer. Chem. Soc., Washington DC 1977) pp. 63–97
9. D. Chandler: *J. Chem. Phys.* **68**, 2959 (1978)
10. P. G. Bolhuis, D. Chandler, C. Dellago, P. L. Geissler: *Annu. Rev. Phys. Chem.* **53**, 291 (2002)
11. C. Dellago, P. G. Bolhuis, P. L. Geissler: *Adv. Chem. Phys.* **123**, 1 (2002)
12. C. J. Cerjan, W. H. Miller: *J. Chem. Phys.* **75**, 2800 (1981)
13. J. P. K. Doye, D. J. Wales: *Z. Phys. D* **40**, 194 (1997)
14. G. Henkelman, G. Jóhannesson, H. Jónsson: 'Methods for Finding Saddle Points and Minimum Energy Paths'. In *Progress in Theoretical Chemistry and Physics.* ed. by S. D. Schwartz (Kluwer Academic Publishers, Dordrecht 2000)
15. A. F. Voter: *Phys. Rev. Lett.* **78**, 3908 (1997)
16. D. J. Bicout, A. Szabo: *Protein Science* **9**, 452 (2000)
17. U. Gasser, E. R. Weeks, A. Schofield, P. N. Nursey, D. A. Weitz: *Science* **292**, 258 (2000)
18. S. Auer, D. Frenkel: *Nature* **409**, 1020 (2001)
19. C. Dellago, P. G. Bolhuis, F. S. Csajka, D. Chandler: *J. Chem. Phys.* **108**, 1964 (1998)
20. C. Dellago, P. G. Bolhuis, D. Chandler: *J. Chem. Phys.* **108**, 9263 (1998)
21. P. G. Bolhuis, C. Dellago, D. Chandler: *Faraday Discuss.* **110**, 421 (1998)
22. P. G. Bolhius, C. Dellago, D. Chandler: *Proc. Natl. Acad. Sci. USA* **97**, 5877 (2000)
23. P. L. Geissler, D. Chandler: *J. Chem. Phys.* **113**, 9759 (2000)
24. J. P. Garrahan, D. Chandler: *Phys. Rev. Lett.* **89**, 035704 (2002)
25. P. G. Bolhuis, C. Dellago, D. Chandler: *Faraday Discussion* **110**, 421 (1998)
26. P. L. Geissler, C. Dellago, D. Chandler, J. Hutter, M. Parrinello: *Chem. Phys. Lett.* **321**, 225 (2000)
27. D. M. Ceperley: *Rev. Mod. Phys.* **67**, 279 (1995)
28. M. Eigen, L. De Maeyer: *Z. Elektrochem.* **59**, 986 (1959)
29. W. C. Natzle, C. B. Moore: *J. Phys. Chem.* **89**, 2605 (1985)
30. R. Car, M. Parrinello: *Phys. Rev. Lett.* **55**, 2471 (1985)
31. B. L. Trout, M. Parrinello: *J. Phys. Chem.* **103**, 7340 (1999)
32. B. L. Trout, M. Parrinello: *Chem. Phys. Lett.* **288**, 343 (1998)
33. N. Agmon: *Israel J. Chem.* **39**, 493 (1999)
34. M. Tuckerman, K. Laasonen, M. Sprik, M. Parrinello: *J. Chem. Phys.* **103**, 150 (1995)
35. M. M. Klosek, B. J. Matkowsky, Z. Schuss: *Ber. Bunsenges. Phys. Chem.* **95**, 331 (1991); V. Pande, A. Y. Grosberg, T. Tanaka, E. I. Shakhnovich: *J. Chem. Phys.* **108**, 334 (1998)
36. P. L. Geissler, C. Dellago, D. Chandler: *J. Phys. Chem. B* **103**, 3706 (1999)
37. C. Dellago, P. G. Bolhuis, D. Chandler: *J. Chem. Phys.* **110**, 6617 (1999)
38. C. J. Geyer, E. A. Thompson: *J. Am. Stat. Assoc.* **90**, 909–920 (1995)
39. T. J. H. Vlugt, B. Smit: *Phys. Chem. Comm.* **2**, 1 (2001)
40. R. Elber, J. Meller, R. Olender: *J. Phys. Chem. B* **103**, 899 (1999)
41. V. Zaloj, R. Elber: *Comp. Phys. Comm.* **128**, 118 (2000)
42. L. R. Pratt: *J. Chem. Phys.* **85**, 5045 (1986)

12 The Stochastic Difference Equation as a Tool to Compute Long Time Dynamics

Ron Elber, Avijit Ghosh, and Alfredo Cárdenas

Department of Computer Science, Cornell University, Ithaca, NY 14850

12.1 Introduction

Molecular Dynamics (MD) simulations provide an atomically detailed description of complex systems on a wide range of temporal and spatial scales. Despite numerous successes and many insightful observations, a clear limitation of the MD approach is its short time scales. Routine simulations of complex and large molecular systems at the atomically level of detail are restricted to nanoseconds. This time scale is far too short to address many interesting processes in biophysics, such as conformational transitions, transport phenomena and reactions.

The focus of this manuscript is on a recently developed methodology [1–3] that enables the calculation of approximate MD trajectories at extended times scales. The method was already applied to investigate numerous systems [2–4], and we computed trajectories at times of nanoseconds [2], microseconds [3], and "milliseconds" [4]. The millisecond trajectories are highly approximate, since a very significant fraction of the motions was filtered out (see Sect. 12.3.3). Nevertheless, they provide a view of the reaction pathway that is useful in interpretations of experimental data [4].

We use the term MD for a simulation technique that solves the classical equations of motion *at the atomic level of detail.* We assume that an atomically detailed potential is available (we use empirical potentials, but other approaches can be used as well). The dynamics on the energy surface is described by classical mechanics (Newton's law). The discussion is limited to dynamical models and differential equations that directly follow from microscopic parameters. For example, this is to be contrasted with the Langevin equation that requires a phenomenological friction constant. The friction constant, which significantly affects the dynamics, is not a microscopic parameter. As a result there is no clear limiting procedure that systematically leads from the results of the model (e.g. the Langevin equation) to the solution of the mechanical equations of motion.

12.2 Molecular Dynamics

We start with a brief review of existing well-established approaches, while emphasizing the current limitations.

12.2.1 Initial Value Formulation

The initial value formulation of classical dynamics (The Newton's equations of motion) is the most widely used numerical procedure:

$$M\frac{d^2X}{dt^2} = -\nabla U \tag{12.1}$$

Throughout the text a lower case (e.g., x, y, or z) denote scalar variables, while uppercase variables (e.g. X, Y, Z) denote vectors. In the above formula M is the mass matrix, X the coordinate vector, t the time and U the potential energy. A successful (and simple) algorithm to solve (12.1) is the Verlet algorithm [5]. In the "velocity" form, it reads:

$$\begin{aligned} X_{i+1} &= X_i + V_i \Delta t - \left(\Delta t^2/2\right) M^{-1} dU/dX_i \\ V_{i+1} &= V_i - \left(\Delta t/2\right) M^{-1} \left[dU/dX_i + dU/dX_{i+1}\right] \end{aligned} \tag{12.2}$$

A trajectory is obtained after specifying two initial conditions, the coordinates, $X\,(t=0) \equiv X_0$, and the velocity vector, $V\,(t=0) \equiv V_0$. The size of the time step is restricted since steps larger than a few femtoseconds result in numerical instabilities. To obtain long time dynamics many small time steps of size Δt are required (e.g. to reach a few nanoseconds millions of steps are required). The necessity of using small time steps is the major obstacle in computations of long time dynamics with the initial value formulation.

12.2.2 A Boundary Value Formulation in Time

Another well established formulation of classical mechanics (which is, of course, equivalent) is based on a boundary value problem. We seek a stationary solution of a functional of the path, S [6]

$$S\left[X\left(t'\right)\right] = \int\limits_0^t L \cdot dt'$$

$$L = \frac{1}{2}\left(\frac{dX}{dt'}\right)^T M \left(\frac{dX}{dt'}\right) - U\left(X\right) \tag{12.3}$$

In (12.3) the two end points, $X\,(0)$ and $X\,(t)$, are held fixed and so is the total time t. It is possible (in principle) to solve (12.3) numerically by discretizing the integral and computing the stationary discrete path (S now is a function of the set $\{X_i\}_{i=1}^N$)

$$S\left[\{X_i\}_{i=1}^N\right] = \Delta t \cdot \sum_{i=0,..,N} \frac{1}{2\Delta t^2}\left(X_{i+1} - X_i\right)^T M \left(X_{i+1} - X_i\right) - U\left(X_i\right) \tag{12.4}$$

Solving (12.4), (i.e. finding a trajectory that makes S stationary), is an alternative to solving the initial value problem (12.2). Equation (12.4) has a certain philosophical appeal since it provides a global solution of the whole path instead of (12.2) that provides a sequence of local solutions in time. There is a hope that having a global view of the classical trajectory will yield more robust and stable solutions as a function of step size.

A support for the "philosophical appeal" is the application of an integral in (12.4) while in (12.2) we use derivatives. Numerical estimates of integrals are in general more accurate and more stable compared to estimates of derivatives. On the other hand, computations of the whole path are more expensive than the calculation of one temporal slice of the trajectory at a time. The computational efforts are larger in the boundary value formulation by at least a factor of N (where N is the number of time slices) compared to the initial value approach. To make the global approach computationally attractive (assuming that it does work) the gain in step size must be substantial.

The hope is then that (12.4) may be a useful alternative to initial value solvers if *approximate* long time trajectories are desirable. However, the use of (12.4) "as is" is problematic. One problem is that even in (12.4) we need an estimate of (first order) derivatives of the coordinates with respect to time. Here, the estimate is based on a finite difference. The finite difference estimate is of poor quality as the time step increases, and it leads to numerical instabilities even in the integral formulation.

Another problem in computing a classical trajectory with (12.4) is that the stationary condition on S does not imply a minimum or a maximum; finding a saddle point can be significantly harder. Finally, if we are after large time steps, the functional above can change from being a minimum to being a maximum as a function of the step size. This is not a desired property for a function to be minimized!

To exemplify the above problems with a simple example, it is instructive to use one-dimensional harmonic oscillator:

$$S\left[\{x_j\}_{j=1}^{N}\right] = \Delta t \cdot \sum_{j=0,..,N} \left(\frac{1}{2}M\frac{(x_{j+1}-x_j)^2}{\Delta t^2} - \frac{K}{2}x_j^2\right) \tag{12.5}$$

The stationary condition on the function S is

$$\frac{1}{\Delta t}\partial S/\partial x_k = \frac{M}{\Delta t^2}\left[(x_k - x_{k+1}) + (x_k - x_{k-1})\right] - Kx_k = 0$$

$$\left(\frac{2M}{\Delta t^2} - K\right)x_k - \frac{M}{\Delta t^2}\left(x_{k+1} + x_{k-1}\right) = 0 \tag{12.6}$$

Guessing a solution of the type $x_k = x_0 \exp\left[-ik\omega\right]$ $(i = \sqrt{-1})$, we have

$$\left(\frac{2M}{\Delta t^2} - K\right) - \frac{2M}{\Delta t^2}\cos\left(\omega\right) = 0$$

$$\left(1 - \frac{K \cdot \Delta t^2}{2M}\right) = \cos\left(\omega\right) \tag{12.7}$$

It is obvious that we obtain a stability condition that is not much different from the stability condition of the initial value equation. If Δt is larger than $2 \cdot \sqrt{M/K}$, the solution grows exponentially and is numerically unstable. Hence, in the straightforward boundary value formulation of classical mechanics, we gain very little in terms of stability and step size compared to the solution of the initial value differential equation. The difficulty is not in the "philosophical" view (global or local) but in the estimate of the derivative, which is approximated by a local finite difference expression.

It is also possible to demonstrate a shift from a maximum to a minimum for S. The diagonal elements of the second derivative matrix of the action, $\partial^2 S/\partial x_k^2 = \left(\frac{2M}{\Delta t^2} - K\right)$, are changing their sign. They are positive at small Δt and become negative at sufficiently large Δt.

Of course, it is not reasonable to expect that oscillations with a frequency ν will be reproduced accurately with a time step Δt larger than π/ν. However, there is a difference between an accurate representation and being blown out of the roof. An attractive alternative (if possible) is the removal of modes that change significantly on a time scale shorter than Δt.

Consider for example a two dimensional harmonic oscillator. Instead of (12.5) we now have:

$$S\left[\{x_j, y_j\}_{j=1}^{N}\right]$$

$$= \Delta t \cdot \sum_{j=1,..,N} \left(\frac{1}{2}M\frac{(x_{j+1} - x_j)^2}{\Delta t^2} + \frac{1}{2}M\frac{(y_{j+1} - y_j)^2}{\Delta t^2} - \frac{K_x}{2}x_j^2 - \frac{K_y}{2}y_j^2\right)$$

$$K_x \gg K_y \tag{12.8}$$

The frequency along the x direction, $\sqrt{K_x/M}$, is set to be much larger than the frequency of the oscillation along the y coordinate, $\sqrt{K_y/M}$. However, we cannot increase the time step, Δt. The use of a time step appropriate for the y direction will cause numerical instability in the x direction, and in the overall solution. In the above trivial case simple filtering of x ($x_j = 0$ for all j) enables exact solution of y. Of course, in (12.8) there is no coupling between the two coordinates. If coupling is important, freezing high frequency modes can provide (at best) only an approximate solution.

It is possible to use multiple sizes of time steps to integrate separately along x and y axes [7–9]. However, this procedure requires the prior identification of the fast and the slow modes, i.e. we need to know which modes to integrate with small time steps and for which modes it is possible to employ the more economic larger time steps.

The identification of fast coordinates can be difficult in simulations of condensed phases. Some of the fast modes are bond or angle vibrations that can be identified and integrated separately. However, other fast modes are transient. They are fast for a short duration of time and slow otherwise. The transient fast modes are collisions, e.g. two atoms that are close and feel strong repulsive forces due to excluded volume interactions. The relevant degree of freedom (the distance between the atoms) is a fast mode during the collision event and a slow mode before or after the short collision period. The fundamental complication in the treatment of these modes is the "identity crisis" of these fast/slow coordinates.

Earlier studies suggest a special treatment to a collision coordinate as a fast coordinate that is turned on and off [10]. Transformations of collision coordinates to forms more appropriate for strongly or weakly interacting coordinates were used in the framework of a mean field approximation. However, tracking down collision events and treating them in a special way is computationally expensive. It cannot be done in practice for more than a few collision events at the same time. Since the number of collisions at a given time slice is proportional to the number of atoms in the system, it is difficult to come with a general scalable tracking scheme that will be independent of the system size. Moreover, as the size and the density of the system increase, three and four body collisions (that are not considered in the above scheme [10]) may be relevant as well.

Our goals are therefore two:

(a) We seek a "stable" treatment of the fast modes (on the scale of Δt) and approximate description of slow modes. We hope that the simplification done for the rapid displacements will still produce sound description of the slow motions.

(b) We seek a formulation in which we will not need to identify (to begin with) what are the slow and the fast degrees of freedom. Hence we seek an automated "stabilizing" algorithm.

To achieve (a) and (b) a new model is required.

We note that we cannot expect to do better than the goals outlined above when a large time step is used. Without detailed small-time-step integration, it is indeed impossible to follow fast motions. So "stabilizing" the modes with frequencies higher than $\pi/\Delta t$ is the best we could hope for.

Before the formulation of an alternative model, we consider yet another boundary value formulation that will be of considerable interest to us later on.

12.2.3 A Boundary Value Formulation in Length

A classical trajectory is a curve in Q dimensional space (where Q is the number of degrees of freedom). The discussion so far considers the parameterization of the curve as a function of time, $X(t)$. However, this parameterization is clearly not unique, and we may choose to distribute the points along the curve in other ways, and more specifically in a computationally more convenient way. Another well established protocol in classical mechanics, which we consider here, parameterizes the curve, $X(l)$, as a function of the path length, l [6].

It is convenient at this point to change to mass weighted coordinates $Z = \sqrt{M}X$. The Lagrangian in (12.3) is now modified to

$$L = \frac{1}{2}\left(\frac{dZ}{dt}\right)^2 - U(Z)$$

The usual formulation of the action as a function of length has fixed end points, fixed total energy and variable length and time.

$$S_L\left[Z\left(l'\right)\right] = \int_{Z(0)}^{Z(l)} \sqrt{2(E-U)} \cdot dl' \tag{12.9}$$

The total energy in (12.9) is denoted by E. One of the advantages of (12.9) with respect to (12.3) is that the total time of the trajectory is an output (versus an input). The energy can be estimated from equilibrium considerations. For example, the total kinetic energy $K \equiv E - U$ may be set to $Qk_BT/2$, where Q is the number of degrees of freedom. The total time can be recovered from the stationary path with length parameterization as

$$t = \int_{Z(0)}^{Z(l)} \frac{dl'}{\sqrt{2(E-U)}} \ .$$

A discrete version of (12.9) can be optimized in a similar way to (12.4).

$$S_L\left[\{Z_j\}_{j=1}^N\right] = \sum_{j=1,..,N-1}\left(\frac{1}{\sqrt{2}}\left[\sqrt{E-U\left(Z_j\right)}+\sqrt{E-U(Z_{j+1})}\right]|Z_j - Z_{j+1}|\right) \tag{12.10}$$

The first and the Nth coordinate sets are fixed. Another advantage (besides switching from constant time to constant energy) is the elimination of time derivatives. A finite difference estimate remains (the length element, $\Delta l' \cong |Z_j - Z_{j+1}|$). However, since the configurations are equally distributed along the path, the length element behaves better than the finite difference estimate of the velocity. For example, the distance between two points provides a lower bound to the true length of the path. No such bound is available when estimating velocity.

The action, S_L, in (12.10) is not necessarily a minimum making the computation of the stationary path a non-trivial challenge. We have tried to use (12.10) in a straightforward way but finally gave up [11]. A variant on (12.10) within the framework of the stochastic difference equation (Sect. 12.3.6) is at present our most promising approach.

To conclude the discussion on the length dependent action we note that an initial value differential equation as a function of length also exists [6] and is

$$\frac{d^2 Z}{dl^2} = -\frac{1}{2(E-U)}\left(\nabla U - (\nabla U \cdot \hat{t})\hat{t}\right) \tag{12.11}$$

where $\hat{t}$ is a unit vector in the direction of the trajectory. This second order differential equation resembles the Newton's equation of motion. It includes a force component only in the direction perpendicular to the path and an effective "mass" of $2(E-U)$.

12.3 The Stochastic Difference Equation

12.3.1 Stochastic Difference in Time: Definition

The expression, which we derive below, leads to an action and to a stationary (minimum) condition on the classical path. The optimal path is a discrete approximation to a classical trajectory. Interestingly, in the integral limit (an infinitesimal time step), the action below was used already by Gauss (!) to compute classical trajectories [12]. At variance with Gauss we keep a finite Δt.

Despite the similarity to the Gauss approach to classical mechanics, there is a key difference between the classical actions described above, and the corresponding action of the Stochastic Difference Equation. The classical actions are deterministic mechanical models; the SDE is a non-deterministic approach that is based on stochastic modeling of the numerical errors.

Computer simulations are (obviously) using a finite time step. Consider a finite difference approximation to the Newton's equations of motion (12.1)

$$M\frac{X_{i+1} + X_{i-1} - 2X_i}{\Delta t^2} + \frac{dU}{dX_i} = \varepsilon_{i+1} \tag{12.12}$$

The new feature in (12.12) is the use of an error vector, ε_{i+1}. Even if the trajectory we have at hand $\{X_i\}_{i=1}^{N}$ is exact, the left hand side of (12.12) will not be zero. This is since a finite difference is used to approximate the second derivative of the coordinate vector as a function of time. In straightforward Molecular Dynamics simulations the time step, Δt, is taken to be small with the hope that the errors (which we denoted by ε_{i+1}) can be neglected. Usually we do not know what the errors are, since we do not have the exact trajectory. Therefore ignoring the errors for small Δt s seems like a reasonable idea. In the SDE approach we attempt to model these errors. This allows us (in principle) the use of larger time steps.

It is important to emphasize that the errors so calculated are with respect to the **exact trajectory**. The computational procedure for the errors is as follows: We have an **exact** solution of the Newton's equation of motion, $\{X_i\}_{i=1}^{N}$, (an analytic or numerical solution). The exact solution is plugged into (12.12) to produce estimates of the errors due to the use of an approximate formula (finite difference estimate of second derivatives). The errors so obtained are therefore not the deviation of an approximate solution from the exact solution. Here the errors estimate the accuracy of a finite difference formula, tested on the exact trajectory. Of course if we solve (12.12) we will have a trajectory that is different from the exact solution, and we may use (12.12) to measure this difference.

The distribution is a property of the exact trajectories. If we generate an approximate trajectory based on the finite difference formula, we should generate error distribution that is consistent with what we know about the true solution.

Note also that the error is evaluated at the upper edge of the interval. The use of the edge for the error calculation is for computational convenience (as discussed below) and should not affect the results. There is an ambiguity about the placement of the error vector that adds interesting complications to the derivation below, and led to some unnecessary anxiety by other researchers in the field. Since we are developing a new statistical model we have the liberty of defining our model at our convenience. Why this choice is indeed convenient will become clearer in Sect. 12.3.2.

We have performed numerous numerical experiments on the properties of the errors. Some of these studies are described in reference [2]. We have performed a few more for the present article. A few experiments are presented below

Based on the numerical experiments we suggest the following two basic assumptions to be used in the error modeling:

(i) The errors are considered stochastic variables, not correlated in time. Or, more explicitly:

$$\langle \varepsilon_i \rangle = 0;\ \langle \varepsilon_i \varepsilon_j \rangle = C\delta_{ij} \tag{12.13}$$

(ii) The probability density of the norms of the error vectors is assumed Gaussian.

$$P(\varepsilon_i) = \sqrt{1/2\pi\sigma^2}\exp[-\varepsilon_i^2/2\sigma^2] \tag{12.14}$$

Assumption (ii) is for convenience only. In fact the formalism can be used for any functional form of the probability density of errors provided that the first assumption is satisfied.

We have no "proof" that the above two assumptions are true. All we have are numerical experiments on systems that vary from a dipeptide to a small solvated protein. Our results suggest that the above assumptions are sound

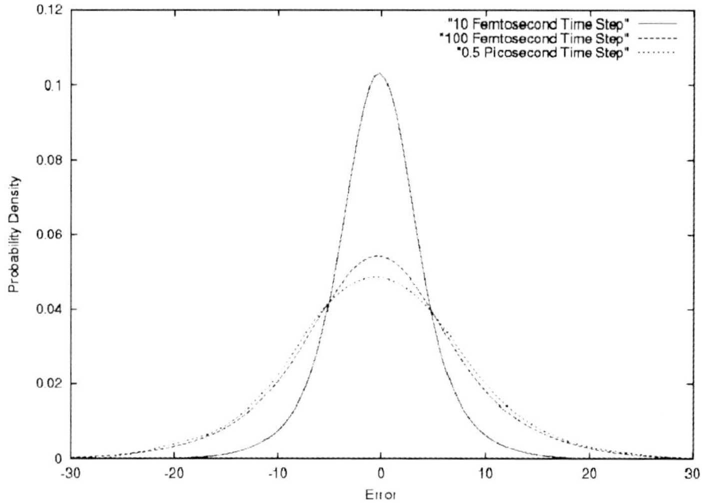

Fig. 12.1. Distribution of norms of the error vectors computed by the finite difference formula (12.12) from exact trajectories of valine dipeptide. The dipeptide was initially equilibrated at 300K. The largest errors are significant and are of the same order of magnitude as the forces.

for sufficiently large time step. We note that at the limit of small Δt s, in which we obtain "an almost" exact result, the (much smaller) errors become correlated. For the statistical assumption to be valid the time step needs to be sufficiently large so that correlations will decay rapidly. A few numerical experiments for different step sizes are presented in Fig. 12.2.

Are the results of the numerical experiments surprising? Let us examine first the second assumption and assume for the moment that the correlation is lost rapidly, is the normal distribution a surprise? It is not. It is a simple demonstration of the Central Limit Theorem (CLT). For sufficiently large systems, and after ensemble average, the addition and averaging of the (uncorrelated) elements of the error vector lead to a normal distribution. Note also that the first and second moments of the errors are bound if the coordinates of the exact trajectory are bound (as they are in practical condensed phase simulations). Therefore the conditions for the application of the CLT are satisfied.

Is it a surprise that the correlations diminished rapidly? If Δt is very small then the two terms, the estimate for the second derivative of the coordinate with respect to time and the force are comparable (along the exact trajectory). As Δt is made larger the force contribution (and the errors) are made larger as well. At very large Δt (maximum errors) the error correlations become the correlation of the forces **computed at time separation** Δt (without short time integration). The forces are changing significantly on the time scale of Δt and their discrete average $C_{dis}(\tau) =$

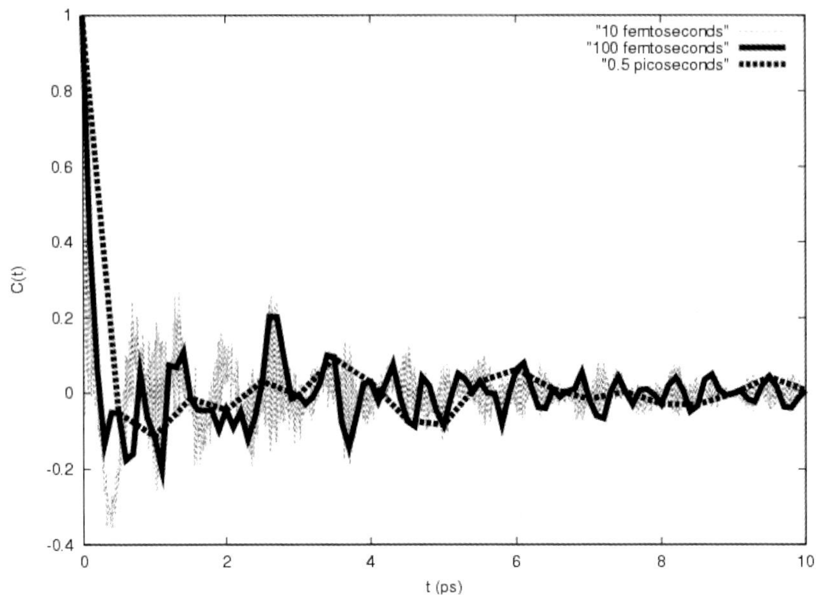

Fig. 12.2. The correlation of errors $\langle(\varepsilon(t)\cdot\varepsilon(0))/(\varepsilon(0)\cdot\varepsilon(0))\rangle$ estimated from exact trajectories of valine dipeptide and (12.12) for three different time steps.

$(1/N)\sum_i F(\tau+i\Delta t)\cdot F(i\Delta t)$ decays significantly faster than the corresponding integral $C_{int}(\tau)=(1/t')\int_0^{t'} F(t)\cdot F(t+\tau)dt$. The integral is known from statistical mechanics to decay quite rapidly. It is sometimes used to estimate the friction kernel and the memory function of the generalized Langevin equation. In the Langevin formulation the extreme view is taken that the correlation function is $C_{int}(\tau)\approx C_0\delta(\tau)$. This is somewhat similar to our (weaker) assumption for $C_{dis}(\tau)$. It is important to emphasize, however, that our model is different from the Langevin equation, which is a stochastic **differential** equation. Our model has no noise at the limit of small time steps in which the numerical errors approach zero. The "noise" we introduce is numerical. Our noise is not attempting to model extra degrees of freedom (we maintain all the degrees of freedom), or to produce different ensembles. It is a noise introduced when we approximate a differential equation by a finite difference formula.

An obvious limitation of the above argument is that it applies only for "sufficiently" large systems and a "sufficiently" large time step. How large is sufficient requires numerical experiments. In our experience, the dynamics of valine dipeptide with 42 degrees of freedom and a time step of 10 femtoseconds already shows the desired properties (Figs. 12.1 and 12.2). Considerably "nicer" Gaussian curves were obtained for yet larger systems (folding of C peptide [2]). It is therefore pointless to test the above assumption on model systems that are not ergodic even if we love them very much (like the one dimensional harmonic oscillator). The conditions for applying the CLT are clearly not satisfied in these cases.

A final comment on the model: in the applications that were pursued so far [1–4] we assume that the variance, σ^2, is a time independent scalar. There is no theoretical or computational restriction to make it so, and a potential extension to the model may make the variance a time dependent tensor. The current choice is based on insufficient data to fit, rather than on true conviction of simplicity.

12.3.2 A Stochastic Model for a Trajectory

If the above model of errors is accepted we can proceed to examine the statistical properties of trajectories. For example, what is the probability of obtaining a sequence of errors, $\{\varepsilon_i\}_{i=1}^{N}$ in N time slices of an exact trajectory? Since the errors are assumed independent of each other we have

$$\bar{P}\left(\{\varepsilon_i\}_{i=1}^{N}\right)\prod_j d\varepsilon_j = \prod_i P(\varepsilon_i)d\varepsilon_i \qquad (12.15)$$

If the errors are zero, we obtain the most probable trajectory within the framework of the stochastic difference equation. This trajectory is **not** exact and is within a distance σ from the exact trajectory (12.12). What are the approximations made? In Sect. 12.3.3 we argue that the approximate trajectory is a solution of the slow modes in the system where the high frequency modes are "stabilized", or filtered out.

Focusing on (12.15), it is more useful to write the probability in terms of coordinates (instead of errors).

$$\bar{P}\left(\{\varepsilon_i\}_{i=1}^{N}\right)\prod_j d\varepsilon_j = \prod_i P\left(\varepsilon_i \equiv M\frac{X_i + X_{i-2} - 2X_{i-1}}{\Delta t^2} + \frac{dU}{dX_{i-1}}\right) \times$$

$$\times \left[\prod_k dX_k\right] \det\left[J_{ij} \equiv \frac{d\varepsilon_i}{dX_j}\right] \qquad (12.16)$$

The determinant at the right side is the Jacobian of transformation from the error vector to the coordinate vector. More explicitly, the Jacobian of the transformation is:

$$\det\left[\frac{d\varepsilon_i}{dX_j}\right] = \begin{bmatrix} \partial\varepsilon_1/\partial X_1 & \partial\varepsilon_1/\partial X_2 & \partial\varepsilon_1/\partial X_3 & \ldots \\ \partial\varepsilon_2/\partial X_1 & \partial\varepsilon_2/\partial X_2 & \partial\varepsilon_2/\partial X_3 & \ldots \\ \partial\varepsilon_3/\partial X_1 & \partial\varepsilon_3/\partial X_2 & \partial\varepsilon_3/\partial X_3 & \ldots \\ \ldots & \ldots & \ldots & \ldots \end{bmatrix}$$

$$= \begin{bmatrix} M/\Delta t^2 & 0 & 0 & \dots \\ \left[\frac{-2M}{\Delta t^2} + \frac{d^2U}{dX_1^2}\right] & M/\Delta t^2 & 0 & \dots \\ M/\Delta t^2 & \left[\frac{-2M}{\Delta t^2} + \frac{d^2U}{dX_2^2}\right] & M/\Delta t^2 & \dots \\ \dots & \dots & \dots & \dots \end{bmatrix} = \left(\frac{M}{\Delta t^2}\right)^N \tag{12.17}$$

Expressing the determinant in (12.17) we find that only the diagonal terms remain, since the upper off-diagonal part of the determinant is zero. The final result is coordinate independent. This is the most convenient choice and the motivation behind the error placement at the edge of the interval. Placing the error at the center will introduce the second derivatives of the potential to the Jacobian that will make it coordinate dependent and more expensive to compute.

The ambiguity in the choice of the Jacobian is well known from path integral studies [13]. There the choice is made based on the physics that we wish to present. There is no "correct" or "wrong" choice before a concrete physical model is introduced. Our freedom in defining the errors we just created, allow us to make the most convenient choice.

Note that the boundary conditions implicitly written into the determinant requires the knowledge of (fixed) X_{-1} and X_{N+1}. Our model as outlined below is leading, at the limit of small Δt, to a fourth order differential equation in time that requires four initial or boundary values. In our experience fixing only pair of coordinates (optimizing also the velocities at the boundaries) affects very little the overall results when a large time step is used.

The probability density of a trajectory can then be written:

$$\begin{aligned} P\left(\{X\}_{i=1}^N\right) &= A\prod_i \exp\left[\frac{\varepsilon_i^2}{2\sigma^2}\right] \\ &= A\prod_i \exp\left[-\frac{1}{2\sigma^2}\left(M\frac{X_{i+1}+X_{i-1}-2X_i}{\Delta t^2}+\frac{dU}{dX_i}\right)^2\right] \\ &= A\exp\left[-\frac{1}{2\sigma^2}\sum_i\left(M\frac{X_{i+1}+X_{i-1}-2X_i}{\Delta t^2}+\frac{dU}{dX_i}\right)^2\right] \\ &= A\exp\left[-\frac{1}{2[\sigma^2\cdot\Delta t]}\sum_i \Delta t\left(M\frac{X_{i+1}+X_{i-1}-2X_i}{\Delta t^2}+\frac{dU}{dX_i}\right)^2\right] \end{aligned} \tag{12.18}$$

where A is the normalization factor that is coordinate independent. The last trick of dividing and multiplying by Δt is not necessary from a computational viewpoint but makes the sum approach a limit at small Δt. We define now an "action" S_{SDET} for the Stochastic Difference Equation in Time. A trajectory that minimizes S_{SDET} is an approximation to the true classical trajectory.

$$S_{SDET}\left(\{X_i\}_{i=1}^N\right) \equiv \sum_i \Delta t \left(M \frac{X_{i+1}+X_{i-1}-2X_i}{\Delta t^2} + \frac{dU}{dX_i} \right)^2$$

$$S_{SDET} \underset{\substack{\Delta t \to 0 \\ N \to \infty}}{\rightarrow} \int_0^{(N-1)\Delta t} dt' \left(M \frac{d^2X}{dt'^2} + \frac{dU}{dX} \right)^2 \equiv S_G \qquad (12.19)$$

The last limiting expression is the Gauss action, S_G, for classical mechanics. It clearly has a minimum that satisfies the equations of motion (when the action is zero). The action is non-negative, which makes it easier to identify the true minimum. The non-negativity is an important difference from the classical action formulation that we introduced at the beginning and makes the calculations with the S_{SDET} and S_G significantly more stable. As we show in the next section (Sect. 12.3.3) the approximate trajectories that are produced by optimization of S_{SDET} are stable. They have been "stabilized" by filtering out frequency components higher than $\pi/\Delta t$.

A complication we should keep in mind when comparing S_G to the usual classical action is that the Newtonian trajectory is not the only stationary solution of the Gauss action. A standard variation of (12.19) leads to a fourth order differential equation and hence to two more solutions (that are related by a time reversal operation) in addition to the true classical trajectory. An example was discussed in details in reference [1] (see discussions and equations (15) and (16) of reference [1]). The good news is that the true trajectory is still the global minimum (when the action is zero), which is a clear computational guideline. However, the possibility of being trapped at a wrong minimum solution exists.

12.3.3 "Stabilizing" Long Time Trajectories, or Filtering High Frequency Modes

In this section we attempt to address the question: what does an optimal S_{SDET} trajectory, computed with a large time step, mean? We consider the simplest trajectory with just one error term, $\{\varepsilon(X_1, X_2, X_3)\}$ (X_1 and X_3 are fixed), which we attempt to optimize. The action S_{SDET} takes the simple form $S_{SDET} = \Delta t \cdot \varepsilon^2$. To appreciate the consequences of this approximation in which only a single intermediate structure describes the trajectory we consider an exact alternative. The action of the stochastic difference equation, S_{SDET}, becomes exact at the limit of the Gauss action, S_G. But then we need to perform an integral over a continuous trajectory, $X(\tau)$, for which we require a representation.

$$S_G = \int_0^{2\cdot\Delta t} \varepsilon^2[X(\tau)]d\tau \qquad (12.20)$$

We write the exact trajectory, $X(\tau)$, as a sum of two terms: (i) a parabolic approximation and (ii) a Fourier series. The argument below can possibly made simpler (without the use of the Fourier series). However, we wanted to demonstrate the filtering effect. The parameters for the parabolic interpolation are extracted from the three coordinate sets: $\{X_1, X_2, X_3\}$, such that the parabola will pass through the three points. The Fourier series takes care of the deviation of the parabolic trajectory from the exact path. We write

$$X(\tau) = a\tau^2 + b\tau + c + \sum_n \alpha_n \sin(\omega_n \tau)$$

$$\begin{aligned} a &= (X_1 + X_3 - 2X_2)/(2 \cdot \Delta t^2) \\ b &= (4X_2 - 3X_1 - X_3)/(2 \cdot \Delta t) \\ c &= X_1 \\ \omega_n &= \frac{n\pi}{2 \cdot \Delta t} \qquad n = 1, 2, 3, ... \end{aligned} \tag{12.21}$$

The condition on ω is a result of the fixed boundaries at X_1 and X_3, forcing the Fourier series to vanish at the edges of the interval $\tau \in [0, 2 \cdot \Delta t]$. The α_n parameters are determined by minimization of S_G similarly to the determination of the coordinate vector $\{X_2\}$. With the exact representation of the trajectory, the next step is to perform the integral that defines the Gauss action in (12.20). It would be great if the integral could have been performed analytically. The remaining task would have been the optimization of the exact action as a function of all the parameters. So far we only used approximate discrete functionals. Unfortunately, the present case is not different from previous attempts. The force, which in general is a non-linear function of the coordinates, makes it impossible to do the integration exactly.

Instead of trying the impossible (searching for an exact expression of S_G) we consider an intermediate case. We use the exact representation of the trajectory, but employ the same approximation of the force that we use in the S_{SDET} algorithm (12.19). We evaluate the force only once at the middle of the time interval (i.e. at $\tau = \Delta t$) and set it to depend only on X_2. The force will depend only on X_2 if the Fourier series will vanish at that time, which implies that $\omega_n = \frac{n\pi}{\Delta t}$ $n = 1, 2, 3, ...$ in contrast to the general formulation in (12.21).

We then examine what are the implications of this approximation on the high frequency components of the trajectory ($\omega \geq \pi/\Delta t$). Substituting

(12.21) into (12.20), we have

$$S_G = \int_0^{2\cdot\Delta t} d\tau \left(c - \sum_n \alpha_n \omega_n^2 \sin(\omega_n \tau) + \frac{dU(X(\tau))}{dX} \right)^2$$

$$S_G \approx \bar{S}_G = \int_0^{2\cdot\Delta t} d\tau \left(c - \sum_n \alpha_n \omega_n^2 \sin(\omega_n \tau) - F(X_2) \right)^2 \tag{12.22}$$

The approximation we made is to use $F(X_2)$, force that depends only on the time independent parameter X_2 instead of $-dU(X(\tau))/dX$. This is exactly the same approximation that we use in S_{SDET}. We have

$$\bar{S}_G = (c - F(X_2))^2 \cdot 2 \cdot \Delta t - 2 \cdot (c - F(X_2)) \sum_n \alpha_n \omega_n^2 \int_0^{2\cdot\Delta t} d\tau [\sin(\omega_n \tau)]$$

$$+ \sum_{n,n'} \alpha_n \alpha_{n'} \omega_n^2 \omega_{n'}^2 \int_0^{2\cdot\Delta t} d\tau [\sin(\omega_n \tau) \sin(\omega_{n'} \tau)] \tag{12.23}$$

Since the integration is over complete periods of the trigonometric functions we have

$$\bar{S}_G = \left[2 \cdot (c - F(X_2))^2 + \sum_n \alpha_n^2 \omega_n^4 \right] \cdot \Delta t \tag{12.24}$$

The optimal coefficients of the Fourier series are obtained by minimizing the $\bar{S}_G$ as a function of α_n, which is trivial:

$$\partial \bar{S}_G / \partial \alpha_n = 2 \cdot \Delta t \cdot \omega_n^4 \cdot \alpha_n = 0 \quad \rightarrow \quad \alpha_n = 0 \tag{12.25}$$

The conclusion from this small exercise is that using our approximate discrete functional results in the disappearance of all the high frequency modes in the system (faster than $\pi/\Delta t$). This is a feature that guarantees the **stability** of the solution in contrast to the previous formulation of functionals in classical mechanics. In previous formulations the rapid motions lead to the exponential growth in the coordinates if the time step was too large. The filtering effect was demonstrated in reference [1] for the simple two-dimensional harmonic oscillator. Numerous numerical examples of more complex systems could be found in other publications. In the present article we demonstrate the filtering for the length parameterization discussed later (Fig. 12.7).

The filtering may seem trivial since it is a direct result of the implementation of the algorithm. It is therefore useful to contrast it with the behavior of initial value solvers. There the high frequency modes (that are not removed) contribute to exponential "blow-up" of the solution.

The effect of filtering high frequency modes on the slow modes is not obvious and requires experimentation. While the studies in references [1], and the computational examples in Sect. 12.4 look promising we are not done yet. More experimentation and experience will clarify the limits of the present approach, and when it is expected to fail. Below we describe one known failure.

While stabilizing the solution by filtering out high frequency modes may sound like a good idea, there is an important class of problems for which the approximate trajectory misses crucial information. Consider a model with Q degrees of freedom that moves in a potential of the form (a system with two wells)

$$U(X) = ax_1^2 - A\exp\left[-x_1^2\right] + \sum_{i=2,\ldots,Q} kx_i^2 + \gamma x_1 x_i \qquad \gamma \ll a, k \quad (12.26)$$

At total energies significantly larger than A, transitions between the wells should be observed. If in addition the average energy (per degree of freedom) is smaller than A, the dynamics is activated. The system remains for a substantial length of time at one of the wells, where it executes (rapid) oscillations with a time scale δt_1. Rarely, (a time scale of τ_2), another rapid motion occurs; a transition from one well to the other at a time scale δt_2. Both high frequency motions, the oscillations in the well, and the rapid jumps between the wells, are filtered out when a large step, Δt, is used ($\tau_2 \gg \Delta t \gg \delta t_1, \delta t_2$). The observed S_{SDET} trajectory will have the system at rest at one well followed by a sudden appearance of the system at the second well without intermediate points (note $\Delta t \gg \delta t_2$). Hence, not only we removed the local quasi-equilibrium and fast small-amplitude vibrations in the wells, but we also filtered out rare transitional events with large spatial amplitudes. Clearly, we are usually interested (at least) in the dynamics of the transition.

A fix to the excessive filtering in the framework of S_{SDET} is not obvious. Rare and large spatial transitions can be detected by sudden and significant changes in the system coordinates in a single Δt step. This specific time interval can be divided into smaller time slices to obtain a refined dynamic picture. However, such refinements require clear separation of time scales, and identification of relevant modes. Note that one of the major advantages of S_{SDET} is the automated filtering of all rapid modes with no need to specify them in advance. The price we pay is the restriction of the present approach to diffusive motions over many low barriers, or over a few identifiable wells separated by high barriers.

12.3.4 Weights of Trajectories and Sampling Procedures

At this point we may continue in one of two directions. We may use a single approximate trajectory at the neighborhood of the exact trajectory; i.e.

the trajectory that was obtained by the minimization of the discrete action. Alternatively, we recognize that the exact trajectory deviates from the optimal trajectory by errors distributed normally. We may sample errors (and plausible trajectories) from the appropriate distribution of coordinates in the neighborhood of the trajectory with filtered high frequency modes. The sampling in the neighborhood of the optimized trajectory should add to one (we approximate one trajectory):

$$\int P(\{X\}_{i=1}^{N}) \prod_{k=2,\dots,N} dX_k = \int A \exp\left[-\frac{S_{SDET}}{2[\sigma^2 \cdot \Delta t]}\right] \prod_{k=2,\dots,N} dX_k = 1 \tag{12.27}$$

where A is the normalization constant. The above expression suggests that the weight of a single trajectory with fixed boundaries is $\exp\left[-\frac{S_{SDET}}{2\sigma^2 \Delta t}\right]$. This weight opens the way for the use of Molecular Dynamics or Monte Carlo procedure to sample **probable solutions** to the boundary value problem. We note that the exact trajectory should be within a distance $\sigma\sqrt{\Delta t}$ from the most probable trajectory with a large time step, Δt, underlining the need for trajectory sampling.

Define the vector $Y \equiv [X_1, X_2, ..., X_N]$ that includes the complete discrete approximation for the trajectory. The action, S_{SDET}, is a function of Y. To generate the trajectory distribution with the above weight, we need to create a canonical distribution with S_{SDET} for an energy function and $2\sigma^2\Delta t$ replacing the usual thermal energy ($k_B T$). In our code MOIL [14] we implemented a Molecular Dynamics protocol that solves the following Newton-like equation:

$$\frac{d^2 Y}{d\eta^2} = -\nabla S_{SDET} \tag{12.28}$$

The parameter η is a fictitious time and the fictitious masses are set uniformly to 1. The solution of (12.28) conserves the total energy of the system where S_{SDET} is the potential. To obtain a temperature of $2\sigma^2\Delta t$ we scale the velocities, $dY/d\eta$, periodically by a single factor λ [15]

$$\lambda \cdot \sum_i \left(\frac{dy_i}{d\eta}\right)^2 = N \cdot Q \cdot (2\sigma^2 \cdot \Delta t) \tag{12.29}$$

The scaling factor λ is chosen to satisfy (12.29). The number of time slices is N and Q is the number of degrees of freedom in a single time slice. The code in MOIL is producing complete trajectories in any of the η steps following the dynamics of (12.28) and (12.29). The trajectories so obtained have the appropriate Boltzmann-like weight.

We note that the code is running in parallel in a very efficient manner. Implementations for loosely coupled clusters of PCs are available for LINUX and Windows operating systems. In practice, the scaling with the number of processors (linear), and the load balancing are excellent [3]. This is of course not surprising since the parallelization protocol is exceedingly simple and consists of assigning different time slices to different CPUs [3]. Only nearest-neighbor communications between the CPUs are required.

The above argument for the weight of trajectories holds for sampling trajectories with fixed end points. The argument becomes subtler if we wish to vary the starting and ending points and sample alternative trajectories. In straightforward classical mechanics all trajectories with the same energy have the same weight. In the present approximate formulation the total time (and not the energy) is made fixed. The weight of different classical trajectories (different boundary conditions) in S_{SDET} can be estimated only approximately. There are two basic assumptions we have to make: the first is on the total energy, and the second is on the kinetic energy.

In the first assumption we set the total energy, E, to be a constant during the whole trajectory. This is correct in a true classical trajectory, but is not exact in an S_{SDET} path. In the same sense that we have errors in our coordinates, we are likely to have errors in the energy as well. Nevertheless, this is a useful constraint to have and is likely to make the trajectories more realistic. Using the first assumption the trajectory weight is (for a thermal system with an inverse temperature β), $\exp\left[-\beta E - \frac{S_{SDET}}{2\sigma^2 \Delta t}\right]$.

Note that we change our philosophy here. If we allow different initial conditions while keeping the total time fixed, we also must allow different energies. We maintain, however, the same energy for one trajectory. The total energy is then written as a sum of kinetic and potential energies ($K = E - U$) and computed for every time slice during the trajectory. The calculation of the energy for each time slice instead of the starting point makes the functional more symmetric, and it should not matter if the energy is indeed conserved. The functional becomes:

$$S_{SDET}^{\beta} \equiv -\beta E - \frac{S_{SDET}}{2\sigma^2 \Delta t} = -\frac{1}{2\sigma^2 \Delta t} \sum_i \left[2\sigma^2 \cdot (\beta/N)(K_i + U_i) + \varepsilon_i^2\right] \Delta t \qquad (12.30)$$

Estimating the total energy, E, of the current S_{SDET}^{β} trajectory is however difficult. The large time step that we employ makes it difficult to estimate time derivatives of the type dX/dt (and the corresponding kinetic energy). The paths we computed do not have enough information to estimate the kinetic and therefore the total energy. We are therefore making a (second) assumption. This time the assumption is on the kinetic energy.

The second assumption is that the **average** kinetic energy is roughly thermal. We assume that

$$\sum_i (\beta/N) K_i \cong Q/2 \qquad (12.31)$$

This assumption is not bad at all for sufficiently large microcanonical systems and long trajectories. It is consistent with the numerical observation of the similarity between canonical and microcanonical ensembles for systems with a few hundreds coupled degrees of freedom.

Using (12.31) we note that the kinetic energy is making no contribution to the weight since the constant factor will disappear in the normalization. The log of the weight of a trajectory with arbitrary starting and ending points is therefore reduced to

$$S_{SDET}^{\beta} \equiv -\beta E - \frac{S_{SDET}}{2\sigma^2 \Delta t} = -\frac{1}{2\sigma^2 \Delta t} \sum_i [2\sigma^2 \cdot (\beta/N) U_i + \varepsilon_i^2] \Delta t \quad (12.32)$$

Note that (12.32) is leading to a very similar trajectory sampling protocol as outlined in (12.28) and (12.29). We only need to change S_{SDET} by the modified action, S_{SDET}^{β} of (12.32).

Of the above two assumptions, (i) constant total energy E, and (ii) thermal average kinetic energy, the first one is the more difficult to justify. It would be nice if we could enforce the energy conservation and still maintain the simple properties of the S_{SDET} formulation. The next formulation that we shall discuss (the Stochastic Difference Equation in Length, S_{SDEL}, Sect. 12.3.6) fixes the energy and is therefore a more natural procedure to sample alternate initial conditions. There is no need to enforce energy conservation as this property is already built in.

12.3.5 Mean Field Approach, Fast Equilibration and Molecular Labeling

So far we discussed algorithms that with the addition of more computational resources (more time slices) approach the exact answer. It is useful at this point to introduce one physically based approximation that reduces significantly the computational resources required. At least we can have it as an option when the computational resources are limited. Perhaps a more significant advance (for us) was the ability to solve the problem of molecular labeling and proper solvent sampling.

The molecular labeling problem is as follows. Consider a solvated system (e.g. a protein immersed in a box of water). To compute a S_{SDET} path we need to specify the initial and the final coordinate sets, X_1 and X_N. Some of the coordinates are the spatial locations of water molecules. The coordinates are required by classical mechanics, true; but the exact labeling of the different water molecules creates a huge labeling degeneracy. All the permutation of the water molecules will create identical trajectories. Moreover, a slight perturbation in the solvent coordinate will create alternative trajectories that we are not interested in. Our prime interest is in the dynamics of the protein and less in the dynamics of the water molecules.

Here we proposed a physically based approximation to get around this problem. We separate the coordinate set X into two domains: X^{slow} and

X^{fast}, where the *"slow"* and *"fast"* are with respect to the rate of approaching equilibrium. For example, we argue that the translation and rotation degrees of freedom of a bulk water molecule relax to equilibrium more rapidly than the protein dihedral angles $(\phi, \psi)_i$ of an amino acid i.

We assume separation of (equilibration) time scale that can be done only if the properties of the system are reasonably well understood. Proceeding with the example of a solvated protein, we set the water coordinates to be X^{fast}, and the protein coordinates to be X^{slow}. Consider a time step, Δt, that is significantly longer than the relaxation time to equilibrium of the fast part but is still slow on the time scale of relaxation to equilibrium of the slow part.

In the above-mentioned example, the time scale of one hundred picoseconds is probably in that range. It is significantly longer than the local orientation and translation relaxation of the water molecules but too short to allow complete relaxation of the protein dihedral angles. If such a time step, Δt, is used in S_{SDET} calculation it eliminates the need to follow the explicit dynamics of the water molecules. On this time scale, in a single step, the water molecules will already relax to equilibrium (with a "frozen" configuration of the slow protein). Their explicit dynamics will become irrelevant.

The consequences of the above picture for the S_{SDET} calculation are: Instead of following the explicit dynamics of all the degrees of freedom in X, we follow the explicit dynamics only of X^{slow} and we thermally average the action (for each time slice) over the X^{fast} coordinates.

$$\langle S_{SDET} \rangle_{X^{fast}} = \sum_i \Delta t \left\langle \left(M^{slow} \frac{X_{i+1}^{slow} + X_{i-1}^{slow} - 2X_i^{slow}}{\Delta t^2} + \frac{dU}{dX_i^{slow}} \right)^2 \right\rangle_{X^{fast}} \tag{12.33}$$

In the above average only the force depends on the fast coordinates. So we are required to perform averages of the type $\left\langle \frac{dU}{dX_i^{slow}} \right\rangle_{X^{fast}}$ and $\left\langle \left(\frac{dU}{dX_i^{slow}} \right)^2 \right\rangle_{X^{fast}}$.

These averages are performed in practice by short molecular dynamics trajectories for the fast components while keeping the slow components fixed at their current time slice configuration. The average, $\langle G(X^{fast}, X^{slow}) \rangle_{X^{fast}}$, of the function, $G(X^{fast}, X^{slow})$, over X^{fast} is computed as follows:

$$\langle G(X^{slow}, X^{fast}) \rangle_{X^{fast}} = \frac{1}{\delta t} \int_0^{\delta t} G(X^{slow}, X^{fast}(t)) \cdot dt$$

$$M^{fast} \frac{d^2 X^{fast}(t)}{dt^2} = -\nabla U(X^{fast}; X^{slow}) \tag{12.34}$$

The molecular dynamics trajectory for X^{fast} is computed at fixed slow coordinates and for time duration, δt, that is significantly smaller than Δt.

In the calculations of folding of C peptide [2], of a conformational transition in hemoglobin [3], and of ion migration through the gramicidin channel [Koneshan Siva and Ron Elber, submitted], the above averaging was used for a selected set of slow coordinates. In these calculations we average the fast coordinates for only tens of steps between sequential optimization steps of the slow coordinates. While more extensive averaging of the fast coordinates could help, this is what we can afford at present.

We also note that there is more than one choice of a function to average. Besides the specific choice made in (12.33), S_{SDET}, it is also possible to average (over the fast coordinates) the weight, $\exp\left[-S_{SDET}/2\sigma^2 \cdot \Delta t\right]$, or the force, $-dU/dX$. Of the three possibilities the last choice is equivalent to generating a potential of mean force prior to the calculations of dynamics. The direct use of a potential of mean force for peptides and proteins is another direction that we currently pursue [16].

12.3.6 Stochastic Difference in Length

The stochastic difference equation in length is conceptually similar to the stochastic difference in time. We therefore do not repeat all of the arguments and discussions above. Rather we briefly list the main formulas and focus on the differences between the two approaches.

Instead of starting from the Newton's equation, we use the actions. In the time formulation, we obtain the equation of motion by requiring that the action is stationary, i.e., $\delta S/\delta X(\tau) = 0$, or in the discrete approximation to the path $\{\partial S/\partial X_i = 0\}_{i=1}^{N}$ (see (12.4)). The S_{SDET} action can be written in that case as

$$S_{SDET} = \sum_i \left(\frac{1}{\Delta t} \partial S/\partial X_i \right)^2 \Delta t \tag{12.35}$$

We are using a similar approach to define the analogous action for the Stochastic Difference Equation in Length, S_{SDEL} (S_L is defined in (12.9))

$$S_{SDEL} = \sum_i \left(\frac{1}{\Delta l_{i,i+1}} \partial S_L/\partial Z_i \right)^2 \Delta l_{i,i+1} \tag{12.36}$$

The optimization of S_{SDEL} is performed subject to the constraint that all the lengths of the path segments, $\Delta l_{i,i+1}$s are the same. This is (of course) equivalent to the requirement that the points are equally distributed along the path. The constraint is conveniently formulated as a penalty function [17]

$$\text{constraint} = \frac{\lambda}{(N-1)Q} \sum_i \left(\Delta l_{i,i+1} - \langle \Delta l \rangle\right)^2 \qquad \langle \Delta l \rangle = \frac{1}{N-1} \sum_i \Delta l_{i,i+1} \tag{12.37}$$

In the above expression λ is a constant that is determined by experimentation. Moreover, since the length element is computed in mass weighted

Cartesian coordinates, $\Delta l_{i,i+1} = |Z_i - Z_{i+1}|$, it is necessary to avoid overall translations and rotations of individual length slices, Z_i. Imposing linear constraints and solving for the corresponding Lagrange's multipliers is a convenient way of removing the rigid body motions [18]. The following constraints are imposed on each length slice

$$\sum_k m_k(R_{ik} - R_{Rk}) = 0$$

$$\sum_k m_k(R_{ik} - R_{Rk}) \times R_{Rk} = 0 \tag{12.38}$$

The k index is running over the atoms in a single length slice i. The vector R_{ik} is of rank 3, and it provides the coordinates of a single atom, m_k is the mass of the k atom. The vector R_{Rk} is a reference coordinate system (the coordinates of the middle intermediate structure).

The errors connected with the length formulation are defined as before. An exact trajectory assesses the accuracy of the finite difference formula. The choice of the finite difference formula to use is biased by the convenience of a constant Jacobian of transformation from the errors to coordinates (see also (12.12))

$$\begin{aligned}[\varepsilon_{i+1}^{(l)}]^T &= \left(\frac{1}{\Delta l_{i,i+1}} \partial S_L/\partial Z_i\right)^T \\ &= 2(E - U(Z_i)) \cdot \frac{(Z_{i+1} + Z_{i-1} - 2Z_i)^T}{\Delta l_{i,i-1}^2} \\ &\quad - (dU/dZ_i)^T \cdot (1 - (Z_i - Z_{i-1})(Z_i^T - Z_{i-1}^T)/\Delta l_{i,i-1}^2)\end{aligned} \tag{12.39}$$

The statistical properties of the "length" errors are similar to the statistical properties of the "time" errors. This is demonstrated in the figure below.

The overall similar behavior of the time dependent and the length dependent errors suggest that a related modeling of trajectories can be used in the last case. This is what we have done. It is also possible to show that the high frequency motions (as a function of path length (!)) are filtered out similarly to the removal of rapid motions as a function of time in S_{SDET}. A legitimate question is then why do we need yet another stochastic formulation? What did we gain (or lose) by the alternative representation?

Below we list the good and the bad news:

One of the difficulties in the time formulation of the errors, which was discussed earlier, is the filtering of fast transitional motions, (e.g. a transition over an energy barrier). These are rare and rapid motions that may be of considerable interest. In the time formulation they are removed since they are fast. In the length formulation the situation is different. These motions are *not* of small amplitude, and the constraint of uniform distribution of points along the path (12.37) ensure proper sampling of spatially significant transitions.

A second point in favor is the use of energy conservation explicitly instead of specifying the total time. The energy, as argued before in Sect. 12.3.4, is easier to estimate since it requires only equilibrium observations (like the temperature of the system). It is also nice to have the total time of the process as an output (Sect. 12.2.3) instead as an input like in S_{SDET}.

Moreover the sampling arguments of Sect. 12.3.4 become simpler conceptually, and there is no need to make additional assumptions (beyond the SDE formulation). The energy conservation of the trajectory is already built in. The usual classical mechanics weight applies: Trajectories with the same energy will have the same weight.

There is also some bad news. The integral that determines the time is a weighted sum over the spatial path. Each Δl is the distance between two sequential points and is the shortest path between them. If the step in length, Δl, is large (i.e., only a small number of grid points is used), then the overall length of the path is bound to be shorter than the length of the true trajectory. The total time of the trajectory, which is an integral over the path, and is an important variable, is likely to be too short. A way to

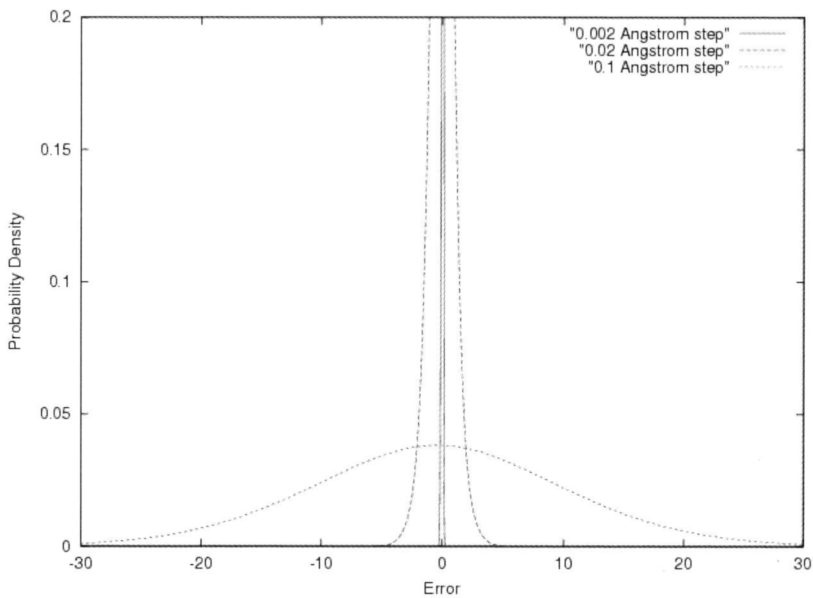

Fig. 12.3a. Distribution of errors extracted from exact trajectories of valine dipeptide and (12.39) at three different length steps. Note that the narrowest distribution corresponds to essentially exact trajectory.

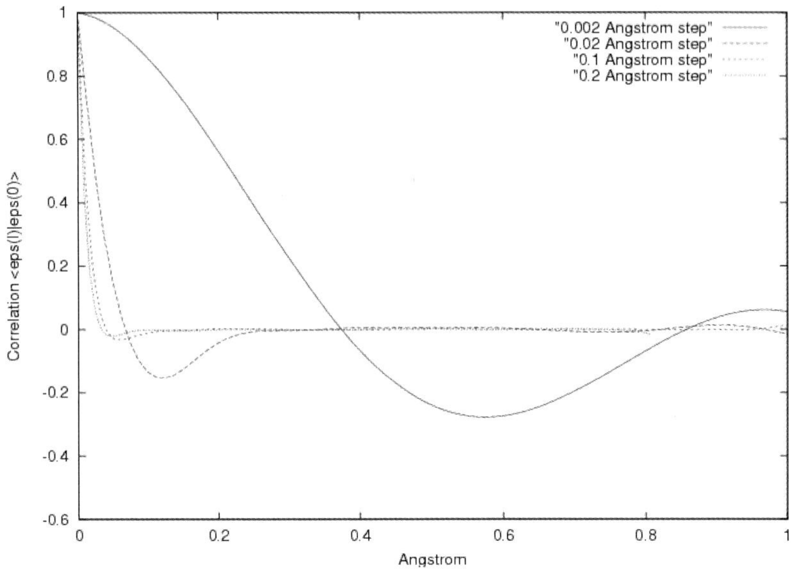

Fig. 12.3b. Correlation function of errors extracted from exact trajectories of valine dipeptide and (12.39) at four different length steps. Note that at the smallest length step (in which the correlation persist for more than one angstrom) the errors are very small and the trajectory is very close to exact. Hence, as we approach the exact result, no filtering of high frequency modes occur, the errors become correlated and the first assumption of the SDE approach is invalid. In practice, it means that there is a minimal size of the step that can be used in the calculation.

get around this problem and obtain a sound estimate of the total time of the trajectory is by statistical refinement, a procedure that is described in the next section.

Another interesting feature of path parameterized by length is the non-uniqueness of the solution. In the length parameterization we fixed the end points and the total energy. If we consider periodic motions, any addition of complete period will return us to the same end points (and of course the same energy). Hence, it is possible to obtain very short (one period) trajectories or infinitely long trajectories with infinitely many repeats of the periodic motions. Depending on our initial guess and the extent of our annealing we may hit trajectories with different total length and time, all of them are legitimate solutions by the conditions we set.

12.3.7 "Fractal" Refinement of Trajectories Parameterized by Length

The argument below is similar in spirit to the estimation of the length of the coast of England [19]. The calculated length depends on the yardstick in which it is measured. The smaller is the yardstick the longer is the observed length. The coast of England is therefore considered to be a fractal, for which the total distance measured as described above, is increasing without limit. For the classical trajectory at hand there is a limit and we define it to be the length element, $\Delta l_{\lim it}$. At this limit the solutions of the initial value differential equation and the boundary value problem agree. It is typically 0.01Å.

The algorithm for trajectory refinement with prime purpose of determining the time of the trajectory is

(i) Initiate: Optimize S_{SDEL} to obtain a path parameterized by length of N intermediate points. We denote the path segment by $\Delta l_{i,i+1}^{(1)}$

(ii) Sample intervals: From the calculated path, sample at random a few pairs of points (intervals), $\{Z_i, Z_{i+1}\}_i$. The number of sampled intervals, ξ, is much smaller than N to ensure computational efficiency. The process below becomes exact when ξ is equal N but we usually do not go that far (Fig. 12.4).

(iii) Refine selected intervals: For each of the ξ intervals compute an interpolating trajectory with N intermediate points (i.e. compute length dependent trajectories for all the $\{Z_i, Z_{i+1}\}_i$ pairs that are used for boundary conditions and sampled in (ii)). The length of the path segment in this refinement is denoted by $\Delta l_{i,i+1}^{(n)}$ where (n) is the index of the refinement cycle.

(iv) Examine the convergence of the newly generated path segments: Convergence is assumed when the path computed with the initial value formulation (12.11) agrees within a threshold to the path created by the optimization of S_{SDEL}. If converged go to (vi)

(v) If convergence was not reached return to (ii): Sampling is now done from the **refined** ξ segments. The sequential points that we sample, $\{Z_i, Z_{i+1}\}$, are from the segments of the segments.

(vi) Convergence was reached. We have an estimate for the value of the time integral (Sect. 12.2.3) for path of length $\{\Delta l_{i,i+1}^{(n)} \equiv |Z_i, Z_{i+1}|\}$. Use this estimate (and estimates from other length intervals) to calculate the time length for an earlier refinement cycle with a length step $\Delta l_{i,i+1}^{(n-1)}$. Repeat until the original length step, $\Delta l_{i,i+1}^{(1)}$, is reached.

After a few refinement cycles (a typical number is 5) the protocol above converges in practice [16]. Path segments calculated with S_{SDEL} or with the initial value formulation ((12.11) with a step size, $\Delta l_{i,i+1}^{(n)}$) are essentially the same (Figs. 12.4 and 12.5)

An intriguing question is how much the spatial distribution of the trajectories changes upon refinement? Do we need to refine the trajectories as described above for all types of studies? Or is it possible to extract useful information from paths with significantly lower resolution? This is a question that will be addressed in the next section on numerical experimentation.

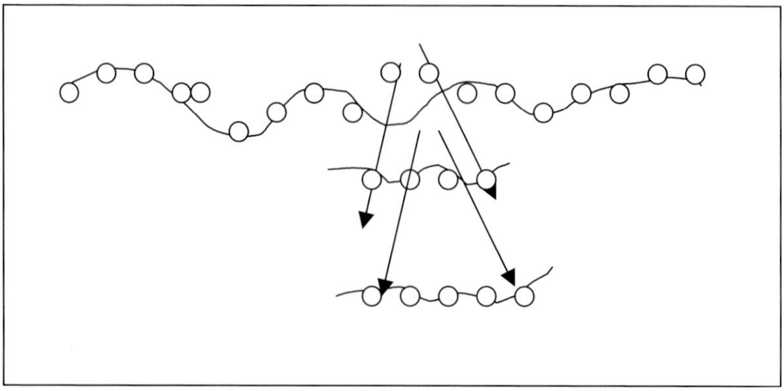

Fig. 12.4a. A schematic drawing of the refinement procedure of an SDEL path. Length segments are sampled and length dependent paths are computed (with more intermediate points) for the selected segments. Convergence is assumed when the criterion discussed in Fig. 12.4.b is satisfied.

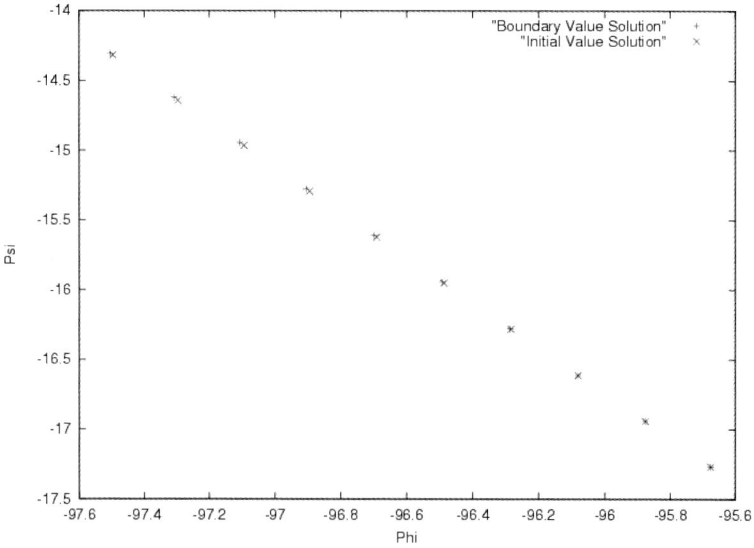

Fig. 12.4b. Comparing the boundary value solution and the initial value formulation for a segment of a refined SDEL path for valine dipeptide. At this level of agreement we considered the refinement process complete.

12.4 Numerical Experiments

The present manuscript is mostly methodological, presenting the conceptual framework behind the new technique of the stochastic difference equation. It is therefore appropriate to discuss numerical examples of small systems for which different aspects can be tested in greater details. On the other hand the

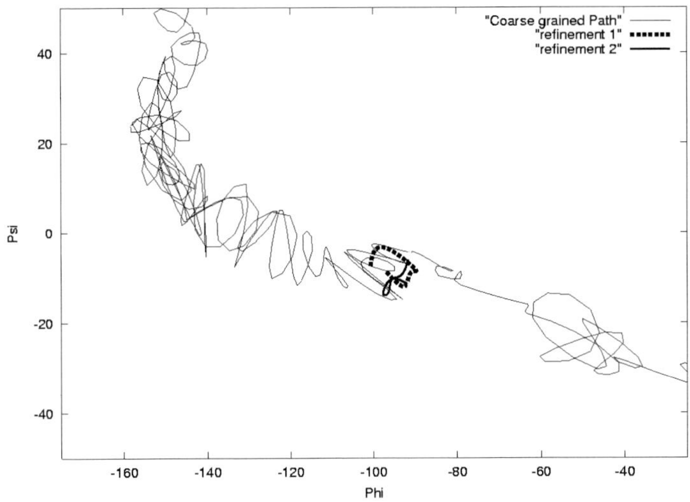

Fig. 12.5. Refinement example of a trajectory along a coarse grained path for valine dipeptide.

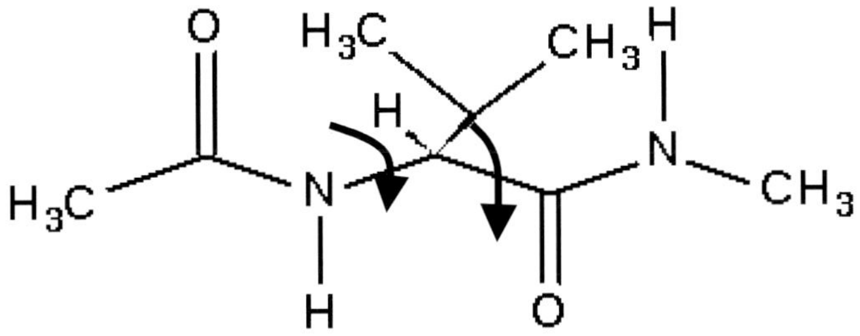

Fig. 12.6. A schematic drawing of valine dipeptide. Note that the only soft degrees of freedom of this system are the rotations (φ, ψ) around the bonds as indicated by the arrows.

numerical examples should be sufficiently complex so that non-trivial effects could be observed. So, despite the fact that the techniques were already applied to investigate much larger systems, we focus here on conformational transitions of smaller systems: dipeptides.

Dipeptides are useful models for conformational transition of proteins. They form "minimal models" on which protein backbone conformational changes can be investigated. A schematic picture of valine dipeptide is shown in Fig. 12.6. An extended atom model (CH_n groups are treated as a point mass) is used.

Glycine dipeptide has 11 extended atoms and valine dipeptide with its larger side chain, 14 atoms. Nevertheless the backbone of the two peptides is essentially identical. It includes only two soft degrees of freedom, the φ, ψ dihedral angles. Other modes are too stiff to be significantly excited at room temperature. It is therefore a common practice to describe the dynamics (and thermodynamics) of dipeptides on a two dimensional φ, ψ energy maps.

The AMBER/OPLS force field is implemented in MOIL [4] and is used throughout the calculations. No cutoffs were used for this small systems and the 1-4 scaling factor were 2 and 8 for electrostatic and van der Waals interactions. No constraints on fast vibration were used. However, the stochastic difference equation filters the bond vibrations anyway. In Fig. 12.7 we compare the energy content of the bond vibrations in S_{SDEL} optimization with different step sizes.

Another question of interest that we can examine in this small system is the dependence of the paths on the number of length slices. In Fig. 12.8 we present a comparison for S_{SDEL} trajectories computed with 80, 320 and 640 number of grid points. The trajectories are shown on (φ, ψ) map, the two relevant degrees of freedom, though the complete 33 degrees of freedom were used in the calculations. The trajectories with different resolution in length cover the same domains in conformation space.

Our trajectories are sampled with the help of simulated annealing protocol. But how can we test that the sampling is appropriate? One measure that can help us to assess the quality of the simulation is the distribution of the orientation of the initial momentum vector. If we sampled effectively the space of initial conditions (by sampling **complete** trajectories) then the

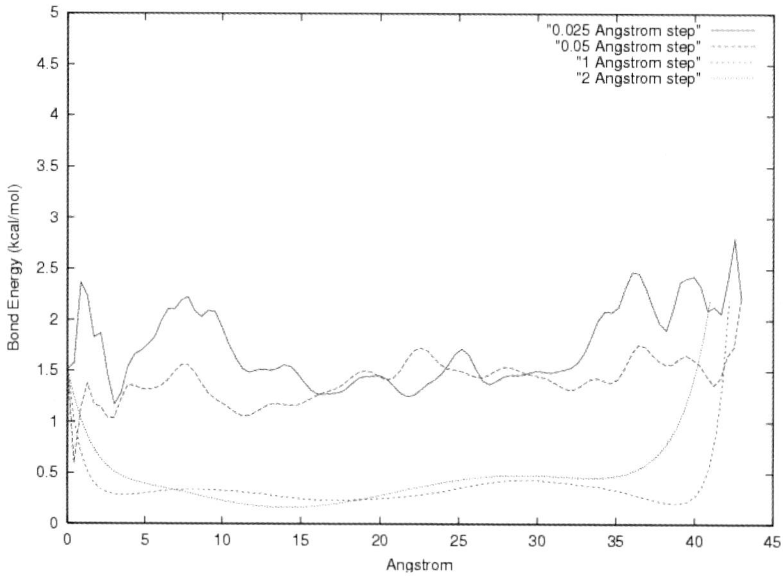

Fig. 12.7. Bond energy distribution along the path for four different sizes of length steps. The data is extracted from SDEL calculations of valine dipeptide. Note the significant reduction of bond energies (filtering) as the step size increases.

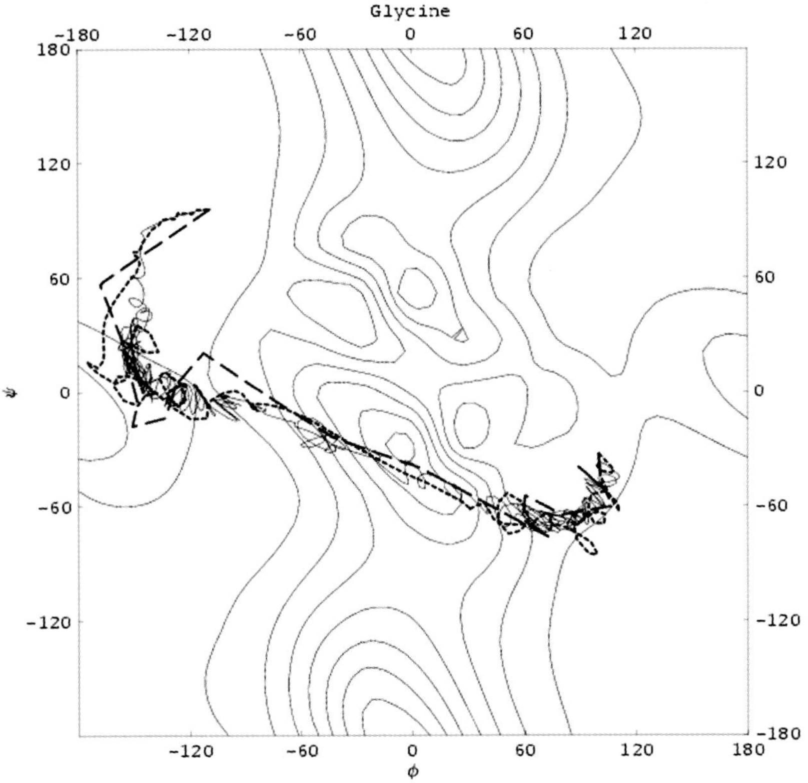

Fig. 12.8. Equatorial to Axial transition for glycine dipeptide using increasing length size with SDEL. Note that the spatial distribution of configurations along the trajectories remains similar at all resolutions. The main difference is the significantly larger density of configurations near minima associated with incubation periods.

momentum vectors should cover all the orientation space. Or alternatively, the initial vectors of the direction of the momentum will behave as random vectors (with norm of one). In Fig. 12.9, we show the distribution function of the scalar products of different (normalized) momentum vectors. The distribution for valine dipeptide is a combination of reactive and non-reactive trajectories. While the distribution is not exactly a Gaussian it is not too far from it...

12.5 Concluding Remarks

We have outlined a new numerical approach to compute approximate long time molecular dynamics trajectories. We have explained the underlying assumptions and the limitations of the present approach as well as its promise. Numerical examples were shown for relatively small system for which detailed

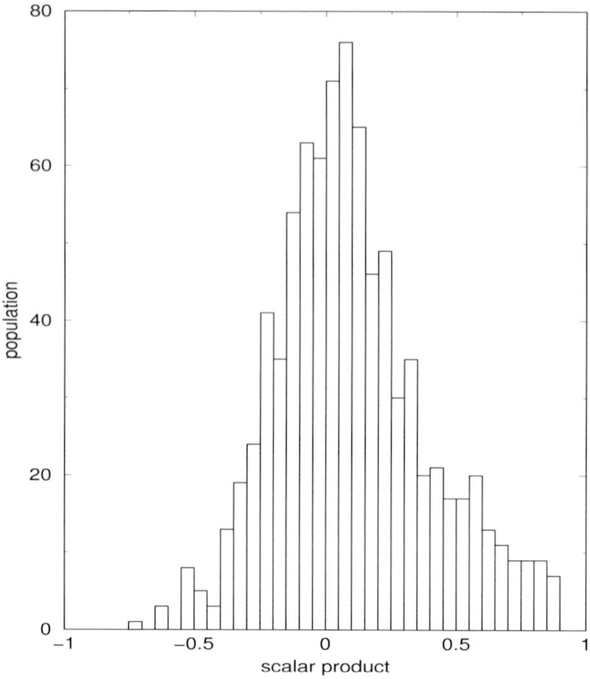

Fig. 12.9. Histogram showing the distribution function of the scalar product of normalized initial momentum vectors for 21 reactive and 21 non-reactive trajectories for valine dipeptide conformational change.

and extensive calculations can be performed. The research described in this article was supported by grants from the NIH GM59796 and the NSF Grant No. 9982524 to Ron Elber.

References

1. R. Olender and R. Elber, "Calculation of classical trajectories with a very large time step: Formalism and numerical examples", J. Chem. Phys., 105, 9299-9315 (1996).
2. R. Elber, J. Meller and R. Olender, "A stochastic path approach to compute atomically detailed trajectories: Application to the folding of C peptide", J. Phys. Chem. B, 103, 899-911 (1999).
3. Veaceslav Zaloj and Ron Elber, "Parallel Computations of Molecular Dynamics Trajectories Using Stochastic Path Approach", Computer Physics Communication, 128, 118-127 (2000).
4. Joost C.M. Uitdehaag, Bart A. van der Veen, Lubbert Dijkhuizen, Ron Elber and Bauke W. Dijkstra, "The enzymatic circularization of a malto-octaose linear chain studied by stochastic reaction path calculations on cyclodextrin glycosyltransferase", Proteins, Structure Function and Genetics, 327-335,43(2001).

5. Verlet L. Physical Review, 98,159(1967)
6. L. D. Landau and E. M. Lifshitz, Mechanics, Butterworth-Heinenann, Oxford, 1993
7. Batcho PF, Case DA, Schlick T, "Optimized particle-mesh Ewald/multiple-time step integration for molecular dynamics simulations", J. Chem. Phys. 115, 4003-4018(2001)
8. Zhou R. H., Harder E, Xu H. F., Berne B. J., "Efficient multiple time step method for use with Ewald and particle mesh Ewald for large biomolecular systems", J. Chem. Phys. 115, 2348-2358(2001)
9. Bishop TC, Skeel RD, Schulten K, "Difficulties with multiple time stepping and fast multipole algorithm in molecular dynamics", J. Comp. Chem. 18, 1785-1791(1997)
10. A. Ulitsky and R. Elber, "The equilibrium of the time dependent Hartree and the Locally Enhanced Sampling approximations: Formal properties, corrections, and computational studies of rare gas clusters", J. Chem. Phys., 98, 3380 (1993).
11. Ghosh A. and Elber R, unpublished
12. Lanczos C. "The variational principles of mechanics", University of Toronto press, 1970
13. Kleinert H., "Path integrals in quantum mechanics, statistics and polymer physics", World Scientific, Singapore, 1995, chapters 18.5 and 18.6
14. Elber R., Roitberg A., Simmerling C., Goldstein R., Li H., Verkhivker G., Keasar C., Zhang J., Ulitsky A. "Moil: A program for simulations of macromolecules", Computer Physics Communication, 91,159-189(1995)
15. J. M. Haile and S. Gupta, J. Chem. Phys. 79,3067(1983)
16. A. Ghosh, A. Cárdenas, and R. Elber, in preparation
17. R. Czerminski and R. Elber, "Self avoiding walk between two fixed points as a tool to calculate reaction paths in large molecular systems", Int. J. Quant. Chem. 24, 167-186 (1990)
18. R. Elber, "Calculation of the potential of mean force using molecular dynamics with linear constraints: An application to a conformational transition in a solvated dipeptide", J. Chem. Phys. 93,4312(1990)
19. Mandelbrot B. M., "The fractal geometry of nature", Freeman, San Francisco, 1977

13 Numerical Simulations of Molecular Systems with Long Range Interactions

Dominique Levesque

Laboratoire de Physique Théorique
Bâtiment 210, Université Paris-Sud
91405 Orsay Cedex, France

13.1 Introduction

Numerical simulations require the summation of long range pair interactions between the charges and/or dipoles of atoms and molecules of many systems such as ionic solutions, molten salts, colloidal suspensions, ... In addition to its infinite range, the Coulomb potential between charges satisfies to the Poisson's equation and Green's theorem and these characteristics induce, in ionic fluids, the well known physical property of charge screening. The interactions between the permanent or induced molecular dipoles are to the origin of the dielectric constants of fluids of polar and polarisable molecules. The aim of simulations being to determine the exact physical properties of atomic and molecular systems, it is needed that, in the numerical simulations, the Coulomb and dipole-dipole interactions are taken into account in such a way that the specific physical properties which derive from their slow decrease and infinite range stay preserved.

It is obvious that this last requirement is strictly incompatible with the use of a cut-off of the Coulomb or dipole-dipole potentials at a distance of the order of the sizes of systems studied in simulations which have typical values of $\sim 1000 - 10000$ Å^3 . Generally in simulations, the systems have periodic boundary conditions. The main problem in the computation of Coulomb or dipole-dipole interactions in finite systems is that the boundary conditions, which, for confined systems, are different in the three spatial directions, are correctly taken into account. First, we discuss the case of the three dimensionnal (3D) systems and, then, that of the confined systems.

13.2 3-D Systems

Considering N atoms or molecules with a charge q_i at a position $\boldsymbol{r}_i$ in a cubic simulation cell of side L with periodic boundary conditions, the coulomb energy of these N charges E_c [1] is

$$E_c = \frac{1}{2} \sum_{i,j=1}^{N}{}' \sum_{n \in Z^3} \frac{q_i \, q_j}{|\boldsymbol{r}_j - \boldsymbol{r}_i + \boldsymbol{n}L|} \, . \tag{13.1}$$

The sum over $\boldsymbol{n}$ (vectors with integer components) corresponds to the interactions between the charges in the simulation cell and the infinite set of their periodic replicas. In the sum over i and j, the $'$ indicates that i must be different of j when $\boldsymbol{n} = 0$. E_c is a conditionnally convergent sum definite only when it is specified how the sum over $\boldsymbol{n}$ is performed. For instance, the sum can be made by adding the terms of increasing modules of $\boldsymbol{n}$; the replicas of the simulation cell in filling the space generate a cubic lattice of identical cells having a quasi-spherical external surface at macroscopic distance surrounded by the vacuum or a dielectic medium of dielectric constant ϵ. This procedure of computation of E_c has been described and studied carefully in the literature [2,3]. It involves the split of the coulomb potential following the identity:

$$\frac{1}{r} = \frac{f(r)}{r} + \frac{1 - f(r)}{r} \tag{13.2}$$

where $r = |\boldsymbol{r}|$ and $f(r)$ is a rapidly decreasing function of r for instance $f(r) = \mathrm{erfc}(\alpha r)$ and the use of the electroneutrality condition $\sum_i q_i = 0$. It is this latter condition which allows to remove the divergent contribution in the sum of (13.1). By the substitution in (13.1) of the identity above, the sum over $\boldsymbol{n}$ is divided in two sums, corresponding to each terms of the right-hand member of (13.2). The first sum is performed in r-space and, after a Fourier transformation, the second sum is made in k-space. These two new sums are rapidly convergent.

The result is

$$\begin{aligned} E_c = & \frac{1}{2} \sum_{i,j=1}^{N}{}' \sum_{n \in Z^3} q_i \, q_j \frac{\mathrm{erfc}(\alpha |\boldsymbol{r}_{ij} + \boldsymbol{n}L|)}{|\boldsymbol{r}_{ij} + \boldsymbol{n}L|} \\ & + \frac{1}{2L^3} \sum_{\boldsymbol{k} \neq 0} \frac{4\pi}{k^2} \exp(-\frac{k^2}{4\alpha^2}) |\sum_j q_j \, e^{i\boldsymbol{k}.\boldsymbol{r}_j}|^2 \\ & - \frac{\alpha}{\sqrt{\pi}} \sum_i q_i^2 + \frac{2\pi}{(1 + 2\epsilon)L^3} |\sum_i q_i \boldsymbol{r}_i|^2 . \end{aligned} \tag{13.3}$$

Equation (13.2) being an identity, the value of E_c is independent of α. However an adequate choice of α allows to suppress the sum on $\boldsymbol{n}$ and to restrict that on $\boldsymbol{k}$ ($\boldsymbol{k} = 2\pi \boldsymbol{n}/L$) to a limited number of terms. These approximations induce a "small" and controlled numerical relative error ($\sim 10^{-4} - 10^{-7}$) in the calculation of E_c. For a "metallic" surrounding medium of the replica lattice $\epsilon = \infty$, the last term, the square of the dipolar electric moment of the charges, disappears ; but when ϵ is finite, some cautions are needed in the calculation of this term which, contrary to the other terms of E_c, is not periodic. In a molecular dynamics simulation, the particle positions at time t, $\boldsymbol{r}_i(t)$, used to compute the dipolar moment must be those of the "image" particles which diffuse from their initial position $\boldsymbol{r}_i(0)$ in the simulation cell through the replica lattice [4].

Some remarks on the calculation of E_c can be done. When $|r_{ij}|$ is computed with a minimum image convention:

$$|\boldsymbol{r}_{ij}|^2 = |x_i - x_j - L\{nint(\frac{x_i - x_j}{L})\}|^2 + |y_i - y_j - L\{nint(\frac{y_i - j}{L})\}|^2$$

$$+ |z_i - z_j - L\{nint(\frac{z_i - z_j}{L})\}|^2 , \tag{13.4}$$

where $nint(a)$ denotes the nearest integer of the decimal number a, the calculation of $|\boldsymbol{r}_{ij}|^2$ needs about 17 elementary "operations" of a standard computer CPU. The second term involving the sum on $\boldsymbol{k}$ is a Fourier series which can be rewritten in order to use a "fast fourier transformation" (FFT) algorithm. The scaling with the number of charges N of the computation time is reduced from N^2 or $N^{3/2}$ (with a good choice of α) to $N \ln N$. The FFT algoritm needs a very carefull numerical implementation in order to calculate E_c with an accuracy of $10^{-6} - 10^{-8}$ [5–7].

Another improvement in the computation of E_c can be made by using a multipole expansion to calculate the terms in r-space [8,9]. It consists to divide the simulation cell volume in m subvolumes (for instance cubic) v_1, ..., v_m, so that the r-space terms of the energy of a charge i are:

$$E_i = \sum_{l=1,m} \sum_{j \in v_l} q_i \, q_j \frac{\mathrm{erfc}(\alpha|\boldsymbol{r}_{ij}|)}{|\boldsymbol{r}_{ij}|} . \tag{13.5}$$

The contribution to E_i of the charges in the subvolume l can be estimated by a Taylor expansion around the position $\bar{\boldsymbol{r}}_l$ of the center of v_l:

$$E_{i,l} \equiv \sum_{j \in v_l} q_i \, q_j \frac{\mathrm{erfc}(\alpha|\boldsymbol{r}_{ij}|)}{|\boldsymbol{r}_{ij}|}$$

$$= q_i \sum_n \frac{1}{n!} \nabla^n_{\bar{\boldsymbol{r}}_l} \frac{\mathrm{erfc}(\alpha|\boldsymbol{r}_i - \bar{\boldsymbol{r}}_l|)}{|\boldsymbol{r}_i - \bar{\boldsymbol{r}}_l|} \sum_j q_j \, \{\boldsymbol{r}_j - \bar{\boldsymbol{r}}_l\}^n , \tag{13.6}$$

which needs the computation of the distances between the charges $j \in v_l$ and $\bar{\boldsymbol{r}}_l$. From the fact that these distances are identical for all the charges i, it results an important decrease of the computation time of E_c. An efficient implementation of the "multipole" method is described in [9] where it is shown that, for $N > 10000$, the computation of E_c by this method is significantly faster than by using the standard Ewald method.

The finite size and boundary conditions of the simulation cell can be taken into account in more physical way by considering that the charges interact by the potential solution of the Poisson's equation in such a volume [10]. Since, as it was known, the potential of a point charge is not defined in a closed space, it is needed to introduce "pseudo-charges" which are defined by the association of a point charge q and a uniform neutralizing background filling the simulation cell, i.e:

$$\rho(r) = q\delta(r) \Longrightarrow \bar{\rho}(r) = q[\delta(r) - \frac{1}{L^3}] . \tag{13.7}$$

The potential $\phi(r)$ produced by $\bar{\rho}(r)$ is the solution of the equation

$$\Delta\phi(r) = 4\pi\bar{\rho}(r) \tag{13.8}$$

with von Neuman boundary conditions which is given by

$$\phi(r) = \frac{q}{L^3} \sum_{\boldsymbol{k}\neq 0} \frac{4\pi}{k^2} e^{i\boldsymbol{k}.\boldsymbol{r}} . \tag{13.9}$$

By using the identity:

$$\frac{1}{k^2} = \frac{1 - \exp(-k^2/4\alpha^2)}{k^2} + \frac{\exp(-k^2/4\alpha^2)}{k^2} \tag{13.10}$$

in (13.9) defining the potential of one charge, it is easy to recover the previous formula for the Coulomb energy E_c of N charges (Eq. (13.3)), for $\epsilon = \infty$, by taking the Fourier transform of the contribution to E_c associated to the first term of the right-hand member of (13.10). This approach to the computation of Coulomb energy opens the possibility to consider simulation cells which are different from a cubic or parallelepipedic volume with periodic boundary conditions. One of these possibilities consists in confining the charged particles on the "surface" of a four dimensionnal (4-D) sphere (a hypersphere) [11,12]. Since the surface of a hypersphere (a 3-D volume) is finite, it is needed to use "pseudo-charges":

$$\bar{\rho}(\mathbf{M}) = q[\delta(\mathbf{M}_o - \mathbf{M}) - \frac{1}{V_H}] , \tag{13.11}$$

where $V_H = 2\pi^2 R^3$ (R, 4-D radius of the hypersphere). $\mathbf{M}_o$ and $\mathbf{M}$ are 4-D vectors which define the positions of the "pseudo-charge" and a point on the surface of the hypersphere, having its center at the origin of the 4-D coordinate referential. The distance between two charges located at $\mathbf{M}_i$ and $\mathbf{M}_j$ is the length of the geodesic linking these positions

$$r_{ij} = R \arccos(\frac{\mathbf{M}_i.\mathbf{M}_j}{R^2}) \equiv R\,\psi_{ij} . \tag{13.12}$$

The potential $\phi(\mathbf{M})$ generated by $\bar{\rho}(M)$ is given by the solution of:

$$\Delta_S\phi(\mathbf{M}) = 4\pi\bar{\rho}(\mathbf{M}) \tag{13.13}$$

where Δ_S is the Laplacian operateur on the surface of a hypersphere [11]. This solution is

$$\phi(\mathbf{M}) = \frac{q}{\pi R}[(\pi - \psi)\cot(\psi) - \frac{1}{2}]\,. \tag{13.14}$$

From this potential, the coulomb energy of N charges located on the hypersphere surface is given by:

$$E_c = \frac{1}{2}\sum_{i,j\neq 1}^{N} \frac{q_i q_j}{\pi R}[(\pi - \psi_{ij})\cot(\psi_{ij}) - \frac{1}{2}]\,. \tag{13.15}$$

The computation of this energy is very straightforward from

$$\cos(\psi_{ij}) = \frac{1}{R^2}[x_i x_j + y_i y_j + z_i z_j + t_i t_j] \tag{13.16}$$

where x_i, y_i, z_i and t_i are the four coordinates of $\mathbf{M}_i$. From the previous equation, it is easy to estimate that the computation of the distance between the charges i and j along a geodesic on the hypersphere surface need about 7 operations. This number of operations must be compared to the 17 operations which are needed to calculate the distance between two charges with the minimum image convention. It give an considerable interest to the hypersphere method due to its excellent computational efficiency for systems with $N \sim$ 1000-10000. Detailed comparisons of the simulation results of charged systems located on a 4-D hypersphere surface or in a 3-D volume with periodic boundary conditions have been done in [12,13]. The simulations of polar fluids and electrolyte solutions can be realised on a 4-D hypersphere surface by deriving from the charge-charge potential $\phi(\mathbf{M})$ the charge-dipole and dipole-dipole potentials [13]. Simulation data are presented in Table 13.1 and 13.2, for the restricted primitive model (RPM) and the RPM in a fluid solvent of hard spheres of diameter σ with a permanent dipole μ):

Table 13.1. Comparison between the "Ewald" method and hypersphere method: results of simulations of the restricted primitive model ($N/2$ hard spheres of diameter σ with a charge q and $N/2$ hard spheres of diameter σ with a charge $-q$) in a volume V. T, ρ, U and P are, respectively, the temperature, density, pressure and internal energy of the considered thermodynamic state of RPM.

$q^2/k_BT\sigma$	$\rho = N\sigma^3/V$	U/k_BT	$P/\rho k_BT$	N	Method
2	0.67	-0.732(3)	4.97(2)	512	"Hypsph."
2	0.67	-0.738(3)	5.06(4)	256	"Ewald"
40	0.67	-20.90(2)	1.98(3)	256	"Hypsph."
40	0.67	-20.86(5)	1.96(3)	128	"Ewald"

In many interesting physical conditions it is relevant to represent a complex molecular or colloidal suspension made of charged molecules or colloidal

Table 13.2. Comparison between the "Ewald" method and hypersphere method: results of simulations of a ionic solutions made of a RPM ($N_i/2$ hard spheres of diameter σ with a charge q and $N_i/2$ hard spheres of diameter σ with a charge $-q$) solvated in a dipolar hard sphere fluid of N_d hard spheres of diameter σ with a permanent dipole μ in a volume V. T, $N = N_i + N_d$, U_{ii}, U_{id} and U_{dd} are, respectively, the temperature, the total number of particles, the internal energies resulting of the charge-charge, charge-dipole and dipole-dipole interactions.

N_i	$q^2/k_BT\sigma$	N_d	μ^2/k_BT	U_{ii}/k_BT	U_{di}/k_BT	U_{dd}/k_BT	$N\sigma^3/V$	Method
54	8	810	3.13	-1.47(4)	-2.85(3)	-2.87(5)	0.6	"Hypsph."
54	8	810	3.13	-1.51(2)	-2.88(2)	-2.78(5)	0.6	"Ewald"

particles, solvent molecules, co-ions and counter-ions by a system of molecules or colloidal particles interacting by a screened potential (Yukawa potential) $Z_e \exp(-\kappa r)/r$ where Z_e is an effective charge and $1/\kappa = \lambda_\kappa$ the damping length. An identical approximation can be also relevant for the interactions bewteen the ions of a plasma screened by the free ectrons.

In simulations where λ_κ is larger than L the side of the simulation cell, it is possible to take into account the tail of the Yukawa interaction by methods similar to those used for the Coulomb interaction.

Considering a cubic volume of side L with periodic boundary conditions, the Yukawa potential generated by a particle at the center of this volume is the solution of the Helmotz's equation

$$(\Delta - \kappa^2)\phi_Y(r) = 4\pi\hat{\delta}(r) \tag{13.17}$$

where $\hat{\delta}(r)$ is

$$\hat{\delta}(r) = \sum_{n \in Z^3} \delta(\boldsymbol{r} + \boldsymbol{n}L) = \frac{1}{L^3}\sum_k e^{\mathrm{i}\boldsymbol{k}.\boldsymbol{r}} . \tag{13.18}$$

The potential $\phi_Y(r)$ is equal to

$$\phi_Y(r) = \frac{4\pi}{L^3}\sum_k \frac{e^{\mathrm{i}\boldsymbol{k}.\boldsymbol{r}}}{k^2 + \kappa^2} . \tag{13.19}$$

By using the identity

$$\hat{\delta}(r) = \hat{\delta}(r) - \lambda f(r) + \lambda f(r) . \tag{13.20}$$

where

$$f(r) = (\frac{\alpha^2}{\pi})^{3/2} e^{-(\alpha r)^2} \tag{13.21}$$

and

$$\lambda = \exp(-\kappa^2/4\alpha^2) , \tag{13.22}$$

it is possible by substituting the Fourier transform of this identity in (13.18), to rewrite $\phi_Y(r)$ in a form which takes into account the tail of the Yukawa interaction and the effect of the periodic boundary conditions. This form which, as expected is similar to that the Ewald potential, is given

$$E_Y = \frac{1}{2}\sum_{i,j}^{N} q_i\, q_j \,\{\sum_{\boldsymbol{n}} \frac{\mathrm{erfc}\,(\alpha|\boldsymbol{r}_i - \boldsymbol{r}_j + \boldsymbol{n}L| + \kappa/(2\alpha))\,\exp(\kappa|\boldsymbol{r}_i - \boldsymbol{r}_j + \boldsymbol{n}L|)}{2|\boldsymbol{r}_i - \boldsymbol{r}_j + \boldsymbol{n}L|}$$

$$+ \frac{\mathrm{erfc}\,(\alpha|\boldsymbol{r}_i - \boldsymbol{r}_j + \boldsymbol{n}L| - \kappa/(2\alpha))\,\exp(-\kappa|\boldsymbol{r}_i - \boldsymbol{r}_j + \boldsymbol{n}L|)}{2|\boldsymbol{r}_i - \boldsymbol{r}_j + \boldsymbol{n}L|} - \frac{4\pi}{\kappa^2\, L^3}\}$$

$$+ \frac{2\pi}{L^3}\sum_{\boldsymbol{k}\neq 0} \frac{\exp[-(k^2+\kappa^2)/(4\alpha^2)]}{k^2+\kappa^2}\,|\sum_{i=1,N} q_i \exp\,(\mathrm{i}\boldsymbol{k}.\boldsymbol{r}_i)\,|^2$$

$$+ \frac{1}{2}\{-\frac{2\alpha}{\sqrt{\pi}}\exp(-\kappa^2/(4\alpha^2)) + \kappa\mathrm{erfc}(\kappa/(2\alpha))$$

$$+ 4\pi\frac{\exp(-\kappa^2/(4\alpha^2)) - 1}{\kappa^2\, L^3}\}\sum_{i=1,N} q_i^2\,, \qquad (13.23)$$

where α is a convergence parameter similar to that in the Ewald expression of E_c [14].

13.3 Confined Systems

When the systems of charged or dipolar particles are confined on a 2-D surface, the two methods discussed above for the 3-D systems can be used to take into account correctly the long range contribution of the Coulomb interactions with minor modifications of the expressions of E_c given in (13.3) [15,16].

The case of systems confined in volumes finite along one of the coordinates has been considered by using the Ewald [17–23] and hypersphere methods [24,25]. In a volume of square base $L \times L$ having a finite extension h along the axis z and periodic boundary conditions in the two directions x and y, the coulomb energy of N charges is:

$$E_{2D} = \frac{1}{2}\sum_{i,j=1}^{N}{}'\sum_{n\in Z^2} \frac{q_i\, q_j}{|\boldsymbol{r}_j - \boldsymbol{r}_i + \boldsymbol{n}L|}\,, \qquad (13.24)$$

where the vectors $\boldsymbol{n}$ have only x and y components.

A procedure similar to that used for the systems with 3-D periodic boundary conditions (spliting of the $1/r$ potential and using a recipe to perform the sum on the two dimensionnal lattice of the replica cells), gives

$$E_{2D} = \frac{1}{2} \sum_{i,j=1}^{N} \sum_{n \in Z^2} {}' q_i \, q_j \frac{\mathrm{erfc}(\alpha|\boldsymbol{t}_{ij} + \boldsymbol{n}L, z_{ij}|)}{|\boldsymbol{t}_{ij} + \boldsymbol{n}L, z_{ij}|}$$

$$+ \frac{\pi}{2L^2} \sum_{i,j=1}^{N} q_i \, q_j \sum_{\boldsymbol{k} \neq 0} \frac{e^{i\boldsymbol{k}.\boldsymbol{t}_{ij}}}{k} [\exp(k z_{ij}) \, \mathrm{erfc}(\frac{k}{2\alpha} + \alpha z_{ij})$$

$$+ \exp(-k z_{ij}) \, \mathrm{erfc}(\frac{k}{2\alpha} - \alpha z_{ij})]$$

$$- \frac{\alpha}{\sqrt{\pi}} \sum_i q_i^2 - \frac{\sqrt{\pi}}{L^2} \sum_{i,j=1}^{N} q_i \, q_j [\frac{1}{\alpha} \exp(-\alpha^2 z_{ij}^2) + \sqrt{\pi} z_{ij} \mathrm{erfc}(\alpha z_{ij})] \tag{13.25}$$

where $\boldsymbol{t}_{ij}$ is a 2-D vector with components equal to $x_i - x_j$ and $y_i - y_j$ and $|\boldsymbol{t}_{ij} + \boldsymbol{n}L, z_{ij}|$ the distance between the particle i and the particle j or one of its replicas.

The computation of the real space term can be reduced, by an adequate choice of the parameter α, to that of the term $\boldsymbol{n} = 0$. However the term in the k-space involves a sum on the pair of charges that it is not possible to rewrite without new approximations as a sum of order N as it is made for the 3-D systems with periodic boundary conditions identical in the three spatial directions (cf. Eq. (13.3)).

On the surface of a 4-D hypersphere, a 3-D volume similar to a volume of finite extension in the direction z with periodic boundary conditions in the direction x and y, is a domain comprised between two parallel "planes" located symmetrically on the opposite sides of the "equator" of the hyperphere. In this confined domain, the charged particles interact by the pair potential given in (13.14). As in the 3-D case, the computational efficiency of the hypersphere method is obvious compared to the complexity of the "Ewald" expression in (13.25). The validity of the ' hypershere" approach for the computation of the properties of confined systems of charged molecules has been carefully checked by A. Delville and his co-workers [24,25].

Several approximation schemes have been developped to overcome the computational complexity of (13.25). Such a sheme has been proposed by Hautman and Klein [26], it starts with the identity

$$\frac{1}{r} = (\frac{1}{r} - \sum_{n=0,m} \frac{a_n z^{2n}}{t^{2n+1}}) + \sum_{n=0,m} \frac{a_n z^{2n}}{t^{2n+1}} \tag{13.26}$$

where $\boldsymbol{r} = \{x, y, z\} = \{\boldsymbol{t}, z\}$ and a_n are the coefficients of the Taylor expansion of $1/r$ in powers of z/t. $1/t^{2n+1}$ is rewritten as:

$$\frac{1}{t^{2n+1}} = \frac{h_n(t,\alpha)}{t^{2n+1}} + \frac{1 - h_n(t,\alpha)}{t^{2n+1}} \tag{13.27}$$

where $h_n(t,\alpha)/t^{2n+1} = \nabla^{2n}(h_0(t,\alpha)/t)/(a_n 2n!)$ and $h_0(t,\alpha) = \mathrm{erf}(t/\alpha)$ a function which for an adequate choice of α and increasing t goes rapidly to 1.

By subsituting these two identities in (13.24) and using the "Ewald" procedure, it is possible to obtain an expression of E_{2D} similar to that found in the case of 3-D periodic boundary conditions: $E_{2D} = V_S + V_L$ where V_S is a term, computed in r-space, taking into account the contribution of the interactions at short distance and V_S, computed in k-space, taking into account the contribution of the interactions at long distances:

$$V_L = \frac{\pi}{L^2} \sum_{i,j=1}^{N} q_i\, q_j \sum_{n=0,m} a_n z_{ij}^{2n} \sum_{k \neq 0} \frac{f_n(k,\alpha)}{k^{2n-1}} e^{\mathrm{i}\boldsymbol{k}.\boldsymbol{t}_{ij}}$$

$$+ \frac{1}{2} \sum_{i,j=1}^{N} q_i\, q_j \sum_{n=0,m} a_n z_{ij}^{2n} C_n(\alpha) + (\text{constant terms}) \tag{13.28}$$

with, for $k \neq 0$

$$f_n(k,\alpha) = \frac{1}{k^{1-2n}} \frac{1}{2\pi} \int \mathrm{d}t\, e^{\mathrm{i}\boldsymbol{k}.\boldsymbol{t}} \frac{h_n(t,\alpha)}{t^{2n+1}} \tag{13.29}$$

and

$$C_n(\alpha) = \frac{1}{L^2} \int \mathrm{d}t\, \frac{h_n(t,\alpha) - \delta_{n,0}}{t^{2n+1}} \,. \tag{13.30}$$

The real space contribution V_S is

$$V_S = \frac{1}{2} \sum_{i \neq j=1}^{N} q_i\, q_j \sum_{n \in Z^2} \frac{1}{|\boldsymbol{r}_j - \boldsymbol{r}_i + \boldsymbol{n}L|} - \sum_{n'=0,m} a_{n'} z_{ij}^{2n'} \frac{h_{n'}(|\boldsymbol{t}_{ij} + \boldsymbol{n}L|, \alpha)}{|\boldsymbol{t}_{ij} + \boldsymbol{n}L|^{2n'+1}}$$

$$+ \quad (\text{constant terms})\,. \tag{13.31}$$

For the computation of V_S, α is chosen so that the sum on n is reduced to its first term. In V_L it is clear that this expression can be rewritten by the inversion of the order of the sums on i, j, n and k in such a way that the sum on the charge pairs is obtained by computing products: $|\sum_i q_i z_i^n \exp(\mathrm{i}\boldsymbol{k}.\boldsymbol{t}_i) \cdot$ $\cdot \sum_j q_j z_j^m \exp(-\mathrm{i}\boldsymbol{k}.\boldsymbol{t}_j)|$ where n and m are powers of z_i or z_j. The calculation of E_{2D} with these expressions of V_S and V_L, when the typical value of the ratios $z_{ij}/t_{ij} \sim h/L$ is small, can be made by using $m \leq 3$ and need a computation time similar to that of E_c for a 3-D systems. A very accurate estimate of E_{2D} needs to take into account in the developments in power of z_{ij} of (13.28) and (13.31) the terms up to the order of z_{ij}^6 [27].

Leckner and Grønbeck-Jensen [28,29] have used a different method to supress the sum on the charge pairs in E_{2D}. In this method, the series associated to the electric field between two charges derived from (13.24) is rewritten by using an integral representation of $1/|\boldsymbol{r}_j - \boldsymbol{r}_i + \boldsymbol{n}L|^3$. The integration of this new expression of the electric field gives the potential V_{ij} between two charges i and j located, for instance, in a slab of sides L_x, L_y and L_z with periodic boundary conditions in the x and y spatial directions:

$$V_{ij}(x_{ij}, y_{ij}, z_{ij}) = 4\frac{q_i q_j}{L_y} \sum_{m=1,\infty} \cos(2\pi m \frac{y_{ij}}{L_y})$$

$$\sum_{k=-\infty,\infty} K_0[2\pi m \{(\frac{L_x}{L_y})^2 (\frac{x_{ij}}{L_x} + k)^2 + (\frac{z_{ij}}{L_y})^2\}^{1/2}]$$

$$- \frac{q_i q_j}{L_y} \ln[\cosh(2\pi \frac{z_{ij}}{L_x}) - \cos(2\pi \frac{x_{ij}}{L_x})] - \frac{q_i q_j}{L_y} \ln 2 \,. \quad (13.32)$$

at this potential must be added constant contributions v_i and v_j which represent the interaction between the charges i and j with their periodic replicas. v_i is given by

$$v_i = \frac{q_i^2}{L_y}[4 \sum_{m=1,\infty} \sum_{k=1,\infty} K_0(2\pi m k \frac{L_x}{L_y}) + C + \ln(\frac{1}{4\pi}\frac{L_x}{L_y})] \,. \quad (13.33)$$

where C is the Euler constant. There is an equivalent expression of V_{ij} where the coordinates x_{ij} and y_{ij} are exchanged. However in order to be accurate for all possible values of the coordinate of x_{ij}, y_{ij} and z_{ij}, these expressions of V_{ij} must be computed by using $m \sim 30 - 100$ and $k \simeq 3$ [30]. These constraints are not favorable to the computational efficiency of the method.

A very simple method for the simulations of charged systems, in a slab of thickness h in the z direction and sides L_x and L_y in x and y directions with L_x and $L_y >> h$, is to locate the slab to the centre of a simulation cell having a dimension L_z much larger than the lateral dimensions L_x and L_y and obviously h. This simulation cell is considered to have periodic boundary conditions in the three directions of space. The charges are located between the slab surfaces perpendicular to the z-axis and to the z coordinates $z = \pm h/2$. Since $L_z >> L_x$ and L_y, the interactions between charges located in the replica of the slab are negligible and E_{2D} can be accurately computed from the expression of E_c, easily generalised to the case of a parallelepipedic simulation cell [31–33].

13.4 Conclusion

In order to preserve, the physical specific properties of systems of charged and/or polar atoms or molecules induced by the long range of the charge-charge, charge-dipole or dipole-dipole interactions, it is essential in numerical simulations realized in volumes with periodic boundary conditions to use an Ewald expression of the energy. An efficient and accurate substitute to this method is the hyperspere method described above which, however, is difficult to use for systems of complex molecules due to geometrical constraints induced by curvature effects.

In confined systems, at the notable exception of the hypersphere method, a very accurate estimate of E_{2D} stays difficult to achieve [34–36]. However it has been recently established [27] that ,as it is expexted, all methods mentionned here when they are carefully implemented give identical results, then the choice between these methods must be made on criterions such accuracy, computational efficiency and fast convergence to thermodynamic limits.

References

1. P. Ewald, Ann. Phys. **64**, 253 (1921)
2. S.W. de Leeuw, J.W. Perram, E.R. Smith: Proc. R. Soc. Lond. A **373**, 27 (1980)
3. M.P. Allen, D.J. Tildesley: *Computer Simulation of liquids* (Clarendon, Oxford 1987)
4. J.M. Caillol: J. Chem. Phys. **101**, 6080 (1994)
5. C. Sagui, T.A. Darden: Annu. Rev. Biophys. Biomol. Struct. **28**, 155 (1999)
6. C. Sagui, T.A. Darden: J. Chem. Phys. **114**, 6578 (2001)
7. M. Desermo, C. Holm: J. Chem. Phys. **109**, 7678 (1998)
8. M. Challacombe, C. White, M. Head-Gordon: J. Chem. Phys. **107**, 10131 (1997)
9. Zhong-Hui Duan, R. Krasny: J. Chem. Phys. **113**, 3492 (2000)
10. S.G. Brush, H.L. Sahlin, E. Teller: J. Chem. Phys. **45**, 2102 (1996)
11. J.M. Caillol: J. Chem. Phys. **96**, 1455 (1992)
12. J.M. Caillol, D. Levesque: J. Chem. Phys. **94**, 597 (1991)
13. J.M. Caillol, D. Levesque: J. Chem. Phys. **96**, 1477 (1992)
14. G. Salin, J.M. Caillol: J. Chem. Phys. **113**, 10459 (2000)
15. H. Totsuji: Phys. Rev. A bf 17, 2827 (1979)
16. G. T. Gao, X.C. Zeng, W. Wang: J. Chem. Phys. **106**, 3311 (1997)
17. D. Parry, Surf. Sci. **49**, 433 (1975)
18. S.Y. Liem, J.H.R. Clarke: Mol. Phys. **92** , 19 (1997)
19. S.W. de Leeuw, J.W. Perram: Mol. Phys. **37**, 1313 (1979)
20. E.R. Smith: Mol. Phys. **65**, 1089 (1988)
21. Y.J. Rhee, J.W. Halley, J. Hautman, A. Rahman: Phys. Rev. B **40**, 36 (1989)
22. A.T. Clark, T.J. Madden and P.W. Warren: Mol. Phys. **87**, 1063 (1996)
23. A. Grzybowski, E. Gwóźdź, A. Bródka: Phys. Rev. B **61**, 6706 (2000)
24. A. Delville, R.J.M. Pellenq, J.M. Caillol: J. Chem. Phys. **106**, 7275 (1997)
25. R.J.M. Pellenq, J.M. Caillol, A. Delville: J. Phys. Chem. B **101**, 8584 (1997)
26. J. Hautman, M.L. Klein: Mol. Phys. **87**, 379 (1992)

27. M.Mazars, J.M.Caillol, J.J.Weis, D.Levesque: Cond. Matter Phys. **4**, 697 (2001)
28. J. Lekner: Physica A **157**, 826 (1989); **176**, 485 (1991)
29. N. Grønbech-Jensen, G. Hummer, K.M. Beardmore: Mol. Phys. **92**, 941 (1997)
30. M. Mazars: J. Chem. Phys. **115**, 2955 (2001)
31. J.C. Shelley, G. N. Patey: Mol. Phys. **88**, 385 (1996)
32. E. Spohr: J. Chem. Phys. **107**, 6342 (1997)
33. In-Chui Yeh, M.L. Berkowitz: J. Chem. Phys. **111**, 3155 (1999)
34. A. Arnold, C. Holm: cond-mat/0202265
35. A. Arnold, J. de Joannis, C. Holm: cond-mat/0202399
36. J. de Joannis, A. Arnold, C. Holm: cond-mat/0202400

Part IX

Perpectives in *ab initio* MD

14 New Developments in Plane-Wave Based *ab initio* Calculations

Glenn J. Martyna[1] and Mark E. Tuckerman[2]

[1] IBM TJ Watson Lab, PO Box 218, Yorktown Heights, NY 10598
[2] Chemistry Department and Courant Institute, New York University, NY, NY 10003

Abstract. New developments in plane-wave based ab initio calculations are reviewed. First, a general framework for electrostatic calculations is presented that allows 3-dimensional periodic systems (solids, liquids), 2-dimensional periodic systems (surfaces), and non-periodic systems (clusters) to be treated properly and easily (*J. Chem. Phys.* **110**, 2810-2821 (1999) and *J. Chem. Phys.* **115**, 3531 (2001)). Second, a general method for describing systems in which the electrons are assumed to be localized in a single small region of space embedded within a large chemically inert bath, is discussed (*J. Chem. Phys.* **116** 5351 (2002)) that significantly enhances the ability of plane-wave based techniques to study reactions in biological systems.

14.1 Introduction

Plane-wave based *ab initio* molecular dynamics simulation studies have been successfully employed to gain insight into many different types of physicals problems [1–6]. Indeed due to *ab initio* molecular dynamics' unique ability to treat bond breaking and bond making, new areas have become available for detailed atomic-level examination. These areas span a wide range from enzymes catalysis, to surface catalysis, to geochemical systems exemplified by minerals under high temperature and pressure, among many others [1–6]. Much of the recent progress in *ab initio* molecular dynamics applications has been spurred by methodological advances as well as powerful parallel computers[7] whose speed can be expected to increase exponentially following Moore's law.

In this article, two recent advances in *ab initio* molecular dynamics technology are reviewed. The first involves generalizing the treatment of periodic boundary conditions with the plane wave basis so that surfaces, wires, and clusters can be studied as efficiently as fluids and solids[8,9]. Thus, transport phenomena in surface catalysis and molecular beam studies, can be modeled without loss of generality. Second, mixed *ab initio*/empirical force-field simulation studies, calculations in which one part of the system is treated using a fully *ab initio* description and another part is treated using an empirical description, have become an important tool to model biological processes such as enzyme catalysis[11]. Therefore, it is shown how to extend plane-wave based *ab initio* methods to model systems in which the electrons are assumed to be localized in a single spatial area, the active site, within a large chemically inert bath, the rest of the enzyme and the surrounding solvent (water) with high efficiency (in order N log N)[12].

The article is organized as follows: A methods section is presented in which the formulae for the two techniques are described. Next, a results section is given wherein the formulae are employed to generate results on both model and realistic systems. The realistic systems include the proton transport at an ice surface and the HCA II enzyme solvated in liquid water.

14.2 Methods

In this section, a generalized reciprocal space based treatment of long range interaction at clusters[8], on surfaces[9] and in solids/liquids, appropriate for use with plane-wave basis sets and Ewald summation is presented. Given the generalized expressions for the long range forces, the techniques required to model systems in which the electrons are assumed to be localized in a single spatial area within a bath using a plane wave basis set are described.

14.2.1 Clusters, Surfaces and Solids/Liquids

Consider a probability density, $n(\mathbf{r})$ enclosed by a three dimensional parallelepiped. Such a function can be expanded in a plane-wave basis set without loss of generality,

$$n(\mathbf{r}) = \frac{1}{V}\sum_{\hat{\mathbf{g}}} \bar{n}(\mathbf{g})\exp(i\mathbf{g}\cdot\mathbf{r}) \tag{14.1}$$

$$\bar{n}(\mathbf{g}) = \int_{D(\mathbf{h})} d\mathbf{r}\, \exp(-i\mathbf{g}\cdot\mathbf{r})n(\mathbf{r}) = V\int_{-\frac{1}{2}}^{\frac{1}{2}}\int_{-\frac{1}{2}}^{\frac{1}{2}}\int_{-\frac{1}{2}}^{\frac{1}{2}} ds_a ds_b ds_c \exp(i2\pi\hat{\mathbf{g}}\cdot\mathbf{s})n(\mathbf{hs})$$

where the domain of integration is restricted to the volume defined by the parallelepiped denoted by $D(\mathbf{h})$. Here, $\mathbf{h}$ is the matrix whose columns are the Cartesian components of the three vectors describing the parallelepiped, $\{\mathbf{a},\mathbf{b},\mathbf{c}\}$, $V = \det(\mathbf{h})$ is the volume, $\hat{\mathbf{g}}$ is the vector of integers $\{\hat{g}_a, \hat{g}_b, \hat{g}_c\}$, $\mathbf{g} = 2\pi\hat{\mathbf{g}}\mathbf{h}^{-1}$ is the reciprocal lattice vector, and $\mathbf{r} = \mathbf{hs}$.

The above result holds in a cluster, along a wire, on a surface or in a solid/liquid. That is, for a cluster, $n(\mathbf{r})$ is assumed to vanish everywhere on the boundary of the parallelepided, while in liquid/solid, the simply density matches on the boundary. For a surface (wire), the density matches on four (two) of the six faces of the parallelepiped and vanishes on the other two (four). Thus, in order to study surfaces, it is useful to examine the where the **a** and **b** axes are taken to lie in the x-y plane and the **c** axis is taken to lie on the z-axis, i.e. the cell angles are taken to be $\alpha = \beta = 90, \gamma \neq 90$. This implies a monoclinic cell of the form

$$\mathbf{h} = \begin{pmatrix} h_{11} & h_{12} & 0 \\ h_{21} & h_{22} & 0 \\ 0 & 0 & L_c \end{pmatrix} = \begin{pmatrix} \mathbf{h}_s & \mathbf{0} \\ \mathbf{0} & L_c \end{pmatrix} \quad , \tag{14.2}$$

reciprocal lattice vectors of the form, $\mathbf{g} = \{\mathbf{g}_s, g_c\}$ with $g_c = 2\pi\hat{g}_c/L_c$, $\mathbf{g}_s = 2\pi\hat{\mathbf{g}}_s\mathbf{h}_s^{-1}$,$\hat{\mathbf{g}}_s = \{\hat{g}_a, \hat{g}_b\}$, and real space vectors of the form $\mathbf{r} = \{\mathbf{r}_s, z\}$ with $\mathbf{r}_s = \{x, y\}$. Surfaces are accurately described if the **c**-boundary height, L_c, is taken large enough that the density vanishes on the **c**-boundary i.e. $L_c > L_t$ where L_t is the surface thickness. Similarly, for a cluster it is useful to consider an orthorhombic cell, $\alpha = \beta = \gamma = 90$ where the **a**,**b** and **c** boundary heights L_a, L_b, and L_c,

$$\mathbf{h} = \begin{pmatrix} L_a & 0 & 0 \\ 0 & L_b & 0 \\ 0 & 0 & L_c \end{pmatrix} \tag{14.3}$$

respectively, are large enough to enclose the density as described above.

Next, consider a system with density $n(\mathbf{r})$ subject to scalar potential $\phi(\mathbf{r})$. The average potential energy taken over the density is

$$\langle\phi\rangle = \frac{1}{2}\int_{D(\mathbf{h})} d\mathbf{r} \int_{D(\mathbf{h})} d\mathbf{r}'\; n(\mathbf{r})n(\mathbf{r}') \sum_{\hat{s}_a,\hat{s}_b,\hat{s}_c} \phi(\mathbf{r} - \mathbf{r}' + \mathbf{h}\hat{\mathbf{S}}). \tag{14.4}$$

For a liquid or solid, the sum is over the infinite array of periodic images along each parallelepiped axis, **a**, **b** and **c** and $\hat{\mathbf{S}} = \{\hat{s}_a, \hat{s}_b, \hat{s}_c\}$ is a vector of integers while for a cluster, $\hat{\mathbf{S}}$ is restricted to the point, $\{0, 0, 0\}$, only (no images!). In contrast, for a surface, $\hat{\mathbf{S}} = \{\hat{s}_a, \hat{s}_b, 0\}$ is taken; there are no images along **c**.

14.2.1.1 Solids/Liquids

In solids and liquids, the standard case, the Poisson summation formula

$$\sum_{\hat{s}_a,\hat{s}_b,\hat{s}_b} F(\mathbf{hS}) = \left[\frac{1}{\det \mathbf{h}}\right] \sum_{\hat{\mathbf{g}}''} G(\hat{\mathbf{g}}''\mathbf{h}^{-1}) \tag{14.5}$$

$$G(\hat{\mathbf{g}}''\mathbf{h}^{-1}) = \int_{-\infty}^{\infty} d\mathbf{r}'' F(\mathbf{r}'') \exp[-i2\pi\hat{\mathbf{g}}''\mathbf{h}^{-1}\mathbf{r}''] = \tilde{F}(\mathbf{g})$$

can be directly inserted into the energy[13] and the result simplified to yield

$$\langle\phi\rangle = \left[\frac{1}{2\det \mathbf{h}}\right] \sum_{\hat{\mathbf{g}}} \int_{D(\mathbf{h})} d\mathbf{r} \int_{D(\mathbf{h})} d\mathbf{r}'\; n(\mathbf{r})n(\mathbf{r}')\tilde{\phi}(\mathbf{g}) \exp(i\mathbf{g}\cdot[\mathbf{r} - \mathbf{r}'])$$

$$= \left[\frac{1}{2\det \mathbf{h}}\right] \sum_{\hat{\mathbf{g}}\neq 0} |\bar{n}(\mathbf{g})|^2 \tilde{\phi}(\mathbf{g}) \tag{14.6}$$

Note, the definition, $\mathbf{g} = 2\pi\hat{\mathbf{g}}\mathbf{h}^{-1}$, given above, has been employed and $\tilde{\phi}(\mathbf{g})$ is the three dimensional Fourier transform of the scalar potential. Finally, the point, $\mathbf{g} = \{0, 0, 0\}$, has been eliminated from sum. This term, described in detail in [14,13], depends on the boundary conditions taken at infinity for long range potentials. If the potential, $\phi(\mathbf{r})$, is Coulomb's law (the usual case), then neglecting $\mathbf{g} = 0$ is equilivalent to surrounding the infinite array of periodic images by a sphere of metal.

14.2.1.2 Clusters

In order to generalize, naturally, the formalism to treat clusters, the "1st periodic image" or "nearest periodic image" form of the potential energy is introduced. The 1st image form of the scalar potential, $\phi^{(1)}(\mathbf{r} - \mathbf{r}')$, can be expressed as $\phi(\mathbf{r} - \mathbf{h}\hat{\mathbf{S}}^{(1)})$ where the three elements of the 1st image, $\hat{\mathbf{S}}^{(1)} = \{\hat{S}_a, \hat{S}_b, \hat{S}_c\}$ with $\hat{S}_\alpha^{(1)}$, are chosen to minimize the distance along each cell axis (given in $\mathbf{h}$). This is equivalent to forcing the scalar potential to assume the periodicity of cell. The first image form of the potential energy exactly reduces to the energy of cluster given in (14.4) **if** the axis of the orthorhombic parallelepiped enclosing the cluster, $\{L_a, L_b, L_c\}$, are taken to be at least twice the diameter of the cluster along each of the three directions rather than simply large enough for the density to vanish on the surface of the parallelepiped. In this way, the nearest image between points in the cell **with non-zero density** is $\hat{\mathbf{S}}^{(1)} \equiv 0$ as stated below Eq. (14.4).

It is, now, a straightforward to write a simple expression for the energy of cluster in reciprocal space which only assumes that simulation parallelepiped is taken suffiently large,

$$
\begin{aligned}
\langle \phi^{(1)} \rangle &= \frac{1}{2} \int_{D(\mathbf{h})} d\mathbf{r} \int_{D(\mathbf{h})} d\mathbf{r}'\, n(\mathbf{r}) n(\mathbf{r}') \phi^{(1)}(\mathbf{r} - \mathbf{r}'). \\
&= \left[\frac{1}{2 \det \mathbf{h}}\right] \int_{D(\mathbf{h})} d\mathbf{r} \int_{D(\mathbf{h})} d\mathbf{r}'\, n(\mathbf{r}) n(\mathbf{r}') \sum_{\hat{\mathbf{g}}} \bar{\phi}(\mathbf{g}) \exp(i\mathbf{g} \cdot (\mathbf{r} - \mathbf{r}')) \\
&= \left[\frac{1}{2 \det \mathbf{h}}\right] \sum_{\hat{\mathbf{g}}} |\bar{n}(\mathbf{g})|^2 \bar{\phi}(\mathbf{g})
\end{aligned}
\tag{14.7}
$$

that the first image reduces to the cluster energy. Here, the fact that $\phi^{(1)}(\mathbf{r})$ is a periodic function has been used by expressing the potential in terms of its plane-wave or Fourier Series expansion,

$$
\bar{\phi}(\mathbf{g}) = \int_{D(\mathbf{h})} d\mathbf{r}\, \exp(-i\mathbf{g} \cdot \mathbf{r}) \phi(\mathbf{r}) \tag{14.8}
$$

where $\phi(\mathbf{r})$ is the scalar potential of interest. This is in contrast to the result for solids/liquids which contains the Fourier transform of the potential.

Given the similarity between the expressions for long range interactions in cluster and solids/liquids, it is natural to accentuate these similarities and rewrite the cluster energy as

$$\langle \phi^{(1)} \rangle = \left[\frac{1}{2 \det \mathbf{h}} \right] \sum_{\hat{\mathbf{g}} \neq 0} |\bar{n}(\mathbf{g})|^2 \left[\tilde{\phi}(\mathbf{g}) + \hat{\phi}^{(\mathrm{screen})}(\mathbf{g}) \right] + \left[\frac{1}{2 \det \mathbf{h}} \right] \bar{\phi}(0) |\bar{n}(0)|^2 \tag{14.9}$$

Here, the "screening function",

$$\hat{\phi}^{(\mathrm{screen})}(\mathbf{g}) = \bar{\phi}(\mathbf{g}) - \tilde{\phi}(\mathbf{g}) \tag{14.10}$$

acts to "screen" or cancel the interactions with the infinite array of periodic images that arises from $\tilde{\phi}(\mathbf{g})$ (see (14.6)). Now, the solid/liquid expression naturally emerges by simply setting the screening function to zero and omitting the $\mathbf{g} = 0$ term. Again, (14.9) is only valid if the cell is taken appropriately large as described above (the edges of the rhombohedral cell, $\{L_a, L_b, L_c\}$, at least twice the diameter of the cluster in the corresponding directions).

For the two most important cases, Coulomb's law, $\phi(\mathbf{r}) = 1/|\mathbf{r}|$ and "softened" Coloumb's law, $\phi(\mathbf{r}) = \mathrm{erf}(\alpha_{ewd}|\mathbf{r}|)/|\mathbf{r}|$, the Fourier Transforms can be determined analytically, $\tilde{\phi}(\mathbf{g}) = 4\pi/|\mathbf{g}|^2$ and $\tilde{\phi}(\mathbf{g}) = 4\pi \exp[-|\mathbf{g}|^2/4\alpha_{ewd}^2]/|\mathbf{g}|^2$, respectively. However, the Fourier series expansion must be evaluated numerically. This can easily be accomplished by writing

$$\begin{aligned} \bar{\phi}(\mathbf{g}) &= 4\pi \int_0^{r_c} dr r^2 \phi(r) \kappa(r, r_c, \lambda) \left[\frac{\sin(gr)}{gr} \right] \\ &\quad + \int_{D(\mathbf{h})} d\mathbf{r} \, \exp(-i\mathbf{g} \cdot \mathbf{r}) \phi(\mathbf{r}) [1 - \kappa(r, r_c, \lambda)] \\ &= \chi_1(\mathbf{g}) + \chi_2(\mathbf{g}) \end{aligned} \tag{14.11}$$

where a cutoff radius, $r_c = (1/2)\mathrm{MIN}(L_a, L_b, L_c)$, and a smooth switching function $\kappa(r, r_c, \lambda)$ which is unity for $r \leq r_c - \lambda$ and zero for $r \geq r_c$ has been employed to divide the integral into two parts. The first integral, $\chi_1(\mathbf{g})$ is one dimensional (radial) and can be evaluated, numerically to high accuracy, quickly and accurately. The second integral, denoted $\chi_2(\mathbf{g})$, is three dimensional but can be evaluated accurately and efficiently using a 3-Dimensional

Fast Fourier Transform,

$$\begin{aligned}\chi_2(\mathbf{g}) &= \int_{D(\mathbf{h})} d\mathbf{r}\, \exp(-i\mathbf{g}\cdot\mathbf{r})\phi(\mathbf{r})[1-\kappa(r,r_c,\lambda)] \\ &= V \int_{-\frac{1}{2}}^{\frac{1}{2}}\int_{-\frac{1}{2}}^{\frac{1}{2}}\int_{-\frac{1}{2}}^{\frac{1}{2}} ds_a ds_b ds_c \exp(-i2\pi\hat{\mathbf{g}}\cdot\mathbf{s})\phi^{(\mathrm{long})}(\mathbf{hs}) \\ &\approx \left[\frac{V(-1)^{j+k+m}}{JKM}\right]\left\{\sum_{u=0}^{J-1}\sum_{v=0}^{K-1}\sum_{w=0}^{M-1} \exp(-i2\pi\hat{\mathbf{g}}\cdot\hat{\mathbf{s}})\phi^{(\mathrm{long})}(\mathbf{h}[\hat{\mathbf{s}}+\mathbf{a}])\right\}\end{aligned} \tag{14.12}$$

where $\phi^{(\mathrm{long})}(\mathbf{r}) = \phi(\mathbf{r})[1-\kappa(r,r_c,\lambda)]$, $\{J,K,M\}$ is the number of equally spaced grid points assigned to descretize $\{s_a,s_b,s_c\}$, respectively, $\hat{\mathbf{s}} = \{u/J, v/K, w/M\}$, $\mathbf{a} = \{-1/2,-1/2,-1/2\}$, and $\hat{\mathbf{g}} = \{j,k,m\}$ are reciprocal lattice indices. That is, the term in the curly brackets is the unscaled 3-Dimensional Fast Fourier Transform of the function, $\phi^{(\mathrm{long})}(\mathbf{h}[\hat{\mathbf{s}}+\mathbf{a}])$. In a realistic calculation, the screening function is evaluated once, at the beginning of a run and stored. That is, the required $\mathbf{g}$ are discrete and the upper limit of reciprocal space sums are truncated in the usual way.

14.2.1.3 Surfaces

From the above discussion, it is clear that the long range interactions on surfaces can be placed on a similar footing. First,, the nearest image form of the potential energy can be invoked along the $\mathbf{c}$-direction so that $\phi^{(1,\mathbf{c})}(\mathbf{r}) = \phi(\mathbf{r}+S_c^{(1)}\mathbf{c})$ with $S_c^{(1)}$ chosen to minimize the distance along $\mathbf{c}$-axis of the monoclinic cell. That is, for surfaces, the $\mathbf{c}$-axis is along z-direction and $\phi^{(1,\mathbf{c})}(\mathbf{r})$ is a periodic function of z. Second, the infinite sum over images along the $\mathbf{a},\mathbf{b}$ axis of the monoclinic cell, can be treated using Poisson summation.

Following Ref.[9], the two dimensional generalization of the Poisson summation formula,

$$\sum_{\hat{s}_a,\hat{s}_b} F(\mathbf{h}_s\mathbf{S}_s) = [\det\mathbf{h}_s]^{-1}\sum_{\hat{\mathbf{g}}_s''} G(\hat{\mathbf{g}}_s''\mathbf{h}_s^{-1}) \tag{14.13}$$

$$G(\hat{\mathbf{g}}_s''\mathbf{h}_s^{-1}) = \int_{-\infty}^{\infty} dx'' \int_{-\infty}^{\infty} dy'' F(\mathbf{r}_s'')\exp[-i2\pi\hat{\mathbf{g}}_s''\mathbf{h}_s^{-1}\mathbf{r}_s'']$$

can then be combined with (14.4) to yield

$$\langle\phi\rangle = \frac{1}{2\det(\mathbf{h}_s)}\int_{D(\mathbf{h})} d\mathbf{r}\int_{D(\mathbf{h})} d\mathbf{r}'\; n(\mathbf{r})n(\mathbf{r}')\sum_{\hat{\mathbf{g}}_s}\exp[i\mathbf{g}_s\cdot(\mathbf{r}_s-\mathbf{r}'_s)]\tilde{\phi}_s(\mathbf{g}_s;z-z')$$

$$\tilde{\phi}_s\left(\mathbf{g}_s;z''\right) = \int_{-\infty}^{\infty} dx'' \int_{-\infty}^{\infty} dy''\; \phi(x''\mathbf{i}+y''\mathbf{j}+z''\mathbf{k})\exp(-i\mathbf{g}_s\cdot\mathbf{r}''_s) \quad (14.14)$$

Here, $\hat{\mathbf{g}}_s$ is the vector of integers, $\{\hat{g}_a,\hat{g}_b\}$, $\mathbf{g}_s = 2\pi\hat{\mathbf{g}}_s\mathbf{h}_s^{-1}$ is the in-plane reciprocal lattice vector, $\mathbf{S}_s = \{\hat{s}_a,\hat{s}_b\}$, $\mathbf{r}_s = \{x,y\}$, $\tilde{\phi}_s(\mathbf{g}_s;z'')$ is the two dimensional Fourier Transform of the potential at fixed z'', and $\mathbf{i}$, $\mathbf{j}$, and $\mathbf{k}$ are the unit vectors in the x, y, and z directions, respectively. In the limit the $\mathbf{c}$-boundary height, L_c, is taken sufficiently large compared to the surface thickness, L_t, the first or nearest image form of the potential energy function along the $\mathbf{c}$-direction is invoked. This form is valid if $L_c > 2L_t$ for discrete density functions (point particles) and $L_c > \sqrt{2}L_t$ for smoothly decaying density functions. The first image form is generated by introducing the plane-wave expansion of the $(z-z')$ dependence of the function[8],

$$\chi(\mathbf{g}_s;z-z') \equiv \frac{1}{\det(\mathbf{h}_s)}\sum_{\hat{\mathbf{g}}_s}\exp[i\mathbf{g}_s\cdot(\mathbf{r}_s-\mathbf{r}'_s)]\tilde{\phi}_s(\mathbf{g}_s;z-z') \quad (14.15)$$

to yield,

$$\langle\phi^{(1)}\rangle = \frac{1}{2V}\sum_{\hat{\mathbf{g}}_s,\hat{g}_c}\int_{D(\mathbf{h})} d\mathbf{r}\int_{D(\mathbf{h})} d\mathbf{r}'\; n(\mathbf{r})n(\mathbf{r}')\exp[i\mathbf{g}\cdot(\mathbf{r}-\mathbf{r}')]\hat{\phi}(\mathbf{g}) \quad (14.16)$$

$$\hat{\phi}(\mathbf{g}) = \int_{-L_c/2}^{L_c/2} dz'' \int_{-\infty}^{\infty} dx'' \int_{-\infty}^{\infty} dy''\phi(\mathbf{r}'')\exp(-i\mathbf{r}''\cdot\mathbf{g}),$$

where $g_c = 2\pi\hat{g}_c/L_c$ is the out-of-plane reciprocal lattice vector and the superscript "(1)" explicitly indicates the first image form. The plane wave expansion employed to generate (14.16) is valid because the quantized complex exponentials, $\exp(-ig_c z)$, form a complete orthonormal set. Last, the plane-wave expansion of the density can be inserted into (14.16) and the resulting expression simplified yielding[9]

$$\langle\phi^{(1)}\rangle = \frac{1}{2V}\sum_{\hat{\mathbf{g}}\neq 0}|\bar{n}(\mathbf{g})|^2\hat{\phi}(\mathbf{g}) + \frac{1}{2V}|\bar{n}(0)|^2\hat{\phi}(0)$$

$$= \frac{1}{2V}\sum_{\hat{\mathbf{g}}\neq 0}|\bar{n}(\mathbf{g})|^2\left[\tilde{\phi}(\mathbf{g}) - \hat{\phi}^{(\mathrm{screen})}(\mathbf{g})\right] + \frac{1}{2V}|\bar{n}(0)|^2\hat{\phi}(0) \quad (14.17)$$

For surfaces, convenient closed form expressions for $\hat{\phi}(\mathbf{g})$, can be derived for two important cases, the pure Coulomb potential, $\phi^{(Coul)}(\mathbf{r}) = 1/|\mathbf{r}|$, and the long range part of the Coulomb potential employed in Ewald summation, $\phi^{(Coul,Ewd)}(\mathbf{r}) = \mathrm{erf}(\alpha_{ewd}|\mathbf{r}|)/|\mathbf{r}|$,

$$\hat{\phi}^{(Coul)}(\mathbf{g}) = \frac{4\pi}{g^2} - \frac{4\pi}{g^2}\left\{\cos\left(\frac{g_c L_c}{2}\right)\exp\left(-\frac{g_s L_c}{2}\right)\right\} \tag{14.18}$$

$$\hat{\phi}^{(Coul,Ewd)}(\mathbf{g},\alpha_{ewd}) = \frac{4\pi}{g^2}\exp\left[-\frac{g^2}{4\alpha_{ewd}^2}\right] - \frac{4\pi}{g^2}\left\{\cos\left(\frac{g_c L_c}{2}\right)\right.$$
$$\times\left[\exp\left(-\frac{g_s L_c}{2}\right) - \frac{1}{2}\exp\left(-\frac{g_s L_c}{2}\right)\mathrm{erfc}\left(\frac{\alpha_{ewd}^2 L_c - g_s}{2\alpha_{ewd}}\right)\right.$$
$$\left. - \frac{1}{2}\exp\left(\frac{g_s L_c}{2}\right)\mathrm{erfc}\left(\frac{\alpha_{ewd}^2 L_c + g_s}{2\alpha_{ewd}}\right)\right]$$
$$\left. + \exp\left(-\frac{g^2}{4\alpha_{ewd}^2}\right) Re\left[\mathrm{erfc}\left(\frac{\alpha_{ewd}^2 L_c + i g_c}{2\alpha_{ewd}}\right)\right]\right\}.$$

Here, $g^2 = |\mathbf{g}|^2$ and $g_s^2 = |\mathbf{g}_s|^2$. Note, the expressions in (14.18) can be easily decomposed into the form $\hat{\phi}^{(Coul)}(\mathbf{g}) = \tilde{\phi}^{(Coul)}(\mathbf{g}) + \hat{\phi}^{(\mathrm{screen,Coul})}(\mathbf{g})$ and $\hat{\phi}^{(\mathrm{Coul,Ewd})}(\mathbf{g},\alpha_{ewd}) = \tilde{\phi}^{(\mathrm{Coul,Ewd})}(\mathbf{g},\alpha_{ewd}) + \hat{\phi}^{(\mathrm{screen,Coul})}(\mathbf{g},\alpha_{ewd})$ where the tilde denotes the three dimensional Fourier Transform of the potential.

14.2.1.4 Wires

For the wire case, it will be assumed that the y and z directions are nonperiodic and that the cell matrix corresponds to an orthorhombic cell, i.e., $\mathbf{h}$ is diagonal with $h_{11} = L_a$, $h_{22} = L_b$, and $h_{33} = L_c$. Then, the contribution of long range potential to the screening function becomes:

$$\hat{\phi}(\mathbf{g}) = \int_{-L_c/2}^{L_c/2} dz \int_{-L_b/2}^{L_b/2} dy \int_{-\infty}^{\infty} dx\, \phi(\mathbf{u})e^{-i\mathbf{g}\cdot\mathbf{u}} \tag{14.19}$$

Again, for the Coulomb potential split according to $1/r = \mathrm{erf}(\alpha r)/r + \mathrm{erfc}(\alpha r)/r$, the integrals can be done partially analytically, with the result for Ewald summation

$$\hat{\phi}^{(Coul,Ewd)}(\mathbf{g}) = \frac{4\pi}{g^2}\exp\left[-\frac{g^2}{4\alpha_{ewd}^2}\right]$$
$$-\frac{4\pi}{g^2}\left[\exp\left[-\frac{g^2}{4\alpha_{ewd}^2}\right] E(\alpha_{ewd}, L_b, g_b)E(\alpha_{ewd}, L_c, g_c)\right)$$

$$+ \cos\left(\frac{g_b L_b}{2}\right) \frac{4\sqrt{\pi}}{\alpha_{ewd} L_b} \exp\left(-g_c^2/4\alpha_{ewd}^2\right) I(\alpha_{ewd}, L_b, L_c, g_c)$$

$$+ \cos\left(\frac{g_c L_c}{2}\right) \frac{4\sqrt{\pi}}{\alpha_{ewd} L_c} \exp\left(-g_b^2/4\alpha_{ewd}^2\right) I(\alpha_{ewd}, L_c, L_b, g_b)\Bigg] \tag{14.20}$$

where

$$I(\alpha, L_1, L_2, g) = \int_0^{\alpha L_1/2} dx\, x e^{-g_a^2 L_1^2/16x^2} e^{-x^2} E\left(\frac{2x}{L_1}, L_2, g\right) \tag{14.21}$$

and

$$E(\lambda, L, g) = \text{erf}\left(\frac{\lambda^2 L + ig}{2\lambda}\right) \tag{14.22}$$

The one-dimensional integrals in (14.21) are well suited to be performed by Gaussian quadrature techniques.

In the limit $\alpha_{ewd} \to \infty$, a simplified partially analytical expression is available:

$$\hat{\phi}^{(Coul)} \longrightarrow \frac{4\pi}{g^2} - \frac{4\pi}{g^2}\left[\cos\left(\frac{g_b L_b}{2}\right) \frac{J(g_c, g_a, L_c, L_b)}{L_b}\right.$$

$$\left. + \quad \cos\left(\frac{g_c L_c}{2}\right) \frac{J(g_b, g_a, L_b, L_c)}{L_c}\right] \tag{14.23}$$

where

$$J(g_1, g_2, , L_1, L_2) = \frac{4}{\pi} \int_0^{L_1/2} dx\, e^{ig_1 x} \sqrt{\theta(x, g_2, L_2)} K_1\left(\sqrt{4\theta(x, g_2, L_2)}\right) \tag{14.24}$$

$$\theta(x, g, L) = \frac{g^2 L^2/16}{1 + 4x^2/L^2} \tag{14.25}$$

and $K_1(z)$ is a modified Bessel function. The complete formalism for the wire case will be discussed in detail in a forthcoming publication [10].

14.2.1.5 Summary

In summary, the energy of a solid, surface or cluster takes on the general form

$$\langle \phi \rangle = \frac{1}{2V} \sum_{\hat{\mathbf{g}} \neq 0} |\bar{n}(\mathbf{g})|^2 \hat{\phi}(\mathbf{g}) + \frac{1}{2V} |\bar{n}(0)|^2 \hat{\phi}(0)$$

$$= \frac{1}{2V} \sum_{\hat{\mathbf{g}} \neq 0} |\bar{n}(\mathbf{g})|^2 \left[\tilde{\phi}(\mathbf{g}) - \hat{\phi}^{(\text{screen})}(\mathbf{g})\right] + \frac{1}{2V} |\bar{n}(0)|^2 \hat{\phi}(0) \tag{14.26}$$

in reciprocal space. Care must be taken to properly evaluate the screening function and to ensure that the cell is taken large enough in clusters, wires and surfaces for the nearest image form to converge to desired result.

14.2.1.6 Application to Ewald Summation

The new formalism can be straightforwardly applied to Ewald summation of the Coloumb energy for a system of point particles the total Coulomb energy is

$$\phi_{tot}^{(\mathrm{Coul})}(\mathbf{R}) = \frac{1}{2}\sum_{\hat{\mathbf{S}}}\sum_{ij}{}^{'} \frac{q_i q_j}{|\mathbf{r}_{ij} + \mathbf{h}\hat{\mathbf{S}}|} \tag{14.27}$$

$$\phi_{tot}^{(\mathrm{Coul})}(\mathbf{R}) = \frac{1}{2}\sum_{\hat{\mathbf{S}}}\sum_{ij} q_i q_j \frac{\mathrm{erf}(\alpha_{ewd}|\mathbf{r}_{ij} + \mathbf{h}\hat{\mathbf{S}}|)}{|\mathbf{r}_{ij} + \mathbf{h}\hat{\mathbf{S}}|} - \frac{\alpha_{ewd}}{\pi^{1/2}}\sum_i q_i^2$$
$$+ \frac{1}{2}\sum_{\hat{\mathbf{S}}}\sum_{ij}{}^{'} q_i q_j \frac{\mathrm{erfc}(\alpha_{ewd}|\mathbf{r}_{ij} + \mathbf{h}\hat{\mathbf{S}}|)}{|\mathbf{r}_{ij} + \mathbf{h}\hat{\mathbf{S}}|}$$

Here, the sum over i, j is over all particles in the system, the prime indicates that $i \neq j$ when $\hat{\mathbf{S}} = 0$, $\mathbf{r}_i$ is the position of particle, i, and q_i its charge, and the Ewald decomposition of the Coulomb potential ($1/r = [\mathrm{erf}(\alpha_{ewd} r) + \mathrm{erfc}(\alpha_{ewd} r)]/r$) has been employed [13–15]. Introducing the Fourier expansion coefficient of the particle charge density

$$S(\mathbf{g}) = \sum_k q_k \exp(-i\mathbf{g}\cdot\mathbf{r}_k) \tag{14.28}$$

the potential energy is

$$\phi_{tot}^{(\mathrm{Coul})}(\mathbf{R}) = \frac{1}{2V}\sum_{\hat{\mathbf{g}}\neq 0}\left[|S(\mathbf{g})|^2 W(\mathbf{g})\right] + \frac{1}{2V}\hat{\phi}^{(\mathrm{screen,Coul})}(0)|S(0)|^2$$
$$-\frac{\alpha_{ewd}}{\pi^{1/2}}\sum_i q_i^2 + \frac{1}{2}\sum_{\hat{\mathbf{S}}}\sum_{ij}{}^{'} q_i q_j \frac{\mathrm{erfc}(\alpha_{ewd}|\mathbf{r}_{ij} + \mathbf{h}\hat{\mathbf{S}}|)}{|\mathbf{r}_{ij} + \mathbf{h}\hat{\mathbf{S}}|} \tag{14.29}$$

$$W(\mathbf{g}) = \left[\frac{4\pi}{g^2}\right]\exp\left(-\frac{g^2}{4\alpha_{ewd}^2}\right) + \hat{\phi}^{(\mathrm{screen,Coul})}(\mathbf{g}) \tag{14.30}$$

In the Ewald form, the two contributions to the Coulomb energy, the reciprocal space sum and the real space sum can both be truncated. The former because the kernel, $W(\mathbf{g})$, approaches zero exponentially quickly at large $\mathbf{g}$ and the latter complementary error function, $\mathrm{erfc}(\alpha_{ewd} r)$ approaches zero exponentially quickly at large distance. Thus, the reciprocal space term can be evaluated in order N log N using Particle Mesh Ewald techniques in the standard way[16] where N is the number of particles. The real space term can be summed in order N using link cell methods or link cell techniques combined with Verlet neighbor lists[15]. The subtleties of the efficiency Particle Mesh Ewald evaluation of the reciprocal space term for surfaces are described in detail in Ref.[9].

14.2.1.7 Application to Plane-Wave Based Density Functional Theory

The general expression for the energy, Eq. (14.17) can be used in conjunction with plane-wave based density functional theory. In the Kohn-Sham formulation [19,20] of density functional theory, the electron density is expanded in a set of orbitals, $\{\psi_i(\mathbf{r})\}$,

$$n(\mathbf{r}) = \sum_{i=1}^{n_{\rm occ}} f_i |\psi_i(\mathbf{r})|^2 \tag{14.31}$$

where $n_{\rm occ}$ is the number of occupied orbitals and the f_j are the occupation numbers[21]. The energy functional is then given by

$$E[n] = T_s[\{\psi_i\}] + E_{\rm H}[n] + E_{\rm xc}[n] + E_{\rm ext}[n], \tag{14.32}$$

where T_s is the kinetic energy of a system of noninteracting electrons, $E_{\rm H}$ is the Hartree energy, $E_{\rm xc}$ is the exchange and correlation energy [21] and $E_{\rm ext}$, represents the interaction of the electrons with the nuclei. If a plane wave basis is employed to minimize the functional, Eq. (14.32), subject to the orthogonality constraint ($\langle\psi_j|\psi_k\rangle = \delta_{jk}$), atomic pseudopotentials are typically used to eliminate the core electrons. Pseudopotentials divide the external potential, $E_{\rm ext}[n]$, into a long range local part, $E_{\rm loc}[n]$, and a short range, non-local, angular moment-dependent part [22]. Thus, the Hartree and local pseudopotential terms contain the only long-range potential interactions present in (14.32).

In plane-wave based calculations at the Γ-point, the orbitals and, hence, the density are expanded according to[2,6]

$$\psi_i(\mathbf{r}) = \frac{1}{\sqrt{V}} \sum_{\mathbf{g}} \bar{\psi}_i(\mathbf{g}) \exp(i\mathbf{g}\cdot\mathbf{r})$$

$$n(\mathbf{r}) = \frac{1}{V} \sum_{\mathbf{g}} \bar{n}(\mathbf{g}) \exp(i\mathbf{g}\cdot\mathbf{r}) \tag{14.33}$$

in a finite basis, $E_{cut}^{(\rm orbitals)} = \hbar^2 g_{max}^2/2m_e$ and $E_{cut}^{(\rm density)} = 4E_{cut}^{(\rm orbitals)}$. Expressions for the Hartree and local external energies can be written in a reciprocal space form following (14.17):

$$E_{\rm H} = \frac{e^2}{2V}\sum_{\hat{\mathbf{g}}}{}'|\bar{n}(\mathbf{g})|^2\left[\frac{4\pi}{|\mathbf{g}|^2}+\hat{\phi}^{(\rm screen,Coul)}(\mathbf{g})\right]$$

$$+\left(\frac{e^2}{2V}\right)\hat{\phi}^{(\rm screen,Coul,0)}\bar{n}(0) \qquad (14.34)$$

$$E_{\rm loc}[n] = \frac{1}{V}\sum_{\hat{\mathbf{g}}}{}'\sum_{I=1}^{N}\bar{n}(-\mathbf{g})\exp(-i\mathbf{g}\cdot\mathbf{R}_I)\left[\tilde{\phi}_{{\rm loc},I}(\mathbf{g})-eq_I\hat{\phi}^{(\rm screen,Coul)}(\mathbf{g})\right]$$

$$+\frac{1}{V}\sum_{I=1}^{N}\bar{n}(0)\left[\tilde{\phi}^{(0)}_{{\rm loc},I}-eq_I\hat{\phi}^{(\rm screen,Coul,0)}\right], \qquad (14.35)$$

provided the cell matrix $\mathbf{h}$ is appropriately chosen for the given periodicity. Here, q_I is the charge on the I$^{\rm th}$ nuclei and $\tilde{\phi}_{{\rm loc},I}(\mathbf{g})$ is the Fourier Transform of the short range part of the electron-nuclear pseudopotential with the I$^{\rm th}$ nuclei In addition, the prime indicates that the $\mathbf{g}=0$ term is eliminated and the term, $\tilde{\phi}^{(0)}_{{\rm loc},I}$ is the non-singular (short range) part of the local pseudopotential at $\mathbf{g}=0$. Equations (14.34) and (14.35) demonstrate that the modifications necessary to treat generally long range interactions are small and can be implemented easily.

14.2.2 Dual Length Scale Approach

It is clear that the standard expressions for the Hartree and local external energies given in (14.34) and (14.35), respectively, only possesses a single length scale. A second length scale can be introduced by first rewriting the real space expressions for these two energies using the identity $\mathrm{erf}(\alpha r)+\mathrm{erfc}(\alpha r)=1$,

$$E_{\rm H}[n] = \left\{\frac{e^2}{2}\sum_{\hat{\mathbf{S}}}\int_{D(\mathbf{h})}d\mathbf{r}\int_{D(\mathbf{h})}d\mathbf{r}'\,\frac{n(\mathbf{r})n(\mathbf{r}')\mathrm{erfc}(\alpha|\mathbf{r}-\mathbf{r}'+\mathbf{h}\hat{\mathbf{S}}|)}{|\mathbf{r}-\mathbf{r}'+\mathbf{h}\hat{\mathbf{S}}|}\right\}$$

$$+\left\{\frac{e^2}{2}\sum_{\hat{\mathbf{S}}}\int_{D(\mathbf{h})}d\mathbf{r}\int_{D(\mathbf{h})}d\mathbf{r}'\,\frac{n(\mathbf{r})n(\mathbf{r}')\mathrm{erf}(\alpha|\mathbf{r}-\mathbf{r}'+\mathbf{h}\hat{\mathbf{S}}|)}{|\mathbf{r}-\mathbf{r}'+\mathbf{h}\hat{\mathbf{S}}|}\right\}$$

$$= E_{\rm H}^{(\rm short)}[n]+E_{\rm H}^{(\rm long)}[n] \qquad (14.36)$$

$$E_{\rm loc}[n] = \left\{ \sum_{\hat{\mathbf{S}}} \sum_{I=1}^{N} \int_{D(\mathbf{h})} d\mathbf{r}\ n(\mathbf{r}) \left[\phi_{{\rm loc},I}(\mathbf{r}-\mathbf{R}_I+\mathbf{h}\hat{\mathbf{S}}) + \frac{eq_I {\rm erf}(\alpha|\mathbf{r}-\mathbf{R}_I+\mathbf{h}\hat{\mathbf{S}}|)}{|\mathbf{r}-\mathbf{R}_I+\mathbf{h}\hat{\mathbf{S}}|} \right] \right\}$$

$$- \left\{ \sum_{\hat{\mathbf{S}}} \sum_{I=1}^{N} \int_{D(\mathbf{h})} d\mathbf{r}\ n(\mathbf{r}) \left[\frac{eq_I {\rm erf}(\alpha|\mathbf{r}-\mathbf{R}_I+\mathbf{h}\hat{\mathbf{S}}|)}{|\mathbf{r}-\mathbf{R}_I+\mathbf{h}\hat{\mathbf{S}}|} \right] \right\}$$

$$= E_{\rm loc}^{\rm (short)}[n] + E_{\rm loc}^{\rm (long)}[n]. \tag{14.37}$$

Here, the first term in the curly brackets in each equation is short range while the second term is long range. Note, both $\phi_{{\rm loc},I}(\mathbf{r})$ and $-eq_I{\rm erf}(\alpha r)/r$ approach $-eq_I/r$, assymptotically where q_I is the charge on I^{th} ion core. In the limit $\alpha V^{1/3} >> 1$, the sum over images in the first term of each expression (i.e. the short range parts) can be truncated at the first or nearest image with exponentially small error.

In order to proceed, it will be assumed that the electrons are localized in a particular region of the large cell described by $\mathbf{h}$ which can be enclosed in a small cell, described by $\mathbf{h}_s$, centered at the point, $\mathbf{R}_c$. That is, the orbitals and, hence, electron density are taken to vanish on the surface of $\mathbf{h}_s$. Furthermore, it is assumed, for simplicity, that the a_s, b_s and c_s axes of $\mathbf{h}_s$ are parallel to the a, b and c axes of $\mathbf{h}$ such that $\mathbf{h}^{-1}\mathbf{h}_s = \mathbf{D}$, a diagonal matrix. Thus, we can define,

$$\psi_j(\mathbf{r}_s+\mathbf{R}_c) = \psi_{j,s}(\mathbf{r}_s)$$

$$n(\mathbf{r}_s+\mathbf{R}_c) = n_s(\mathbf{r}_s) \tag{14.38}$$

where, the $\mathbf{r}_s$ span the small cell and can be expressed as $\mathbf{r}_s = \mathbf{h}_s\mathbf{s}$ with $0 \le s_\alpha \le 1$ and, both, $\psi_j(\mathbf{r}) \equiv 0$ and $n(\mathbf{r}) \equiv 0$ for $\mathbf{r}_s = \mathbf{r} - \mathbf{R}_c$ outside the domain of $\mathbf{h}_s$. The orbitals and the electron density can be expanded in a plane wave basis set that spans the small cell, only,

$$\psi_{j,s}(\mathbf{r}_s) = \frac{1}{\sqrt{V_s}} \sum_{\hat{\mathbf{g}}_s} \bar{\psi}_{j,s}(\mathbf{g}_s) \exp(i\mathbf{g}_s\cdot\mathbf{r}_s)$$

$$n_s(\mathbf{r}_s) = \frac{1}{V_s} \sum_{\hat{\mathbf{g}}_s} \bar{n}_s(\mathbf{g}_s) \exp(i\mathbf{g}_s\cdot\mathbf{r}_s) \quad , \tag{14.39}$$

where $\mathbf{g}_s = \mathbf{h}_s^{-1}\hat{\mathbf{g}}_s$, the vector of integers, $\hat{\mathbf{g}}_s = \{g_{a,s}, g_{b,s}, g_{c,s}\}$, indexes the small reciprocal space and $V_s = \det \mathbf{h}_s$ is the volume of the small cell. The plane wave energy cutoff is taken to be $E_{cut}^{(\rm short)}$ (with the cutoff on the density $4E_{cut}^{(\rm short)}$).

Given that the electron density is localized in the small cell, the short range components of the Hartree and local pseudopotential energies can be evaluated straightforwardly,

$$E_{\rm H}^{(\rm short)}[n] = \frac{e^2}{2}\int_{D(\mathbf{h}_s)} d\mathbf{r}\int_{D(\mathbf{h}_s)} d\mathbf{r}'\,\frac{n_s(\mathbf{r})n_s(\mathbf{r}')\mathrm{erfc}(\alpha|\mathbf{r}-\mathbf{r}'|)}{|\mathbf{r}-\mathbf{r}'|}$$

$$= \frac{e^2}{2V_s}\sum_{\hat{\mathbf{g}}_s}{}'\,\bar{n}_s(-\mathbf{g}_s)\bar{n}_s(\mathbf{g}_s)\left[\frac{4\pi}{g_s^2}\right]\left[1-\exp\left(-\frac{g_s^2}{4\alpha^2}\right)\right] + \frac{e^2\pi}{2V_s\alpha^2}|n_s(0)|^2 \quad (14.40)$$

$$E_{\rm loc}^{(\rm short)}[n] = \sum_{J=1}^{N_s}\int_{D(\mathbf{h}_s)} d\mathbf{r}\; n_s(\mathbf{r})\left[\phi_{{\rm loc},J}(\mathbf{r}-\mathbf{R}_J+\mathbf{R}_c) + \frac{eq_J\mathrm{erf}(\alpha|\mathbf{r}-\mathbf{R}_J+\mathbf{R}_c|)}{|\mathbf{r}-\mathbf{R}_J+\mathbf{R}_c|}\right]$$

$$= \frac{1}{V_s}\sum_{\hat{\mathbf{g}}_s}{}'\sum_{J=1}^{N_s}\bar{n}_s^*(\mathbf{g}_s)\exp(-i\mathbf{g}_s\cdot[\mathbf{R}_J-\mathbf{R}_c]) \;\times$$

$$\times\;\left[\tilde{\phi}_{{\rm loc},J}(\mathbf{g}_s) + \frac{4\pi e q_J}{g_s^2}\exp\left(-\frac{g_s^2}{4\alpha^2}\right)\right]$$

$$+\;\frac{1}{V_s}\sum_{J=1}^{N_s}\bar{n}_s(0)\left[\tilde{\phi}_{{\rm loc},J}^{(0)} - \frac{eq_J\pi}{\alpha^2}\right]. \quad (14.41)$$

where the J sum runs over the N_s ions within the small cell, the $\hat{\mathbf{g}}_s$ sum runs over the large reciprocal-space grid of the small cell and $\mathbf{R}_c$ is position of the small cell inside the large. Since the full system is not periodic on $\mathbf{h}_s$ but on $\mathbf{h}$, Eqs. (14.40-14.41) will only yield the correct short range energy if $\alpha V_s^{1/3} >> 1$ and $n(\mathbf{r}_s)$ vanishes on the small cell boundary. The non-local pseudopotential energy is short range and is assumed to be evaluated within the small cell (only, considering the N_s ions in the small cell and using the small cell reciprocal space). Similarly, the exchange correlation and the electronic kinetic energies can also be evaluated in the small cell using standard techniques[2,6].

Next, the expressions for the long range portions of the Hartree and local pseudopotential energies must be formulated. This can be accomplished by expanding the electron density localized in the small cell in terms of the plane waves of the large cell. This expansion is permitted because the electron density, localized in the small cell, obeys periodic boundary conditions in the large cell (i.e. it is zero on the surface of $\mathbf{h}$). Thus,

$$E_{\mathrm{H}}^{(\mathrm{long})}[n] = \frac{e^2}{2}\sum_{\hat{\mathbf{S}}}\int_{D(\mathbf{h})}d\mathbf{r}\int_{D(\mathbf{h})}d\mathbf{r}'\,\frac{n(\mathbf{r})n(\mathbf{r}')\mathrm{erf}(\alpha|\mathbf{r}-\mathbf{r}'+\mathbf{h}\hat{\mathbf{S}}|)}{|\mathbf{r}-\mathbf{r}'+\mathbf{h}\hat{\mathbf{S}}|}$$

$$= \frac{e^2}{2V}\sum_{\hat{\mathbf{g}}}{}'\,\bar{n}(-\mathbf{g})n(\mathbf{g})\left[\frac{4\pi}{g^2}\exp\left(-\frac{g^2}{4\alpha^2}\right)+\hat{\phi}^{(\mathrm{screen,Coul})}(\mathbf{g})\right]$$

$$+\left(\frac{e^2}{2V}\right)\left[\hat{\phi}^{(\mathrm{screen,Coul})}(0)-\frac{\pi}{\alpha^2}\right]|n(0)|^2 \tag{14.42}$$

$$E_{\mathrm{loc}}^{(\mathrm{long})}[n] = -\sum_{\hat{\mathbf{S}}}\sum_{I=1}^{N}\int_{D(\mathbf{h})}d\mathbf{r}\;n(\mathbf{r})\left[\frac{eq_I\mathrm{erf}(\alpha|\mathbf{r}-\mathbf{R}_I+\mathbf{h}\hat{\mathbf{S}}|)}{|\mathbf{r}-\mathbf{R}_I+\mathbf{h}\hat{\mathbf{S}}|}\right]$$

$$= -\frac{e}{V}\sum_{\hat{\mathbf{g}}}{}'\,\bar{n}^*(\mathbf{g})S(\mathbf{g})\left[\frac{4\pi}{g^2}\exp\left(-\frac{g^2}{4\alpha^2}\right)+\hat{\phi}^{(\mathrm{screen,Coul})}(\mathbf{g})\right]$$

$$-\frac{e}{V}\bar{n}_s(0)S(0)\left[\hat{\phi}^{(\mathrm{screen,Coul})}(0)-\frac{\pi}{\alpha^2}\right]. \tag{14.43}$$

where

$$S(\mathbf{g}) = \sum_I q_I\exp(i\mathbf{g}\cdot\mathbf{R}_I) \tag{14.44}$$

is the atomic structure factor and

$$\bar{n}(\mathbf{g}) = \int_{D(\mathbf{h})}d\mathbf{r}\exp[-i\mathbf{g}\cdot\mathbf{r}]n(\mathbf{r}) \tag{14.45}$$

$$= \int_{D(\mathbf{h}_s)}d\mathbf{r}_s\exp[-i\mathbf{g}\cdot\mathbf{r}_s]n(\mathbf{r}_s+\mathbf{R}_c)$$

$$= \int_{D(\mathbf{h}_s)}d\mathbf{r}_s\exp[-i\mathbf{g}\cdot(\mathbf{r}_s-\mathbf{R}_c)]n_s(\mathbf{r}_s)$$

are the plane wave expansion coefficients of the electron density in the reciprocal space of the large cell, $\mathbf{g} = \mathbf{h}^{-1}\hat{\mathbf{g}}$. The integral in (14.45) can be extended to cover the domain described by the large cell without loss of generality because $n(\mathbf{r}_s+\mathbf{R}_c)\equiv 0$ outside of the small cell. Note, $\bar{n}(\mathbf{g}) = \bar{n}_s(\mathbf{g}_s)$ if $\mathbf{h}_s\equiv\mathbf{h}$ and $\mathbf{R}_c = 0$. Methods for the efficient evaluation of (14.45) and, hence, (14.42) and (14.43) are developed below.

First, it is clear from the long range/short range decomposition of the Hartree and local pseudopotential energies that a different plane wave cutoff can be introduced to treat each part. That is, one cutoff, $E_{cut}^{(\mathrm{short})}$, can be used to evaluate the short range components of the energy, (14.40) and (14.41), and another, $E_{cut}^{(\mathrm{long})}$ can be used to evaluate the long range components, (14.42) and (14.43). While the long range/short range decomposition is, general, it is expected that the short range contributions will be obtained by integration over functions that rapidly vary spatially while the long range contributions will be obtained by integration over a slowly varying function. Therefore, the short range energy contributions must be evaluated using a large reciprocal space cutoff (i.e. the standard $E_{cut}^{(\mathrm{density,short})} = 4E_{cut}^{(\mathrm{short})}$). In contrast, the long range part can be evaluated, in reciprocal space, using a small cutoff, $E_{cut}^{(\mathrm{long})} << E_{cut}^{(\mathrm{short})}$. Thus, by splitting the electronic energy into two parts, large gains in efficiency are possible.

Next, consider the case that the number of particles in the small cell, N_s and the small cell volume, V_s), are much less than their large cell counterparts ($N_s << N$ and $V_s << V$) as would be the case for a large, chemically inert bath surrounding a chemically active subsystem. The computational cost of evaluating the short range local pseudopotential and short range Hartree, exchange correlation, non-local pseudopotential and the electronic kinetic energy as well as the overlap matrix, $\langle\psi_{j,s}|\psi_{i,s}\rangle$, scales like $\sim N_s^3$. The computational cost of evaluating the long range part of the Hartree and local pseudopotential energies depends on the computational cost of evaluating the atomic charge density, $\mathrm{S}(\mathbf{g})$, and the plane wave expansion of the density in the large cell (see Eq. (14.45)). Since the atomic charge density can be evaluated in $N \log\ N$ using Particle Mesh Ewald techniques [16], if (14.45) could also be evaluated in $N \log\ N$, the computational cost of the method would then be $N \log\ N$ at fixed $\mathbf{h}_s$ and N_s. (The present approach yields a linear scaling method because the number of plane waves increases linearly with particle number at fixed particle density and plane wave cutoff).

In order to achieve linear scaling, the electron density must be interpolated from the small cell where it is described by a plane wave expansion with a large cutoff, $E_{cut}^{(\mathrm{short})}$, to the large cell where it is described by a plane wave expansion with a small cutoff, $E_{cut}^{(\mathrm{long})}$. First, consider the Fourier components of the density

$$\bar{n}(\mathbf{g}) = \int\limits_{D(\mathbf{h})} d\mathbf{r} \exp[-i\mathbf{g}\cdot\mathbf{r}] n(\mathbf{r}). \tag{14.46}$$

If $n(\mathbf{r})$ can be expressed in a finite plane wave basis,

$$n(\mathbf{r}) \equiv \frac{1}{V} \sum_{\hat{g}_a=-P_a/2+1}^{P_a/2} \sum_{\hat{g}_b=-P_b/2+1}^{P_b/2} \sum_{\hat{g}_c=-P_c/2+1}^{P_c/2} \exp(i\mathbf{g}\cdot\mathbf{r})\bar{n}(\mathbf{g}), \tag{14.47}$$

then the Fourier coefficients can also be determined (exactly) from a discrete sum over a real space grid

$$\bar{n}(\mathbf{g}) \equiv \frac{V}{P_a P_b P_c} \sum_{\hat{s}_a=0}^{P_a-1} \sum_{\hat{s}_b}^{P_b-1} \sum_{\hat{s}_c=0}^{P_c-1} e^{-2\pi i \hat{g}_a \hat{s}_a / P_a} e^{-2\pi i \hat{g}_b \hat{s}_b / P_b} e^{-2\pi i \hat{g}_c \hat{s}_c / P_c} n(\mathbf{hs}) \tag{14.48}$$

Here, P_a, P_b, and P_c are both the number of reciprocal lattice points along each direction and the number of points discretizing the $\mathbf{a}, \mathbf{b}, \mathbf{c}$ axes of the cell, and $s_\alpha = \hat{s}_\alpha / P_\alpha$. Importantly, Eq. (14.48) and its inverse, Eq. (14.47), can be evaluated using a three dimensional Fast Fourier Transforms (3D-FFT) in order $N \log\, N$. A spherical cutoff is introduced in reciprocal space by simply assuming that $n(\mathbf{r})$ is described by a basis in which $\bar{n}(\mathbf{g}) \equiv 0$ when $\hbar^2 |\mathbf{g}|^2 / 2m_e > E_{cut}$.

Next, consider a function, $f(\mathbf{r})$ with plane wave expansion coefficients,

$$\bar{f}(\mathbf{g}) = \int_{D(\mathbf{h})} d\mathbf{r}\ \exp[-i\mathbf{g}\cdot\mathbf{r}] f(\mathbf{r})$$

$$= V \int_0^1 ds_a \int_0^1 ds_b \int_0^1 ds_c e^{-2\pi i \hat{g}_a s_a} e^{-2\pi i \hat{g}_b s_b} e^{-2\pi i \hat{g}_c s_c} f(\mathbf{hs}). \tag{14.49}$$

that can be described on a finite reciprocal space (cf. Eq. (14.48)). In order to express the plane wave expansion coefficients, accurately, in terms of a sum over an arbitrary set of equally spaced discrete points in real space (as opposed to the continuous integrals given in (14.49) or the discretization required by (14.48), it useful to introduce the Euler exponential spline[17,18]

$$\exp\left(\frac{2\pi i \hat{g}_\alpha u}{\tilde{P}_\alpha}\right) = d_m(\hat{g}_\alpha, \tilde{P}_\alpha) \sum_{\hat{s}=-\infty}^{\infty} M_m(u - \hat{s}) \exp\left(\frac{2\pi i \hat{g}_\alpha \hat{s}}{\tilde{P}_\alpha}\right) + \mathcal{O}\left(\frac{2|\hat{g}_\alpha|}{\tilde{P}_\alpha}\right)^m \tag{14.50}$$

$$d_m(\hat{g}_\alpha, \tilde{P}_\alpha) = \frac{\exp\left(2\pi i (m-1)/\tilde{P}_\alpha\right)}{\left[\sum_{j=0}^{m-2} M_m(j+1) \exp\left(2\pi i \hat{g}_\alpha j / \tilde{P}_\alpha\right)\right]}$$

where $\hat{s}$ is an integer, u is a real number, m is the spline order assumed to be even and the $M_m(u)$ are the Cardinal B splines

$$M_2(u) = 1 - |u-1| \tag{14.51}$$

$$M_m(u) = \left[\frac{u}{m-1}\right] M_{m-1}(u) + \left[\frac{m-u}{m-1}\right] M_{m-1}(u-1)$$

$$M_m(u) \neq 0 \qquad 0 < u < m$$

$$M_m(u) = 0 \qquad u \leq 0, u \geq m$$

$$\sum_{\hat{s}=\infty}^{\infty} M_m(u-\hat{s}) = 1$$

$$\frac{dM_m(u)}{du} = M_{m-1}(u) - M_{m-1}(u-1)$$

Inserting the Euler exponential spline into (14.49) yields a well defined approximation to $\bar{f}(\mathbf{g})$,

$$\bar{f}(\mathbf{g}) \approx \left[V d_m^*(\hat{g}_a, \tilde{P}_a) d_m^*(\hat{g}_b, \tilde{P}_b) d_m^*(\hat{g}_c, \tilde{P}_c)\right] \tag{14.52}$$
$$\times \sum_{\hat{s}_a=0}^{\tilde{P}_a-1} \sum_{\hat{s}_b=0}^{\tilde{P}_b-1} \sum_{\hat{s}_c=0}^{\tilde{P}_c-1} e^{-2\pi i \hat{g}_a \hat{s}_a/\tilde{P}_a} e^{-2\pi i \hat{g}_b \hat{s}_b/\tilde{P}_b} e^{-2\pi i \hat{g}_c \hat{s}_c/\tilde{P}_c} f^{(\mathrm{conv})}(\mathbf{hs})$$

where

$$f^{(\mathrm{conv})}(\mathbf{hs}) = \int_0^1 ds'_a \int_0^1 ds'_b \int_0^1 ds'_c \sum_{k_a=-\infty}^{\infty} \sum_{k_b=-\infty}^{\infty} \sum_{k_c=-\infty}^{\infty} f(\mathbf{hs}') \tag{14.53}$$
$$\times M_m([s'_a - k_a]\tilde{P}_a - \hat{s}_a) M_m([s'_b - k_b]\tilde{P}_b - \hat{s}_b) M_m([s'_c - k_c]\tilde{P}_c - \hat{s}_c).$$

is the interpolation of $f(\mathbf{r})$ onto the discrete real space grid defined by $s_\alpha = \hat{s}_\alpha/\tilde{P}_\alpha$ and $0 \leq \hat{s}_\alpha \leq \tilde{P}_\alpha - 1$.

Equation (14.52) can be evaluated using a 3D-FFT in order $N \log N$ provided the function, $f^{(\mathrm{conv})}(\mathbf{hs})$, defined on the discrete real space, can be constructed in a computationally efficient manner. In addition, Eq. (14.52) is smooth and possesses $m-2$ continuous derivatives. Note, if $\tilde{P}_a > m+1$ then each point in the continuous space, $\{s'_a, s'_b, s'_c\}$, is mapped to m^3 unique points on the discrete grid indexed by $\{\hat{s}_a, \hat{s}_b, \hat{s}_c\}$ due to the finite support of the $M_m(p)$ (see Eq. (14.51)).

It is now a simple matter to generate a computationally efficient and well defined approximation to the Fourier coefficients, $\bar{n}(\mathbf{g})$, of an electron density

$n(\mathbf{r})$ that is assumed to be nonzero only in the small cell described by $\mathbf{h}_s$. First, given that $\bar{n}_s(\mathbf{g}_s)$, defined in (14.39), exists on a finite reciprocal space, the identity given in (14.48) holds. Thus, the discrete form of the density can be inserted into (14.53) and the integrals performed using trapezoidal rule integration with loss of generality to yield the desired interpolation from the small cell to the large cell,

$$n^{(\mathrm{conv})}(\mathbf{hs}) = \left[\frac{V_s}{V}\right]\left[\frac{1}{P_{a,s}P_{c,s}P_{c,s}}\right]$$
$$\times \sum_{\hat{s}'_a=0}^{P_{a,s}-1}\sum_{\hat{s}'_b=0}^{P_{b,s}-1}\sum_{\hat{s}'_c=0}^{P_{c,s}-1}\sum_{k_a=-\infty}^{\infty}\sum_{k_b=-\infty}^{\infty}\sum_{k_c=-\infty}^{\infty} n_s(\mathbf{h}_s\mathbf{s}')$$
$$\times M_m([s'_a + S_{a,s} - k_a]\tilde{P}_a - \hat{s}_a)M_m([s'_b + S_{b,s} - k_b]\tilde{P}_b - \hat{s}_b)$$
$$\times M_m([s'_c + S_{c,s} - k_c]\tilde{P}_c - \hat{s}_c). \tag{14.54}$$

Here, $\{P_{a,s}, P_{b,s}, P_{c,s}\}$ are defined by the size of the small cell reciprocal space (through the cutoff, $E_{cut}^{(\mathrm{short})}$), $s'_\alpha = \hat{s}'_\alpha/P_{\alpha,s}$, $\mathbf{S}_s = \mathbf{h}^{-1}\mathbf{R}_c$, and $V_s/V = \det \mathbf{D}$ while the $\{\tilde{P}_a, \tilde{P}_b, \tilde{P}_c\}$ are defined by the size of the large cell reciprocal space (through the cutoff, $E_{cut}^{(\mathrm{long})}$).

The desired plane wave expansion of the density, $\bar{n}(\mathbf{g})$, is constructed by inserting $n^{(\mathrm{conv})}(\mathbf{hs})$ into (14.52) and performing a 3D-FFT. Note, in the limit, $\tilde{P}_a = P_{a,s}$, $\tilde{P}_b = P_{b,s}$, $\tilde{P}_c = P_{c,s}$ or $E_{cut}^{(\mathrm{short})} = E_{cut}^{(\mathrm{long})}$, and $\mathbf{h} = \mathbf{h}_s$, then $\bar{n}_s(\mathbf{g}_s) \equiv \bar{n}(\mathbf{g})$ because (14.48) is exact for a finite reciprocal space and the Euler exponential splines are exact at the knots[17,18]. Importantly, Eq. (14.54) can be evaluated in order $N_s m^3$ **and** the (dense) discrete real space grid spanning the small cell, $\mathbf{h}_s$, and the (sparse) discrete real space grid spanning the large cell, $\mathbf{h}$, need not be commensurate. In addition, the separable form of the $M_m(p)$, which is a consequence of the choice $\mathbf{h}^{-1}\mathbf{h}_s = \mathbf{D}$, allows the required $M_m(p)$ to be evaluated independently in order $mN_s^{1/3}$. Thus, the overall computational cost of constructing $\bar{n}(\mathbf{g})$ is $N \log\, N$ (dominated by the FFT). Finally, the resulting $\bar{n}(\mathbf{g})$ (i.e. obtained by inserting (14.54) into (14.52)) is continuously differentiable with respect to the expansion coefficients of the orbitals, $\bar{\psi}_{j,s}(\mathbf{g}_s)$, defined in (14.39).

14.3 Results

In this section, model and realistic surfaces and clusters are examined, first, in order to demonstrate the effectiveness of the new long range forces techniques. Next, applications of the new hybrid *ab initio*-empirical force field technique to model and realistic systems are given.

14.3.1 Clusters

The efficacy of the plane-wave based expressions for the energy in clusters is shown for a series of model and realistic problems.

14.3.1.1 Hartree and Local Pseudopotential Energies for a Model Density

In order to examine the efficacy of the cluster methodology, a Gaussian density $n(\mathbf{r})$,

$$n(\mathbf{r}) = \left(\frac{\kappa^2}{\pi}\right)^{3/2} \exp(-\kappa^2 r^2) \tag{14.55}$$

$$\bar{n}(\mathbf{g}) = \exp\left(-\frac{g^2}{4\kappa^2}\right).$$

is first examined. We choose to treat the large system cell size limit, $\kappa L >> 1$, where L is the side of the cell. In this limit the Hartree and local external energy, Eqs. (14.34) and (14.35), are given by

$$E_{\mathrm{H}} = \frac{\kappa e^2}{\sqrt{2\pi}}$$

$$E_{\mathrm{loc}} = -\left[\frac{4}{\pi}\right]^{1/2} \kappa e^2 \tag{14.56}$$

where the local potential is just Coulombs Law, $V_{loc}(\mathbf{r}) = -e^2/r$. Comparisons to the exact results above and those given by (14.34) and (14.35), respectively, are presented in Tables 14.1 and 14.2. Accurate results are obtained at relatively small box edge, $\kappa L \geq 15$. Note, that if the screening function is omitted, very large errors result.

14.3.1.2 Water Molecule and Hydronium Ion

In Tables 14.3 and 14.4, the behavior of the *total energy difference* between an H_2O molecule and an H_3O^+ ion as a function of box size for LDA [23] and GG-LDA [24,25], respectively, are given. Uniform convergence with box size at fixed plane-wave cutoff is observed. In addition, although different functionals and pseudopotentials are used in the calculations presented in the two tables, the total energy difference is reproduced accurately (within 70 Kelvin). At fixed box edge, $L = 9$ Å , the energy difference convergences rapidly with cutoff. For example, under LDA, ΔE(180 Ry) = 0.268921 Hartree, ΔE(240 Ry) = 0.269400 Hartree and ΔE(300 Ry) = 0.269497 Hartree. Note, energy differences between systems with different total charge are ill defined when the screening function is neglected [14,13]. Indeed, the results indicate that neglecting the screening function increases the error, markedly.

Table 14.1. Convergence of the Hartree and local external energies for the model density of (14.55) using $\kappa = 2$ Å $^{-1}$. The exact energies are $E_{\mathrm{H}} = 0.422222$ Hartree and $E_{\mathrm{loc}} = -1.194225$ Hartree. The double asterisk indicates that the screening function has been set to zero.

L (Å)	$\alpha_{conv}L$	E_{H} (Hartree)	ΔE_{H}(Kelvin)	E_{loc} (Hartree)	ΔE_{loc}(Kelvin)
6	3	0.422678	144.0	-1.194382	-49.6
6	4	0.422446	70.7	-1.194230	-1.6
6	5	0.422436	67.5	-1.194226	-0.6
8	3	0.422295	23.1	-1.194282	-17.8
8	4	0.422226	1.3	-1.194226	-0.6
8	5	0.422225	1.3	-1.194225	0.0
8**	5	0.322194	-29376.0	-1.00735	59008.4
10	3	0.422252	9.4	-1.194225	0.0
10	4	0.422222	0.1	-1.194225	0.0
10	5	0.422222	0.0	-1.194225	0.0

Table 14.2. Convergence of the Hartree and local external energies for the model density of (14.55) using $\kappa = 1$ Å $^{-1}$. The exact energies are $E_{\mathrm{H}} = 0.211111$ Hartree and $E_{\mathrm{loc}} = -0.5971125$ Hartree. The double asterisk indicates that the screening function has been set to zero.

L (Å)	$\alpha_{conv}L$	E_{H} (Hartree)	ΔE_{H}(Kelvin)	E_{loc} (Hartree)	ΔE_{loc}(Kelvin)
10	5	0.211873	240.6	-0.597142	-9.4
14	5	0.211124	4.1	-0.597112	-0.4
14	5	0.211124	4.1	-0.597112	-0.4
14**	5	0.158094	-16741.5	0.490472	33674.2
18	5	0.211111	0.0	-0.597112	0.0

14.3.2 Surface Ewald Summation

Here, the accuracy and effectiveness of plane-wave based expressions for the energy on surfaces are demonstrated for a series of model and realistic problems.

14.3.2.1 Model BCC Surface

The Ewald surface summation, Eqs. (14.29) and Eqs. (14.28), was tested by considering an out-of-plane bcc surface of point changes (positive charges in the body center and negative charges on the corner of the cube above and below the z=0 plane – see Table 14.5). In all cases, rapid convergence to the exact result (obtained via explicit Madelung summation) is observed as a function of magnitude of the reciprocal space cutoff (g_{cut}). However,

Table 14.3. Convergence of the total energy difference between a water molecule and a hydronium ion for different box lengths using a plane-wave cutoff of $E_{cut} =$ 180 Ry and $\alpha_{\rm conv} L = 7$. The H_2O molecule was given a O-H bond length of r_{OH}0.99 Å and a HOH bond angle of θ_{HOH}=105.8 degrees. The hydronium was given a O-H bond length of $r_{OH} = 0.98$ Å and a HOH bond angle of θ_{HOH}=116.7 degrees. Exchange and correlation were treated with the local density approximation and BHS pseudopotentials were employed to treat core electrons. The quantity $\Delta\Delta E = \Delta E(L) - \Delta E(12$ Å $)$. The double asterisk indicates that the screening function has been set to zero.

L (Å)	$E(H_2O)$ (Hartree)	$E(H_3O^+)$ (Hartree)	ΔE (Hartree)	$\Delta\Delta E$ (Kelvin)
6	-17.180726	-17.450848	0.270122	382.
8	-17.181933	-17.450895	0.268962	16.
8**	-17.182263	-17.549832	0.367569	31154.
10	-17.181977	-17.450888	0.268911	0.
12	-17.182030	-17.450942	0.268912	0.

Table 14.4. Convergence of the total energy difference between a water molecule and a hydronium ion for different box lengths using a plane-wave cutoff of $E_{cut} =$ 140 Ry and $\alpha_{\rm conv} L = 7$. Geometries of the H_2O and H_3O^+ molecules are as specified in the previous table. Exchange and correlation were treated within the GGA using the B-LYP functional, and Troullier-Martins type pseudopotentials were employed to treat core electrons. The quantity $\Delta\Delta E = \Delta E(L) - \Delta E(12$ Å $)$. The double asterisk indicates that the screening function has been set to zero.

L (Å)	$E(H_2O)$ (Hartree)	$E(H_3O^+)$ (Hartree)	ΔE (Hartree)	$\Delta\Delta E$ (Kelvin)
6	-17.158608	-17.428996	0.270388	394.
8	-17.160274	-17.429403	0.269129	3.
8**	-17.157504	-17.528353	0.370849	34912.
10	-17.160263	-17.429405	0.269142	1.
12	-17.160260	-17.429399	0.269139	0.

evaluating the surface energy within the supercell approximation (setting the screening function to zero), yields poor results.

Next, the bcc surface was modified by replacing each negative point charge by a Gaussian charge density $n(\mathbf{r}) = (\kappa^2/\pi)^{3/2}\exp(-\kappa^2|\mathbf{r}|^2)$ of width, $\kappa = 4$ Å. The total negative charge density was expanded in a plane wave basis and the total lattice energy evaluated as a function of **c**-boundary height, L_c and plane wave basis set size by combining Eqs. (14.29-14.28) with Eqs. (14.34-14.35). In Table 14.6, the lattice energy with the self energy of the charge density, $N_{\rm cell}\kappa/\sqrt{2\pi}$, removed, analytically, is presented. Excellent agreement with the exact results is obtained. The supercell approximation (setting the

Table 14.5. The Coulomb energy per unit cell for the out-of-plane bcc surface with a unit cell length of $L_s = 2$Å for a different number of unit cells, $M \times M \times 1$ ($\phi^{(Coul,exact)}$ = -0.16400). L_c is the length of the box in the non-periodic direction and g_{cut} is the reciprocal-space cutoff value The star indicates a calculation carried out using three dimensional periodicity ($\hat{\phi}^{(\mathrm{screen,Coul})}(\mathbf{g}, \alpha_{ewd}) \equiv 0$).

M	L_c(Å)	α_{ewd}(Å^{-1})	g_{cut}(Å^{-1})/2π	ϕ_{coul}(Hartree)	$\Delta\phi$(Kelvin)
1	4.0	6.0	1.5	-0.10451	18789
			3.0	-0.16265	426
			6.0	-0.16400	0
2	4.0	3.0	1.5	-0.11685	14889
			3.0	-0.16398	6
			6.0	-0.16400	0
3	6.0	2.0	1.0	-0.13256	9928
			2.0	-0.16400	0
			4.0	-0.16400	0
*2.0	4.0	6.0	6.0	-0.37171	-65590
*4.0	4.0	3.0	6.0	-0.37171	-65590
*6.0	6.0	2.0	4.0	-0.30254	-43748

screening function to zero), again, converges rather slowly with out-of-plane boundary height, L_c.

14.3.2.2 Ice Surface with a Defect

An ice surface with a H_3O^+/OH^- defect was studied using an *ab initio* approach (valence electrons plus ion cores). The defect was created by performing an inter-layer proton transfer followed by an adjacent intra-layer proton transfer. Thus, the ion-pair is separated by a single layer and bridged by a single water molecule to which the H_3O^+ donates a hydrogen bond and from which the OH^- accepts a hydrogen bond (see Fig. 14.1). The electronic structure was treated within the generalized gradient approximation (GGA) using the B-LYP exchange and correlation function [24,25]. Core electrons were treated using the norm-conserving pseudopotentials of Troullier and Martins [26]. Convergence of the total energy of this system as a function of out-of-plane distance, L_c, at a plane wave cutoff of $E_{cut}^{(\mathrm{orbitals})}$=80 Ry is given in Table 14.7. Comparison to the supercell approximation (neglecting the screening function) to the new technique at L_c = 20 Å shows that while the new method is well converged, the supercell approximation generates a substantial error in the energetics.

Table 14.6. The Coulomb energy per unit cell for an out-of-plane bcc surface with gaussian charge density, $n(\mathbf{r}) = (\kappa^2/\pi)^{3/2}\exp(-\kappa^2|\mathbf{r}|^2)$, $\kappa = 4.0$ Å^{-1} in the body center of a $2 \times 2 \times 1$ system of length, $L_s = 4$Å ($\phi^{(Coul,exact)}$ = -0.08200 without gaussian self energy). A plane wave basis set with an energy cutoff, $E_{\rm cut}^{\rm (orbitals)}$, was used to describe the gaussian. The star indicates a calculation carried out using three dimensional periodicity (i.e $\hat{\phi}^{\rm (screen,Coul)}(\mathbf{g},\alpha_{ewd}) \equiv 0$).

L_c(Å)	$E_{\rm cut}(Ry)$	E(Hartree)	ΔE(Kelvin)
4	25	-0.16170	-25167
	50	-0.10897	-8517
	100	-0.10768	-8109
	200	-0.10768	-8109
6	25	-0.13352	-16269
	50	-0.08324	-392
	100	-0.08200	0
	200	-0.08200	0
8	25	-0.13436	-16534
	50	-0.08328	-378
	100	-0.08200	0
	200	-0.08200	0
*6	100	-0.21951	-43422
*8	100	-0.18588	-32803
*10	100	-0.16512	-26247
*20	100	-0.12356	-13124
*32	100	-0.10797	-8201
*40	100	-0.10278	-6562

Table 14.7. Convergence of the total energy (electronic energy plus the ion-ion Coulomb interaction) for the ice surface described in the text, in a 9.01 Å ×7.81 Å ×L_c cell, at fixed plane wave cutoff, $E_{\rm cut}^{\rm (orbitals)}$ = 80Ry, and varying box length, L_c. Martins-Troullier pseudopotentials were employed with B-LYP exchange and correlation functionals The quantity ΔE=$E(L_c)$-$E(27)$, and the asterisk indicates a calculation carried out using three-dimensional periodicity (i.e $\hat{\phi}^{\rm (screen,Coul)}(\mathbf{g},\alpha_{ewd}) \equiv 0$)

L_c(Å)	E(Hartree)	ΔE(Kelvin)
10	-273.07260	444150
15	-274.47337	1820
20	-274.47906	20
20*	-274.48209	-940
27	-274.47912	0

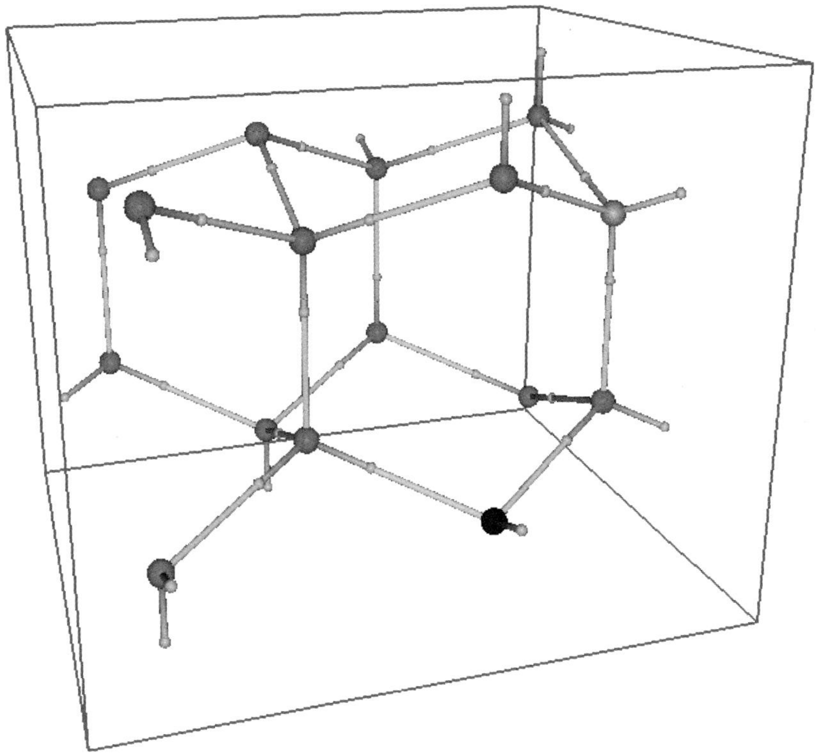

Fig. 14.1. An ice surface with an H_3O^+/OH^- defect generated via an inter-layer proton transfer followed by an adjacent intra-layer proton transfer. The ion-pair is separated by a single layer and bridged by a single water molecule to which the H_3O^+ donates a hydrogen bond and from which the OH^- accepts a hydrogen bond.

14.3.3 Mixed *ab initio*/Empirical Force Fields

Next, the efficacy of the novel mixed *ab initio*/empirical force field technique is tested. Two systems, a single "*ab initio*" solvated by 63 "empirical" water molecules and an HCA II enzyme solvated by $\approx$ 9000 "empirical" water molecules are considered.

14.3.3.1 Neat Water

A system of 64 water molecules in a cubic box, subject to periodic boundary conditions, is examined. Although relatively small, a 64 molecule system, has been shown to be a reasonable model of liquid water. Therefore, a system of 64 empirical TIP3P model water molecules was equilibrated at the state point $\{\rho = 1g/cm^3, T = 300K\}$ for 1 nanosecond of simulation time. One TIP3P water molecule was then replaced by an *ab initio* water molecule (8 valence electrons and three "*ab initio* atoms" or "ions"). The electrons were allowed to interact with the TIP3P model molecules and the H^+ and O^{6+}

ions via pseudopotentials fit by the authors and standard Troullier-Martins pseudopotentials [26], respectively. In addition, the ions were permitted to interact with the TIP3P molecules via electrostatic and van der Waals forces. The B-LYP model, exchange-correlation density functional[24,25] was used in all the calculations, in conjunction with a small cell plane wave cutoff of $E_{cut}^{(short)} = 100$ Ry. In Table 14.8, the convergence of the electronic energy is shown as a function of the large cell plane wave cutoff and Cardinal B-spline interpolation order. As can be seen from the table, rapid convergence with both parameters is achieved. In addition, the small cell edge, L_s, can be taken to be much less than the large cell edge without loss of accuracy. Note, Fig. 14.2, shows the radial distribution function of a 50/50 mixture of TIP3P-*ab initio* water molecules, indicating the model is reasonable.

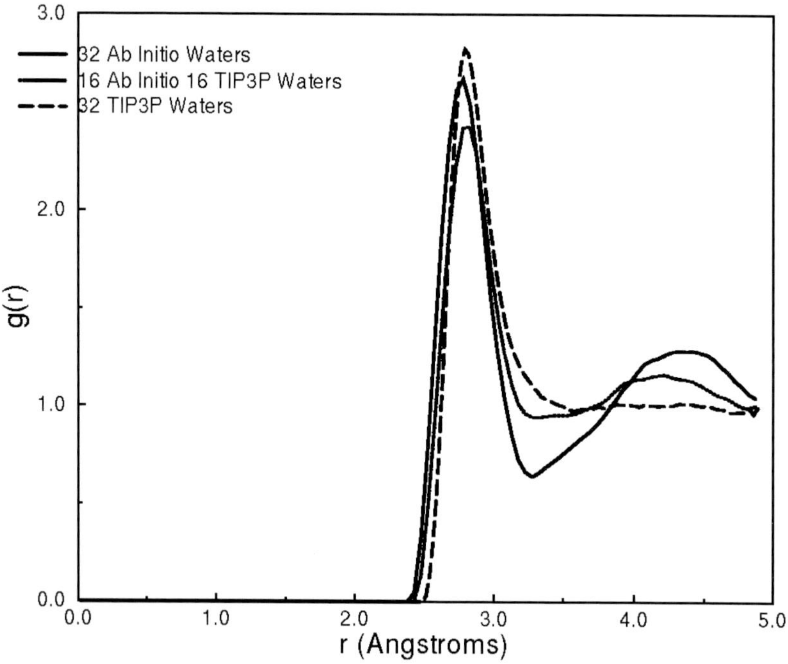

Fig. 14.2. The oxygen-oxygen radial distribution function for a system consisting of 16 TIP3P water molecules and 16 *ab initio* molecules at the state point $\{\rho = 1g/cm^3, T = 300K\}$ is presented (dotted line). Comparisons are made to distribution functions generated using fully *ab initio* (solid line) and a fully empirical model (dashed line).

Table 14.8. The total electronic energy of a single *ab initio* water molecule immersed in a bath of TIP3P molecules as a function of large cell plane cutoff and spline interpolation order. The large cell size was fixed by the state point, $L_l = 12.43$Å, on edge. The small cell cutoff was fixed at 100Ry. The electrostatic division parameter was set to $\alpha = 6/L_s$ and $\Delta E_{tot} = E_{tot} - E_{tot}^{(std)}$ where $E_{tot}^{(std)} = -20.28767$ is the result of a standard calculation with $L_s = L_l = 12.43$Å.

L_s (Å)	$E_{cut}^{(\mathrm{long})}$ (Rydberg)	m	E_{tot} (Hartree)	ΔE_{tot} (Kelvin)
6	4	4	-20.27980	2480
		6	-20.28127	2020
		8	-20.28133	2002
	8	4	-20.28108	2080
		6	-20.28133	2000
		8	-20.28134	2000
8	4	4	-20.28684	260
		6	-20.28718	150
		8	-20.28718	150
	8	4	-20.28711	175
		6	-20.28718	150
		8	-20.28718	150
9	4	4	-20.28773	-20
		6	-20.28790	-70
		8	-20.28790	-70
	8	4	-20.28787	-60
		6	-20.28790	-70
		8	-20.28790	-70

14.3.3.2 HCA II in Water

Last, an HCA II enzyme solvated in liquid water, is considered. In detail, the 260-residue HCA-II enzyme (complete with catalytic zinc), was solvated by 8,859 waters, for a total of 30,649 atoms. Clearly, a fully *ab initio* treatment of such a large system is not feasible, at present. However, a hybrid model, wherein only the catalytic zinc, the side-chains of active site residues, HIS 94, HIS 96, HIS 119, THR 199, GLU 106 and the five water molecules in the active site are treated using an *ab initio* description, can be studied.

Thus, 320 valence electrons of 80 atoms (see Fig. 14.3) are treated at an *ab initio* level while the remainder of the system is treated using the empirical CHARMM22 all-atom parameter force field which includes TIP3P water model[27]. Briefly, the electrons are assumed to interact with "*ab initio*" atoms via standard Troullier-Martins pseudopotentials [26] and with "empirical atoms" via pseudopotentials fit by the authors (see also [28,29]), and the B-LYP, density functional[24,25] was employed to treat exchange and correlation. *Ab initio* atoms (ion-cores) were permitted to interact with neighboring "empirical atoms" via appropriate bond, bend, torsion, one-four, van der Waals and Coulomb forces, parameterized, by the authors. The parameters were obtained by enforcing good agreement between mixed models, fully empirical models and fully *ab initio* models of relevant fragments. For example,

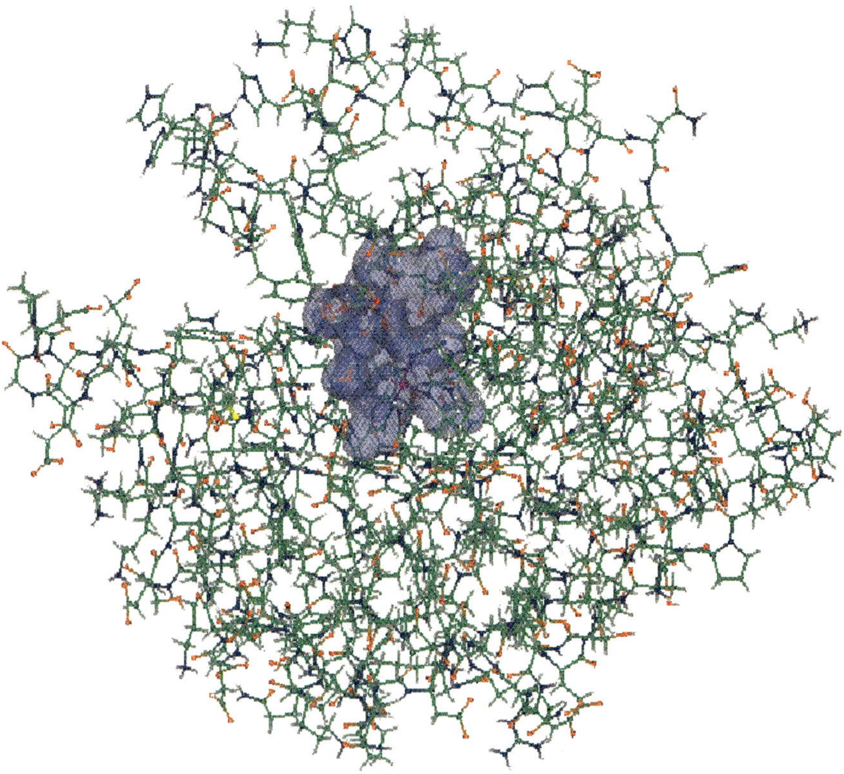

Fig. 14.3. Human carbonic anhydrase treated using the mixed *ab initio*/empirical force field based approach described in the text. The full enzyme wherein the wire frame represents atomic sites and the blue cloud represents the electron density of the valence electrons associated with "*ab initio* atoms".

Table 14.9. The total electronic energy of the active site of HCA II immersed in a bath of TIP3P molecules and CHARMM22 model amino acid residues as a function of large cell plane cutoff and spline interpolation order. The large cell size is fixed by the state point, 66.7Å, on edge. The small cell size was fixed at 18Å on edge and the small cell cutoff was fixed at 70Ry. The electrostatic division parameter was set to be $\alpha = 9/L_s$ and the accuracy measure is defined to be $\Delta E_{tot} = E_{tot}(E_{cut}^{(\mathrm{long})}, m) - E_{tot}(4, 8)$.

$E_{cut}^{(\mathrm{long})}$ (Rydberg)	m	E_{tot} (Hartree)	ΔE_{tot} (Kelvin)
0.5	6	-2329.31984	9200
	8	-2329.33018	5900
2	6	-2329.34896	32
	8	-2329.34905	3
4	6	-2329.34905	3
	8	-2329.34906	0

the minimum energy geometry of hybrid model $CH_3CO - (HIS) - NHCH_3$ deviates at most 2 degrees in the bend angles and 0.02Å in the bond lengths from the standards (CHARMM22 and fully *ab initio* treatments as appropriate).

The HCA II/water system described above was prepared by taking the crystallographic configuration of the enzyme (PDB identification label, "1RAY") [30] and immersing it in TIP3P water. Next, a 1 ns constant temperature molecular dynamics calculation was performed using a fully empirical treatment[27]. This was followed by a 1 ns constant pressure molecule dynamics calculation. At this point, the hybrid model was introduced. In Table 14.9, the convergence of the electronic energy for a representative configuration taken from the simulation of the hybrid model, is shown versus the large cell, plane wave cutoff and the Cardinal B-spline interpolation order. As in previous example, accurate energies are obtained for low spline orders and plane wave cutoffs. The results, therefore, indicate that large complex systems can be studied efficiently using the new methodology.

14.4 Conclusion

In this review article, recent advances in plane-wave based *ab initio* molecular dynamics methodology were discussed. In particular, a generalized treatment of electrostatic interactions that permits surfaces and clusters to be modeled as easily and naturally as fluids and solids was given[8,9]. In this way, many important processes in systems with restricted periodicity can now be examined. Thus, catalytic behavior at surfaces, and reactions in clusters, can be modeled effectively. Next, an extension of plane-wave based *ab initio* methods to model systems in which the electrons are assumed to be localized in a single spatial area, within a large chemically inert bath, in a computational efficient, accurate and controlled manner, was given[12]. An application to a mixed *ab initio*/empirical force-field simulation study of a full 30,000 atom system, the enzyme, HCA-II solvated by approximately 9000 water molecules, with 320 valence electrons of 80 atoms treated at an *ab initio* level of detail, was presented. These new methods and the results that they enable, demonstrate the tremendous ability of *ab initio* MD calculations to impact important areas of physics, chemistry, materials science and biology.

Acknowledgements

This research was supported by NSF - CHE - 0121367 and IBM (G.J.M.), Research Corporation RI0218 and NSF CHE - 98 - 75824 (M.E.T.) and EIA-0081307 (G.J.M. and M.E.T.).

References

1. R. Car and M. Parrinello. *Phys. Rev. Lett.*, 55:2471–2474, (1985).
2. G. Galli and M. Parrinello. *Computer Simulation in Materials Science*, 3:283, (1991).
3. M.C. Payne, M.P. Teter, D.C. Allen, T.A. Arias, and J.D. Joannopoulous. *Rev. Mod. Phys.*, 64:1045, (1992).
4. M.E. Tuckerman, D. Marx, M.L. Klein, and M. Parrinello. *Science*, 275:817, 1997.
5. M. Diraison, M.E. Tuckerman, and G.J. Martyna. *J. Chem. Phys.*, 110:1096, (1999).
6. D.K.Remler and P.A. Madden. *Mol. Phys.*, 70:921, (1990).
7. F. Allen et al. *IBM Syst. J.*, 40:310, (2001).
8. G.J. Martyna and M.E. Tuckerman. *J. Chem. Phys.*, 110:2810, (1999).
9. P. Minary, M.E. Tuckerman, K.A. Pihakari, and G.J. Martyna. *J. Chem. Phys.*, 116:5351, (2002).
10. P. Minary, M. E. Tuckerman, and G. J. Martyna. *J. Chem. Phys.* (to be submitted).
11. A. Warshel and M. Levitt. *J. Mol. Biol.*, 103:227, (1976).
12. D.A. Yarne, M.E. Tuckerman, and G.J. Martyna. *J. Chem. Phys.*, 115:3531, (2001).
13. J.P. Hansen. "MD dynamics simulation of coulomb systems" in *MD simulation of statistical mechanical systems*. North Holland Physics, Amsterdam, (1986).
14. S.W. deLeeuw, J.W. Perram, and E.R. Smith. *Proc. R. Soc. London A*, 373:27, 1980.
15. M.P. Allen and D.J. Tildesley. *Computer Simulations of Liquids*. Claredon Press, Oxford, (1989).
16. U. Essmann, L. Perera, M.L. Berkowitz, T. Darden, H. Lee, and L.G. Pedersen. *J. Chem. Phys.*, 103:8577, 1995.

17. C.K. Chu. *An Introduction to Wavelets.* Academic Press, Boston,MA, (1992).
18. I.J. Schoenberg. *Cardinal Spline Interpolation.* Society for Industrial and Applied Math, Philadelphia,PA, (1973).
19. W. Kohn and L.J. Sham. *Phys. Rev.*, 140:A1133, 1965.
20. P. Hohenberg and W. Kohn. *Phys. Rev.*, 136:B86, 1964.
21. S. Lundqvist and N.H. March, editors. *Theory of Inhomogeneous Electron Gas.* Plenum Press, New York, (1983).
22. G. Bachelet, D. Hamann, and M. Schluter. *Phys. Rev. B*, 26:4199, (1982).
23. J.P. Perdew and A. Zunger. *Phys. Rev. B.*, 23:5048, (1981).
24. A.D. Becke. *Phys. Rev. A*, 38:3098, (1988).
25. C. Lee, W. Yang, and R.G. Parr. *Phys. Rev. B*, 37:785, (1988).
26. N. Troullier and J.L. Martins. *Phys. Rev. B*, 43:1993, (1991).
27. A. MacKerell Jr., D. Bashford, M. Bellott, R.L. Dumbrack, J.D. Evanseck, M.J. Field, S. Fischer, H. Guo, S. Ha, D. Joseph-McCarthy, L. Kcuhnir, K. Kuczera, F.T.K. Lau, C. Mattos, S. Michnick, T. Ngo, D.T. Nguyen, B. Prodhom, W.E. Reiher III, B. Roux, M. Schlenkrich, J.C. Smith, R. Stote, J. Straub, M. Watanabe, J. Wiorkiewicz-Kuczera, D. Yin, and M. Karplus. *J. Phys. Chem. B*, 102:3586, 1998.
28. Y. Zhang, T. Lee, and W. Yang. *J. Chem. Phys.*, 110:46, (1999).
29. J. Gao, P. Amara, C. Alhambra, and M. J. Field. *J. Phys. Chem. A*, 102:4714, (1998).
30. B.M. Jonsson, K. Hakansson, and A. Liljas. *Febs. Lett.*, 322:186, (1993).

15 Time and Length Scales in *ab initio* Molecular Dynamics

Ursula Röthlisberger[1], Michiel Sprik[2], and Jürg Hutter[3]

[1] Institute of Molecular and Biological Chemistry, Swiss Federal Institute of Technology, EPFL, CH-1015 Lausanne, Switzerland
[2] Chemistry Department, University of Cambridge, Cambridge CB2 1EW, United Kingdom
[3] Physical Chemistry Institute, University of Zurich, CH-8057 Zurich, Switzerland

Abstract. Time and length scales accessible to *ab initio* molecular dynamics simulations are necessarily limited. This is the price one pays for an "in principle" unbiased description of chemical reactivity. We address the problem of times scales focusing on the determination of the reaction path and mechanism of chemical reactions using advanced sampling techniques. Further in the discussion about the calculation of thermochemical constants both, time and length scale problems are covered. Finally, new techniques for extending system sizes with density functional theory are presented.

15.1 Introduction

Ab initio molecular dynamics (AIMD) [1] is gradually becoming a well established tool for the study of physical [2], chemical [3] and biochemical [4] systems. This method has enabled direct simulations of chemical reactions in extended and complex environments. However, typical characteristic time windows that can be covered with first-principles Car-Parrinello simulations are currently limited to few tens of picoseconds, in vast contrast to realistic time scales of dynamical processes in chemical and biological systems that often extend into the ms-hours range. Hence, rare conformational transition or reactive events are rarely observed spontaneously during such MD runs.

Ab initio molecular dynamics simulations have made huge progress in recent years. New developments in algorithms and more powerful computers make it possible to simulate larger systems for longer times. Nevertheless, there is the need for simulations of even larger systems as there is the need to be able to go beyond the time scales accessible by todays methods. The simulation of the electronic structure of biomolecules (especially DNA), systems with long range interactions (strongly ionic systems), electrochemical system, or the behavior of nano structures are of major importance. However, systems of this size will not be accessible with traditional algorithms used in electronic structure theory. These algorithms have a scaling behavior that ranges from N^3 to N^8 for independent particle models to highly accurate couple cluster theory, respectively, where N denotes the system size.

In the contribution of Rothlisberger in Sect. 15.2 the time scale problem is addressed with a focus on the reaction path and mechanism of chemical reactions. In Sect. 15.3 Sprik discusses the time and length scale problems involved in the calculation of thermochemical constants. In the final Sect. 15.4 new techniques for extending system sizes are reviewed by Hutter.

15.2 Overcoming the Time Scale Barrier: Enhanced Sampling Techniques for *ab initio* Molecular Dynamics Simulations

15.2.1 Time Scale Limitations in *ab initio* Molecular Dynamics Simulations

The development of new methods that can significantly extend the time scale, or enhance the sampling efficiency of MD simulations is a longstanding topic of research and several schemes have been proposed for this purpose.[5–11] Among the particularly powerful approaches in this respect is the introduction of well-controlled modifications of the potential energy surface (PES) designed in such a way as to lower activation free energies and increase the rate for activated events [11]. Within transition state theory, one can estimate that for a decrease in potential depth by V_{bias} the rate constant increases exponentially by $\approx e^{\beta V_{bias}}$. The same exponential acceleration can be expected in the convergence of the free energy. It is possible to obtain the corrected results for the original system (with Hamiltonian H and inverse temperature *beta*) from the properties of the biased ensemble (described with H' and *beta*$'$). The thermodynamic average of a function $f(p,q)$ can be written as

$$< f(p,q) >_{\beta H} = < f(p,q) e^{\beta' H' - \beta H} >_{\beta' H'} \frac{1}{< e^{\beta' H' - \beta H} >_{\beta' H'}} \tag{15.1}$$

This equation shows that the average of the function f of the system with Hamiltonian H ($< f(p,q) >_{\beta H}$) can be obtained as a function of the averages of two different functions, $f(p,q)e^{\beta' H' - \beta H}$ and $e^{\beta' H' - \beta H}$, of a system based on the Hamiltonian H'. If H' is chosen in an appropriate way the new averages $< f(p,q)e^{\beta' H' - \beta H} >_{\beta' H'}$ and $< e^{\beta' H' - \beta H} >_{\beta' H'}$ converge faster than the original one. The technique will work efficiently as long as sufficiently many configurations sampled with the Hamiltonian H' are significant also for the system with Hamiltonian H, so that the reweighting factor $e^{\beta' H' - \beta H}$ is non-vanishing during a significant part of the trajectory. An appropriate choice of the bias V_{bias} is therefore essential for the efficiency of the method. In the following, we shall give an example for an atomic as well as for an electronic bias potential, which have both proven to be particularly useful in the context of AIMD and mixed quantum mechanical/molecular mechanical (QM/MM)-AIMD [12] simulations.

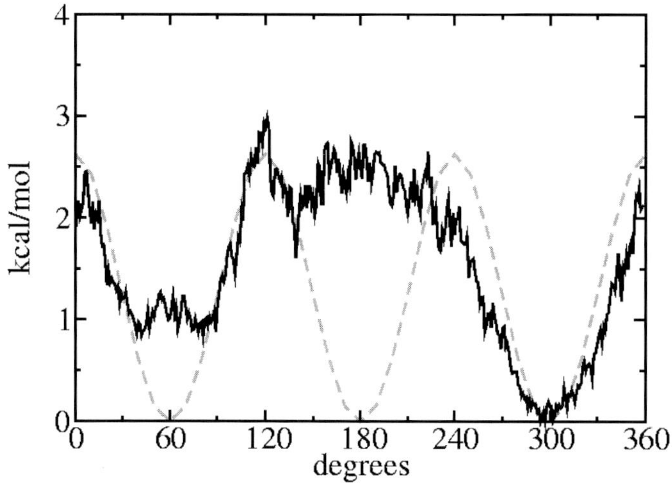

Fig. 15.1. Free energy profile for methyl group rotation in ethane, as obtained from an unbiased AIMD run at 500 K (full line). The dashed curve is the potential energy calculated on a grid of 10 degrees and is given for comparison

15.2.2 The Use of Classical Force Fields as Bias Potentials for an Enhanced Sampling of Conformational Transitions

A straightforward way to construct suitable bias potentials, especially appropriate for the enhanced sampling of conformational transitions, is the use of selected terms of a classical force field. [13] The configurational space can be restricted by including only specific terms (such as e.g. torsional potentials). Clearly, if the force field is sufficiently close to the ab initio results then the resulting potential ($V' = V + V_{bias}$) will be nearly flat, and the trajectory can visit all points in space without having to cross potential energy barriers. It is important to point out that the converged values for the sampled properties do not depend on the classical force field, and are thus of *ab initio* quality. Only the rate of convergence depends on the quality of the force field. Furthermore, this approach is inexpensive in both computational time, and implementation effort.

As an illustrative example, we show in Fig. 15.1 the one dimensional free energy profile for rotation of the methyl groups in ethane. The barrier for rotation[14], that is the energy difference between the eclipsed and the staggered conformation, is ~2.6 kcal/mol. The severe time scale barrier inherent to ab initio MD is clearly apparent. Even at 500 K, where ethane can often be regarded as a free rotator, direct ab initio molecular dynamics fails to sample the three symmetry equivalent minima homogeneously. Through the application of the classical bias method presented here these shortcomings can be fully overcome. Here, the classical bias potential was constructed in the form of a torsional potential $\frac{1}{2}V_0(1 - cos(3\phi))$. The distribution of the

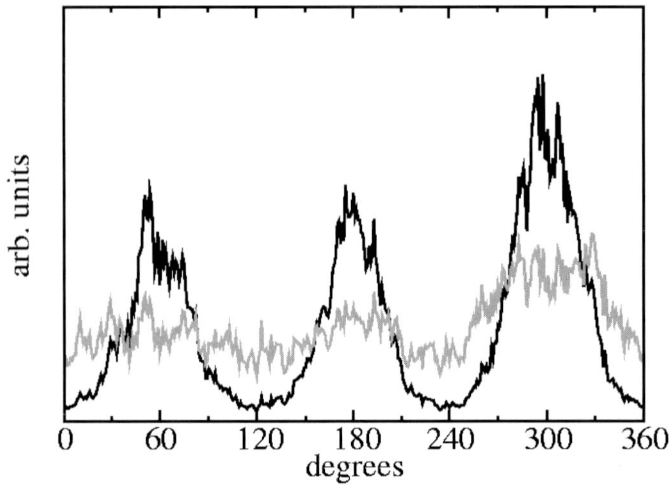

Fig. 15.2. Probability density (gray) and corrected probability density (black) distribution for the torsional angle from the biased run

torsional angle sampled during the simulation is shown in Fig. 15.2, together with the corrected distribution as obtained using (15.1). Clearly, the three equivalent minima are now sampled almost equally. The free energy (F) is obtained from $F(x) = -kT\ln(P(x))$, where P is the sampled distribution. This free energy profile is shown in Fig. 15.3 together with the potential energy profile. By inspecting the deviation from three-fold symmetry we can

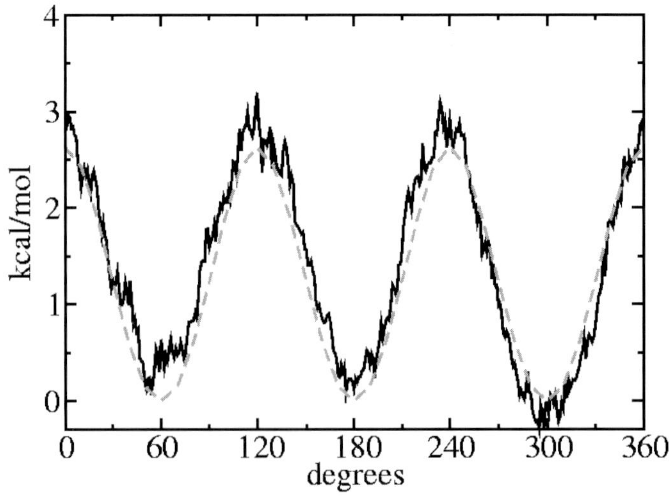

Fig. 15.3. Free energy profile from the biased run (full line) in comparison with the potential energy (dashed line)

estimate that the accuracy is better than 0.5 kcal/mol over the whole profile. Thus, the statistical error is certainly smaller than the error due to the approximations made to the exchange and correlation functional within DFT, and the neglect of the quantum nature of the hydrogens. A analogous strategy has been applied for a determination of the two dimensional free energy surface for dihedral transitions in peroxynitrous acid (ONOOH) [13].

15.2.3 Finite Electronic Temperatures as Electronic Bias Potentials

First-principles molecular dynamics simulations are especially valuable for complex systems in which a reaction coordinate is not known *a priori* and may involve a nontrivial combination of multiple degrees of freedom. Unfortunately, this approach is currently limited by the very short time window that can be simulated which restricts its application to the spontaneous observation of fast processes with low activation barriers. Usual simulation times are 10-100 ps, i.e. orders of magnitude shorter than what would be needed in order to observe chemical reactions with rates in the millisecond to hours range implying typical activation barriers of the order of 10 - 30 kcal/mol. At present, such cases can only be studied if some information about a possible reaction pathway is known so that it is possible to constrain the system along an *a priori* chosen coordinate in order to guide it over the activation barrier [5,6]. Since this information is not always available, and an inappropriately chosen coordinate can even be misleading, it is of primary importance to develop methods that are able to explore possible reaction pathways and potentially reactive configurations in an efficient and unprejudiced way. We have recently proposed a novel method that is capable to specifically enhance the sampling of rare reactive events in chemical systems [15]. A lot of information about the intrinsic chemical reactivity of a system is directly encoded in its electronic structure. This simple fact has been used by chemists for decades and lies at the roots of the overwhelming success of frontier orbital theory [16]. Within DFT, this intimate link between electronic structure and chemical properties can be formalized in the concepts of reactivity indices based on chemical hardness and softness [17]. Within a frozen orbital picture, the highest o ccupied orbital(s) (HOMO(s)) determine the electron-donor properties of a reactive system, whereas the lowest occupied orbital(s) (LUMO(s))are relevant for its electron-acceptor behavior and a linear combination of HOMOs and LUMOs is relevant in the context of radical reactions [17]. This suggests that a method that uses the electronic structure to construct a bias potential could be especially efficient in exploiting chemically relevant regions of phase space without having to resort to *a priori* assumptions about likely reaction pathways based on atomic coordinates. The simple electronic bias potential presented in Ref. [15] exploits the electronic structure in a straightforward way by simulating the system at an increased electronic temperature using the DFT free energy functional introduced by Mermin [18]. A simple form of this free energy functional is given by[20]

$$F[\{R_I\},T] = \min_{\{\psi_i\},\{f_{ij}\}} \sum_{ij} f_{ji} \int dr \left(\psi_i^* \hat{T} + \hat{V}_{ext}\psi_j\right) + E_{Hxc}[n,T] - TS[\{f_{ij}\}]. \tag{15.2}$$

Here E_{Hxc} is the Hartree and the exchange and correlation, S is the entropy term given by $-\mathrm{tr}(f\ln(f) + (1-f)\ln(1-f))$. The solution of the equations reduces to assigning occupation numbers (f_i) to the Kohn-Sham single particle states wavefunction ψ_i using the Fermi-Dirac distribution i.e. $f_i = 1/(e^{\beta(\epsilon_i-\mu)}+1)$ where ϵ_i is the eigenvalue of the wavefunction (ψ_i), and converging the wavefunctions self-consistently. The free energy can be used as a potential energy surface for the dynamics of the ions, and recent advances in computational methodology made it possible to use the free energy functional for MD.[19–21] Through the introduction of a finite electronic temperature, electron density is transferred from the highest lying donor orbitals of the entire reactive system to its lowest lying acceptor orbitals, i.e. its lowest-energy virtual states. Thus, for a fixed electronic temperature the free energy will be lowered more for a system with a small gap than for systems with a large gap. We can thus consider the difference between the finite temperature (FT) potential (V_{FT}) and the zero temperature (ZT) potential (V_{ZT}) to be a bias potential ($V_{bias} = V_{FT} - V_{ZT}$) that enhances the exploration of regions with small electronic gaps. This particular choice of electronic bias potential can be further rationalized using concepts from chemical reactivity theory, such as the maximum hardness principle [22–26] which states that the ground state of any system has a maximal hardness. This implies, that through the admixture of virtual excited states the system becomes softer, i.e. more reactive towards soft, orbital-driven reactions. The electronic bias potentials we present here can therefore be expected to work best for this type of reactions whereas typical hard reactions are electrostatically driven and can often be accelerated in more obvious ways. In order to examine the efficiency of this approach in a quantitative manner, we calculated energy profiles along simple reaction coordinates for a set of prototypical chemical reactions in gas and condensed phase (including bond dissociation, proton and electron transfer as well as isomerization reactions) and compared the barrier heights at finite electronic temperature with the 0 K results. In all cases, we found that setting the electronic temperature such that $f_{LUMO} \approx 0.05$ gave a significant lowering of the barriers while leaving the overall location of the minima unchanged. An example is given in Table 15.1 for the dissociation of the peroxy bond in ONOOH. As demonstrated in Table 15.1, substantial gains in sampling efficiency (e.g. up to 10 orders of magnitude!) can be achieved in this way. This simple method works surprisingly well in the case of orbital-driven chemical reactions that are dominated by the HOMO and LUMO orbitals for which impressive acceleration factors of several orders of magnitude can be achieved in a straightforward manner. This approach should be considered as

Table 15.1. O-O bond dissociation in ONOOH. ΔE is the lowering in dissociation energy in kcal/mol, Enhancement is an estimate of the corresponding increase in sampling efficiency for a simulation at room temperature, or, equivalently, the increase in the rate constant at room temperature ($\approx exp(\frac{\Delta E}{kT})$)

Temperature (K)	ΔE (kcal/mol)	Enhancement
1000	0	1
2000	1	10
4000	8	5e+06
5000	11	8e+09

a first step towards more advanced electronic bias methods that make direct use of the chemical information inherent in the electronic structure.

15.3 Computation of Acid Dissociation Constants

15.3.1 Time and Length Scales in Aqueous Chemistry

Liquids, by nature, require longer runs than solids or surfaces, since their stability depends on dynamical fluctuations. The liquids of interest to use here are aqueous solutions. The crucial time scale in these systems is the ≈ 3 ps lifetime of a hydrogen bond (for a review see for example Ref. [27]). Related to this is the $\approx 1-2$ ps hopping time for excess protons in solutions at non-neutral pH[27]. These are fairly favourable numbers from the point of view of *ab initio* MD studies of chemical reactions in aqueous solution. It means that a run in the order of 5-10 ps might just be long enough to allow for resolvation of reactants after a change of charge or polarity during a reaction. Also redistribution of surrounding excess protons is possible in this length of time. Generation of trajectories of a duration of 10 ps for systems of 30-50 water molecules plus one or two small solute molecules are now standard and no longer represent the huge computational effort it did ten years ago when the first simulations on 32 molecules of liquid water were carried out[28].

The opening up of this computational "window" is exploited by an increasing number of *initio* MD studies of elementary reactions in aqueous chemistry (see for example Refs. [29] - [39]). Reactions involving transport of excess protons[29,30] and also dissociation of strong acids[31] occur spontaneously on this time scale, and could be studied in all microscopic detail for the first time. However, while the solvent response falls (more or less) in the *ab initio* time window, almost all reactions of interest involve some formation or rearrangement of chemical bonds of the reactive solutes which is either endothermic or activated and, hence, impossible to observe in 10 ps. Special sampling techniques are needed which force these rare events to occur. The challenge, of course, is to have the system find a trajectory in phase space as close as possible to the true, but unknown, reaction path. A further condition, of particular importance in liquids, is that these methods can provide us with an estimate of the free energy profile, since in a fluctuating finite temperature environment transition paths are not unique.

A variety of techniques has been developed for this purpose some of which are discussed in detail elsewhere in the present proceedings. Early *ab initio* MD applications relied on control of some predetermined order parameter by mechanical constraints[5]. While computational efficient, the coordinate driven approach has the drawback that it can easily produce inaccurate or even erroneous results if the reaction coordinate is a poorly described by the chosen order parameter. An alternative, special to *ab initio* MD method, is increasing the electronic temperature[15]. The best solution to the rare event sampling problem is perhaps the transition path sampling method, which was recently developed by the Berkeley group of David Chandler[8,9]). This method is free of any dynamical bias. Only specification of a initial (reactant) and final (product) state is required (see also Ref. [10]). A pioneering *ab initio* MD application in the condensed phase of this method can be found in Ref. [38].

15.3.2 Determination of Free Energy Profiles

A free energy profile $\Delta w\left(q\right)$ describes the variation of free energy with an order parameter q (not necessarily identical to the reaction coordinate). $\Delta w\left(q\right)$ is related to the work performed during reversible changes of the order parameter

$$\Delta w\left(q\right) = w\left(q\right) - w\left(q_0\right) = -\int_{q_0}^{q} dq' f\left(q'\right) \qquad (15.3)$$

where $f\left(q\right)$ is the mean force and q_0 is the value of q at some reference state[40]. Equation (15.3) forms the basis of the thermodynamic integration method for determination of relative free energy. What makes it a practical computational scheme in MD simulation is the relationship between mean force $f\left(q\right)$ and the mechanical force f_q exerted on the order parameter when held fixed at a specified value q'. $f\left(q\right)$ can be estimated by the time average of f_q

$$f\left(q'\right) = \left\langle f_q \right\rangle_{q=q'} \qquad (15.4)$$

where the subscript attached to the brackets denotes a constrained average. If, moreover, the method of constraints is used f_q can be obtained from the corresponding Lagrange multiplier[6].

Clearly for determination of activation energies it is crucial that the order parameter not only distinguishes initial and final states but also gives an accurate representation of the reaction path connecting these two states. For some elementary reactions this condition can be met, to a good approximation, by simple geometric order parameters such as distance[32,35,36], dihedral angles[33,34] and coordination numbers[37] of selected atoms. However, even in these seemingly simple chemical processes rearrangements of the environment (e.g. conformation and solvent degrees of freedom) can cause complications and the results of coordinate driven methods need to be interpreted with caution[9]. On the other hand, if we are only interested in relative free energies of reactant and products the demands on the transition path are less stringent. In principle any reversible path linking reactant and product could be used. In practice reversibility can be difficult to achieve along an unstable path. Therefore, for thermochemical calculations, the method of constraining approximate reaction coordinates remains an attractive option which may well have advantages in terms of computational efficiency.

15.3.3 Statistical Thermodynamics of Gas-Phase Equilibria

In thermochemical studies the ultimate aim is estimation of the equilibrium constant of a reaction, which can be directly compared to experiment. The equilibrium constant K_c for a dissociation reaction $AB \rightarrow A + B$ is defined as

$$K_c = \frac{[\mathrm{A}]\,[\mathrm{B}]}{c^{\mathrm{o}}\,[\mathrm{AB}]} \tag{15.5}$$

where $[X]$ denotes the equilibrium concentration of component $X = A, B, AB$, i.e. $[X] = N_X/V$ if N_X is the number of particles of species X and V is the volume of the container. $c^{\mathrm{o}} = 1 \;\; \mathrm{mol\ dm}^{-3}$ is the concentration in the standard state. Assuming the mixture has been produced by dissociation of an initially pure sample of N dimers, K_c can also be written in terms of a dissociation fraction fraction $\alpha = N_A/N = N_B/N$ (hence, $N = N_A + N_{AB}$). This expression, which is needed later, is

$$K_c = \frac{\alpha^2}{1-\alpha}\frac{N}{Vc^{\mathrm{o}}} \tag{15.6}$$

How to compute K_c (or alternatively α) from the relative free energy Δw obtained by thermodynamic integration? To see the relation between these quantities we first consider gas-phase dissociation. The equilibrium will lie strongly on the left(reactant side) because of the large energy required to break the dimer bond. In the gas-phase this energy is simply the difference $\Delta\epsilon$ between the (ground-state) energies ϵ_X of the dimer and the atomic fragments:

$$\Delta\epsilon = \epsilon_{\mathrm{A}} + \epsilon_{\mathrm{B}} - \epsilon_{\mathrm{AB}} > 0 \tag{15.7}$$

The textbook statistical thermodynamics expression for the dissociation constant is

$$K_c = \frac{v^{\mathrm{o}}}{\Lambda_\mu^3}\left(q_{vib}\,q_{rot}\right)^{-1} e^{-\Delta\epsilon/k_BT} \tag{15.8}$$

where $v^{\mathrm{o}} = 1/c^{\mathrm{o}} = 1660\ \mathrm{\AA}^3$. The small exponent for the endothermic dissociation process is multiplied by a prefactor consisting of the (inverse) vibrational and rotational partition function of the AB dimer indicated by q_{vib}

respectively q_{rot} and a translational contribution depending on the thermal wavelength associated with the *reduced* mass

$$\Lambda_\mu = \sqrt{\frac{h^2}{2\pi\mu k_B T}} \tag{15.9}$$

Typical values for the dimer partition functions are $q_{vib} \approx 1$ and $q_{rot} \approx 50$. Thermal wavelengths of atoms, on the other hand, are very small $\Lambda_\mu \approx 0.01 - 0.1$Å and therefore $\Lambda_\mu^3 \ll v^{\text{o}}$, leading to substantial enhancement of the prefactor. The explanation for this effect, favouring the product, is the huge increase in entropy due to the creation of an extra unbound translational degree by the dissociation. The result is a significant reduction of the standard dissociation (Gibbs) free energy

$$\frac{\Delta G^{\text{o}}}{k_B T} = -\ln K_c = \frac{\Delta\epsilon}{k_B T} - \ln\frac{v^{\text{o}}}{\Lambda_\mu^3} + \ln q_{vib} + \ln q_{rot} \equiv \ln 10 \times pK_{diss} \tag{15.10}$$

The last identity defines the pK for dissociation. Measured in terms of a contribution to the pK the translational entropy term in (15.10) can be as large as several units.

15.3.4 Reversible Work and Equilibrium Constants

In order to generalize this fundamental result from the statistical thermodynamics of gas-phase reactions to solutions we first notice that in solution the role of $\Delta\epsilon$ is taken over by the reversible work Δw (Eq. (15.3)) for breaking up the dimer and separating the two fragments A and B, which now can also be ionic. Δw contains already an entropy contribution, but this is the entropy related to the solvation of reactants and products and should be distinguished from the translational entropy term appearing in (15.10). This term, arising from the partition function from a free quantum particle, cannot be straightforwardly transferred to an interacting particle in a dense liquid. One way to circumvent this problem is by a thermodynamic (Born-Haber) cycle, taking the dimer out of solution to vacuum, and then, after dissociation, reinserting the product atoms back into solution(see for example [41]). This requires computation of the solvation free energies of all components involved, which is relatively easy to do when the solvent is modelled by a reaction field, as is the case in many quantum chemistry code, but is a very difficult calculation in the small periodic model systems used in *ab initio* MD.

More in the spirit of atomistic simulation (MD or Monte Carlo) is the purely classical expression for the dissociation equilibrium constant derived by David Chandler in his textbook on statistical mechanics (DC)[40]. This expression, which is valid in the gas-phase and liquid alike, is based on the reversible work theorem: The theorem states that the reversible work $w(r)$

for bringing particles together from infinity to a distance r apart can be used as an effective potential to generate the radial distribution function $g(r)$

$$g(r) = e^{-\beta w(r)} \tag{15.11}$$

with $\beta = 1/k_B T$. The function $w(r)$ in (15.11) is often referred to as the potential of mean force (PMF). It is equal to the mean force integral of (15.3) w.r.t to r provided $w(r_0) = 0$ (see below). However, the radial distribution function (RDF) can be interpreted as a (conditional) probability distribution for distance. Thus, by introducing a distance criterion R_c for bonding

$$|\mathbf{r}_A - \mathbf{r}_B| = r \begin{cases} \leq R_c \; \textit{bonded AB dimer reactant} \\ > R_c \; A + B \textit{ dissociation product} \end{cases} \tag{15.12}$$

the equilibrium constant for association (dimerization) can be directly obtained as the integral of the RDF

$$K_c^{-1} = c^{\text{o}} \int_0^{R_c} dr\, 4\pi r^2 g(r) \tag{15.13}$$

(For more details see DC pag. 212). In accordance with (15.5) concentration is measured in units of standard hence the factor c^{o} in (15.13). Substitution of (15.11) gives

$$K_c^{-1} = c^{\text{o}} \int_0^{R_c} dr\, 4\pi r^2 e^{-\beta w(r)} \tag{15.14}$$

which relates the equilibrium constants to the PMF. To verify that (15.14) indeed accounts for the creation of translational entropy we consider the simple example of a model gas-phase dimer bound by a square well potential

$$v(r) = \begin{cases} -\Delta\epsilon \;\; r \leq R_c \\ \;\;\; 0 \;\;\;\; r > R_c \end{cases} \tag{15.15}$$

In the low density limit the PMF can be replaced by the pair potential. Inserting (15.15) into (15.14) we find

$$K_c = \frac{3v^{\text{o}}}{4\pi R_c^3}\, e^{-\Delta\epsilon/k_B T} \tag{15.16}$$

Similar to (15.8) the standard volume per particle $v^{\text{o}} = 1/c^{\text{o}} = 1660\ \text{Å}^3$ is compared to a molecular "small" volume, namely the volume $V_c = 4/3\pi R_c^3$ of the sphere with radius R_c. However, since in general $V_c \gg \Lambda_\mu^3$ the classical prefactor in (15.16) will be much smaller than the quantum prefactor in (15.8).

15.3.5 Controlled Dissociation in a Small Box

The formalism of Sect. 15.3.4 reduces the problem of time scales for computation of equilibrium constants to a question of reversibility. The finite size effects remain, which we will address now. Distances in a periodic cubic cell of size L are limited to a maximum of $R_{max} \approx L/2$. PMF's determined under periodic boundary conditions are, therefore, necessarily relative to the value at R_{max}

$$\Delta w(r) = w(r) - w(R_{max}) \tag{15.17}$$

Using (15.14) without further corrections yields a "truncated" equilibrium constant

$$\overline{K}_c^{-1} = c_0 \int_0^{R_c} dr\, 4\pi r^2 e^{-\beta \Delta w_{\mathrm{AB}}(r)} = e^{\beta w(R_{max})} K_c^{-1} \tag{15.18}$$

In general $w(R_{max}) < 0$, leading to underestimation of the pK for dissociation

$$\overline{pK}_{diss} = pK_{diss} + \frac{w(R_{max})}{2.303\, k_B T} < pK_{diss} \tag{15.19}$$

At 300 K a pK unit corresponds to 1.4 kcal mol^{-1}. Therefore, for the typical *ab initio* MD cell size of $L \approx 10$ Å errors can be substantial. Interaction between periodic images gives rise to further inaccuracies, distorting the mean force $f(r)$, which by symmetry must vanish at $r \approx L/2$. For heterolytic reactions in a polar solvent involving charge separation, errors can be expected to be especially large. For these reactions estimated pK's can be too low by several pK units.

The systematic bias due to system size can be reduced by modifying Chandler's procedure for computing dissociation constants making explicit use of the "tight" conditions prevalent in small model systems. We will focus on the determination of the pK_a of a weak acid, i.e. the dissociation constant of the reaction AH $\rightarrow$ A^- + H^+. In view of the active participation of the solvent, this reaction is an elementary but important example of application of *ab initio* MD methods to aqueous chemistry. Because of the restrictions in system size the number of AH species is limited to a single molecule, which is subjected to controlled dissociation. This enables us to fix the reference free energy $w(R_{max})$ by normalizing the RDF of the acid H w.r.t its conjugate base A to unity

$$\frac{1}{V}\int_V d\mathbf{r}\, e^{-\beta \Delta w(r)} = 1 \tag{15.20}$$

where $V = L^3$ is the volume of the cell. The rationale behind (15.20) is that there is only one excess proton, which can only be in the MD cell and nowhere else. Because of this constraint on the number and position of leaving protons, the dissociation fraction α can be directly obtained from the coordination number computed from the normalized RDF of (15.20).

$$\alpha(R_c) = 1 - \frac{\int_0^{R_c} dr\, 4\pi r^2 e^{-\beta \Delta w(r)}}{\int_0^{R_{max}} dr\, 4\pi r^2 e^{-\beta \Delta w(r)}} \tag{15.21}$$

R_c is again the critical distance for bonding defined in (15.12). To simplify the integration we have approximated the integral over the cubic volume in (15.20) by in an integral over a sphere of radius R_{max}. The dissociation fraction can be converted to a equilibrium constant using (15.6) giving for the pK_a

$$pK_a = -\frac{1}{2.303}\left\{\ln\left[\frac{\alpha(R_c)^2}{1-\alpha(R_c)}\right] + \ln\frac{v_o}{v_m}\right\} \tag{15.22}$$

The constant $v_m = V/N$ is the volume per base A (whether dissociated or not). In this case $N = 1$. A pK computed according to the scheme of (15.22) is not identical to the result obtained from (15.14). The discrepancies become increasingly significant for R_c approaching R_{max}: whereas the pK_a predicted by (15.14) remains finite, the pK_a of (15.22) diverges since the dissociation fraction of (15.21) by vanishes by construction for $R_c = R_{max}$.

15.3.6 Computation of the Water Dissociation Constant

High on the list of acid dissociation constants to be calculates is the ionization product of liquid water. This quantity is defined as

$$K_w = (c^o)^{-2}\left[H_3O^+\right]\left[OH^-\right] \tag{15.23}$$

K_w corresponds formally to a dissociation constant at unit activity of the acid (=solvent). K_w can be computed, as before, by expressing it in terms of a dissociation fraction

$$K_w = \left(\frac{\alpha(R_c)\,N}{V c^o}\right)^2 \tag{15.24}$$

and approximating $\alpha(R_c)$ by (15.21). Now, however, since all "acid" molecules are equivalent, N is the total number of water molecules in the cell of volume V. Hence, under ambient conditions, $N/(Vc^o) = 55.4$. Substituting in (15.24) and taking the base 10 logarithm we find

$$pK_w = -0.868\left\{\ln\left[\alpha(R_c)\right] + \ln 55.4\right\} \tag{15.25}$$

Auto dissociation of liquid water has developed into a popular subject for *initio* MD simulation. Here we mention four recent calculations all based on very similar pseudo-potential plane wave methods but differing in the rare event sampling techniques. Refs. [36] and [37] use mechanical constraint methods to deprotonate a selected water molecule: one of its OH bonds is stretched until it breaks. In the process the proton is transferred to a second hydrogen bonded water molecule which plays the role of acceptor. In Ref. [36] the bond is broken by step wise increase of the length R_{OH}. In Ref. [37] proton release is enforced by reducing the proton coordination n_{H} of a specified H_2O oxygen from two to one. Water auto-dissociation was also the first condensed *ab initio* MD problem to which full scale transition path sampling was applied, revealing a number of details about the cooperative hydrogen bond motion driving the transfer[38]. Ref. [39], is a study of accelerated dissociation under conditions of elevated temperature and pressure.

The free energy profiles obtained by the constraint calculations of Refs. [36] and [37] provide us with unique material to test the procedure for pK computation since these data should be directly comparable (for technical details, such as density functionals, pseudo potentials etc, we refer to the original papers). Setting $r = R_{\mathrm{OH}}$ the $\Delta w\,(R_{\mathrm{OH}})$ of Ref. [36] can be immediately inserted in (15.21) giving an estimate of the dissociation fraction. The corresponding pK_w computed according to (15.25) is plotted as a function of the bonding radius R_c in Fig. 15.4.

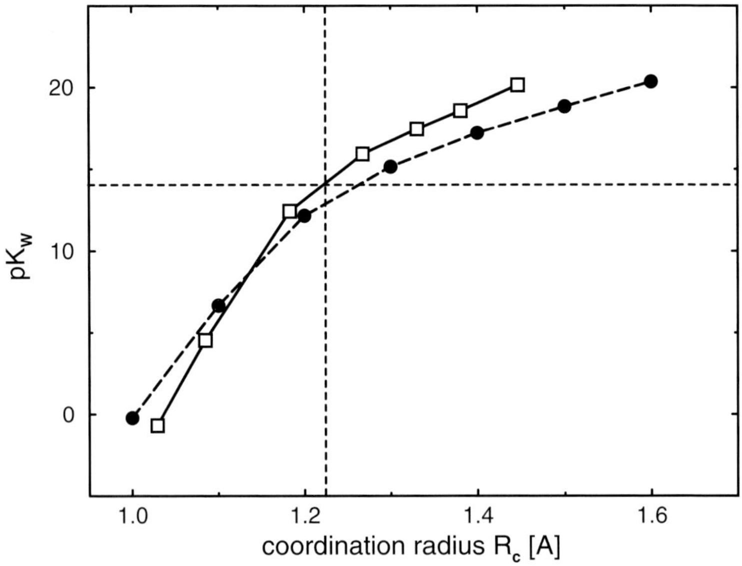

Fig. 15.4. pK for the auto-dissociation of liquid water computed according to (15.25) as a function of the critical radius R_c for OH bonding. Compared are the results for two independent calculations using different constraint methods to enforce bond breaking. Solid circles are based on the free energy profile of Ref. [36] obtained using a OH bond constraint. The data for the open squares were generated using a proton coordination constraint[37]. Both calculations were performed for a 32 water molecule sample in a cubic box of size 9.86Å using the BLYP functional. Crossing dashed lines indicate the value of R_c reproducing the experimental pK_w.

The free energy profile $\Delta w(n_{\mathrm{H}})$ of Ref. [37] for proton coordination number changes can be subjected the same treatment after having been converted to a dependence on OH distance. The straightforward way to do this is to search for the proton H* which is second nearest to the coordination constrained oxygen atom O*. This must be the leaving proton (the nearest proton is the proton that remains). Under a coordination constraint on O* the distance $R_{\mathrm{O^*H^*}}$ is a dynamical quantity (fluctuation in time (see Ref. [37], note that even the identity of H* can change). So, we identify the average over a trajectory at a given fixed value of n_{H} with the order parameter r in (15.21). The corresponding $\Delta w(r)$ function is then obtained by transforming the $\Delta w(n_{\mathrm{H}})$ profile according to the inverse the $r(n_{\mathrm{H}})$ relationship. The pK_w is again determined using (15.25). The result is also displayed in Fig. 15.4. Considering that the two curves in Fig. 15.4 were generated using two rather different constraint schemes the agreement can be seen as a validation of our approach. At this point a comment on the approach of ref. [37] is in order. In this work, which provided the data for the coordination constraint curve in Fig. 15.4, the pK_w is estimated by simply the free energy difference between the equilibrium OH distance and the dissociated state. A value of 13 was obtained. This approach, however, is not correct. The agreement with experiment is presumably the result of fortuitous cancellation of the errors of the size effect and the neglect of the translational entropy of the dissociation product.

As will be evident from Fig. 15.4 the computed pK_W is strongly dependent on R_c making the choice of this parameter critical. In spite of this, the radius of $R_c = 1.23$Å reproducing the experimental $pK_w = 14$ (see Fig. 15.4) is entirely consistent with accepted criteria for maximum extension of OH bond length. This observation suggests a pragmatic solution to the dilemma of the specification of R_c: We simply treat R_c as an adjustable parameter and use the pK_w data of Fig. 15.4 to fix its value. The resulting R_c of 1.23Å can be applied to compute the pK_a of other weak acids involving ionization of hydroxyl groups.

15.3.7 Application to Weak Acids and Evaluation of Method

A first such "calibrated" calculation of acid dissociation constants is reported in Ref. [42]. The weak acid studied there is aqueous $\mathrm{P(OH)_5}$. This molecule is a model for a family of phosphoranes occurring as intermediates in biochemical processes such as RNA hydrolysis. Phosphoranes are compounds containing a central phosphor bonded to five atoms, in this case all oxygen, in an approximately trigonal bipyramidal coordination. An important issue in the chemistry of these unstable molecules is the strength of the second bond formed by the equatorial oxygens relative to the bonds formed by the axial oxygens. This will also affect the relative pKa of hydroxyl groups in these positions, hence this quantity can be used as a probe of bond strength.

Unfortunately, the lifetime of phosphoranes is too short for standard pKa determination. Estimates can only made by indirect methods, such as structure reactivity relations (for more details see Ref. [42]). Independent verification by computational methods can therefore be of considerable help.

The calculation followed the same scheme as applied in the reference study of auto-dissociation of water[37]. In two separate series of calculations an equatorial and axial hydroxyl group of $P(OH_5)$ were deprotonated using the coordination number constraint method[42]. This resulted in two free energy profiles $\Delta w(n_{\mathrm{H}})$ which were transformed to PMF's for OH distance as described in Sect. 15.3.6 and then used to compute the dissociation constant $\alpha(R_c)$ as a function of R_c according to (15.21). Substituting in (15.22) leads to R_c dependent pKa curves similar to Fig. 15.4. The final estimate of pKa is obtained by reading off the value at $R_c = 1.22$Å, the bonding radius fitted to the experimental pKw (see Fig. 15.4). The result is a value of 9.8 for equatorial dissociation compared to 14.2 for axial dissociation, which is in good agreement with the experimental estimates of 8.6 respectively 13.5(for the experiment see also Ref. [42]).

These numbers are encouraging, in fact better than expected in view of the many uncertainties involved. Apart from the density functional that is employed, the main source of error is indeed the limited system size. By treating the bonding radius R_c as an "empirical" parameter adjusting it to the experimental ionization product of water, we have made this fundamental reaction effectively into a reference for the computation of pK's of weak acids. The explanation for the apparent success of this approach in periodic cells as small as used in *ab initio* MD, is not that long range effects are unimportant, but that they are very similar for the acid dissocations considered. Evidently, what distinguishes these reactions is the short range process of breaking the OH bond and the subsequent formation of an $O^- - H_3O^+$ contact ion pair.

This simple justification of our rather pragmatic approach leaves a number of open questions. To mention just one: Treated as a single (highly correlated) solution, the acid solute and its periodic images reach concentrations in the order of 2M. In this regime activity coefficients can deviate from unity by factors two or more. How to account for these effects is not clear. Further test and analysis must bring answers to these and other questions. The example of pKa calculation, however, shows that the use of reference systems is a valid option to compensate for small system size in *ab initio* methods.

15.4 Linear Scaling Electronic Structure Methods for *ab initio* Molecular Dynamics

Ab initio molecular dynamics simulations have made huge progress in recent years. Recently, much work has been done to devise new algorithms that reduce the scaling behavior and the ultimate goal, linear scaling has been achieved for most electronic structure methods. The review of Goedecker [43] gives an overview of the different attempts.

Although most electronic structure methods are now available in a form that allows for linear scaling algorithms there are still huge differences in performance. Combining the electronic structure methods with molecular dynamics is on a regular basis only possible for the independent particle methods. Of those the Kohn–Sham method of density functional theory provides the highest accuracy and widest applicability. In the following sections we will discuss methods and algorithms that allow for an optimal implementation of linear scaling techniques with the Kohn–Sham method.

The Kohn–Sham energy is defined as

$$\mathrm{E}_{\mathrm{KS}} = E_{\mathrm{kin}}[\{\Phi\}] + E_{\mathrm{ext}}[n] + E_{\mathrm{H}}[n] + E_{\mathrm{XC}}[n] \ , \tag{15.26}$$

where the kinetic energy E_{kin} is a function of the Kohn–Sham orbitals $\{\Phi\}$ and the external energy E_{ext}, Hartree energy E_{H}, and exchange and correlation energy E_{XC} are functions of the electron density n. The election density is calculated from the orbitals

$$n(\mathbf{r}) = \sum_i f_i |\Phi_i(\mathbf{r})|^2 \ , \tag{15.27}$$

and the occupation numbers f_i. Using the variational principle and imposing the orthogonality constraint on the orbitals leads to the Kohn–Sham equations (in canonical form)

$$H_{\mathrm{KS}} \Phi_i = \epsilon_i \Phi_i \ . \tag{15.28}$$

As the Kohn–Sham Hamiltonian H_{KS} depends on the orbitals, these equations have to be solved iteratively until self–consistence is achieved.

Expanding the orbitals in a basis set transforms the equation into algebraic form and we are left with the tasks to calculate the Kohn–Sham Hamiltonian in its matrix representation from a given set of orbitals and the search for the eigenvalues and eigenfunctions of this matrix. In the following sections we will discuss methods and algorithms to efficiently perform these two tasks.

15.4.1 Kohn–Sham Matrix Calculation

To be able to calculate the Kohn–Sham matrix in a number of steps proportional to the system size, first it is necessary to find a matrix representation with only O(N) non–vanishing elements. This can be achieved by using a basis set with finite support. There are two basis sets of this type used in linear scaling calculations. Basis sets connected with grids in real space include finite difference methods [44], finite element type methods based on B-splines [45] and plane wave derived basis sets [46]. Another possibility is to use basis sets derived from atomic orbitals. These basis sets are used in quantum chemistry calculations and are very popular with linear scaling approaches [47,48]. The matrix elements to be calculated are

$$H_{\mu\nu} = \langle\mu| - \frac{1}{2}\nabla^2|\nu\rangle + \langle\mu|V_{\text{ext}}(\mathbf{r})|\nu\rangle + \langle\mu|V_{\text{xc}}(\mathbf{r})|\nu\rangle + \frac{1}{2}\sum_{\alpha\beta} P_{\alpha\beta}\langle\mu\nu||\alpha\beta\rangle \ , \tag{15.29}$$

where V_{ext} is the external potential and V_{xc} the exchange and correlation potential. Coulomb integrals are defined by

$$\langle\mu\nu||\alpha\beta\rangle = \int\int \frac{\varphi_\mu(\mathbf{r})\varphi_\nu(\mathbf{r})\varphi_\alpha(\mathbf{r}')\varphi_\beta(\mathbf{r}')}{|\mathbf{r}-\mathbf{r}'|} \, d\mathbf{r} \, d\mathbf{r}' \ . \tag{15.30}$$

There are N^4 integrals of this type, but using screening techniques [49] the asymptotic number of non–vanishing integrals for large systems is N^2. To further reduce the scaling of this part of the calculation, adaptations of fast multipole methods to charge distributions have been developed. [50]

Another approach expands the charge distribution into an auxiliary basis [51]

$$n(\mathbf{r}) = \sum_{\alpha\beta} P_{\alpha\beta}\varphi_\alpha(\mathbf{r})\varphi_\beta(\mathbf{r}) = \sum_k d_k \chi_k(\mathbf{r}) \ . \tag{15.31}$$

This reduces the inherent scaling for the Coulomb integrals to N^3, which can be again lowered to N^2 using integral screening. The advantage is that for small and medium system sizes, where the screening is not yet efficient, considerable speed up can be achieved compared to the original method. [52]

A plane wave expansion of the electronic density can be viewed as a special auxiliary basis set. Plane waves have the advantage of being orthogonal and the expansion coefficients do not depend on the metric used. Further do plane waves describe the Hartree potential derived from the density exactly in the same basis and the calculation of the potential only needs O(N) steps. The major disadvantage is the number of plane waves needed to accurately describe the density. This effectively limits its use to methods where the strongly varying parts of the electron density are either handled by pseudo potentials or taken care of in other ways.

Another advantage of a plane wave expansion is that the long range part of the external potential can be treated together with the other Coulombic terms, greatly reducing the pre factor of the calculation of those matrix elements. In the following the derivation of the electrostatic energy in a suitable form for the treatment with plane waves is outlined.

The electrostatic energy of a system of nuclear charges Z_I at positions $\mathbf{R}_I$ and an electronic charge distribution $n(\mathbf{r})$ consists of three parts: the Hartree energy of the electrons, the interaction energy of the electrons with the nuclei and the internuclear interactions

$$E_{\text{ES}} = \frac{1}{2}\int\int d\mathbf{r} \, d\mathbf{r}' \frac{n(\mathbf{r})n(\mathbf{r}')}{|\mathbf{r}-\mathbf{r}'|} + \sum_I \int d\mathbf{r} \, V_{\text{core}}^I(\mathbf{r})n(\mathbf{r}) + \frac{1}{2}\sum_{I\neq J} \frac{Z_I Z_J}{|\mathbf{R}_I - \mathbf{R}_J|} \ . \tag{15.32}$$

The Ewald method can be used to avoid singularities in the individual terms when the system size is infinite. In order to achieve this a Gaussian core charge distribution associated with each nuclei is defined

$$n_{\mathrm{c}}^{I}(\mathbf{r}) = -\frac{Z_I}{(\mathrm{R}_I^{\mathrm{c}})^3}\pi^{-3/2}\exp\left[-\left(\frac{\mathbf{r}-\mathbf{R}_I}{\mathrm{R}_I^{\mathrm{c}}}\right)^2\right] . \tag{15.33}$$

At this point we redefine the core potential to be the potential of the Gaussian charge distribution of Eq. (15.33)

$$V_{\mathrm{core}}^{I}(\mathbf{r}) = \int d\mathbf{r}'\,\frac{n_{\mathrm{c}}^{I}(\mathbf{r}')}{|\mathbf{r}-\mathbf{r}'|} = -\frac{Z_I}{|\mathbf{r}-\mathbf{R}_I|}\mathrm{erf}\left[\frac{|\mathbf{r}-\mathbf{R}_I|}{\mathrm{R}_I^{\mathrm{c}}}\right] , \tag{15.34}$$

where erf is the error function. This definition is convenient for the further development of the electrostatic energy and leaves an effective external potential that only has short range terms. The interaction energy of this Gaussian charge distributions is added and subtracted from the total electrostatic energy. This allows to combine terms to the electrostatic energy of a total charge distribution $n_{\mathrm{tot}}(\mathbf{r}) = n(\mathbf{r}) + n_{\mathrm{c}}(\mathbf{r})$, where $n_{\mathrm{c}}(\mathbf{r}) = \sum_I n_{\mathrm{c}}^{I}(\mathbf{r})$. The remaining terms are rewritten as a double sum over nuclei and a sum over self–energy terms of the Gaussian charge distributions

$$\begin{aligned} E_{\mathrm{ES}} &= \frac{1}{2}\int\!\!\int d\mathbf{r}\,d\mathbf{r}'\,\frac{n_{\mathrm{tot}}(\mathbf{r})n_{\mathrm{tot}}(\mathbf{r}')}{|\mathbf{r}-\mathbf{r}'|} \\ &+\frac{1}{2}\sum_{I\neq J}\frac{Z_I Z_J}{|\mathbf{R}_I-\mathbf{R}_J|}\mathrm{erfc}\left[\frac{|\mathbf{R}_I-\mathbf{R}_J|}{\sqrt{\mathrm{R}_I^{\mathrm{c}\,2}+\mathrm{R}_J^{\mathrm{c}\,2}}}\right] - \sum_I \frac{1}{\sqrt{2\pi}}\frac{Z_I^2}{R_I^{\mathrm{c}}} , \end{aligned} \tag{15.35}$$

where erfc denotes the complementary error function.

For a periodically repeated system the total energy per unit cell (volume V) is derived from the above expression by using the solution to Poisson's equation in Fourier space for the first term and make use of the quick convergence of the second term in real space. The total charge is expanded in plane waves with expansion coefficients

$$n_{\mathrm{tot}}(\mathbf{G}) = n(\mathbf{G}) - \frac{1}{\mathrm{V}}\sum_I \frac{Z_I}{\sqrt{4\pi}}\exp\left[-\frac{1}{2}G^2 R_I^{\mathrm{c}\,2}\right] e^{i\mathbf{G}\cdot\mathbf{R_I}} . \tag{15.36}$$

This leads to the electrostatic energy for a periodic system

$$E_{\mathrm{ES}} = 2\pi\,\mathrm{V}\sum_{\mathbf{G}\neq 0}\frac{|n_{\mathrm{tot}}(\mathbf{G})|^2}{G^2} + E_{\mathrm{ovrl}} - E_{\mathrm{self}} , \tag{15.37}$$

where

$$E_{\mathrm{ovrl}} = \sum_{I,J}{}' \sum_{\mathbf{L}} \frac{Z_I Z_J}{|\mathbf{R}_I-\mathbf{R}_J-\mathbf{L}|}\mathrm{erfc}\left[\frac{|\mathbf{R}_I-\mathbf{R}_J-\mathbf{L}|}{\sqrt{\mathrm{R}_I^{\mathrm{c}\,2}+\mathrm{R}_J^{\mathrm{c}\,2}}}\right] \tag{15.38}$$

and

$$E_{\mathrm{self}} = \sum_I \frac{1}{\sqrt{2\pi}} \frac{Z_I^2}{R_I^{\mathrm{c}}} \ . \tag{15.39}$$

Here, the sums expand over all atoms in the simulation cell, all direct lattice vectors $\mathbf{L}$, and the prime in the first sum indicates that $I < J$ is imposed for $\mathbf{L} = \mathbf{0}$. The electrostatic potential needed in the calculation of the Kohn–Sham matrix is easily calculated in the plane wave basis

$$V_{\mathrm{H}}(\mathbf{r}) = 4\pi \mathrm{V} \sum_{\mathbf{G}\neq 0} \frac{n_{\mathrm{tot}}(\mathbf{G})}{G^2} e^{i\mathbf{G}\cdot\mathbf{r}} \ . \tag{15.40}$$

Efficient screening can be applied to the calculation of the matrix elements of this potential in real space [47].

The Gaussian–Augmented Plane Wave Method (GAPW) method [53,54] uses a linear combination of atomic orbital expansion of the Kohn–Sham orbitals

$$\Phi_i(\mathbf{r}) = \sum_\alpha c_{\alpha i} \varphi_\alpha(\mathbf{r}). \tag{15.41}$$

However, instead of using the straightforward definition of the electron density, another representation used in the projector–augmented wave (PAW) method [55] is employed

$$n(\mathbf{r}) = \tilde{n}(\mathbf{r}) - \tilde{n}^1(\mathbf{r}) + n^1(\mathbf{r}), \tag{15.42}$$

where $\tilde{n}$ is smooth and distributed over all space, and

$$n^1(\mathbf{r}) = \sum_A n_A^1(\mathbf{r} - \mathbf{R}_A) \tag{15.43}$$

$$\tilde{n}^1(\mathbf{r}) = \sum_A \tilde{n}_A^1(\mathbf{r} - \mathbf{R}_A) \tag{15.44}$$

are sums of atom–centered contributions. The individual densities are constructed such that n_A^1 and $\tilde{n}_A^1$ cancel outside a spherical radius of atom A and inside this radius $\tilde{n}$ is equal to the atom–centered contribution $\tilde{n}_A^1$. Based on this rewriting of the electron density the method used by Blöchl for the PAW method can be applied to redefine Hartree and exchange and correlation energies as sums of independent contributions from the smooth density $\tilde{n}$ and the atom–centered charge distributions.

The conditions on the different parts of the density are not strictly imposed but only approximately using projector methods. [55] Products of functions from a second basis $\{g\}$, derived from the original basis $\{\varphi\}$ are used to expand the atom–centered densities. The relation of the two basis sets is governed by matrices $D_{a\mu}^A$ and $\tilde{D}_{a\mu}^A$, where a denotes a function g_a at atom A and μ any function φ_μ from the original basis. Details on how to construct the basis functions $\{g\}$ and the matrices D are given in Refs. [53,54]. Finally, the smooth density is represented by an auxiliary basis set of plane waves

$$\tilde{n}(\mathbf{r}) = \frac{1}{\mathrm{V}} \sum_{|\mathbf{G}|<\mathrm{G_c}} \tilde{n}(\mathbf{G}) e^{i\mathbf{G}\cdot\mathbf{r}} \ , \tag{15.45}$$

where V is the volume of the periodic computational box. The plane wave expansion is limited to functions with a kinetic energy smaller than a cut-off value $\mathrm{E_c} = 1/2\mathrm{G_c}^2$. With this one can write the final definition of the electronic density in the GAPW formulation

$$n(\mathbf{r}) = \frac{1}{\mathrm{V}} \sum_{|\mathbf{G}|<\mathrm{G_c}} \tilde{n}(\mathbf{G}) e^{i\mathbf{G}\cdot\mathbf{r}} - \sum_{A} \sum_{ab\in A} \sum_{\mu\nu} \tilde{D}^A_{a\mu} P_{\mu\nu} \tilde{D}^A_{b\mu} g_a(\mathbf{r}) g_b(\mathbf{r})$$
$$+ \sum_{A} \sum_{ab\in A} \sum_{\mu\nu} D^A_{a\mu} P_{\mu\nu} D^A_{b\mu} g_a(\mathbf{r}) g_b(\mathbf{r}) \ , \tag{15.46}$$

where all quantities are either analytic functions of the basis set $\{\varphi\}$ or the density matrix P. The matrices D^A are sparse, as only few basis function close to atom A have a non–vanishing contribution. Therefore the one–center terms of the density can be calculated in a number of operations that does not depend on system size, making the total calculation of all atom–centered densities linear scaling. The plane wave expansion of the smooth density is calculated in real space where an efficient screening can be applied due to the locality of the basis functions. [47,53] A fast Fourier transform is finally used to calculated $\tilde{n}(\mathbf{G})$, making the total calculation of the density a $N \log N$ method (N being the system size).

The Kohn–Sham matrix derived from the GAPW energy functional can be calculated from several different contributions.

1. Integrals involving two basis functions and a local operator (e.g. kinetic energy or screened external potential). These integrals can be calculated analytically, provided that a basis of Gaussians is used. Linear scaling is easily achieved by screening of integrals. A small pre factor in the scaling law is ensured by the short range of all operators.
2. Contributions from the exchange and correlation functional of the atom–centered densities. These integrals are calculated using numerical quadrature. The number of integrals of this type is proportional to the number of atoms and therefore linear in system size.
3. Integrals involving the Hartree and exchange and correlation potential derived from the smooth density $\tilde{n}$. These potentials are available in the plane wave basis. In its real space representation efficient screening can be applied for these integrals [47,53] and linear scaling can be achieved.

The accuracy of the GAPW method has been tested on many systems [53,54]. The scaling behavior for the construction of the Kohn–Sham matrix is shown

Table 15.2. Timings in seconds for the construction of the Kohn–Sham matrix for zeolite systems and linear alkane chains. Number of atoms N_{atom}, number of electrons N_{elec}, number of basis functions N_{bas}, calculations performed on a IBM 397 workstation.

System	N_{atom}	N_{elec}	N_{bas}	CPU time
$Si_{24}O_{48}$	72	384	936	110
$Si_{48}O_{96}$	144	768	1872	226
$Si_{72}O_{144}$	216	1152	2808	346
$Si_{96}O_{192}$	288	1536	3744	466
$Si_{120}O_{240}$	360	1920	4680	611
$C_{60}H_{122}$	182	482	1084	176
$C_{80}H_{162}$	242	682	1444	243
$C_{100}H_{202}$	302	802	1804	302

for different systems in Table 15.2. As expected one finds for a three dimensional periodic system (zeolite) as well as for linear alkane chains an almost linear scaling of the CPU time for the construction of the Kohn–Sham matrix.

15.4.2 Wavefunction Optimization; Solving the Kohn–Sham Equations

Once the Kohn–Sham matrix has been calculated the task left to do is to solve the Kohn–Sham equations. These equations for a fixed potential are equivalent to other independent particle models like the Hartree–Fock or tight–binding Hamiltonians. The equations can be summarized by

$$\mathbf{HC} = \mathbf{CE} \ , \tag{15.47}$$

where $\mathbf{H}$ is the Hamiltonian matrix in an orthogonal basis, $\mathbf{E}$ the Lagrange multiplier matrix and $\mathbf{C}$ the orbital coefficients. This is an eigenvalue problem that scales with the third power of the system size. Many approaches have been devised to reduce the pre factor of the scaling law for solving (15.47). For example the Car–Parrinello method [1] can be used to solve (15.47) with a scaling of $\mathrm{N}_{\text{occ}}^2\mathrm{N}_{\text{bas}}$, where N_{occ} is the number of occupied orbitals and N_{bas} the number of basis functions. If the number of occupied orbitals is much smaller than the number of basis functions, e.g. in plane wave calculations, this results in a dramatic decrease of the pre factor compared to a direct diagonalization scheme. Similar improvements can be achieved by other iterative diagonalization methods.

However, for systems with more than about 100 atoms, solving the Kohn–Sham equations even with the best diagonalization method becomes the bottleneck of the electronic structure calculation. It was therefore important to develop new methods that allow to find solutions to (15.47) with scaling laws better than cubic. In the last ten years several such methods have been proposed (see Ref [43] for a review). Some methods make use of the invariance of the energy functional to unitary transformations of the occupied orbitals by searching for a special set of localized orbitals. Other methods avoid the orbital picture all together and directly search for the density matrix. One set of linear scaling methods is based on the observation that the density matrix is a functional of the Kohn–Sham Hamiltonian

$$\mathbf{P} = f(\mathbf{H}) \ . \tag{15.48}$$

The matrix function f is then approximated either by a Chebychev polynomial [56] or a rational function approximation based on a Cauchy integral [57]. Another approach was put forward by Li et al. [58]. They define an extended energy functional

$$\Omega = \mathrm{Tr}(3P^2 - 2P^3)(H - \mu I) \ . \tag{15.49}$$

The density matrix is obtained by minimizing (15.49) with respect to P. This method is variational and no constraint has to be imposed. The first part of the functional ensures the idem-potency of the density matrix and the chemical potential μ has to be adjusted such that the total number of electrons is correct. This method has recently been combined with the Car–Parrinello algorithm [59]. O(N) algorithms alone do not yield linear scaling. However, in large molecules we can take advantage of matrix sparsity, i.e., the fact that many elements of a matrix are zero or below a certain threshold. For large systems, where the number of significant elements scales linearly with the size of the system, the cost of all sparse matrix operations will scale linearly.

The sparsity of the density matrix, depends on several factors. We can write the density matrix as,

$$P_{ij} = \sum_{kl} S_{ik}^{-1} Q_{kl} S_{lj}^{-1} \ , \tag{15.50}$$

$$Q_{kl} = \int\!\!\int drdr' \, \varphi_k(r) P(r,r') \varphi_l(r') \ , \qquad S_{ij} = \int dr \, \varphi_i(r) \varphi_j(r) \ ,$$

where $P(r,r')$ denotes the one-particle density operator and Q_{kl} is its matrix representation with respect to the atomic basis functions. From (15.50) it is clear that the sparsity of P depends on the decay properties of $P(r,r')$ as well as on the sparsity of S^{-1}.

Theoretical models of periodic solids suggest that the locality of $P(r,r')$ is related to the band gap $\Delta\epsilon$. For the case of an insulator, the one-particle density operator decays asymptotically as an exponential [60]

$$P(r,r') \approx \exp\left(-\sqrt{\Delta\epsilon}|r - r'|\right) \ . \tag{15.51}$$

While the decay behavior of $P(r,r')$ is a fundamental property of the respective quantum mechanical system, this is certainly not true for the sparsity of

S^{-1}. The determining quantities are kind and size of the basis set chosen for the calculation. In the following we will concentrate on Gaussian-type basis functions, since they are almost exclusively used in ab-initio quantum chemistry. Extended basis sets are often needed in DFT calculations to achieve converged results. These extended basis sets, expecially if they include diffuse functions may often lead to near linear dependencies. Procedures used to eliminate the dependencies can not be applied in the linear scaling context as they destroy the local character of the basis. For a more systematic treatment of the consequences of extended basis sets, it is useful to consider the following definition. A well conditioned overlap matrix is one containing no linear or near linear dependencies. Its condition number, the ratio of its largest and smallest eigenvalue, is small. Using this definition, it can be shown [61], that a localized and well conditioned overlap matrix leads to a similarly localized inverse and finally via (15.50) to a localized density matrix. Unfortunately the reverse conclusion is also true, ill conditioned overlap matrices caused by large nonorthogonal basis sets destroy the sparsity of the density matrix. Since this is the crucial point determining whether a O(N) method is faster than traditional electronic structure calculations, linear scaling methods are not used when large nonorthogonal basis sets are required. Standard methods used in quantum chemistry to deal with ill conditioned overlap matrices are not applicable in linear scaling calculations. These algorithms rely on the eigenfunctions of the overlap matrix which are non local. However, in the context of their energy renormalization group method, Baer and Head-Gordon [62] have devised an algorithm that produces a localized basis with given condition number.

In the following we will present another method that constructs a well conditioned basis on the fly. To achieve this a small fraction of the flexibility of the basis set is lost.

Recently, Lee and Head-Gordon introduced the polarized atomic orbital method [63,64] to construct small basis sets optimized in the molecular environment. A flexible PAO basis set with a dimension typically equal to the size of a minimal basis, is formed from atom-centered linear combinations of a larger set of atomic orbitals. While atom-optimized minimal basis sets, e.g. of the STO-nG type, are often performing poorly in practical calculations, the PAO's derive their flexibility from the fact that they can adapt to the molecular environment by the admixture of higher angular momentum functions. From a computational point of view, PAO's have the attractive feature of greatly reducing the number of independent variables to be determined during a density update. This offers the possibility for studying large systems currently intractable with larger than minimal basis sets, while retaining a considerable amount of flexibility in the basis set. Besides, being considerably less memory and CPU intensive, the PAO method is also useful in combination with linear scaling density update methods. As seen in the last section a major drawback of linear scaling approaches are numerical instabilities resulting from ill-conditioned overlap matrices encountered when large nonorthogonal basis sets are used. PAO's provides a solution to this problem at the cost of a slight reduction of the flexibility of the basis set. One finds that the condition number of the PAO overlap matrix is almost independent from the condition number of the underlying extended basis set and consequently no numerical instabilities are encountered (see Table 15.3).

Table 15.3. Logarithm of the condition number of the overlap matrix for different basis sets computed from full SCF and PAO-SCF calculations.

	Basis	MIN	DZ	DZSP	TZDP
10-glycine	SCF	2.83	9.39	9.60	10.87
	PAO-SCF	—	2.08	2.15	2.12
$(H_2O)_{30}$	SCF	2.26	6.18	6.63	9.01
	PAO-SCF	—	1.70	1.87	1.85
C_{60}	SCF	3.37	13.25	13.40	14.56
	PAO-SCF	—	2.49	2.49	2.41
$C_{19}H_{28}O_2$	SCF	3.24	10.58	10.90	12.71
	PAO-SCF	—	2.38	2.46	2.48

PAOs are variationally optimized during an iterative SCF calculation. They are defined as a linear transformation of an underlying extended basis set. The transformation is strictly limited to atom-centered linear combinations of basis functions. In other words, only basis functions from the same atom are allowed to mix,

$$\tilde{\varphi}_i(\mathbf{r}) = \sum_j B_{ji}\varphi_j(\mathbf{r}) \; , \tag{15.52}$$

where i and j belong to the same atom. Here, φ denotes an atomic orbital basis functions, B is the transformation from the extended to the PAO basis set and the tilde is used to distinguish quantities related to the PAO basis set. B is a strictly atom-centered block diagonal matrix. Using (15.52), it is straightforward to compute the one-particle density matrix in the extended basis as a function of the PAO density matrix and the transformation matrix B,

$$P = B\tilde{P}B^T \; . \tag{15.53}$$

In the following we describe a scheme that builds on existing methods for solving the Hartree–Fock or Kohn–Sham equations for the determination of the PAO orbitals.

We split the optimization process into two parts. Optimization of the transformation matrix is, at each step, accompanied by an optimization of the density matrix (DM) in the current PAO basis. The calculation of the density matrix is performed in the PAO basis and can be done either by

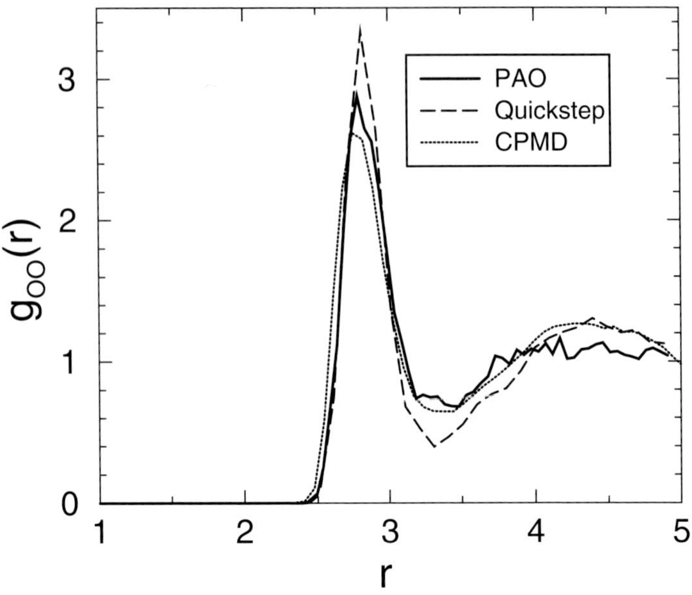

Fig. 15.5. O-O pair correlation function of 32 water molecules obtained from PAO-MD (PAO) compared to a MD simulation using the same program from ref. [54] (Quickstep) and to a Car-Parrinello MD from ref. [67] (CPMD).

conventional methods (e.g. diagonalization) or linear scaling methods mentioned in previous sections. Thus, this scheme allows for great flexibility and requires only minor changes in existing implementations. During the MO/DM update we transform both the MO's/DM and the Hamiltonian matrix into an orthonormal basis. The transformation matrix is updated by minimizing a constraint energy functional using a conjugate gradient scheme.
The density matrix in the PAO basis is calculated by minimizing the energy functional Ω given in matrix representation

$$\Omega = \mathrm{Tr}[\tilde{P}B^T HB] + \mathrm{Tr}[\tilde{\Lambda}(\tilde{P} - \tilde{P}B^T SB\tilde{P})] \;, \tag{15.54}$$

where $\tilde{\Lambda}$ is the Lagrangian multiplier matrix corresponding to the idempotency constraint.

Given a transformation matrix B, the next step is to compute a variational density matrix, either from converged MO's or by solving directly for the PAO density matrix without any explicit construction of orbitals. If we transform the Hamiltonian matrix and the overlap matrix into the PAO basis, $\tilde{H} = B^T HB$ and $\tilde{S} = B^T SB$, Eq. (15.54) can be rewritten and we have to minimize,

$$\Omega = \mathrm{Tr}[\tilde{P}\tilde{H}] + \mathrm{Tr}[\tilde{\Lambda}(\tilde{P} - \tilde{P}\tilde{S}\tilde{P})] \;, \tag{15.55}$$

which leads to a generalized eigenvalue problem, $\tilde{H}\tilde{C} = \tilde{S}\tilde{C}\tilde{\Lambda}$ or can be solved directly for the PAO density matrix using linear scaling methods. However, both procedures are only in the minimal basis.

In the optimization cycle for the PAO basis the derivative of Ω, as defined in (15.54), with respect to the degrees of freedom of B are needed. A convenient way to parameterize B is through a exponential function of a skew–Hermitian matrix [65], where only the matrix elements connecting the PAO and the excluded subspace are non zero.

Using the results above we have now a scheme to solve for the PAO's by iterating the following steps:

1. Choose an initial guess for B.
2. Compute the Hamiltonian matrix and the overlap matrix in the minimal basis,
   $\tilde{H} = B^T H B$ and $\tilde{S} = B^T S B$.
3. Determine a variational density matrix, i.e. $\frac{\partial \Omega}{\partial \tilde{P}} = 0$, by minimizing (15.55) by either one of the following methods
   (a) Solve $\tilde{H}\tilde{C} = \tilde{S}\tilde{C}\tilde{\Lambda}$ and compute $\tilde{P} = \tilde{C}\tilde{C}^T$.
   (b) Use a linear scaling density update method.
4. Calculate the gradient, $\frac{\partial \Omega}{\partial B}$, and update B, by e.g. a conjugate gradient method.
5. If convergence is achieved compute the density matrix in the extended basis,
   $P = B\tilde{P}B^T$, otherwise continue with step 2.

The scheme described above contains the solution of a Roothaan-type equation in step 3. However, this has to be done only in the PAO basis, reducing the work considerably both in a diagonalization based method as also in linear scaling methods.

Figure 15.6 shows the results from test calculations on linear alkane chains using diagonalization and different linear scaling algorithms together with the PAO approach. For large enough system sizes the linear scaling approaches are better performing than the diagonalization. However, break even points for the different methods are at rather large system sizes. This is due to the fact that the diagonalization scheme profits even more from the PAO method than the linear scaling methods. Namely, the pre factor is reduced by the third power of the ratio of the extended to the PAO basis set size.

The algorithms described can also be used to calculate forces which then allow to perform molecular dynamics simulations with linear scaling methods. These algorithmic improvements, together with the continuing development of faster computer hardware, makes it possible to extend molecular dynamics simulations to larger systems and longer time scales.

Acknowledgments

U.R. would like to thank J. VandeVondele for his important contributions to this work.

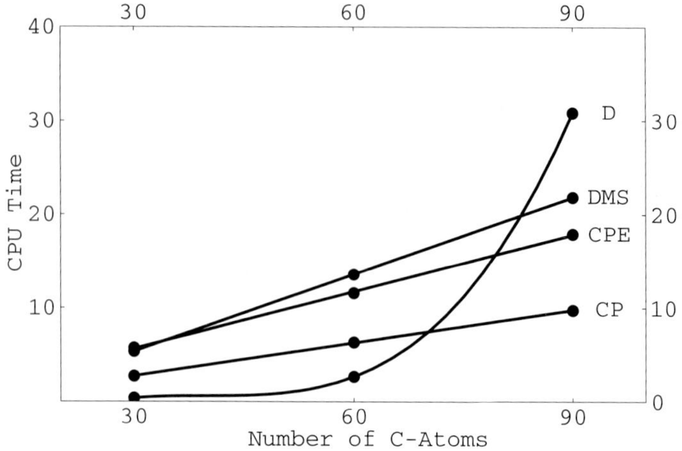

Fig. 15.6. CPU time requirements [s] of a density matrix update step for the linear alkane chains $C_{30}H_{62}$, $C_{60}H_{122}$ and $C_{90}H_{182}$ using diagonalization (D), Chebychev polynomial expansion (CPE) [56], canonical purification of the density matrix (CP) [66] and conjugate gradient density matrix search (DMS) [58]. All density update procedures are in terms of the PAO minimal basis.

References

1. R. Car, and M. Parrinello, Phys. Rev. Lett., **55**, 2471, (1985).
2. see e.g. C. Cavazzoni, G.L. Chiarotti, S. Scandalo, E. Tosatti, M. Bernasconi, M. Parrinello: Science **283**, 44 (1999)
3. U. Rothlisberger: '15 Years of Car-Parrinello Simulaions in Physics, Chemistry and Biology'. In: *Computational Chemistry: Reviews of Current Trends*, ed. by J. Leszczynski (World Scientific, Singapore 2001) pp.33–68
4. P. Carloni, U. Rothlisberger: 'Simulations of Enzymatic Systems: Perspectives from Car-Parrinello Molecular Dynamics Simualations'. In: *Theoetical Biochemistry - Processes and Properties of Biological Systems*, ed. by L. Eriksson (Elsevier Science 2001) pp.215
5. E. A. Carter, G. Ciccotti, J. T. Hynes, and R. Kapral, Chem. Phys. Lett., **156**, 472, (1989).
6. M. Sprik and G. Ciccotti, J. Chem. Phys., **109**, 7737, (1998).
7. A. Amadei, A.B.M. Linsen, H.J.C. Berendsen: Proteins: Structure, Function and Genetics **17**, 412 (1993); B. Space, H. Rabitz, A. Askar: J. Chem. Phys. **99**, 9070 (1993); N. Nakajima, H. Nakamura, A. Kidera: J. Phys. Chem. B **101**, 817 (1997); M. Marchi, P. Ballone: J. Chem. Phys. **110**, 3697 (1999)
8. C. Dellago, P. G. Bolhuis, F. S. Csajka, D. J. Chandler, J. Chem. Phys., **108**, 1964, (1998).
9. P. G. Bolhuis, C. Dellago, D. J. Chandler, Faraday Discuss., **110**, 42, (1998).
10. D. Passerone and M. Parrinello, Phys. Rev. Lett., **87**, 108302, (2001).
11. G.M. Torrie, J.P. Valleau: J. Chem.Phys. **66**, 1402 (1977); K. Kuczera: J. Comp. Chem. **17**, 1726 (1996); Y. Wang, K. Kuczera: Theor. Chem. Acc. **101**, 274 (1999); C. Bartels, M. Karplus: J. Comp. Chem. **18**, 1450 (1997);

J. Apostolakis, P. Ferrara, A. Caflisch: J. Chem. Phys. **110**, 2099 (1999); C. Bartels, M. Schaefer, M. Karplus: J. Chem . Phys. **111**, 8084 (1999); S. Crouzy, J. Naudry, J.C. Smith, B. Roux: J. Comp. Chem. **20**, 1644 (1999); T. Huber, A.E. Torda, W.F. vamGunsteren: J. Comput.-Aided Mol. Design **8**, 695 (1994); H. Grubmüller: Phys. Rev. E **52**, 2893 (1995); A.F. Voter: J. Chem. Phys. **106**, 4665 (1997); A.F. Voter: Phys. Rev. Lett. **78**, 3908 (1997); M.M. Steiner, P.-A. Genilloud, J.W. Wilkins: Phys. Rev. B **57**, 10236 (1998); X.G. Gong, J.W. Wilkins: Phys. Rev. B **59**, 54 (1999)

12. A. Laio, J. VandeVondele, U. Rothlisberger: J. Chem. Phys. **116**, 6941 (2002); M. Colombo, L. Guidoni, A. Laio, A. Magistrato, P. Maurer, S. Piana, U. Röhrig, K. Spiegel, M. Sulpizi, J. VandeVondele, M. Zumstein, and U. Rothlisberger: CHIMIA **55**,11-17 (2002); A. Laio, J. VandeVondele, U. Rothlisberger: J. Phys. Chem. B **106**,7300 (2002);
13. J. VandeVondele, U. Rothlisberger: J. Chem. Phys. **113**, 4863 (2000)
14. Density Functional calculations with the LDA functional, a basis set of plane waves with a kinetic energy cutoff of 60Ry, Norm-Conserving non local pseudo potentials and a face centered cubic cell of edge a = 10A. The calculated barrier is in good agreement with the experimental value (2.75 kcal/mol) given in Ref. G. Herzberg, 'Molecular spectra and molecular structure' (Krieger Publishing Company, Malabar, Florida 1991).
15. J. VandeVondele, U. Rothlisberger: J. Am. Chem. Soc. **124**,8163 (2002);
16. K. Fukui, H. Fujimoto, 'Frontier orbitals and reaction paths' (World Scientific, Singapore, 1997)
17. R.G. Parr, W. Yang, 'Density-Functional Theory of Atoms and Molecules' (Oxford University Press, New York, 1989)
18. N. D. Mermin: Phys. Rev. **137**, A1441 (1965).
19. A. Alavi, J. Kohanoff, M. Parrinello and D. Frenkel: Phys. Rev. Lett. **73**, 2599 (1994).
20. N. Marzari, D. Vanderbilt, and M. C. Payne: Phys. Rev. Lett. **79**, 1337 (1997).
21. J. VandeVondele and A. De Vita: Phys. Rev. B **60**, 13241 (1999).
22. R. G. Pearson: J. Chem. Educ. **64**, 561 (1987).
23. R. G. Pearson: Acc. Chem. Res. **26**, 250 (1993).
24. R. G. Parr, Z. Zhou: Acc. Chem Res. **26**, 256 (1993).
25. P. K. Chattaraj, H. Lee, and R. G. Parr: J. Am. Chem. Soc. **113**, 1855 (1991).
26. R. G. Parr and P. K. Chattaraj: J. Am. Chem. Soc. **113**, 8154 (1991).
27. N. Agmon, Chem. Phys. Lett., **244**, 456, (1995).
28. K. Laasonen, M. Sprik, M. Parrinello, and R. Car, J. Chem. Phys., **99**, 9080, (1993).
29. M. Tuckerman, K. Laasonen, M. Sprik, and M. Parrinello, J. Phys. Chem., **99**, 5749, (1995); J. Chem. Phys., **103**, 150, (1995).
30. D. Marx, M. E. Tuckerman, J. Hutter, and M. Parrinello, Nature, **397**, 601, (1999).
31. K. E. Laasonen and M. L. Klein, J. Phys. Chem. A, **101**, 98, (1997).
32. E. J. Meijer and M. Sprik, J. Am. Chem. Soc., **120**, 6345, (1998).
33. K. Doclo and U. Roethlisberger, J. Phys. Chem. A, **104**, 6464, (2000).
34. P. Carloni, M. Sprik, and W. Andreoni, J. Phys. Chem. B, **104**, 823, (2000).
35. D. Aktah and I. Frank, J. Am. Chem. Soc., (2002) (in press).
36. B. L. Trout and M. Parrinello, Chem. Phys. Lett., **288**, 343, (1998); J. Phys. Chem. B, **103**, 7340, (1999).

37. M. Sprik, Chem. Phys., **258**, 139, (2000).
38. P. Geissler, C. Dellago, D. Chandler, J. Hutter, and M. Parrinello, Science, **291**, 2121, (2001).
39. E. Schwegler, G. Galli, F. Gygi, and R. Q. Hood, Phys. Rev. Lett., **87**, 265501, (2001).
40. D. Chandler, *"Introduction to Modern Statistical Mechanics"*, Oxford University Press, 1987.
41. G. Schüürmann, M. Cossi, V. Barone, and J Tomasi, J. Phys. Chem. A, **102**, 6706, (1998).
42. J. E. Davies, N. L. Doltsinis, A.JL. Kirby, C. D. Roussev, and M. Sprik M., J. Am. Chem. Soc., J. Am. Chem. Soc. 124, 6594 (2002)
43. S. Goedecker, Rev. Mod. Phys., **71** 1085 (1999)
44. J. R. Chelikowsky, N. Trouiller, Y. Saad, Phys. Rev. Lett., **72**, 1240, (1994).
45. E. Hernández, M. J. Gillan, and C. M. Goringe, Phys. Rev. B, **55**, 13485 (1997).
46. C. K. Gan, P. D. Haynes, and M. C. Payne, Phys. Rev. B, art. no. 205109
47. G. Lippert, J. Hutter and M. Parrinello, Mol.Phys.,
48. P. Ordejón, E. Artacho, and J. M. Soler, Phys. Rev. B, **53**, R10441 (1996).
49. H. Horn, H. Weiss, M. Häser, M. Ehrig, and R. Ahlrichs, J. Comp. Chem., **12**, 1058, (1991).
50. C. White, B. Johnson, P. Gill, and M. Head–Gordon, Chem. Phys. Lett., **230**, 8, (1994).
51. H. Sambe and R. H. Felton, J. Chem. Phys., **62**, 1122, (1975).
52. K. Eichkorn, O. Treutler, H. Ohm, M. Häser, and R. Ahlrichs, Chem. Phys. Lett., **240**, 283, (1995).
53. G. Lippert, J. Hutter and M. Parrinello, Theor. Chem. Acc., **103**, 124, (1999).
54. M. Krack and M. Parrinello, Phys. Chem. Chem. Phys., **2**, 2105, (2000).
55. P. Blöchl, Phys. Rev. B, **50**, 17953, (1994).
56. S. Goedecker and M. Teter, Phys. Rev. B, **51**, 9455 (1995).
57. S. Goedecker, J. Comp. Phys., **118**, 261 (1995).
58. X. P. Li, R. W. Nunes, and D. Vanderbilt, Phys. Rev. B, **47**, 10891 (1993).
59. H. B. Schlegel, J. M. Millam, S. S. Iyengar, G. A. Voth, A. D. Daniels, G. E. Scuseria, M. J. Frisch, J. Chem. Phys., **114**, 9758, (2001).
60. W. Kohn, Chem. Phys. Lett., **208**, 167 (1993).
61. P. E. Maslen, C. Ochsenfeld, C. A. White, M. S. Lee, and M. Head-Gordon, J. Phys. Chem. A, **102**, 2215 (1998).
62. R. Baer and M. Head-Gordon J. Chem. Phys., **109**, 10159, (1998).
63. M. S. Lee and M. Head-Gordon, J. Chem. Phys., **107**, 9085, (1997).
64. M. S. Lee and M. Head-Gordon, Computers and Chemistry, **24**, 295, (2000).
65. J. Hutter, M. Parrinello, and S. Vogel, J. Chem. Phys., **101**, 3862 (1994).
66. A. H. R. Palser and D. E. Manolopoulos, Phys. Rev. B, **58**, 12704 (1998).
67. M. Sprik, J. Hutter and M. Parrinello, J. Chem. Phys., **105**, 1142, (1996).

Part X

Quantum Simulations

16 A Statistical Mechanical Theory of Quantum Dynamics in Classical Environments

Raymond Kapral[1] and Giovanni Ciccotti[2]

[1] Chemical Physics Theory Group, Department of Chemistry, University of Toronto, Toronto, ON M5S 3H6, Canada
[2] INFM and Dipartimento di Fisica, Università "La Sapienza", Piazzale Aldo Moro, 2, 00185 Roma, Italy

16.1 Introduction

Since it is difficult to simulate the quantum dynamics of large, complex many-body systems, one is led to construct a statistical mechanical description of matter based on a mixture of quantum and classical dynamics. Many physically interesting systems may be partitioned into subsystems where certain degrees of freedom must necessarily be treated quantum mechanically, while others behave classically to a high degree of accuracy. Examples of systems with these characteristics are familiar and include proton and electron transfer processes and systems with electronic degrees of freedom coupled to heavy nuclei. In these cases it is useful to construct a quantum-classical dynamics that not only accounts for the quantum and classical dynamics of the two isolated subsystems but also describes their interaction. [1–3] The most widely used approaches are based on surface-hopping schemes where the coupling between the two subsystems induces quantum transitions. [4–7]

The primary interest is in the computation of quantities such as expectation values of dynamical variables or transport coefficients usually determined from integrals of time correlation functions. Consequently it is not sufficient to simply focus on the development of quantum-classical dynamics; instead one must formulate the statistical mechanics of such systems in order to provide a route to the computation of these observables. In this chapter we develop a scheme for carrying out quantum-classical evolution of many-body systems and, having established the nature of this dynamics, formulate a statistical mechanics for such systems and devise schemes for the computation of expectation values in this quantum-classical world.

We begin the presentation with a brief overview of quantum statistical mechanics in Sect. 16.2 where the response function and autocorrelation function expressions for transport properties are given. This section also introduces the partial Wigner representation and formulates quantum statistical mechanics in this form. The passage to quantum-classical dynamics is considered in Sect. 16.3 and the forms of the quantum-classical Liouville equation for the density matrix and dynamical variables are presented. Since quantum-classical dynamics has some unusual features, its nature is discussed in Sect. 16.4. The evolution equation for the density matrix is expressed in an adiabatic basis and the evolution of the density matrix is determined in terms of an ensemble of surface-hopping trajectories. Section 16.5 carries out an analogous analysis for the evolution of a dynamical variable. The quantum-classical form of the canonical equilibrium density matrix is the topic of Sect. 16.6. A knowledge of this quantity is necessary for the computation of equilibrium statistical mechanical properties. Nonequilibrium statistical mechanics in the quantum-classical world is formulated in Sect. 16.7. This section gives the expressions for quantum-classical transport properties and time correlation functions. Properties of the correlation functions are discussed and comparisons with full quantum mechanics are made. Section 16.8 gives an example of the calculation of the expectation value of an observable for a many-body system: the spin-boson model. The conclusions of the chapter are given in Sect. 16.9.

16.2 Quantum Dynamics and Statistical Mechanics

The time evolution of the quantum mechanical density matrix $\hat{\rho}$ is given by the von Neumann equation,

$$\frac{\partial \hat{\rho}(t)}{\partial t} = -\frac{i}{\hbar}[\hat{H}, \hat{\rho}(t)] , \tag{16.1}$$

where $\hat{H}$ is the hamiltonian of the system. Its formal solution is

$$\hat{\rho}(t) = e^{-i\hat{L}t}\hat{\rho}(0) = e^{-i\hat{H}t/\hbar}\hat{\rho}(0)e^{i\hat{H}t/\hbar} , \tag{16.2}$$

with $i\hat{L} = (i/\hbar)[\hat{H}, \;]$ the quantum Liouville operator. In the Heisenberg picture of quantum mechanics, the time evolution of a dynamical variable $\hat{B}$ is given by

$$\frac{d\hat{B}(t)}{dt} = \frac{i}{\hbar}[\hat{H}, \hat{B}(t)] , \tag{16.3}$$

whose formal solution is

$$\hat{B}(t) = e^{i\hat{L}t}\hat{B} = e^{i\hat{H}t/\hbar}\hat{B}e^{-i\hat{H}t/\hbar} . \tag{16.4}$$

Usually, one is not simply interested in the time evolution of such quantities but rather in statistical mechanical quantities like the average values of observables, dynamical properties or transport coefficients defined in terms of time integrals of correlation functions. The average value of a dynamical variable is given by

$$\overline{B(t)} = \mathrm{Tr}\hat{B}\hat{\rho}(t) = \mathrm{Tr}\hat{B}(t)\hat{\rho}(0) , \tag{16.5}$$

where we have used the fact that the time dependence can be transferred from the density matrix to the operator using cyclic permutations under the trace. Thus, as is well known, one may either evolve a given initial density matrix backward in time and compute the average of $\hat{B}$ using the time evolved density matrix to find $\overline{B(t)}$ or evolve the operator forward in time and average over the initial value of the density matrix.

In quantum mechanical systems in thermal equilibrium, we are often interested in the calculation of transport properties such as diffusion coefficients or rate constants. The microscopic forms for such transport coefficients can be obtained by applying linear response theory to a system in equilibrium subjected to an external force or by monitoring the decay of fluctuations about the equilibrium state. In linear response theory, it is assumed that a time dependent external force $F(t)$ couples to an operator $\hat{A}^\dagger$. Then the system hamiltonian takes the form

$$\hat{H}(t) = \hat{H} - \hat{A}^\dagger F(t) , \tag{16.6}$$

and the evolution equation for the density matrix reads

$$\begin{aligned}\frac{\partial \hat{\rho}(t)}{\partial t} &= (i\hbar)^{-1}[\hat{H}(t), \hat{\rho}(t)] , \\ &= -(i\hat{L} - i\hat{L}_A F(t))\hat{\rho}(t) ,\end{aligned} \tag{16.7}$$

where $i\hat{L}_A \equiv (i/\hbar)[\hat{A}^\dagger,\ ]$. The adjoint of $\hat{A}$ is denoted by $\hat{A}^\dagger$.

Assuming the system was in thermal equilibrium in the distant past, the solution of this equation to linear order in the external force is [8]

$$\hat{\rho}(t) = \hat{\rho}_e^Q + \int_{-\infty}^t dt'\ e^{-i\hat{L}(t-t')} i\hat{L}_A \hat{\rho}_e^Q F(t') . \tag{16.8}$$

Here $\hat{\rho}_e^Q = Z_Q^{-1}\exp(-\beta\hat{H})$ is the canonical equilibrium density matrix and $Z_Q = \mathrm{Tr}\exp(-\beta\hat{H})$ is the partition function. The response of the system to the external force may be determined by computing the average value of an operator $\hat{B}$ using the density matrix at time t,

$$\begin{aligned}\overline{B(t)} = \mathrm{Tr}\hat{B}\hat{\rho}(t) &= \int_{-\infty}^t dt'\ \mathrm{Tr}\hat{B}e^{-i\hat{L}(t-t')} i\hat{L}_A \hat{\rho}_e^Q F(t') \\ &= \frac{i}{\hbar}\int_{-\infty}^t dt'\ \mathrm{Tr}\hat{B}(t-t')[\hat{A}^\dagger, \hat{\rho}_e^Q]F(t') \\ &= \frac{i}{\hbar}\int_{-\infty}^t dt'\ \mathrm{Tr}[\hat{B}(t-t'), \hat{A}^\dagger]\hat{\rho}_e^Q F(t') \equiv \int_{-\infty}^t dt'\ \phi_{BA}(t-t')F(t') .\end{aligned} \tag{16.9}$$

For simplicity, the operator $\hat{B}$ was assumed to have zero average value in equilibrium. The last line in (16.9) defines the response function

$$\phi_{BA}(t) = \langle \frac{i}{\hbar}[\hat{B}(t), \hat{A}^\dagger]\rangle_Q , \tag{16.10}$$

where the angle brackets denote a quantum canonical equilibrium average, $\langle \cdots \rangle_Q = \mathrm{Tr} \cdots \hat{\rho}_e^Q$.

The response function may be written in a equivalent form by using the quantum mechanical operator identity, [8]

$$\frac{i}{\hbar}[\hat{A}^\dagger, \hat{\rho}_e^Q] = \int_0^\beta d\lambda \, \hat{\rho}_e^Q \, \dot{\hat{A}}^\dagger(-i\hbar\lambda) \,, \tag{16.11}$$

in the second line of (16.9) to obtain

$$\phi_{BA}(t) = \int_0^\beta d\lambda \; \mathrm{Tr} \, \dot{\hat{A}}^\dagger(-i\hbar\lambda)\hat{B}(t)\hat{\rho}_e^Q \,. \tag{16.12}$$

If we choose $\hat{B} = \dot{\hat{A}} \equiv \hat{f}_A$, the flux corresponding to the operator $\hat{A}$, the response has the form of a macroscopic law and the response function is proportional to the flux autocorrelation function

$$\phi_{\dot{A}A}(t) = \langle \frac{i}{\hbar}[\hat{f}_A(t), \hat{A}^\dagger] \rangle_Q = \int_0^\beta d\lambda \; \mathrm{Tr} \hat{f}_A^\dagger(-i\hbar\lambda)\hat{f}_A(t)\hat{\rho}_e^Q \equiv \beta \langle \hat{f}_A^\dagger ; \hat{f}_A(t) \rangle_Q \,. \tag{16.13}$$

The last equality defines the Kubo transformed correlation function. A simple transport property λ_A in quantum mechanics is proportional to the time integral of the flux autocorrelation function,

$$\lambda_A \propto \int_0^\infty dt \; \langle \frac{i}{\hbar}[\hat{f}_A(t), \hat{A}^\dagger] \rangle_Q \propto \int_0^\infty dt \; \langle \hat{f}_A ; \hat{f}_A(t) \rangle_Q \,. \tag{16.14}$$

The quantum mechanical correlation functions satisfy time translation symmetry,

$$\langle \hat{f}_A ; \hat{f}_A(t) \rangle_Q = \langle \hat{f}_A(\tau) ; \hat{f}_A(t+\tau) \rangle_Q \,, \tag{16.15}$$

as can be verified by using the explicit form of the canonical equilibrium density matrix and cyclic permutations under the trace.

16.2.1 Mixed Representation of Quantum Statistical Mechanics

To obtain an alternative description of the quantum statistical mechanics of the system, we partition it into two subsystems: the first subsystem contains n particles with masses m and coordinate operators $\hat{q}$; the second subsystem comprises N particles with masses M and coordinate operators $\hat{Q}$.

The hamiltonian operator may be written as

$$\hat{H} = \frac{\hat{P}^2}{2M} + \frac{\hat{p}^2}{2m} + \hat{V}(\hat{q}, \hat{Q}) \,, \tag{16.16}$$

where $\hat{p}$ and $\hat{P}$ are momentum operators and $\hat{V}(\hat{q}, \hat{Q})$ is the total potential energy. We employ a condensed notation such that $\hat{q} = (\hat{q}_1, \hat{q}_2, \ldots \hat{q}_{3n})$ and $\hat{Q} = (\hat{Q}_1, \hat{Q}_2, \ldots \hat{Q}_{3N})$, with an analogous notation for $\hat{p}$ and $\hat{P}$.

The alternative description we wish to consider is based on a partial Wigner transformation [9] of the density matrix with respect to the subset of Q coordinates, [10]

$$\hat{\rho}_W(R,P) = (2\pi\hbar)^{-3N} \int dz e^{iP\cdot z/\hbar} \langle R - \frac{z}{2}|\hat{\rho}|R + \frac{z}{2}\rangle \, . \tag{16.17}$$

In this representation the quantum Liouville equation is,

$$\begin{aligned} \frac{\partial \hat{\rho}_W(R,P,t)}{\partial t} &= -\frac{i}{\hbar}\left((\hat{H}\hat{\rho})_W - (\hat{\rho}\hat{H})_W\right) \\ &= -\frac{i}{\hbar}\left(\hat{H}_W e^{\hbar\Lambda/2i}\hat{\rho}_W(t) - \hat{\rho}_W(t)e^{\hbar\Lambda/2i}\hat{H}_W\right) \, , \end{aligned} \tag{16.18}$$

where the partially Wigner transformed Hamiltonian is

$$\hat{H}_W(R,P) = \frac{P^2}{2M} + \frac{\hat{p}^2}{2m} + \hat{V}_W(\hat{q},R) \, , \tag{16.19}$$

and Λ is the negative of the Poisson bracket operator,

$$\Lambda = \overleftarrow{\nabla}_P \cdot \overrightarrow{\nabla}_R - \overleftarrow{\nabla}_R \cdot \overrightarrow{\nabla}_P \, . \tag{16.20}$$

The direction of an arrow indicates the direction in which the operator acts. To obtain this equation we used the definition of the partial Wigner transform of an observable,

$$\hat{A}_W(R,P) = \int dz e^{-iP\cdot z/\hbar} \langle R + \frac{z}{2}|\hat{A}|R - \frac{z}{2}\rangle \, , \tag{16.21}$$

and the fact that the partial Wigner transform of a product of operators is [11]

$$(\hat{A}\hat{B})_W(R,P) = \hat{A}_W(R,P)e^{\hbar\Lambda/2i}\hat{B}_W(R,P) \, . \tag{16.22}$$

We may rewrite the quantum Liouville equation in a more compact form [12]

$$\begin{aligned} \frac{\partial \hat{\rho}_W(R,P,t)}{\partial t} &= -\frac{i}{\hbar}\left(\overrightarrow{H}_\Lambda \hat{\rho}_W(t) - \hat{\rho}_W(t)\overleftarrow{H}_\Lambda\right) \, , \\ &\equiv -i\hat{L}_W\hat{\rho}_W(t) \equiv -(H_W, \hat{\rho}_W(t))_Q \, . \end{aligned} \tag{16.23}$$

by defining the quantum Liouville operator and quantum Lie bracket. In these equations we have defined the right ($\overrightarrow{H}_\Lambda$) and left ($\overleftarrow{H}_\Lambda$) acting operators,

$$\overrightarrow{H}_\Lambda = \hat{H}_W(R,P)e^{\hbar\Lambda/2i} \, , \qquad \overleftarrow{H}_\Lambda = e^{\hbar\Lambda/2i}\hat{H}_W(R,P) \, . \tag{16.24}$$

The second equality in (16.23) defines the quantum Liouville operator $i\hat{L}_W$ in the partial Wigner representation while the third defines the associated Lie bracket $(H_W, \ )_Q$. More generally the Lie bracket of two partially Wigner transformed operators is defined as

$$(\hat{A}_W, \hat{B}_W)_Q = \frac{i}{\hbar}\left(\overrightarrow{A}_\Lambda \hat{B}_W - \hat{B}_W \overleftarrow{A}_\Lambda\right)$$

$$= \frac{i}{\hbar}\left(\hat{A}_W e^{\hbar\Lambda/2i}\hat{B}_W - \hat{B}_W e^{\hbar\Lambda/2i}\hat{A}_W\right) . \qquad (16.25)$$

Here the $\overrightarrow{A}_\Lambda$ and $\overleftarrow{A}_\Lambda$ operators are defined as in (16.24) with the replacement $\hat{H}_W \to \hat{A}_W$.

The formal solution of (16.23) is

$$\hat{\rho}_W(R,P,t) = e^{-i\overrightarrow{H}_\Lambda t/\hbar}\hat{\rho}_W(R,P,0)e^{i\overleftarrow{H}_\Lambda t/\hbar} ,$$

$$= e^{-i\hat{L}_W t}\hat{\rho}_W(R,P,0) . \qquad (16.26)$$

A similar set of equations may be written for the evolution of any quantum operator $\hat{A}$. In the Wigner representation these equations and their solutions, respectively, take the form,

$$\frac{d\hat{A}_W(R,P,t)}{dt} = i\hat{L}_W\hat{A}_W(R,P,t) = (H_W(R,P), \hat{A}_W(R,P,t))_Q , \qquad (16.27)$$

and

$$\hat{A}_W(R,P,t) = e^{i\hat{L}_W t}\hat{A}_W(R,P) = e^{i\overrightarrow{H}_\Lambda t/\hbar}\hat{A}_W(R,P)e^{-i\overleftarrow{H}_\Lambda t/\hbar} . \qquad (16.28)$$

We shall drop the dependence of quantities like $\hat{A}_W(R,P)$ on the bath phase space coordinates when confusion is unlikely to arise. However, we stress that the time dependence of the observables cannot be expressed as, e.g., $\hat{A}_W(R(t),P(t))$

We now consider some important properties of products of partially Wigner transformed operators. The Wigner transform of a product of operators satisfies the associative product rule,

$$(\hat{A}\hat{B}\hat{C})_W = \left(\left(\hat{A}_W e^{\hbar\Lambda/2i}\hat{B}_W\right)e^{\hbar\Lambda/2i}\hat{C}_W\right)$$

$$= \left(\hat{A}_W e^{\hbar\Lambda/2i}\left(\hat{B}_W e^{\hbar\Lambda/2i}\hat{C}_W\right)\right) , \qquad (16.29)$$

which may be generalized to products of n operators.

Next, consider a quantum operator $\hat{C} = \hat{A}\hat{B}$ which is the product of two operators. Since the time evolution of $\hat{C}$ may be written as $\hat{C}(t) = \hat{A}(t)\hat{B}(t)$, its partial Wigner transform is

$$\hat{C}_W(t) = \hat{A}_W(t)e^{\hbar\Lambda/2i}\hat{B}_W(t) . \qquad (16.30)$$

The quantum mechanical Lie bracket, either in its original form as $(i/\hbar)[\hat{A},\hat{B}]$ or in its partially Wigner transformed form $(\hat{A}_W,\hat{B}_W)_Q$, satisfies the Jacobi identity,

$$(\hat{A}_W,(\hat{B}_W,\hat{C}_W)_Q)_Q+(\hat{C}_W,(\hat{A}_W,\hat{B}_W)_Q)_Q+(\hat{B}_W,(\hat{C}_W,\hat{A}_W)_Q)_Q=0, \tag{16.31}$$

so that it has the Lie algebraic structure of any true dynamics, quantum or classical.

This general formulation of quantum dynamics reduces to standard descriptions in certain limiting cases. If the Q subsystem is absent, the system comprises only q degrees of freedom and we recover the usual quantum dynamical description in terms of the von Neumann equation (16.1). If one considers the Q dynamics alone without any q subsystem, one has the ordinary Wigner representation of quantum mechanics and all partially Wigner transformed operators become simple phase space functions: $\hat{A}_W(R,P)\to A_W(R,P)$. The classical limit of the quantum Q dynamics, which consists in keeping only terms of order $\hbar^0$ in the evolution operator, is obtained by truncating the power series expression of the exponential operator: $\exp(\hbar\Lambda/2i)=1+\hbar\Lambda/2i$. In this limit the bracket $(H_W,\ )_Q$ reduces to the Poisson bracket $\{H_W,\ \}$, and the Wigner representation of the quantum Liouville equation becomes the classical Liouville equation, $\partial\rho_C/\partial t=\{H_W,\rho_C\}=-iL_C\rho_C(t)$, whose solution may be written as,

$$\rho_C(R,P,t)=e^{-iL_Ct}\rho_C(R,P,0)=\rho_C(R(-t),P(-t),0)\ . \tag{16.32}$$

Having given this brief overview of quantum statistical mechanics, we turn to the central problem of this chapter: the construction of the analogs of these results for quantum-classical systems.

16.3 Quantum-Classical World

As discussed above, we consider a quantum mechanical system partitioned into two subsystems. Now, however, the second subsystem, comprising N particles with masses M and coordinate operators $\hat{Q}$, is taken to represent an environment or bath with $M\gg m$. We wish to study the limit where the bath degrees of freedom may be treated classically but the quantum character of the first subsystem (hereafter referred to as the quantum subsystem) cannot be neglected. The approximation to the full quantum dynamics we want to consider is depicted schematically in Fig. 16.1.

The passage to quantum-classical dynamics is made by first scaling distances in terms of the wavelength appropriate for the mass m particles, $\lambda_m=(\hbar^2/m\epsilon_0)^{1/2}$, where ϵ_0 is a suitable energy unit, scaling the momenta of the light and heavy particles by $p_m=(m\lambda_m/t_0)=(m\epsilon_0)^{1/2}$

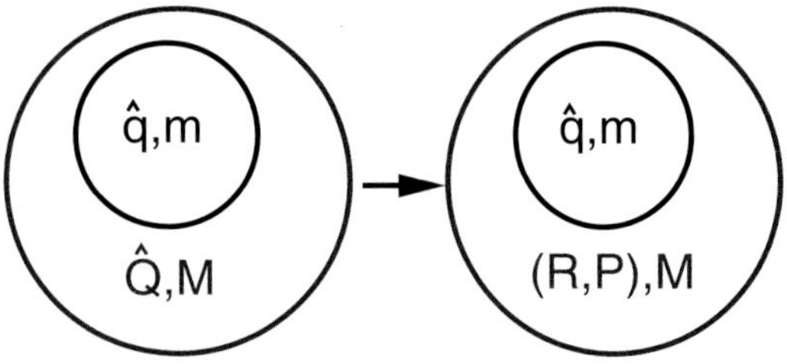

Fig. 16.1. Schematic representation of a quantum system partitioned into two subsystems and its approximation as a quantum subsystem in a classical bath.

and $P_M = (M\epsilon_0)^{1/2}$, respectively, and time by $t_0 = \hbar/\epsilon_0$. [10] The evolution operator in the quantum Liouville equation may then be expanded in $\mu = (m/M)^{1/2}$ and retaining terms to first order in this quantity we find (in the original unscaled variables), [10]

$$\begin{aligned} \frac{\partial \hat{\rho}_W(R,P,t)}{\partial t} &= -\frac{i}{\hbar}[\hat{H}_W, \hat{\rho}_W(t)] + \frac{1}{2}\left(\left\{\hat{H}_W, \hat{\rho}_W(t)\right\} - \left\{\hat{\rho}_W(t), \hat{H}_W\right\}\right) \\ &= -\frac{i}{\hbar}\left(\overrightarrow{\mathcal{H}_\Lambda}\hat{\rho}_W(t) - \hat{\rho}_W(t)\overleftarrow{\mathcal{H}_\Lambda}\right) \\ &\equiv -i\hat{\mathcal{L}}\hat{\rho}_W(t) \equiv -(\hat{H}_W, \hat{\rho}_W(t)) , \end{aligned} \tag{16.33}$$

where we have defined the right and left acting operators, respectively, as

$$\overrightarrow{\mathcal{H}_\Lambda} = \hat{H}_W\left(1 + \frac{\hbar\Lambda}{2i}\right), \qquad \overleftarrow{\mathcal{H}_\Lambda} = \left(1 + \frac{\hbar\Lambda}{2i}\right)\hat{H}_W , \tag{16.34}$$

the quantum-classical Liouville operator $\hat{\mathcal{L}}$ and the quantum-classical bracket as

$$(\hat{A}_W, \hat{B}_W) = \frac{i}{\hbar}\left(\overrightarrow{\mathcal{A}_\Lambda}\hat{B}_W - \hat{B}_W\overleftarrow{\mathcal{A}_\Lambda}\right) , \tag{16.35}$$

where $\overrightarrow{\mathcal{A}_\Lambda}$ is defined as $\overrightarrow{\mathcal{H}_\Lambda}$ in (16.34) with $\hat{H}_W \to \hat{A}_W$. Equation (16.33) is the quantum-classical Liouville equation [10,13–20] whose solution will be discussed later in this chapter. The quantum-classical Liouville equation describes the coupled evolution of these two subsystems. We shall see that as a result of this coupling a purely Newtonian description of the bath dynamics is no longer possible. The quantum-classical equation of motion for a dynamical variable $\hat{B}_W$ can be written in a similar form as

$$\frac{d\hat{B}_W(t)}{dt} = (\hat{H}_W, \hat{B}_W(t)) . \tag{16.36}$$

The formal solutions of (16.33) and (16.36) are [12]

$$\hat{\rho}_W(t) = e^{-i\hat{\mathcal{L}}t}\hat{\rho}_W(0) = \mathcal{S}\left(e^{-i\overrightarrow{\mathcal{H}_\Lambda}t}\hat{\rho}_W(0)e^{i\overleftarrow{\mathcal{H}_\Lambda}t}\right) , \tag{16.37}$$

and

$$\hat{B}_W(t) = e^{i\hat{\mathcal{L}}t}\hat{B}_W = \mathcal{S}\left(e^{i\overrightarrow{\mathcal{H}_\Lambda}t/\hbar}\hat{B}_W e^{-i\overleftarrow{\mathcal{H}_\Lambda}t/\hbar}\right) , \tag{16.38}$$

where the operator $\mathcal{S}$ is needed to prescribe how the left and right acting operators are to be evaluated to yield the evolution determined by the quantum-classical Liouville operator. The presence of the $\mathcal{S}$ operator signals the existence of differences in the formal structures of quantum-classical and quantum dynamics. Quantum-classical dynamics does not possess a Lie algebraic structure like quantum mechanics since the properties in (16.29), (16.30) and (16.31) are violated to some order in $\hbar$. In particular, the Jacobi identity [18,12]

$$(\hat{A}_W, (\hat{B}_W, \hat{C}_W)) + (\hat{C}_W, (\hat{A}_W, \hat{B}_W)) + (\hat{B}_W, (\hat{C}_W, \hat{A}_W)) = \mathcal{O}(\hbar), \tag{16.39}$$

is valid only to terms $\mathcal{O}(\hbar)$.

16.4 Nature of Quantum-Classical Dynamics

In order to gain insight into the nature of quantum-classical dynamics, in this section we show how the evolution of the density matrix can be expressed in terms of an ensemble of trajectories. Equation (16.33) is independent of the basis used to represent the quantum subsystem, and any convenient basis may be chosen to study the evolution. Here, however, we use an adiabatic basis since it provides a fruitful way to analyze the dynamics and carry out simulations. As we shall see below, we are forced to adopt an Eulerian description of the density matrix evolution since the evolution operator cannot be reduced to a streaming operator acting on the "classical" (R, P) coordinates. At each coordinate point R of the classical bath we define the Hamiltonian

$$\hat{h}_W(R) = \frac{\hat{p}^2}{2m} + \hat{V}_W(\hat{q}, R) , \tag{16.40}$$

whose eigenvalue problem

$$\hat{h}_W(R)|\alpha; R\rangle = E_\alpha(R)|\alpha; R\rangle , \tag{16.41}$$

yields the adiabatic states and energies. In this adiabatic basis the density matrix has matrix elements $\rho_W^{\alpha\alpha'}(R, P, t) = \langle\alpha; R|\hat{\rho}_W(R, P, t)|\alpha'; R\rangle$. The density matrix may be written as a vector function with components ρ_W^s by associating an index $s = \alpha\mathcal{N} + \alpha'$ with the pair $(\alpha\alpha')$, where $0 \leq \alpha, \alpha' < \mathcal{N}$ for an $\mathcal{N}$-state quantum subsystem.

Using this notation and introducing a subscript to label different values of s, e.g. $s_k = \alpha_k \mathcal{N} + \alpha'_k$, the quantum-classical Liouville equation has the form [21–23]

$$\frac{\partial \rho_W^{s_j}(R,P,t)}{\partial t} = \sum_{s_k} -i\mathcal{L}_{s_j s_k} \rho_W^{s_k}(R,P,t) \ . \tag{16.42}$$

After some algebra, the matrix elements of the quantum-classical Liouville operator are found to be [10,21]

$$\begin{aligned} -i\mathcal{L}_{s_j s_k} &= -(i\omega_{s_j} + iL_{s_j})\delta_{s_j s_k} + J_{s_j s_k} \\ &\equiv -i\mathcal{L}^0_{s_j s_k} + J_{s_j s_k} \ . \end{aligned} \tag{16.43}$$

The diagonal term $i\mathcal{L}^0_{s_j s_k} = i\mathcal{L}^0_{s_j}\delta_{s_j s_k} = (i\omega_{s_j} + iL_{s_j})\delta_{s_j s_k}$ contains the frequency $\omega_{s_j}(R) = (E_{\alpha_j}(R) - E_{\alpha'_j}(R))/\hbar$ and the classical Liouville operator L_{s_j}

$$iL_{s_j} = \frac{P}{M}\cdot\frac{\partial}{\partial R} + \frac{1}{2}\left(F_W^{\alpha_j} + F_W^{\alpha'_j}\right)\cdot\frac{\partial}{\partial P} \ , \tag{16.44}$$

where $F_W^{\alpha_j} = -\langle \alpha_j; R|\frac{\partial \hat{V}_W(\hat{r},R)}{\partial R}|\alpha_j; R\rangle = -\partial E_{\alpha_j}(R)/\partial R$ is the Hellmann-Feynman force that governs the motion on the adiabatic surface corresponding to the state $|\alpha_j; R\rangle$ of the $\hat{q}$ subsystem. The term $J_{s_j s_k}$ is responsible for non-adiabatic transitions and has the form [10]

$$\begin{aligned} J_{s_j s_k} = &-\frac{P}{M}\cdot d_{\alpha_j \alpha_k}\left(1 + \frac{1}{2} S_{\alpha_j \alpha_k}\cdot\frac{\partial}{\partial P}\right)\delta_{\alpha'_j \alpha'_k} \\ &-\frac{P}{M}\cdot d^*_{\alpha'_j \alpha'_k}\left(1 + \frac{1}{2} S^*_{\alpha'_j \alpha'_k}\cdot\frac{\partial}{\partial P}\right)\delta_{\alpha_j \alpha_k} \ , \end{aligned} \tag{16.45}$$

where $S_{\alpha_j \alpha_k} = (E_{\alpha_j} - E_{\alpha_k}) d_{\alpha_j \alpha_k} (\frac{P}{M}\cdot d_{\alpha_j \alpha_k})^{-1}$ and $d_{\alpha_j \alpha_k} = \langle \alpha_j; R|\frac{\partial}{\partial R}|\alpha_k; R\rangle$ is the non-adiabatic coupling matrix element which determines the non-adiabaticity of the system. Henceforth, we choose a real-valued adiabatic basis so that $d_{\alpha\alpha} = 0$ and $J_{s_j s_k}$ is off-diagonal. Moreover, when the first term on the right hand side of (16.45) is non-zero, the second term is zero, and vice versa. Specifically, the first term is non-zero for values of s_j and s_k such that $s_j - s_k = \pm \ell \mathcal{N}$, while the second term is non-zero only if $s_j - s_k = \pm \ell$, where $1 \leq \ell < \mathcal{N}$. These two conditions restrict the sequences of transitions that can occur. [23]

We can solve the equation for the density matrix formally to give

$$\rho_W^{s_j}(R,P,t) = \sum_{s_k} \left(e^{-i\mathcal{L}t}\right)_{s_j, s_k} \rho_0^{s_k}(R,P) \ , \tag{16.46}$$

where $\rho_0^{s_k}(R,P) \equiv \rho_W^{s_k}(R,P,0)$. Using the form of the Liouville operator $i\mathcal{L}_{s_j, s_k}$ in (16.43), we may use the Dyson identity to write the evolution operator as

$$\left(e^{-i\mathcal{L}t}\right)_{s_j,s_k} = e^{-i\mathcal{L}^0_{s_j}t}\delta_{s_js_k} + \sum_{s_l}\int_0^t dt' e^{-i\mathcal{L}^0_{s_j}(t-t')} J_{s_js_l}\left(e^{-i\mathcal{L}t'}\right)_{s_l,s_k} . \tag{16.47}$$

Equation (16.47) may be substituted into (16.46) and iterated to yield

$$\rho_W^{s_0}(R,P,t) = e^{-i\mathcal{L}^0_{s_0}t}\rho_0^{s_0}(R,P) + \sum_{n=1}^{\infty}\sum_{s_1\ldots s_n}\int_0^{t_0}dt_1\int_0^{t_1}dt_2\ldots\int_0^{t_{n-1}}dt_n$$
$$\times\prod_{k=1}^{n}\left[e^{-i\mathcal{L}^0_{s_{k-1}}(t_{k-1}-t_k)}J_{s_{k-1}s_k}\right]e^{-i\mathcal{L}^0_{s_n}t_n}\rho_0^{s_n}(R,P) , \tag{16.48}$$

where $t_0 \equiv t$. In this representation of the dynamics, the evolution operator $\exp(-i\mathcal{L}^0_{s_j}t)$ determines the evolution between the quantum transitions governed by $J_{s_js_k}$. For a diagonal contribution with $s_j \sim (\alpha_j\alpha_j)$, $\exp(-i\mathcal{L}^0_{s_j}t) = \exp(-iL_{\alpha_j}t)$ is the ordinary classical evolution operator governed by the potential $E_{\alpha_j}(R)$,

$$e^{-iL_{s_j}t}f_{s_j}(R,P) = f_{s_j}(R^t_{0,s_j}, P^t_{0,s_j}) , \tag{16.49}$$

where $f_{s_j}(R,P)$ is a function of the phase space point (R,P). Here $(R^t_{0,s_j}, P^t_{0,s_j})$ is the result of backward evolution to time zero of the phase point (R,P) at time t. For an off-diagonal contribution with $s_j \sim (\alpha_j\alpha'_j)$, the phase factor comes into play and in Ref. [10] we have shown that

$$e^{-(i\omega_{s_j}+iL_{s_j})t}f_{s_j}(R,P) = e^{-i\int_0^t d\tau\omega_{s_j}(R^\tau_{0,s_j},P^\tau_{0,s_j})}e^{-iL_{s_j}t}f_{s_j}(R,P)$$
$$\equiv W_{s_j}(t,0)f_{s_j}(R^t_{0,s_j}, P^t_{0,s_j}) . \tag{16.50}$$

In this case the classical evolution is determined by the mean potential $(E_{\alpha_j}(R) - E_{\alpha'_j}(R))/2$ of the two coherently coupled adiabatic states.

Figure 16.2 shows one of the trajectory segments contributing to the second order term in the density matrix. In this figure we are interested in the value of the $(\alpha\alpha)$ component of the density matrix at phase point (R,P) at time t. The phase point (R,P) is evolved backward in time on the E_α potential energy surface until time t'' where a quantum transition to adiabatic state β occurs. At this time the operator $J_{\alpha\alpha,\alpha\beta}$ acts, for example, to change the state of the second index of the density matrix, $(\alpha\alpha) \to (\alpha\beta)$; a corresponding continuous change occurs in the bath momentum determined by the momentum derivative in the J operator. States α and β are now coherently coupled. The phase point evolves backward in time on the mean of the E_α and E_β potential energy surfaces between time t'' where the quantum transition occurred and time t' where another quantum transition takes place.

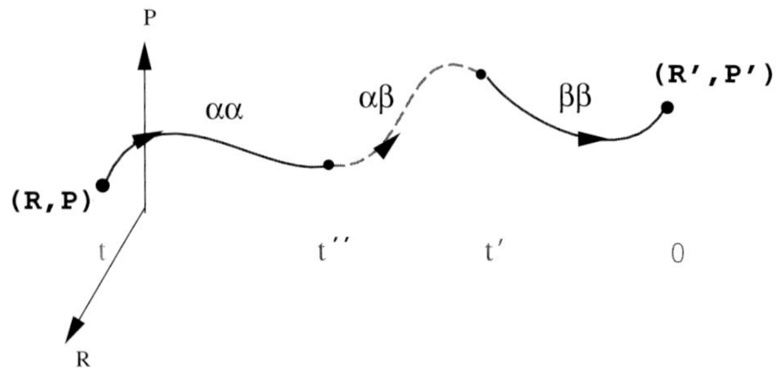

Fig. 16.2. Schematic picture of a trajectory that enters into the computation of the density matrix.

During this time interval the phase factor (using an obvious generalization of the definition in (16.50))

$$W_{\alpha\beta}(t',t'') = e^{-i\int_{t''}^{t'} d\tau\, \omega_{\alpha\beta}(R^{\tau}_{t'',\alpha\beta})} , \tag{16.51}$$

accumulates its value reflecting the coherent evolution of the off-diagonal element of the density matrix. At time t' a second quantum transition, e.g., $(\alpha\beta) \to (\beta\beta)$, occurs. As before the operator $J_{\alpha\beta,\beta\beta}$ determines the nature of this transition and specifies the momentum change in the bath. Due to this second quantum transition the system is once again in a diagonal state $(\beta\beta)$. As a result of this transition back to the diagonal state, no phase factor enters the evolution to time zero on the single adiabatic surface E_β to yield the phase point (R', P').

The density matrix element at time t can be constructed from an ensemble of such "surface-hopping" trajectories where all possible numbers of quantum transitions to all possible intermediate quantum states at all possible intermediate times are considered. This ensemble of trajectories provides an exact solution of the density matrix in the quantum-classical limit.

In order to illustrate the nature of this ensemble we consider a simple example where a two-level quantum subsystem is coupled to a single classical one-dimensional harmonic oscillator. [21,22] Since the classical phase space is two dimensional, we may easily visualize the classical trajectories that contribute to the density matrix evaluation. The hamiltonian $\hat{h}_W(R)$ can be written in terms of the quantum subsystem hamiltonian $(\hat{h}_s)$ plus the bath (V_b) and coupling $(\hat{V}_c)$ potentials as $\hat{h}_W(R) = \hat{h}_s + V_b(R) + \hat{V}_c(\hat{q}, R)$, with $\hat{h}_s = \hat{p}^2/2m + \hat{V}_s$. The eigenvalue problem for $\hat{h}_s$ is $\hat{h}_s|i\rangle = \tilde{\epsilon}_i|i\rangle$, where the space is spanned by the two eigenstates $|1\rangle$ and $|2\rangle$. We take the matrix elements of the coupling potential in this basis to be $\langle i|\hat{V}_c(\hat{q}, R)|j\rangle = \hbar\gamma(R)(1-\delta_{ij})$ where we have assumed that $V_{ii} = 0$. In terms of the diabatic or spin up and spin down states, $|\uparrow> = 2^{-1/2}(|1\rangle + |2\rangle)$ and $|\downarrow> = 2^{-1/2}(|1\rangle - |2\rangle)$, respectively, and taking the energy constant at zero, the hamiltoniam matrix of $\hat{h}_W(R)$ is

$$\mathbf{h}_W = -\hbar\Omega\hat{\sigma}_x + V_b(R)\mathbf{I} + \hbar\gamma(R)\hat{\sigma}_z , \tag{16.52}$$

where $2\hbar\Omega = \tilde{\epsilon}_2 - \tilde{\epsilon}_1$, **I** is the unit matrix and the Pauli matrices are

$$\hat{\sigma}_x = \begin{pmatrix} 0 & 1 \\ 1 & 0 \end{pmatrix}, \quad \hat{\sigma}_z = \begin{pmatrix} 1 & 0 \\ 0 & -1 \end{pmatrix} . \tag{16.53}$$

The solution of the eigenvalue problem for $\mathbf{h}_W$ yields the adiabatic energies and eigenstates. The adiabatic energies are

$$E_{1,2}(R) = V_b(R) \mp \hbar(\Omega^2 + \gamma(R)^2)^{1/2} . \tag{16.54}$$

The adiabatic eigenstates are

$$|1; R\rangle = \left(2(1+G^2)\right)^{-1/2} \left((1+G)|\uparrow\rangle + (1-G)|\downarrow\rangle\right) ,$$

$$|2; R\rangle = \left(2(1+G^2)\right)^{-1/2} \left((G-1)|\uparrow\rangle + (1+G)|\downarrow\rangle\right) , \tag{16.55}$$

with $G(R) = (\gamma(R))^{-1}(-\Omega + (\Omega^2 + \gamma(R)^2)^{1/2})$. The non-adiabatic coupling matrix element is $d_{12} = -d_{21} = -(1+G^2)^{-1}G'$.

An example of the ensemble of trajectories, with up to four non-adiabatic transitions, contributing to the 11 element of the density matrix at phase point (R, P) at a specific time $t = 4.4$ is shown in Fig. 16.3 for this system. [21] In this figure the three solid curves represent deterministic classical evolution on the three potential energy surfaces that enter into the description of the dynamics: the ground state adiabatic surface, $E_1(R)$, (outermost curve labeled 11), the excited state adiabatic surface, $E_2(R)$, (innermost curve labeled 22) and the mean of these two surfaces, $(E_1(R) + E_2(R))/2 = V_b(R)$,

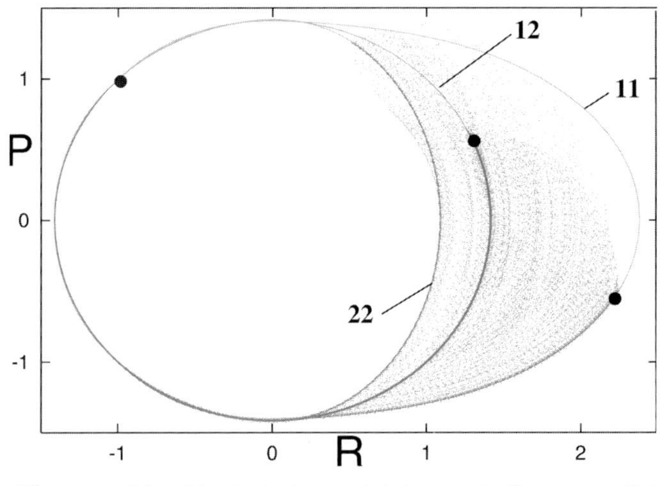

Fig. 16.3. The ensemble of trajectories containing up to four non-adiabatic transitions contributing to the 11-element of the density matrix at phase point $(-1, 1)$ at time t.

which is the same as the bare bath potential energy for this model system with two quantum states (middle curve labeled 12). All members of the ensemble start at phase point $(-1, 1)$ at time t indicated by a heavy dot. As an illustration of how the features of the trajectories can be understood, consider, for example, those trajectories that involve a single non-adiabatic transition between times t and 0. If the non-adiabatic transition $1 \to 2$ takes place at $t = 0^+$, the system will evolve on the ground adiabatic state (outermost curve) for entire trajectory ending at the heavy dot on the $E_1(R)$ surface. If the non-adiabatic transition takes place at $t = t^+$, the system will evolve on the bath potential (middle curve) for its entire history ending at the heavy dot on the $V_b(R)$ curve. If the transition occurs at any time intermediate between these two limits, ($t = 4.4$ and $t = 0$), the trajectory will end at a point on an arc that lies at the end of the high density region between the 12 and 11 curves and connects the two heavy dots on these curves. For these trajectories it is $\rho_0^{12}(R_{t,12}, P_{t,12})$ (and a similar quantity with $(1 \leftrightarrow 2)$) that determines the $\rho_W^{11}(R, P, t)$. A similar analysis can be carried out for trajectories with a larger number of non-adiabatic transitions.

16.5 Time Evolution of Dynamical Variables

In most statistical mechanical applications one is not interested in the evolution of a particular element of the density matrix at a particular phase point but, rather, in expectations of dynamical variables or time correlation functions. It is therefore often more useful to consider the evolution of a dynamical variable instead of the density matrix. Although the methods used to carry out this evolution are similar to those for the density matrix, it is convenient to show how the quantum-classical evolution equation for a dynamical variable can be analyzed.

The equation of motion for the dynamical variable was given in (16.36) and its formal solution, which was given in (16.38) can be written more explicitly as

$$B_W^{s_j}(R, P, t) = \sum_{s_k} \left(e^{i\mathcal{L}t}\right)_{s_j s_k} B_0^{s_k}(R, P) \;, \tag{16.56}$$

where $B_W^{s_j}(R, P, 0) = B_0^{s_j}(R, P)$. Using the form of the Liouville operator $i\mathcal{L}_{s_j s_k}$ in (16.43), we may use a variant of the identity in (16.47) to write the evolution operator as

$$\left(e^{i\mathcal{L}t}\right)_{s_j s_k} = e^{i\mathcal{L}^0_{s_j} t} \delta_{s_j s_k} - \sum_{s_l} \int_0^t dt' e^{i\mathcal{L}^0_{s_j} t'} J_{s_j s_l} \left(e^{i\mathcal{L}(t-t')}\right)_{s_l s_k} \;. \tag{16.57}$$

Substitution of this expression into (16.36), followed by iteration of the resulting equation and the change of variables $\tau_1 = t_1$ and $\tau_i = t_i + \tau_{i-1}$,

$(i > 1)$, yields the result,

$$B_W^{s_0}(R,P,t) = e^{\mathcal{L}^0_{s_0} t} B_0^{s_0}(R,P) + \sum_{n=1}^{\infty}(-1)^n \sum_{s_1 \dots s_n} \int_0^t d\tau_1 \int_{\tau_1}^t d\tau_2 \dots \int_{\tau_{n-1}}^t d\tau_n$$

$$\times \prod_{k=1}^{n} \left[e^{i\mathcal{L}^0_{s_{k-1}}(\tau_k - \tau_{k-1})} J_{s_{k-1}s_k} \right] e^{i\mathcal{L}^0_{s_n}(t-\tau_n)} B_0^{s_n}(R,P) \,. \quad (16.58)$$

The forward-evolved classical trajectory segments are defined in a manner that parallels the earlier discussion for the backward-evolved trajectories. We let

$$(\bar{R}_{s_j,t}, \bar{P}_{s_j,t}) = e^{iL_{s_j} t}(R,P) \,, \quad (16.59)$$

be the trajectory that starts at (R, P) at time 0 and ends at $(\bar{R}_{s_j,t}, \bar{P}_{s_j,t})$ at time t. The action of the evolution operator $e^{i\mathcal{L}^0_{s_j} t}$ on any phase function is

$$e^{i\mathcal{L}^0_{s_j} t} f_{s_j}(R,P) = e^{i\int_0^t d\tau \omega_{s_j}(\bar{R}_{s_j,\tau})} e^{iL_{s_j} t} f_{s_j}(R,P)$$

$$\equiv \mathcal{W}_{s_j}(t,0) f_{s_j}(\bar{R}_{s_j,t}, \bar{P}_{s_j,t}) \,. \quad (16.60)$$

We may write the form of the dynamical variable at time t more explicitly by using the forms of the phase points evolved under quantum-classical dynamics. In (16.45) we saw that J could be written as the sum of two contributions that determine which of the two indices in s_j changes in a non-adiabatic transition. We use a symbol $\kappa = 0, 1$ to denote these two contributions and label the S and d factors with the same symbol. While the dynamics may be carried out using the explicit expression for J as discussed in the previous section, in many instances it is sufficient to write J in the momentum-jump approximation [10]. To present the evolution results in their simplest form, we utilize this approximation in the remainder of this section.

The momentum-jump approximation to J may be constructed in the following way. The operator J involves differential operators of the form $\left(1 + \frac{1}{2} S_{\alpha\beta} \cdot \frac{\partial}{\partial P}\right)$ acting on functions of the classical phase space coordinates. It is possible to write this operator approximately as a "momentum jump" operator whose effect on the momentum is to shift it by some value. In order to carry out this calculation one must account for the fact that $S_{\alpha\beta}$ depends on the momenta. In spite of this dependence, one may introduce a translation operator in a new variable to complete the demonstration. Since $S_{\alpha\beta} = \Delta E_{\alpha\beta} \hat{d}_{\alpha\beta} (\frac{P}{M} \cdot \hat{d}_{\alpha\beta})^{-1}$, with $\Delta E_{\alpha\beta} = E_\alpha - E_\beta$, we may write

$$\left(1 + \frac{1}{2} S_{\alpha\beta} \cdot \frac{\partial}{\partial P}\right) = 1 + \frac{1}{2} \Delta E_{\alpha\beta} M \frac{1}{(P \cdot \hat{d}_{\alpha\beta})} \frac{\partial}{\partial (P \cdot \hat{d}_{\alpha\beta})}$$

$$= 1 + \Delta E_{\alpha\beta} M \frac{\partial}{\partial (P \cdot \hat{d}_{\alpha\beta})^2} \quad (16.61)$$

If we now consider the action of the operator on any function $f(P)$ of the momentum, we have

$$\left(1+\Delta E_{\alpha\beta}M\frac{\partial}{\partial(P\cdot\hat{d}_{\alpha\beta})^2}\right)f(P)\approx e^{\Delta E_{\alpha\beta}M\partial/\partial(P\cdot\hat{d}_{\alpha\beta})^2}f(P)$$

$$= e^{\Delta E_{\alpha\beta}M\partial/\partial(P\cdot\hat{d}_{\alpha\beta})^2}f\left(\hat{d}^{\perp}_{\alpha\beta}(P\cdot\hat{d}^{\perp}_{\alpha\beta})+\hat{d}_{\alpha\beta}\mathrm{sgn}(P\cdot\hat{d}_{\alpha\beta})\sqrt{(P\cdot\hat{d}_{\alpha\beta})^2}\right)$$

$$= f\left(\hat{d}^{\perp}_{\alpha\beta}(P\cdot\hat{d}^{\perp}_{\alpha\beta})+\hat{d}_{\alpha\beta}\mathrm{sgn}(P\cdot\hat{d}_{\alpha\beta})\sqrt{(P\cdot\hat{d}_{\alpha\beta})^2+\Delta E_{\alpha\beta}M}\right). \quad (16.62)$$

In the second line of this equation we have written the momentum vector as a sum of its components along $\hat{d}_{\alpha\beta}$ and perpendicular to $\hat{d}^{\perp}_{\alpha\beta}$, and in the last line we have used the fact that the exponential operator is a translation operator in the variable $(P\cdot\hat{d}_{\alpha\beta})^2$. If the energy difference times the mass $\Delta E_{\alpha\beta}M$ is small, we may expand the square root in the argument of f to obtain,

$$f\left(\hat{d}^{\perp}_{\alpha\beta}(P\cdot\hat{d}^{\perp}_{\alpha\beta})+\hat{d}_{\alpha\beta}\mathrm{sgn}(P\cdot\hat{d}_{\alpha\beta})\sqrt{(P\cdot\hat{d}_{\alpha\beta})^2+\Delta E_{\alpha\beta}M}\right)$$

$$\approx f\left(\hat{d}^{\perp}_{\alpha\beta}(P\cdot\hat{d}^{\perp}_{\alpha\beta})+\hat{d}_{\alpha\beta}(P\cdot\hat{d}_{\alpha\beta})+\frac{1}{2}(P\cdot\hat{d}_{\alpha\beta})^{-1}\Delta E_{\alpha\beta}M\right)$$

$$= f\left(P+\frac{1}{2}S_{\alpha\beta}\right). \quad (16.63)$$

Collecting these results we may write,

$$\left(1+\frac{1}{2}S_{\alpha\beta}\cdot\frac{\partial}{\partial P}\right)f(P)\approx e^{\Delta E_{\alpha\beta}M\partial/\partial(P\cdot\hat{d}_{\alpha\beta})^2}f(P)=f(P+\frac{1}{2}S_{\alpha\beta}). \quad (16.64)$$

Thus, to lowest order in the small parameter $\Delta E_{\alpha\beta}M$ we may write the operators in J as momentum translation (jump) operators.

This approximation may yield useful results beyond its strict domain of validity. Non-adiabatic transitions are likely to occur when adiabatic potential energy surfaces lie close in energy so that $\Delta E_{\alpha\beta}$ is small. In such circumstances the non-adiabatic coupling matrix element $d_{\alpha\beta}$ is typically large. The momentum jump approximation will be valid in such cases provided $P\cdot d_{\alpha\beta}$ is not too small. If $\Delta E_{\alpha\beta}$ is large, i.e. when the approximation fails, the prefactor of $(1+\frac{1}{2}S_{\alpha\beta}\cdot\frac{\partial}{\partial P})$, $P\cdot d_{\alpha\beta}/M$, is typically small and the contributions to the evolution coming from the J factors carry a small weight.

If we then consider the evolution of a phase point using the first-order momentum jump approximation, we can label the evolved phase point with state and κ labels to specify its history. Using this notation, the sequence of bath phase space coordinates at times $\tau_1,\tau_2,\ldots\tau_n$, supposing that one of the components of J acts at each of these times, is

$$(\bar{R}_{s_0,\tau_1}, \bar{P}_{s_0,\tau_1}) = e^{iL_{s_0}(\tau_1)}(R, P)$$

$$(\bar{R}^{\tau_1,\kappa_1}_{s_1,\tau_2}, \bar{P}^{\tau_1,\kappa_1}_{s_1,\tau_2}) = e^{iL_{s_1}(\tau_2-\tau_1)}(\bar{R}_{s_0,\tau_1}, \bar{P}_{s_0,\tau_1} + \frac{S_1^{\kappa_1}}{2})$$

$$\dots$$

$$(\bar{R}^{\{\tau_i,\kappa_i\}}_{s_i,\tau_{i+1}}, \bar{P}^{\{\tau_i,\kappa_i\}}_{s_i,\tau_{i+1}}) = e^{iL_{s_i}(\tau_{i+1}-\tau_i)}(\bar{R}^{\{\tau_{i-1},\kappa_{i-1}\}}_{s_{i-1},\tau_i}, \bar{P}^{\{\tau_{i-1},\kappa_{i-1}\}}_{s_{i-1},\tau_i} + \frac{S_i^{\kappa_i}}{2}) \,. \quad (16.65)$$

Here $\{\tau_i, \kappa_i\} = ((\tau_1, \kappa_1), (\tau_2, \kappa_2), \dots, (\tau_i, \kappa_i))$ labels the history of the choice of the two terms in J.

Using this form we may write the solution in terms of surface-hopping trajectories as

$$\begin{aligned}
B_W^{s_0}(R,P,t) &= \mathcal{W}_{s_0}(t,0) B_0^{s_0}(\bar{R}_{s_0,t}, \bar{P}_{s_0,t}) \\
&+ \sum_{n=1}^{\infty} (-1)^n \sum_{s_1\kappa_1,\dots,s_n\kappa_n} \int_0^t d\tau_1 \int_{\tau_1}^t d\tau_2 \dots \int_{\tau_{n-1}}^t d\tau_n \\
&\times \prod_{k=1}^{n} \left[\mathcal{W}_{s_{k-1}}(\tau_k, \tau_{k-1}) \frac{\bar{P}^{\{\tau_k,\kappa_k\}}_{s_{k-1},\tau_k})}{M} \cdot d^{\kappa_k}_{s_{k-1}\to s_k}(\bar{R}^{\{\tau_k,\kappa_k\}}_{s_{k-1},\tau_k}) \right] \\
&\times \mathcal{W}_{s_n}(t,\tau_n) B_0^{s_n}(\bar{R}^{\{\tau_n,\kappa_n\}}_{s_{n-1},\tau_n}, \bar{P}^{\{\tau_n,\kappa_n\}}_{s_{n-1},\tau_n}) \,.
\end{aligned} \quad (16.66)$$

16.5.1 Equations for Canonical Variables

To gain some appreciation for the nature of quantum-classical evolution of a dynamical variable, we consider the equations of motion for the "classical" canonical variables. Letting $B_0^{\alpha\alpha'} = R\delta_{\alpha\alpha'}$ or $B_0^{\alpha\alpha'} = P\delta_{\alpha\alpha'}$, and using (16.36) we find

$$\frac{dR_W^{\alpha\alpha'}(t)}{dt} = \sum_{\beta\beta'} i\mathcal{L}_{\alpha\alpha',\beta\beta'} R_W^{\beta\beta'}(t) = \sum_{\beta\beta'} \left(e^{i\mathcal{L}t}\right)_{\alpha\alpha',\beta\beta'} \frac{P}{M} \delta_{\beta\beta'} \equiv \frac{P_W^{\alpha\alpha'}(t)}{M} \quad (16.67)$$

$$\frac{dP_W^{\alpha\alpha'}(t)}{dt} = \sum_{\beta\beta'} i\mathcal{L}_{\alpha\alpha',\beta\beta'} P_W^{\beta\beta'}(t) = \sum_{\beta\beta'} \left(e^{i\mathcal{L}t}\right)_{\alpha\alpha',\beta\beta'} F_W^{\beta} \delta_{\beta\beta'} \equiv F_W^{\alpha\alpha'}(t) \,, \quad (16.68)$$

where we have used the fact that $i\mathcal{L}_{\alpha\alpha',\beta\beta'} R = (P/M)\delta_{\alpha\beta}\delta_{\alpha'\beta'}$ and $i\mathcal{L}_{\alpha\alpha',\beta\beta'} P = F_W^{\beta}\delta_{\alpha\beta}\delta_{\alpha'\beta'}$. While the first equation for the time evolution of the position has the same form as Newton's equations of motion, the second equation of motion for the momentum does not. The "force" $F_W^{\alpha\alpha'}(t)$ cannot be expressed simply as function of $R_W^{\alpha\alpha'}(t)$ as may be verified by repeated application of $i\mathcal{L}$ to its initial value. Consequently, to evaluate the canonical variables one must use the techniques described above for a general dynamical variable.

16.6 Quantum-Classical Equilibrium Density

The form of the canonical equilibrium density that appears in the quantum mechanical expressions for transport coefficients is $\hat{\rho}_e^Q = Z_Q^{-1}\exp(-\beta\hat{H})$ and expressed in terms of the partial Wigner transform it can be written as

$$\hat{\rho}_{We}^Q(R,P) = (2\pi\hbar)^{-3N}\int dz e^{iP\cdot z/\hbar}\langle R - \frac{z}{2}|\hat{\rho}_e^Q|R + \frac{z}{2}\rangle \; . \qquad (16.69)$$

The equilibrium density is stationary under full quantum dynamics, either in its original or partial Wigner transformed forms. It is not stationary under quantum-classical dynamics and in this section we discuss the quantum-classical analog of this equilibrium density which satisfies, [12]

$$i\hat{\mathcal{L}}\hat{\rho}_{We} = \frac{i}{\hbar}(\overrightarrow{\mathcal{H}_\Lambda}\hat{\rho}_{We} - \hat{\rho}_{We}\overleftarrow{\mathcal{H}_\Lambda}) = 0 \; . \qquad (16.70)$$

One way to find a solution of this equation is in terms of a power series expansion in $\hbar$. Letting

$$\hat{\rho}_{We} = \sum_{n=0}^{\infty} \hbar^n \hat{\rho}_{We}^{(n)} \; , \qquad (16.71)$$

substituting this expression in (16.70) and grouping by powers of $\hbar$, we obtain the following recursion relations: for $n = 0$,

$$i\left[\hat{H}_W, \hat{\rho}_{We}^{(0)}\right] = 0 \; , \qquad (16.72)$$

and for $n \geq 0$,

$$i\left[\hat{H}_W, \hat{\rho}_{We}^{(n+1)}\right] = \frac{1}{2}\left\{\hat{H}_W, \hat{\rho}_{We}^{(n)}\right\} - \frac{1}{2}\left\{\hat{\rho}_{We}^{(n)}, \hat{H}_W\right\} \; . \qquad (16.73)$$

If a similar set of recursion relations is written for the partial Wigner transform of the full quantum mechanical canonical equilibrium density matrix, one finds that the two set of recursion relations are identical to $\mathcal{O}(\hbar)$. Since the recursion relations permit us to obtain the terms higher order in $\hbar$ from those with lower orders, we have sufficient information to construct the quantum-classical stationary density that is consistent with the quantum mechanical equilibrium density to order $\hbar$.

To compute the equilibrium density, it is convenient for our purposes to consider these recursion relations in an adiabatic basis where they take the form,

$$iE_{\alpha\alpha'}\rho_{We}^{(0)\alpha\alpha'} = 0 \; , \qquad (16.74)$$

$$iE_{\alpha\alpha'}\rho_{We}^{(n+1)\alpha\alpha'} = -iL_{\alpha\alpha'}\rho_{We}^{(n)\alpha\alpha'} + \sum_{\nu\nu'} J_{\alpha\alpha',\nu\nu'}\rho_{We}^{(n)\nu\nu'} . \tag{16.75}$$

One may prove that these equations can be solved to any order in $\hbar$ to obtain the equilibrium density. While it is difficult to find the full solution to any order in $\hbar$, it is not difficult to find the solution analytically to order $\hbar$. It is given by [12]

$$\rho_{We}^{\alpha\alpha'} = \rho_{We}^{(0)\alpha}\Big(\delta_{\alpha\alpha'} - i\frac{P}{M}\cdot d_{\alpha\alpha'}\Big(\frac{\beta}{2}(1+e^{-\beta E_{\alpha'\alpha}})$$
$$+\frac{1}{E_{\alpha\alpha'}}(1-e^{-\beta E_{\alpha'\alpha}})\Big)(1-\delta_{\alpha\alpha'})\Big) + \mathcal{O}(\hbar^2) . \tag{16.76}$$

The utility of this expression for the equilibrium density matrix to $\mathcal{O}(\hbar)$ for the calculation of time correlation functions can be subjected to numerical test.

16.7 Quantum-Classical Time Correlation Functions

We are now in a position to address the problem posed at the begining of this chapter in Sect. 16.2: the nature of the quantum-classical forms for equilibrium time correlation functions and their associated transport coefficients. The more general issue we address is the construction of a nonequilibrium statistical mechanics in a world obeying quantum-classical dynamics. To carry out this program we begin by constructing a linear response theory for quantum-classical dynamics. [12] The formalism parallels that for quantum (or classical) systems. We suppose the quantum-classical system with hamiltonian $\hat{H}_W$ is subjected to a time dependent external force that couples to the observable $\hat{A}_W$, so that the total hamiltonian is

$$\hat{\mathbf{H}}_W(t) = \hat{H}_W - \hat{A}_W^\dagger F(t) . \tag{16.77}$$

The evolution equation for the density matrix takes the form

$$\frac{\partial \hat{\rho}_W(t)}{\partial t} = (i\hbar)^{-1}\left(\overrightarrow{\mathbf{H}}_A(t)\hat{\rho}_W(t) - \hat{\rho}_W(t)\overleftarrow{\mathbf{H}}_A(t)\right) ,$$
$$= -(i\hat{\mathcal{L}} - i\hat{\mathcal{L}}_A F(t))\hat{\rho}_W(t) , \tag{16.78}$$

where $\overrightarrow{\mathbf{H}}_A(t) = \overrightarrow{\mathcal{H}}_A - \overrightarrow{\mathcal{A}}_A^\dagger F(t)$ and $i\hat{\mathcal{L}}_A$ has a form analogous to $i\hat{\mathcal{L}}$ with $\hat{A}_W^\dagger$ replacing $\hat{H}_W$, $i\hat{\mathcal{L}}_A = (\hat{A}_W^\dagger,\ )$. The formal solution of this equation is found by integrating from t_0 to t,

$$\hat{\rho}_W(t) = e^{-i\hat{\mathcal{L}}(t-t_0)}\hat{\rho}_W(t_0)$$
$$+\int_{t_0}^t dt'\, e^{-i\hat{\mathcal{L}}(t-t')} i\hat{\mathcal{L}}_A \hat{\rho}_W(t')F(t') . \tag{16.79}$$

In parallel with the derivation for quantum mechanical systems outlined in Sect. 16.2, we choose $\hat{\rho}_W(t_0)$ to be the equilibrium density matrix, $\hat{\rho}_{We}$. As discussed in Sect. 16.6, $\hat{\rho}_{We}$ is defined to be invariant under quantum-classical dynamics, $i\hat{\mathcal{L}}\hat{\rho}_{We} = 0$. In this case the first term on the right hand side of (16.79) reduces to $\hat{\rho}_{We}$ and is independent of t_0. We may assume that the system with hamiltonian $\hat{H}_W$ is in thermal equilibrium at $t_0 = -\infty$, and with this boundary condition, to first order in the external force, (16.79) is

$$\hat{\rho}_W(t) = \hat{\rho}_{We} + \int_{-\infty}^{t} dt' \, e^{-i\hat{\mathcal{L}}(t-t')} i\hat{\mathcal{L}}_A \hat{\rho}_{We} F(t') \, . \tag{16.80}$$

Then, computing $\overline{B_W(t)} = \mathrm{Tr}' \int dRdP \;\, \hat{B}_W \hat{\rho}_W(t)$ to obtain the response function, we find

$$\begin{aligned}\overline{B_W(t)} &= \int_{-\infty}^{t} dt' \, \mathrm{Tr}' \int dRdP \, \hat{B}_W e^{-i\hat{\mathcal{L}}(t-t')} i\hat{\mathcal{L}}_A \hat{\rho}_{We} F(t') \\ &= \int_{-\infty}^{t} dt' \, \langle (\hat{B}_W(t-t'), \hat{A}_W^{\dagger}) \rangle F(t') \equiv \int_{-\infty}^{t} dt' \, \phi_{BA}^{QC}(t-t') F(t') \, .\end{aligned} \tag{16.81}$$

Thus, the quantum-classical form of the response function is

$$\phi_{BA}^{QC}(t) = \langle (\hat{B}_W(t), \hat{A}_W^{\dagger}) \rangle \, . \tag{16.82}$$

The derivation of linear response theory in the quantum-classical world is completely analogous to that in quantum mechanics. The response function in (16.82) is structurally similar to that in (16.10) with the quantum-classical bracket replacing the quantum Lie bracket, $(i/\hbar)[\ ,\ ] \to (\ ,\ )$, the dynamical variable $\hat{B}_W(t)$ evolves under quantum-classical dynamics instead of full quantum mechanics and the trace over the quantum canonical density matrix is replaced by an average over $\hat{\rho}_{We}$. One may regard these replacements as correspondence rule to transform a quantum response function to the quantum-classical response function.

Knowing the response function, we may derive an expression for a transport property by taking $\hat{B}_W = \dot{\hat{A}}_W = i\mathcal{L}\hat{A}_W \equiv \hat{f}_{AW}$. The quantum-classical analog of the expression for a quantum mechanical transport coefficient in (16.14) is given by

$$\lambda_A \propto \int_0^{\infty} dt \, \langle (\hat{f}_{AW}(t), \hat{A}_W^{\dagger}) \rangle = \int_0^{\infty} dt \, \mathrm{Tr}' \int dRdP \, \hat{f}_{AW}(t) (\hat{A}_W^{\dagger}, \hat{\rho}_{We}) \, . \tag{16.83}$$

In writing the second line of (16.83) we have used cyclic permutations under the trace and integrations by parts. In addition to this direct derivation via linear response theory, this form can also be obtained by applying the correspondence rule to the first expression on the right hand side of (16.14).

At this point we have all the ingredients for the computation of transport properties and expectation values of dynamical variables in a quantum-classical world. The equilibrium time correlation function in (16.83) entails evolution of $\hat{f}_W(t)$ under quantum-classical classical dynamics as discussed in Sect. 16.5, evaluation of the quantum-classical bracket of $\hat{A}_W^\dagger$ and $\hat{\rho}_{We}$, and an integration over the classical phase space coordinates and trace over the quantum states. In this formulation of quantum-classical nonequilibrium statistical mechanics, correlation functions should be computed by ensemble averages of dynamical quantities, as specified by (16.83), rather than by time averages. In addition, we observe that while $\hat{\rho}_{We}$ is not a probability density, $\hat{\rho}_W^{(0)}$ provides a weight function for sampling the phase space points and quantum states for the evaluation of the average.

While this statistical mechanical formulation is complete it is worth remarking that some aspects of the quantum mechanical calculation do not carry over to the quantum-classical world. These concern time translation invariance and alternate forms for the time correlation function expressions for transport coefficients.

The first issue we examine is time translation invariance of the equilibrium time correlation functions. Consider the quantum mechanical response function in (16.10). This function may also be written as

$$\phi_{BA}(t) = \langle \frac{i}{\hbar}[\hat{B}(t), \hat{A}^\dagger]\rangle_Q = \langle \frac{i}{\hbar}[\hat{B}(t+\tau), \hat{A}^\dagger(\tau)]\rangle_Q \, , \tag{16.84}$$

using the form of the canonical equilibrium density matrix and cyclic permutations under the trace. This property is not exactly satisfied by the correlation function in quantum-classical response function (16.82). To see this we may write (16.82) more explicitly as

$$\begin{aligned}\phi_{BA}^{QC}(t) &= \langle (\hat{B}_W(t), \hat{A}_W^\dagger) \rangle \\ &= \frac{i}{\hbar}\Big(\langle \hat{B}_W(t)\,(1+\hbar\Lambda/2i)\,\hat{A}_W^\dagger\rangle - \langle \hat{A}_W^\dagger\,(1+\hbar\Lambda/2i)\,\hat{B}_W(t)\rangle \Big) \, ,\end{aligned} \tag{16.85}$$

Using cyclic permutations under the trace, integration by parts and the fact that $\hat{\rho}_{We}$ is invariant under quantum classical dynamics, one may show that

$$\langle \hat{B}_W(t)\,(1+\hbar\Lambda/2i)\,\hat{A}_W^\dagger\rangle = \langle e^{i\mathcal{L}\tau}(\hat{B}_W(t)\,(1+\hbar\Lambda/2i)\,\hat{A}_W^\dagger)\rangle \, . \tag{16.86}$$

However, the evolution of a composite operator in quantum-classical dynamics cannot be written exactly in terms of the quantum-classical evolution of its constituent operators, but only to terms $\mathcal{O}(\hbar)$. To see this consider the action of the quantum-classical Liouville operator on the composite operator

$\hat{C}_W = \hat{B}_W(1+\hbar\Lambda/2i)\hat{A}_W^\dagger$. We have

$$\begin{aligned} i\hat{\mathcal{L}}\hat{C}_W &= \frac{i}{\hbar}\left(\overrightarrow{\mathcal{H}_\Lambda}\left(\hat{B}_W\left(1+\frac{\hbar\Lambda}{2i}\right)\hat{A}_W^\dagger\right)\right) \\ &\quad -\frac{i}{\hbar}\left(\left(\hat{B}_W\left(1+\frac{\hbar\Lambda}{2i}\right)\hat{A}_W^\dagger\right)\overleftarrow{\mathcal{H}_\Lambda}\right), \\ &= (i\hat{\mathcal{L}}\hat{B}_W)\left(1+\frac{\hbar\Lambda}{2i}\right)\hat{A}_W^\dagger + \hat{B}_W\left(1+\frac{\hbar\Lambda}{2i}\right)(i\hat{\mathcal{L}}\hat{A}_W^\dagger) + \mathcal{O}(\hbar)\,. \end{aligned} \tag{16.87}$$

It follows that

$$\begin{aligned} \hat{C}_W(\tau) = e^{i\hat{\mathcal{L}}\tau}\hat{C}_W &= \left(e^{i\hat{\mathcal{L}}\tau}\hat{B}_W\right)\left(1+\frac{\hbar\Lambda}{2i}\right)\left(e^{i\hat{\mathcal{L}}\tau}\hat{A}_W^\dagger\right) + \mathcal{O}(\hbar) \\ &= \hat{B}_W(\tau)\left(1+\frac{\hbar\Lambda}{2i}\right)\hat{A}_W^\dagger(\tau) + \mathcal{O}(\hbar)\,. \end{aligned} \tag{16.88}$$

Therefore, the quantum-classical correlation function satisfies standard time translation invariance only to $\mathcal{O}(\hbar)$,

$$\phi_{BA}^{QC}(t) = \langle(\hat{B}_W(t), \hat{A}_W^\dagger)\rangle = \langle(\hat{B}_W(t+\tau), \hat{A}_W^\dagger(\tau))\rangle + \mathcal{O}(\hbar)\,, \tag{16.89}$$

although its most strict form, (16.86), is surely satisfied. Consequently, transport properties should be computed using ensemble averages as in (16.83), rather than through time averages assuming ergodicity.

Next, we consider alternate forms for correlations that are commonly used in computations. We saw that the quantum mechanical response function (16.10) could be written in the equivalent form (16.12) using the Kubo identity (16.11). However, the quantum-classical version of the Kubo identity holds only to $\mathcal{O}(\hbar)$, [12]

$$(\hat{A}_W^\dagger, \hat{\rho}_{We}) = \int_0^\beta d\lambda \hat{\rho}_{We}(1+\frac{\hbar\Lambda}{2i})\dot{\hat{A}}_W^\dagger(-i\hbar\lambda) + \mathcal{O}(\hbar)\,. \tag{16.90}$$

If we then write the quantum-classical transport coefficient (16.83) as

$$\lambda_A \propto \int_0^\infty dt\, \mathrm{Tr}' \int dRdP\, \hat{f}_{AW}(t)(\hat{A}_W^\dagger, \hat{\rho}_{We})\,, \tag{16.91}$$

and use (16.90), we find another expression for the transport coeficient in Kubo transformed form,

$$\lambda_A \propto \int_0^\infty dt \int_0^\beta d\lambda \mathrm{Tr}' \int dRdP\, \left(\hat{f}_{AW}^\dagger(-i\hbar\lambda)(1+\frac{\hbar\Lambda}{2i})\hat{f}_{AW}(t)\right)\hat{\rho}_{We} + \mathcal{O}(\hbar), \tag{16.92}$$

Since the quantum-classical form of the Kubo identity is valid only to $\mathcal{O}(\hbar)$, the two forms of the autocorrelation function expressions for the transport coefficient are no longer equivalent. The results of comparisons of computations of both forms of the correlation functions can provide information about the reduction to the quantum-classical limit.

16.8 Simulation Schemes

In this section we give a few technical details concerning the implementation of the development presented above needed to simulate quantum-classical dynamics and evaluate expectation values and time correlation functions.

To simulate quantum-classical evolution one must evaluate the action of the time evolution operator $\exp(i\mathcal{L}t)$ on a dynamical variable. There are many ways to do this. While we have stressed the use of an adiabatic basis, the choice of basis is often determined by the physical application under consideration. One may implement schemes that treat the operator J responsible for non-adiabatic transitions and bath momentum changes either exactly [21,23] or approximately by momentum jump approximations [10,21–23]. Finally, a variety of schemes may be devised for the computation of the evolution operator. [10,21,22,19,23–25]

To illustrate the application of the formalism and techniques described here to a many-body system, we consider the calculation of the average value of an observable $\hat{B}_W$ at time t for the spin-boson system [23],

$$\overline{B_W(t)} = \sum_{\alpha\alpha'} \int dRdP\ B_W^{\alpha\alpha'}(t)\rho_W^{\alpha'\alpha}(R,P,0)\ , \tag{16.93}$$

where $\hat{\rho}_W(R,P,0)$ is the initial value of the density matrix. We have expressed the expectation value in an adiabatic basis. In order to compute such average values, one must sample initial phase space points and quantum states from a weight determined by $\rho_W^{\alpha'\alpha}(R,P,0)$ and evolve $B_W^{\alpha\alpha'}(t)$ according to quantum-classical dynamics. We have already shown in (16.66) how to express the time evolution of $B_W^{\alpha\alpha'}(t)$ in terms of a sequence of surface-hopping trajectories. Consequently, a solution for $\overline{B_W(t)}$ in terms of surface-hopping trajectories may be found using a hybrid Monte Carlo-Molecular Dynamics scheme that combines a numerical implementation of (16.66) with sampling according to $\rho_W^{\alpha'\alpha}(R,P,0)$. [21–23]

In order to complete the calculation of the nth order term in (16.66) we must carry out the sums over the discrete s_k indices and perform the multiple time integrals. Not all the sequences $\{s_0,\dots,s_n\}$ contribute since only a subset $\mathcal{S}$ of the sequences is physically permissible. The allowed sequences have $s_k - s_{k+1} = \pm\ell\mathcal{N}$ or $s_k - s_{k+1} = \pm\ell$ for $1 \le \ell < \mathcal{N}$. One can either count all the elements of $\mathcal{S}$ contributing to (16.66) or estimate the sum through a Monte Carlo sampling of the summand over $\mathcal{S}$. [23]

The time integrals that must be computed are of the type

$$\mathcal{I} = \int_0^{t_0} dt_1 \int_0^{t_1} dt_2 ... \int_0^{t_{n-1}} dt_n f(t_1, t_2, ..., t_n), \tag{16.94}$$

and simplest Monte Carlo method that can be used for their evaluation involves uniform sampling. Alternatively, sampling based on least-discrepancy sequences or other more sophisticated sampling schemes may be used. [23]

16.8.1 Spin–Boson Model

The spin-boson model is one of the most widely studied model systems because it provides a simple description of many physical phenomena and is amenable to analysis. [26–29] It has served as testing ground for many simulation methods [30,31] and it in this spirit which we study it here. The spin-boson model describes a two-level system, with states $\{|\uparrow>, |\downarrow>\}$, bilinearly coupled to a harmonic bath of N oscillators with masses M_j and frequencies ω_j, and has hamiltonian

$$\hat{H} = -\hbar \Omega \hat{\sigma}_x + \sum_{j=1}^{N} \left(\frac{\hat{P}_j^2}{2M_j} + \frac{1}{2} M_j {\omega_j}^2 \hat{R}_j^2 - c_j \hat{R}_j \hat{\sigma}_z \right) . \tag{16.95}$$

The energy gap of the isolated two-state system is $2\hbar\Omega$ and $\hat{\sigma}_x$ and $\hat{\sigma}_z$ are Pauli matrices. The coupling constants c_j and frequencies ω_j in this hamiltonian have been taken from Makri and Thompson [31],

$$c_j = \sqrt{\xi \hbar \omega_0 M_j} \omega_j, \qquad \omega_j = -\omega_c \ln \left(1 - j \frac{\omega_0}{\omega_c} \right) , \tag{16.96}$$

where $\omega_0 = (\omega_c/N)(1 - \exp(-\omega_{max}/\omega_c))$. The spectral density is characterized by the Kondo parameter ξ and frequency ω_c. The parameter ω_{max} is a cut-off frequency. With this parameter choice we have a model of an infinite bath with Ohmic spectral density in terms of a finite number of oscillators.

Taking the partial Wigner transform[9] over the bath degrees of freedom, the hamiltonian becomes

$$\begin{aligned} \hat{H}_W &= -\hbar \Omega \hat{\sigma}_x + \sum_{j=1}^{N} \left(\frac{{P_j}^2}{2M_j} + \frac{1}{2} M_j {\omega_j}^2 {R_j}^2 - c_j R_j \hat{\sigma}_z \right) , \\ &= \sum_{j=1}^{N} \frac{{P_j}^2}{2M_j} - \hbar \Omega \hat{\sigma}_x + V_b(R) + \hbar \gamma(R) \hat{\sigma}_z . \end{aligned} \tag{16.97}$$

which depends on the classical phase space coordinates (R, P) and the spin degrees of freedom. The last three terms in the second line of (16.97) have the same form (16.52) with

$$V_b(R) = \sum_{j=1}^{N} \frac{1}{2} M_j {\omega_j}^2 \hat{R}_j^2 \,, \tag{16.98}$$

and $\gamma(R) = -\sum_{j=1}^{N} c_j R_j$. The adiabatic energies $E_\alpha(R)$, ($\alpha = 1, 2$) and corresponding eigenvectors have the same forms as (16.54) and (16.55). The results will be presented in dimensionless variables, $R'_j = (M_j \omega_c/\hbar)^{1/2} R_j$ and $P'_j = (\hbar M_j \omega_c)^{-1/2} P_j$, and we henceforth drop the primes on the variables and assume that these dimensionless variables are used.

We have assumed that the density matrix at $t = 0$ is uncorrelated so that the subsystem is in state $|\uparrow\rangle$ and bath is in thermal equilibrium,

$$\hat{\rho}(0) = \hat{\rho}_s(0) Z_b^{-1} e^{-\beta \hat{H}_b} \,, \quad \hat{\rho}_s(0) = \begin{pmatrix} 1 & 0 \\ 0 & 0 \end{pmatrix} , \tag{16.99}$$

where Z_b is the bath partition function. The partial Wigner transform of this initial density operator is [9]

$$\hat{\rho}_W(R, P, 0) = \hat{\rho}_s(0) \rho_{bW}(R, P), \tag{16.100}$$

where

$$\rho_{bW}(R, P) = \prod_{i=1}^{N} \frac{\tanh(\beta\omega_i/2)}{\pi} \exp\left[-\frac{2\tanh(\beta\omega_i/2)}{\omega_i} \left(\frac{P_i^2}{2} + \frac{\omega_i^2 R_i^2}{2} \right) \right] . \tag{16.101}$$

The time evolution of the difference in population between the ground and excited states (expectation value of $\hat{\sigma}_z$)

$$\overline{\hat{\sigma}_z(t)} = \mathrm{Tr}' \int dR dP \; \hat{\sigma}_z(t) \hat{\rho}_W(R, P, 0) = \sum_{\alpha\alpha'} \int dR dP \; \sigma_z^{\alpha\alpha'}(t) \rho_W^{\alpha'\alpha}(R, P, 0) \,, \tag{16.102}$$

was computed for this ten-oscillator spin-boson model using the surface-hopping scheme discussed in Sects. 16.4 and 16.5. The calculations were carried out using both the exact form of the J operator as well as its representation in terms of the momentum jump approximation. [23]

The results obtained using the quantum-classical surface-hopping scheme [23] are compared below with the known numerically exact results [31] for this model. Figure 16.4 (left panel) is a plot of $\overline{\sigma_z(t)}$ versus time computed using the surface-hopping algorithm including up to four ($n = 4$) non-adiabatic transitions along with the influence functional results for a Kondo parameter of $\xi = 0.007$. One can see that for the time interval shown our results for $n = 4$ are in complete accord with those of Makri and Thompson. [31]

It is instructive to examine the individual adiabatic and non-adiabatic contributions to the surface-hopping solution as a function of time. These

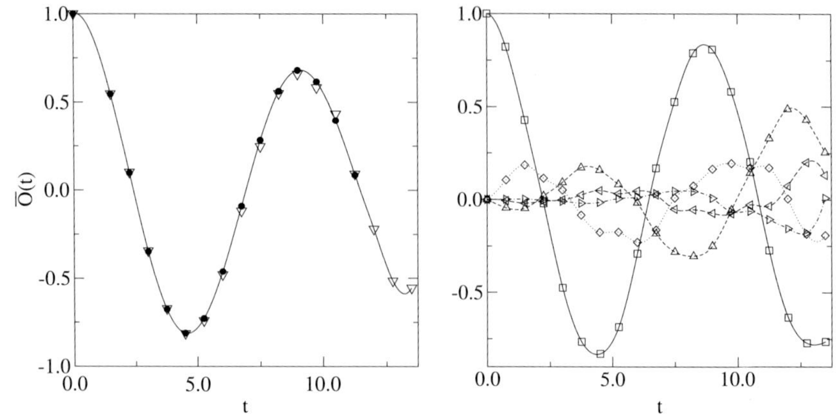

Fig. 16.4. Left panel: $\overline{\sigma_z(t)}$ versus time for $\xi = 0.007$. Influence functional results (filled circle) (see text), surface-hopping results for n up to 4 (down triangle). Right panel: Contributions to $\overline{\sigma_z(t)}$ versus time for $\xi = 0.007$. Individual contributions are: adiabatic dynamics, $n = 0$, (square); non-adiabatic contributions, $n = 1$, (diamond); $n = 2$, (up triangle); $n = 3$, (left triangle); $n = 4$, (right triangle).

results are shown in Fig. 16.4 (right panel). While the coupling to the bath is quite weak and adiabatic dynamics dominates the structure for this value of the Kondo parameter, the dynamics has non-negligible non-adiabatic components. The convergence of the surface-hopping results may also be gauged from an examination of this figure: the third and fourth order contributions are small over the entire time interval studied. Additional details of the simulation method as well as results for stronger coupling may be found in Ref. [23].

16.9 Conclusion

Quantum-classical dynamics leads to viable methods for studying many-body systems where the quantum character of certain degrees of freedom must be taken into account. In this chapter we have presented a formulation of quantum-classical dynamics that accounts for the coupled evolution of quantum and classical subsystems and have given a description of quantum-classical dynamics in terms of an ensemble of "surface-hopping" trajectories. In addition, the nonequilibrium statistical mechanics of such systems was constructed and expressions for expectation values of dynamical variables and transport properties were derived. The statistical mechanical formulation provides formulas for time correlation functions and specifies how they must be evaluated in simulations. Consequently, the results presented here provide one with all the theoretical tools needed to evaluate observables in a quantum-classical world. Further developments of the topics presented here will likely center on extensions of the theory of quantum-classical dynamics, the construction of efficient simulation schemes for quantum-classical evolution and applications to realistic systems with physical interest.

Acknowledgements

This work was supported in part by a grant from the Natural Sciences and Engineering Research Council of Canada. Acknowledgement is made to the donors of The Petroleum Research Fund, administered by the ACS, for partial support of this research. This work was also partly supported by a MIUR COFIN 2000 project. We would also like to thank Steve Nielsen and Donal Mac Kernan for numerous discussions on quantum-classical dynamics and their collaboration in this project.

References

1. P. Pechukas: Phys. Rev. **181**, 166, 174 (1969).
2. M. F. Herman: Annu. Rev. Phys. Chem. **45**, 83 (1994).
3. J. C. Tully: *Modern Methods for Multidimensional Dynamics Computations in Chemistry*, ed. D. L. Thompson, (World Scientific, NY, 1998), p. 34.
4. J.C. Tully: J. Chem. Phys. **93,** 1061 (1990); J.C. Tully: Int. J. Quantum Chem. **25**, 299 (1991); S. Hammes-Schiffer and J.C. Tully: J. Chem. Phys. **101**, 4657 (1994); D. S. Sholl and J. C. Tully, J. Chem. Phys. **109**, 7702 (1998).
5. L. Xiao and D. F. Coker: J. Chem. Phys. **100**, 8646 (1994); D. F. Coker and L. Xiao: J. Chem. Phys. **102**, 496 (1995); H. S. Mei and D.F. Coker: J. Chem. Phys. **104**, 4755 (1996).
6. F. Webster, P. J. Rossky, and P. A. Friesner: Comp. Phys. Comm. **63**, 494 (1991); F. Webster, E. T. Wang, P. J. Rossky, and P. A. Friesner: J. Chem. Phys. **100**, 483 (1994).
7. T. J. Martinez, M. Ben-Nun, and R. D. Levine: J. Phys. Chem. A **101**, 6389 (1997).
8. R. Kubo: J. Phys. Soc. (Japan) **12**, 570 (1957); R. Kubo: Repts. Prog. Phys. **29**, 255 (1966).
9. E. Wigner: Phys. Rev. **40**, 749 (1932).
10. R. Kapral and G. Ciccotti: J. Chem. Phys. **110**, 8919 (1999).
11. K. Imre, E. Özizmir, M. Rosenbaum and P. F. Zwiefel: J. Math. Phys. **5**, 1097 (1967); M. Hillery, R. F. O'Connell, M. O. Scully and E. P. Wigner, Phys. Repts. **106**, 121 (1984).
12. S. Nielsen, R. Kapral and G. Ciccotti: J. Chem. Phys. **115**, 5805 (2001).
13. I. V. Aleksandrov: Z. Naturforsch. **36a**, 902 (1981).
14. V. I. Gerasimenko: Theor. Math. Phys. **50**, 77 (1982); D. Ya. Petrina, V. I. Gerasimenko and V. Z. Enolskii, Sov. Phys. Dokl. **35**, 925 (1990).
15. W. Boucher and J. Traschen: Phys. Rev. D **37**, 3522 (1988).
16. W. Y. Zhang and R. Balescu: J. Plasma Phys. **40**, 199 (1988); R. Balescu and W. Y. Zhang: J. Plasma Phys. **40**, 215 (1988).
17. A. Anderson: Phys. Rev. Lett. **74**, 621 (1995).
18. O. V. Prezhdo and V. V. Kisil: Phys. Rev. A **56**, 162 (1997).

19. C. C. Martens and J.-Y. Fang: J. Chem. Phys. **106**, 4918 (1996); A. Donoso and C. C. Martens: J. Phys. Chem. **102**, 4291 (1998).
20. C. Schütte, preprint SC 99-10 (Konrad-Zuse-Zentrum, 1999).
21. S. Nielsen, R. Kapral and G. Ciccotti: J. Chem. Phys. **112**, 6543 (2000).
22. S. Nielsen, R. Kapral and G. Ciccotti: J. Stat. Phys. **101**, 225 (2000).
23. D. Mac Kernan, G. Ciccotti and R. Kapral: J. Chem. Phys. **106**, (2002).
24. C. Wan and J. Schofield: J. Chem. Phys. **113**, 7047 (2000).
25. M. Santer, U. Manthe, G. Stock: J.Chem. Phys. **114**, 2001 (2001).
26. A. J. Leggett, S. Chakravarty, A. T. Dorsey, M. P. A. Fisher, A. Garg and M. Zwerger: Rev. Mod. Phys. **59**, 1 (1987).
27. U. Weiss: *Quantum Dissipative Systems* (World Scientific, Singapore, 1999).
28. E. B. Davis: *Quantum Theory of Open Systems* (Academic, London, 1976).
29. T. Dittrich, P. Hänggi G.-L. Ingold, B. Kramer G. Schön and W. Zwerger: *Quantum Transport and Dissipation*, (Wiley, New York, 1998).
30. D. E. Makarov and N. Makri: Chem. Phys. Lett. **221**, 482 (1994); N. Makri and D. E. Makarov: J. Chem. Phys. **102**, 4600, 4611 (1995); N. Makri: J. Math. Phys. **36**, 2430 (1995); E. Sim and N. Makri: Comp. Phys. Commun. **99**, 335 (1997): N. Makri: J. Phys. Chem. **102**, 4414 (1998).
31. N. Makri and K. Thompson: Chem. Phys. Lett. **291**, 101 (1998); K. Thompson and N. Makri: J. Chem. Phys. **110**, 1343 (1999); N. Makri: J. Phys. Chem. B **103**, 2823 (1999).

17 The Coupled Electronic–Ionic Monte Carlo Simulation Method

David Ceperley[1,2], Mark Dewing[3], and Carlo Pierleoni[1,4]

[1] CECAM, c/o ENS Lyon, 46 Allée d'Italie, 69364 Lyon (France)
[2] Department of Physics, University of Illinois at Urbana-Champaign, 1110 West Green Street, Urbana, Illinois 61801 (USA)
[3] INTEL, Fox Drive, Champaign, IL
[4] INFM and Department of Physics, University of L'Aquila, Via Vetoio, L'Aquila (Italy)

Abstract. Quantum Monte Carlo (QMC) methods such as Variational Monte Carlo, Diffusion Monte Carlo or Path Integral Monte Carlo are the most accurate and general methods for computing total electronic energies. We will review methods we have developed to perform QMC for the electrons coupled to another MC simulation for the ions. In this method, one estimates the Born-Oppenheimer energy $E(Z)$ where Z represents the ionic degrees of freedom. That estimate of the energy is used in a Metropolis simulation of the ionic degrees of freedom. Important aspects of this method are how to deal with the noise, which QMC method and which trial function to use, how to deal with generalized boundary conditions on the wave function so as to reduce the finite size effects. We discuss some advantages of the CEIMC method concerning how the quantum effects of the ionic degrees of freedom can be included and how the boundary conditions can be integrated over. Using these methods, we have performed simulations of liquid H_2 and metallic H on a parallel computer.

17.1 Introduction

The first computer simulations of a condensed matter system used the simplest inter-atomic potential, the hard sphere interaction[1]. As computers and simulation methods progressed, more sophisticated and realistic potentials came into use, for example the Lennard–Jones potential to describe rare gas systems, the potential functions being parameterized and then fit to reproduce experimental quantities. Both Molecular Dynamics (MD) and Monte Carlo (MC) methods can be used to generate ensemble averages of many-particle systems, MC being simpler and only useful for equilibrium properties.

Inter–atomic potentials originate from the microscopic structure of matter, described in terms of electrons, nuclei, and the Schrödinger equation. But the many-body Schrödinger equation is too difficult to solve directly, so approximations are needed. In practice, one usually makes the one electron approximation, where a single electron interacts with the potential due to the nuclear charge and with the mean electric field generated by all the other electrons. This is done by Hartree–Fock (HF) or with Density Functional Theory (DFT)[2]. DFT is, in principle, exact, but contains an unknown exchange and correlation functional that must be approximated, the most simplest being the Local Density Approximation (LDA) but various improvements are also used.

In 1985, Car and Parrinello introduced their method, which replaced an assumed functional form for the potential with a LDA-DFT calculation done "on the fly"[3]. They did a molecular dynamics simulation of the nuclei of liquid silicon by computing the density functional forces of the electronic degrees of freedom at every MD step. It has been a very successful method, with the original paper being cited thousands of times since its publication. There are many applications and extensions of the Car–Parrinello method[4–7]. The review of applications to liquid state problems by Sprik[6] notes that the LDA approximation is not sufficient for an accurate simulation of water although there are improved functionals that are much more accurate.

Quantum Monte Carlo (QMC) methods have developed as another means for accurately solving the many body Schrödinger equation[8–11]. The success of QMC is to a large extend because electrons are directly represented as particles, so that the electronic exchange and correlation effects can be directly treated. Particularly within the LDA, DFT has known difficulties in handling electron correlation[12].

In the spirit of the Car-Parrinello method, in this paper we describe initial attempts to combine a Classical Monte Carlo simulation of the nuclei with a QMC simulation for the electrons. This we call Coupled Electronic-Ionic Monte Carlo (CEIMC)[13]. As an example of this new method we apply it to warm dense many-body hydrogen. Hydrogen is the most abundant element in the universe, making an understanding of its properties important, particularly for astrophysical applications. Models of the interiors of the giant planets depends on a knowledge of the equation of state of hydrogen[14,15]. Hydrogen is also the simplest element, but it still displays remarkable variety in its properties and phase diagram. It has several solid phases at low temperature, and the crystal structure of one of them (phase III) is not fully known yet. At high temperature and pressure the fluid becomes metallic, but the exact nature of the transition is not known, nor is the melting transition from liquid to solid for pressures above 1 MBar. The present knowledge of the phase diagram of hydrogen is summarized in Fig. 17.1.

Some of the previous QMC calculations have been at high temperature using the restricted Path Integral MC method. This method become computationally inefficient at temperatures a factor of ten lower than the Fermi temperature[16]. At the present time it is not known how to make the PIMC method efficient at the low temperatures needed to calculate interesting portions of the phase diagram. Zero temperature QMC methods have been used for calculations in the ground state[17–19] with full quantum effects used for

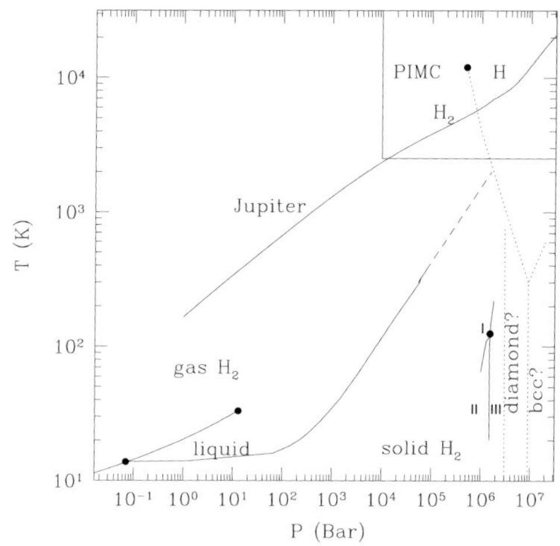

Fig. 17.1. Phase Diagram of hydrogen. Solid lines are experimental determination, dashed line are theoretical estimates. The solid line labelled Jupiter is the set of points (P,T) estimated to exist in the planet. The box shows roughly the domain of applicability of PIMC.

both the electronic and protonic degrees of freedom. In such cases it is hard to ensure that the protonic degrees of freedom are fully converged because of the problem that the electron and protons require two different time scales which differ by three orders of magnitude. In addition, finite temperature effects of the protons are beyond the reach of the method. CEIMC provides a middle way: the electrons are at zero temperature where accurate trial functions are known and the zero variance principle applies, while the protons (either classical or quantum) are at finite temperature and not subjected to the limitations imposed by the electronic time scale.

The electrons are assumed to be in their ground state, both in the Car–Parrinello method and in CEIMC. There are two internal effects that could excite the electrons, namely coupling to nuclear motion and thermal excitations. In the first case, we make the Born–Oppenheimer approximation, where the nuclei are so much more massive than the electrons that the electrons are assumed to respond to nuclear motion instantaneously, and so stay in their ground state. We neglect any occupation of excited states of the electrons due to coupling to nuclear motion. To estimate the effect of thermal excitation in metallic hydrogen, consider a gas of degenerate electrons at a density of $n = 0.0298$ electrons per cubic Bohr (i.e. $r_s = (4\pi n/3)^{-1/3} = 2.0$). This has a Fermi temperature of about 140,000 K. In the molecular hydrogen phase, the gap between the ground state and the first excited state of a hydrogen molecule at the equilibrium bond distance is about 124,000 K. Since our temperatures are well below this, and we are not at too high pressures (since the pressure decreases the gap), the thermal occupation of excited states can be neglected. At higher pressure however, when the electrons becomes delocalized and the system becomes metallic thermal effects can be relevant.

This report brings up-to-date previous work on CEIMC described in ref. [20]. The rest of this paper is as follows. First, we will describe the penalty method to rigorously deal with the noisy QMC estimates of energy differences. Then we will briefly discuss method for computing energy differences. Next, the choice of trial wave function will be discussed. Finally, we put all the pieces of a CEIMC simulation together and discuss preliminary results appropriate to many-body hydrogen.

17.2 The Coupled Electronic-Ionic Monte Carlo Method

First let us recall the basic ideas of Variational Monte Carlo (VMC) and Diffusion Monte Carlo. VMC uses the Metropolis method to sample the ratio of integrals and gives an upper bound to the exact ground state energy.

$$E = \frac{\int dR \, |\psi_T(R)|^2 \, E_L(R)}{\int dR \, |\psi_T(R)|^2} \tag{17.1}$$

where $E_L = (H\psi_T)/\psi_T$ is the local energy. Important features of VMC are that any computable trial function can be used for ψ_T and that the statistical uncertainty vanishes as ψ_T approaches an exact eigenstate.

The second QMC method we apply is diffusion Monte Carlo (DMC) in which the Hamiltonian is applied to the VMC distribution to project out the ground state:

$$\phi(t) = \psi_T e^{-(H-E_T)t} \phi(0)/\psi_T . \tag{17.2}$$

The VMC method, though it can directly include correlation effects, is not sufficiently accurate, at relevant temperatures, as we discuss below. The projection is implemented by a branching, drifting random walk[21] though there are some advantages to working in a time independent framework of ground state path integrals. To maintain a positive function, needed for efficient sampling, the fixed-node approximation is used. Though an uncontrolled approximation, estimates of the resulting error lead to the conclusion[8] that the systematic error of this approximation are small, especially when accurate nodal surfaces are used.

In the CEIMC method we move the protons with a "classical" Monte Carlo and accept or reject to satisfy detailed balance. The Metropolis acceptance formula is

$$A = \min\left[1, \exp(-\Delta)\right] \tag{17.3}$$

where $\Delta = \beta[V(s') - V(s)]$ and $V(s)$ is the BO electronic energy, computed with one of the QMC methods. The QMC simulation will yield a noisy estimate for Δ, which we denote as δ. The exponential in the acceptance ratio is nonlinear, so that $\langle \exp(-\delta)\rangle \neq \exp(\langle -\delta\rangle)$. The noise will introduce a bias into our acceptance ratio formula. Such bias is unacceptable since the main motivation for the CEIMC method is to improve the accuracy beyond what can be achieved with alternative approaches. To avoid this bias in our simulations, we can either run until the noise is negligible, but that is *very* time-consuming, or we can use the penalty method[22] which tolerates noise. We describe this method next.

17.3 The Penalty Method

The basis of the penalty method is to satisfy detailed balance on average by using information about the energy differences. We introduce the "instantaneous" acceptance probability, $a(\delta)$, which is a function of the estimated energy difference. The average acceptance probability is the acceptance probability averaged over the noise,

$$A(s \to s') = \int_{-\infty}^{\infty} d\delta P(\delta; s \to s') a(\delta). \tag{17.4}$$

We need to satisfy detailed balance on average,

$$A(s \to s') = A(s' \to s) \exp[-\Delta] \tag{17.5}$$

If the noise is normally distributed with variance, σ, it has the distribution

$$P(\delta) = (2\sigma^2\pi)^{-1/2} \exp\left[-\frac{(\delta - \Delta)^2}{2\sigma^2}\right]. \tag{17.6}$$

Then a simple solution that satisfies average detailed balance is

$$a(\delta) = \min\left[1, \exp(-\delta - \frac{\sigma^2}{2})\right] \tag{17.7}$$

The extra $-\sigma^2/2$ term causes additional rejections of trial moves due to noise. For this reason, it is called the penalty method.

An important issue is to verify that the energy differences are normally distributed. Recall that if moments of the energy are bounded, the central limit theorem implies that given enough samples, the distribution of the mean value will be Gaussian. Careful attention to the trial function to ensure that the local energies are well behaved may be needed.

In practice, the variance is also estimated from the data, and a similar process leads to additional penalty terms. Let χ be the estimate for σ using n samples. Then the instantaneous acceptance probability is

$$a(\delta, \chi^2, n) = \min[1, \exp(-\delta - u_B)] \tag{17.8}$$

where

$$u_B = \frac{\chi^2}{2} + \frac{\chi^4}{4(n+1)} + \frac{\chi^6}{3(n+1)(n+3)} + \cdots \tag{17.9}$$

Note that as the number of independent samples n gets large, the first term dominates.

The noise level of a system can be characterized by the relative noise parameter, $f = (\beta\sigma)^2 t/t_0$, where t is the computer time spent reducing the noise, and t_0 is the computer time spent on other pursuits, such as optimizing the VMC wave function or equilibrating the DMC runs. A small f means little time is being spent on reducing noise, where a large f means much time is being spent reducing noise. For a double well potential, the noise level that gives the maximum efficiency is around $\beta\sigma \approx 1$, with the optimal noise level increasing as the relative noise parameter increases [22].

We can use multi–level sampling to make CEIMC more efficient [23]. An empirical potential is used to "pre-reject" moves that would cause particles to overlap and be rejected anyway. A trial move is proposed and accepted or rejected based on a classical potential

$$A_1 = \min\left[1, \frac{T(R \to R')}{T(R' \to R)} \exp(-\beta \Delta V_{cl})\right] \tag{17.10}$$

where $\Delta V_{cl} = V_{cl}(R') - V_{cl}(R)$ and T is the sampling probability for a move. If it is accepted at this first level, the QMC energy difference is computed and accepted with probability

$$A_2 = \min\left[1, \exp(-\beta \Delta V_{QMC} - u_B) \exp(\beta \Delta V_{cl})\right] \tag{17.11}$$

where u_B is the noise penalty.

Compared to the cost of evaluating the QMC energy difference, computing the classical energy difference is much less expensive. Reducing the number of QMC energy difference evaluations reduces the overall computer time required. For the molecular hydrogen system, using the pre–rejection technique with a CEIMC–DMC simulation results in a first level (classical potential) acceptance ratio of 0.43, and a second level (quantum potential) acceptance ratio of 0.52. The penalty method rejects additional trial moves because of noise. If these rejections are counted as acceptances (i.e., no penalty method or no noise), then the second level acceptance ratio would be 0.71.

17.4 Energy Differences

In Monte Carlo it is the energy difference between an old position and a trial position that is needed. Using correlated sampling methods it is possible to compute the energy difference with a smaller statistical error than each individual energy. We also need to ensure that that energy difference is unbiased and normally distributed. In this section we briefly discuss several methods for computing that difference.

17.4.1 Direct Difference

The most straightforward method for computing the difference in energy between two systems is to perform independent computations for the energy of each system. Then the energy difference and error estimate are given by

$$\Delta E = E_1 - E_2 \tag{17.12}$$

$$\sigma(\Delta E) = \sqrt{\sigma_1^2 + \sigma_2^2} \tag{17.13}$$

This method is simple and robust, but has the drawback that the error is related to the error in computing a single system. If the nuclear positions are close together, the energy difference is likely to be small and difficult to resolve, since σ_1 and σ_2 are determined by the entire system. Hence the computation time is proportional to the size of the system, not to how far the ions are moved.

17.4.2 Reweighting

Reweighting is the simplest correlated sampling method. The same set of sample points, obtained by sampling $p(R) \propto \psi_1^2$ is used for evaluating both energies. The energy difference is estimated as:

$$\begin{aligned}\Delta E &= E_1 - E_2 \\ &= \frac{\int dR\, \psi_1^2\, E_{L1}}{\int dR\, \psi_1^2} - \frac{\int dR\, \psi_2^2\, E_{L2}}{\int dR\, \psi_2^2} \\ &= \frac{\int dR\, p(R) \left(\frac{\psi_1^2}{p(R)}\right)\, E_{L1}}{\int dR\, p(R) \left(\frac{\psi_1^2}{p(R)}\right)} - \frac{\int dR\, p(R) \left(\frac{\psi_2^2}{p(R)}\right)\, E_{L2}}{\int dR\, p(R)\, \left(\frac{\psi_2^2}{p(R)}\right)}.\end{aligned}$$

Then an estimate of ΔE for a finite simulation is

$$\Delta E \approx \sum_{R_i \in \psi_1^2} \left[\frac{E_{L1}(R_i)}{N} - \frac{w(R_i) E_{L2}(R_i)}{\sum_i w(R_i)}\right] \tag{17.14}$$

where $w = \psi_2^2/p(R)$.

Reweighting works well when ψ_1 and ψ_2 are not too different, and thus have large overlap. As the overlap between them decreases, reweighting gets worse due to large fluctuations in the weights. Eventually, one or a few large weights will come to dominate the sum, the variance in ΔE will be larger than that of the direct method. In addition, the distribution of energy differences will be less normal.

17.4.3 Importance Sampling

In [20] we discussed two-sided sampling: the advantages of sampling the points in a symmetrical way. Here we introduce a similar method, namely the use of importance sampling to compute the energy difference. Importance sampling is conceptually similar to the reweighting described above, however, we optimize the sampling function $p(r)$ so as to minimize the variance of the energy difference. If we neglect sequential correlation caused by the Markov sampling, it is straightforward to determine the optimal function:

$$p^*(R) \propto |\psi_1^2(R)(E_{L1}(R) - E_1) - Q\psi_2^2(R)(E_{L2}(R) - E_2)| \tag{17.15}$$

Here E_1 and E_2 are the energies of the two systems, and $Q = \int \psi_1^2 / \int \psi_2^2$ is the ratio of the normalization of the trial functions. In practice, since these are unknown, one replaces them by a fuzzy estimate of their values, namely we maximize $p^*(R)$ within an assumed range of values of E_1, E_2, Q . A nice feature of the optimal function in (17.15) is that it is symmetric in the two systems leading to correct estimate of the fixed node energy, even when the nodes for the two systems do not coincide. Another advantage, is that the distribution of energy differences is bounded and the resulting energy difference is guaranteed to be normal. This is because the sampling probability depends on the local energy. The use of this distribution with nodes could lead to ergodic problems, but in practice no such difficulty has been encountered in generating samples with p^* using "smart MC" methods.

As another sampling example, we consider a simplification of the optimal distribution, namely $P_s(R) \propto \psi_1^2 + \psi_2^2$. This is quite closely related to the two-sided method used earlier [20]. In this distribution, only a single trajectory is computed, no local energies are needed in the sampling, and the estimation of the noise is a bit simpler.

Shown in Fig. 17.2 is the efficiency computed with the different methods, as a function of the proton step. The curves show that the various correlated methods have roughly the same efficiency, which is independent of the size of the proton move. Correlated methods are more efficient that the direct methods, as long as the proton are moved less than $\approx 0.8\mathring{A}$. The optimal importance sampling has about 10% lower variance than the reweighting. In addition, the estimates are less biased and approach a normal distribution much more rapidly[20]. We used the two-sided method and the importance sampling method for computing energy differences of trial moves with VMC, but only used the direct method with DMC.

17.5 Choice of Trial Wave Function

An essential part of the CEIMC method is the choice of the trial wave function. Variational Monte Carlo (VMC) depends crucially on the trial wave function to find the minimum energy. The trial wave function is also important in DMC, to reduce the variance and the projection time, and for

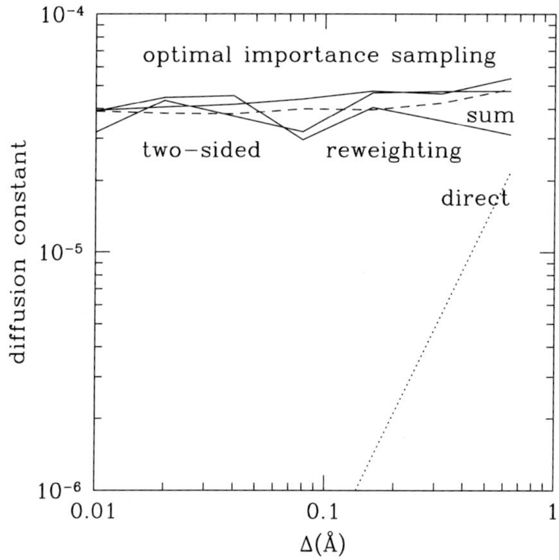

Fig. 17.2. Efficiency versus importance function on a system with $N_e = N_p = 16$ and $r_s = 1.31$. In one system the protons are taken in a sc lattice and in the other they are displaced randomly, with an average displacement of Δ. The diffusion constant is defined as Δ^2/T_{CPU} where T_{CPU} is the computer time needed to calculate the energy difference to an accuracy of 1000 K.

accurate nodal surfaces within the fixed-node method. CEIMC places special demands since optimization of a trial wave function must be done very often and quickly and without direct user control.

A typical form of the variational wave function used in QMC is a Jastrow factor (two body correlations) multiplied by two Slater determinants of one body orbitals.

$$\psi_T = \exp\left[-\sum_{i<j} u(r_{ij})\right] \text{Det}\left(S^{\uparrow}\right) \text{Det}\left(S^{\downarrow}\right) \tag{17.16}$$

The Slater determinant is taken from a mean field calculations such as Hartree–Fock or approximate density functional theory. The cusp condition can be used to well approximate this at short distances and RPA to determine the behavior at large distances[8].

In the molecular phase of hydrogen we estimated that using the orbitals determined from a separate DFT calculation would have been too slow. For the molecular phase we resort a simpler alternative namely we used gaussian single body orbitals, pinned in the center of the molecular bonds. Optimization of the gaussian, one for each of the molecules, took much of the computer time. See [13,20] for a detailed discussion of those results.

For the metallic hydrogen phase, in a previous QMC investigation, Natoli[18] found that simple plane wave nodes are inaccurate by 0.05 eV/atom within the fixed-node approach at the transition to metallic hydrogen ($r_s = 1.31$) necessitating the use of more accurate (LDA) nodes. However it is inconvenient and inefficient to solve the LDA equations for each new position of the ions in CEIMC. In addition, one has to modify the LDA orbitals to take into account the effect of explicit electron-electron correlation. The same problem of disordered ionic configurations arises from zero point motion of the protons in the solid state. In earlier work on molecular hydrogen, we were unable to use high quality LDA orbitals when the molecules were angularly disoriented [19].

We have recently generalized the backflow and three–body wave function to a two component system of electrons and protons at high enough density so that the electrons are delocalized and all the hydrogen molecules are dissociated. For metallic hydrogen, as an element without a core, the formalism leading to the improved wave functions is simplest [24]. These wave functions depend explicitly and continuously on the ionic variables and as a consequence do not have to be reoptimized for movements of the ions. These trial functions are a generalization of the backflow three–body wave functions used very successfully in highly correlated homogeneous quantum liquids: liquid ^{3}He and electron gas. Backflow trial functions show much improvement over the pair product getting approximately 80% of the missing correlation and even more of the energy when done with the fixed-node method. Backflow wave functions utilize the power of the QMC sampling approach: one can calculate properties of such a wave function without changing the algorithm in an essential way, whileas in approaches based on explicit integration, one is limited in the form of the trial function by the ease performing the integration. We will discuss this functions in more detail below.

17.6 Twist Average Boundary Conditions

Almost all QMC calculations in periodic boundary conditions have assumed that the phase of the wave function returns to the same value if a particle goes around the periodic boundaries and returns to its original position. However, with these boundary conditions, delocalized fermion systems converge slowly to the thermodynamic limit because of shell effects in the filling of single particle states. One can allow particles to pick up a phase when they wrap around the periodic boundaries,

$$\Psi(\boldsymbol{r}_1 + L\hat{\boldsymbol{x}}, \boldsymbol{r}_2, \cdots) = \mathrm{e}^{i\theta_x}\Psi(\boldsymbol{r}_1, \boldsymbol{r}_2, \cdots). \tag{17.17}$$

The boundary condition $\theta = 0$ is periodic boundary conditions (PBC), and the general condition with $\theta \neq 0$, twisted boundary conditions (TBC). The use of twisted boundary conditions is commonplace for the solution of the band structure problem for a periodic solid, particularly for metals. In order to calculate properties of an infinite periodic solid, properties must be averaged by integrating over the first Brillouin zone.

For a degenerate Fermi liquid, finite-size shell effects are much reduced if the twist angle is averaged over: twist averaged boundary conditions (TABC). This is particularly important in computing properties that are sensitive to the single particle energies such as the kinetic energy and the magnetic susceptibility. By reducing shell effects, much more accurate estimations of the thermodynamic limit for these properties can be obtained. What makes this even more important is that the most accurate quantum methods have computational demands which increase rapidly with the number of fermions. Examples of such methods are exact diagonalization (exponential increase in CPU time with N), variational Monte Carlo (VMC) with wave functions having backflow and three-body terms [25,26] (increases as N^4), and transient-estimate and released-node Diffusion Monte Carlo methods [27] (exponential increase with N). Methods which can extrapolate more rapidly to the thermodynamic limit are crucial in obtaining high accuracy.

Twist averaging is especially advantageous in combination with CEIMC (i.e. QMC) because the averaging does not necessarily slow down the evaluation of averages, except for the necessity of doing complex rather than real arithmetic. In a metallic system, such as hydrogen at even higher pressure when it becomes a simple metal, results in the thermodynamic limit require careful integration near the Fermi surface because the occupation of states becomes discontinuous. Within LDA this requires "k–point" integration, which slows down the calculation linearly in the number of k-points required. Within QMC such k-point integration takes the form of an average over the (phase) twist of the boundary condition and can be done in parallel with the average over electronic configurations without significantly adding to the computational effort. We typically spawn about 100 distinct QMC processes, run for a fixed time, and then average the resulting properties.

17.7 Fluid Molecular Hydrogen

We now describe our calculations on liquid molecular hydrogen. First of all, we examine the accuracy of several methods for computing total energy. We took several configurations from PIMC simulations at 5000 K at two densities ($r_s = 1.86$ and $r_s = 2.0$), and compared the electronic energy using VMC, DMC, DFT-LDA, and some empirical potentials. The DFT–LDA results were obtained from a plane wave code using an energy cutoff of 60 Rydbergs, and using the Γ point approximation [28]. The empirical potentials we used are the Silvera–Goldman [29] and the Diep–Johnson [30,31]. To these we added the energy from the Kolos [32] intramolecular potential to get the energy as a function of the bond length variations. The Silvera–Goldman potential was obtained by fitting to low temperature experimental data, with pressures up

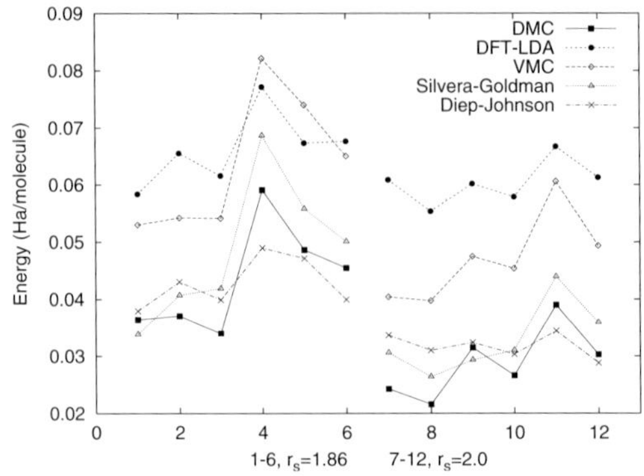

Fig. 17.3. Electronic energy for several configurations computed by various methods. The energy is relative to an isolated H_2 molecule.

to 20 Kbar, and is isotropic. The Diep–Johnson potential is the most recent in a number of potentials for the isolated H_2–H_2 system. It was fit to the results of accurate quantum chemistry calculations for a number of H_2–H_2 configurations and included anisotropic effects in the potential.

The energies relative to an isolated H_2 molecule are shown in Fig. 17.3. The first thing we notice is that the classical potentials are more accurate than VMC or DFT. The Silvera–Goldman does a good job of reproducing the DMC results. Some of the failures of the SG potential can be attributed to the lack of anisotropy. The isolated H_2–H_2 potential (Diep–Johnson) has much weaker interactions, compared with interactions in a denser system.

The PIMC method itself gives an average energy of about 0.07(3) Ha for both densities. Improvements in the fermion nodes and in other aspects of the PIMC calculation appear to lower the energy [16,33,34], although the error bars are still quite large. The PIMC energy is in rough agreement with the DMC energy.

As mentioned above, we used the Silvera–Goldman potential for pre-rejection. As seen in the Fig. 17.3, it resembles the DMC potential even though it lacks anisotropy. Each trial move consisted of moving multiple molecules (usually four). This increased efficiency by amortizing the cost of the QMC calculation. Each molecular move was decomposed into a translation of the center of mass, a rotation of the molecule, and a change in the bond length.

Shown in Tables 17.1–17.2 are CEIMC results at three state points two of which can be compared with the gas gun data of Holmes *et al.* [35]. The pressure is given in Table 17.1 with results from the gas gun experiments, the free energy model of Saumon and Chabrier [36–38], from simulations using the Silvera–Goldman potential, and from our CEIMC simulations. These state points are in the fluid molecular H_2 phase. For the gas gun experiments, the uncertainties in the measured temperatures are around 100–200 K. The experimental uncertainties in the volume and pressure were not given, but previous work indicates that they are about 1–2% [39].

We did CEIMC calculations using VMC or DMC for computing the underlying electronic energy, which are the first such QMC calculations in this range. The simulations at $r_s = 2.1$ and $r_s = 1.8$ were done with 32 molecules, and the simulations at $r_s = 2.202$ were done with 16 molecules. We see that the pressures from VMC and DMC are very similar, and that for $r_s = 2.1$ we get good agreement with experiment.

There is a larger discrepancy with experiment at $r_s = 2.202$. The finite size effects are fairly large, especially with DMC. We also did simulations at $r_s = 2.1$ with 16 molecules and obtained pressures of 0.264(3) Mbar for CEIMC–VMC and 0.129(4) Mbar for CEIMC–DMC. The Silvera–Goldman potential showed much smaller finite size effects than the CEIMC simulations, so we conclude that the electronic part of the simulation is largely responsible for the observed finite size effects.

The energies for all these systems are given in Table 17.2. The energy at $r_s = 2.1$ with 16 molecules for CEIMC–VMC is 0.0711(4) Ha and for CEIMC–DMC is 0.0721(8) Ha. The proton–proton distribution functions comparing CEIMC–VMC and CEIMC–DMC are shown in Fig. 17.4. The VMC and DMC distribution functions look similar, with the first large intramolecular peak around $r = 1.4$ and the intermolecular peak around $r = 4.5$.

Table 17.1. Pressure from simulations and shock wave experiments

r_s	T(K)	Pressure (Mbar)				
		Gasgun	S–C	S–G	CEIMC–VMC	CEIMC–DMC
2.100	4530	0.234	0.213	0.201	0.226(4)	0.225(3)
2.202	2820	0.120	0.125	0.116	0.105(6)	0.10(5)
1.800	3000	-	-	0.528	-	0.357(8)

Table 17.2. Energy from simulations and models, relative to the ground state of an isolated H_2 molecule. The H_2 column is a single thermally excited molecule plus the quantum vibrational KE.

r_s	T(K)	Energy (Ha/molecule)				
		H_2	S–C	S–G	CEIMC–VMC	CEIM-C-DMC
2.100	4530	0.0493	0.0643	0.0689	0.0663(8)	0.0617(2)
2.202	2820	0.0290	0.0367	0.0408	0.0305(8)	0.0334(9)
1.800	3000	0.0311	-	0.0722	-	0.048(1)

The CEIMC–VMC simulations at $r_s = 1.8$ and 3000 K never converged. Starting from a liquid state, the energy decreased during the entire simulation. Visualization of the configurations revealed that they were forming a plane. It is not clear whether it was trying to freeze, or forming structures similar to those found in DFT–LDA calculations with insufficient Brillouin

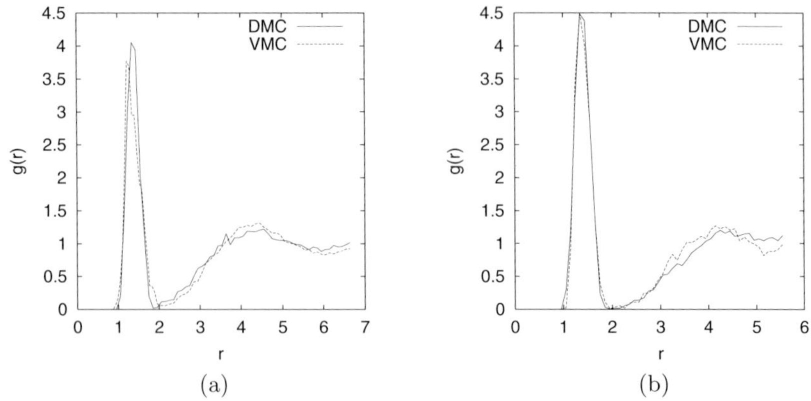

Fig. 17.4. Proton pair distribution, $g(r)$, for **(a)** $r_s = 2.1$ and T=4530 K **(b)** $r_s = 2.202$ and T=2820 K

zone sampling[40,41]. (note that the molecular hydrogen calculations were done at the Γ point.)The CEIMC–DMC simulations did not appear to have any difficulty, so it seems the VMC behavior was due to inadequacies of the wave function.

Hohl *et al.*[40] did DFT–LDA simulations at $r_s = 1.78$ and T=3000 K, which is very close to our simulations at $r_s = 1.8$. The resulting proton-proton distribution functions are compared in Fig. 17.5. The CEIMC distribution has more molecules and they are more tightly bound. The discrepancy between CEIMC and LDA in the intramolecular portion of the curve has several possible causes. On the CEIMC side, it may be due to lack of convergence or to the molecular nature of the wave function, which does not allow dissociation. The shift of the molecular bond length peak can be accounted for because LDA is known to overestimate the bond length of a free hydrogen molecule [40] which would account for the shifted location of the bond length peak. The deficiencies of LDA may account for it preferring fewer and less tightly bound molecules.

17.8 The Atomic–Metallic Phase

In this section, we describe preliminary results for metallic atomic hydrogen from a recent implementation of the method using an improved wave functions including threebody and backflow terms and taking advantage of averaging over the twist angle to minimize size effects.

17.8.1 Trial Wave Function and Optimization

We have seen that an important part of the CPU time is needed in the optimization of the molecular trial wave functions which contain a number of vari-

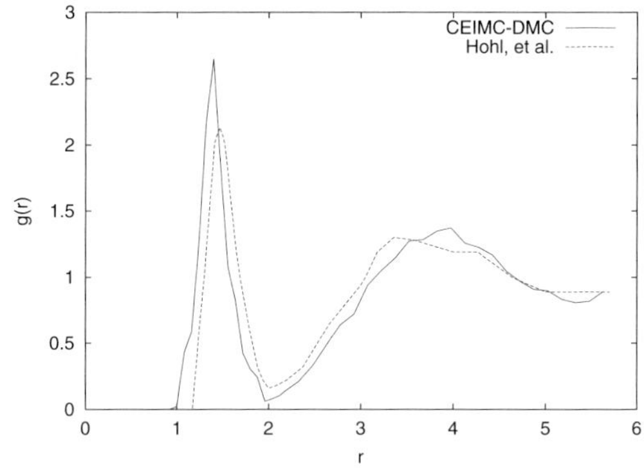

Fig. 17.5. The proton pair distribution function, $g(r)$, near $r_s = 1.8$ and T=3000K.

ational parameters proportional to the number of molecules and which need to be optimized individually for each protonic steps. A major improvement in the efficiency of the method can be achieved by using more sophisticated wave functions, namely analytic functions in terms of the proton positions which move with the protons and which depend on few variational parameters (about 10, regardless of the number of particles). Moreover, one can explore the possibility of optimizing the variational parameters only once at the beginning of the calculation, either on ordered or disordered protonic configurations, and using the optimized wave function during the simulation. In this section, we consider hydrogen at densities at which molecules are dissociated ($r_s \leq 1.31$) and the system is metallic. We will therefore avoid the complications arising from the presence of bound states (either molecular or atomic). In this case one can show that improved wave functions with respect to the simple Slater-Jastrow form includes backflow and threebody terms between electrons and protons [24] in a very similar fashion as the ones used by Kwon et al.[25,26] for the electron gas. We assume a trial wave function of the form

$$\Psi_T(\boldsymbol{R}) = det(e^{i\mathbf{k}_i \cdot \mathbf{x}_j}) exp\left(-\sum_{i<j}^{N} \tilde{u}(r_{ij}) - \frac{\lambda_T}{2} \sum_{l=1}^{N} \mathbf{G}(l) \cdot \mathbf{G}(l) \right) \tag{17.18}$$

where

$$\mathbf{x}_i = \mathbf{r}_i + \sum_{j \neq i}^{N} \eta(r_{ij})(\mathbf{r}_i - \mathbf{r}_j) \tag{17.19}$$

$$\mathbf{G}(l) = \sum_{i \neq l}^{N} \xi(r_{li})(\mathbf{r}_l - \mathbf{r}_i) \tag{17.20}$$

$$\tilde{u}(r) = u(r) - \lambda_T \xi^2(r) r^2 \tag{17.21}$$

with

$$\eta(r) = \lambda_b exp[-(r - r_b)^2 / w_b^2] \tag{17.22}$$

$$\xi(r) = exp[-(r - r_T)^2 / w_T^2] \tag{17.23}$$

and $u(r)$ is an optimized version of the RPA pseudopotential [17].

In what follows we only consider the effect of electron–proton backflow and electron–proton–proton three body terms, while the electronic part of the wave function is of the simple Slater–Jastrow form. To establish the goodness of this wave function for metallic hydrogen we perform VMC and DMC calculation for 16 protons on a bcc lattice at $r_s = 1.31$. In Table 17.3 we compare the energy and the variance of the local energy of this wave function with data obtained with the simple Slater-Jastrow wavefunction and with an improved wavefunction in which a determinant of single body orbital from a separate LDA calculation has been used [18]. From these results we infer that the nodes of the new wavefunction are as accurate as the LDA nodes.

Table 17.3. $r_s = 1.31$. Energy and variance for 16 protons in the bcc lattice. SJ, SJ3B and LDA indicate optimized Slater–Jastrow, optimized Slater–Jastrow+three–body+backflow, and LDA nodes respectively. Energies per particle are in Rydbergs.

	E_{VMC}	σ^2_{VMC}	E_{DMC}
SJ	-0.4754(2)	0.0764(9)	-0.4857(1)
SJ3B	-0.4857(2)	0.0274(2)	-0.4900(1)
LDA	-0.4870(10)		-0.4890(5)

Having established that our wavefunction at $r_s = 1.31$ is as good as the most accurate wavefunction used for metallic hydrogen so far, we continue our study at slightly higher density, namely $r_s = 1$. It can be shown that the above form of the wavefunction is obtained using perturbation theory from the high density limit and we expect that its accuracy improves for decreasing r_s.

We first perform a number of optimizations of the trial wave function. Beside the RPA e–p Jastrow, we consider an extra 2 body (e–p) gaussian term with two variational parameters (λ_e and w_e). In Table 17.4 we report the values of the variational parameters obtained minimizing a linear combination of the local energy and its variance over a set of different protons and electrons configurations. Typically 1000 configurations have been used, by saving a configuration after 10 or 20 protonic steps during a previous run. We also studied the relative importance of the different terms in the

Table 17.4. $r_s = 1, T = 5000\,K, N_p = N_e = 16$. Optimized values of the variational parameters for the VMC trial function. The values are obtained minimizing local energy and variance for 1000 different equilibrium configurations.

	λ_b	r_b	w_b	λ_T	r_T	w_T	λ_e	w_e	E(a.u.)
SJ	–	–	–	—	–	–	–	–	-0.117(1)
SJE	—	—	—	–	–	–	0.06167	0.9497	-0.1180(4)
SJB	-0.60824	-1.3726	1.44822	–	–	–	–	–	-0.1207(4)
SJEB	-0.45828	-0.60202	0.91273	–	–	–	-0.0874	1.7002	-0.1227(5)
SJE3B	-0.4671	-0.6217	1.0193	-2.4676	-1.0917	3.0029	-0.0844	1.5130	-0.1238(2)

trial wavefunction by performing calculations with partially improved wave functions. In Fig. 17.6 we compare the pair correlation functions for the various calculations in Table 17.4. No significant difference is observed in the electron–electron and in the proton–proton pair correlation functions among different forms of the trial function. We can see that the cooperative effects of the optimized Jastrow factor and the backflow term are responsible of an enhancing of electron-proton correlation as seen in the $g_{ep}(r)$. Inclusion of three–body terms lowers the energy but does not change the pair correlations.

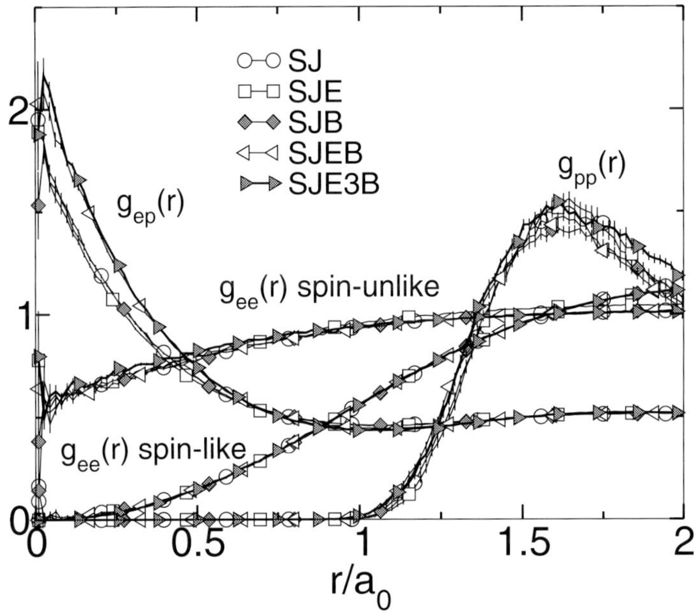

Fig. 17.6. $r_s = 1, T = 5000\,K, N_e = N_p = 16$ spin unpolarized. Pair correlations functions with various trial wave functions. The entries in the legend corresponds to the entries in Table 17.4. The $g_{ep}(r)$ have been shifted downward by 0.5 for sake of clarity.

In principle, the optimization study could be repeated at each temperature needed for the CEIMC simulation. It is therefore of practical interest to investigate the transferability at finite temperature of wavefunctions optimized for the lattice configurations of protons.

An additional ingredient discussed above and crucial for a metallic system are finite size effects. It has been shown recently [42] that the very irregular behavior of the energy versus N observed in the presence of a Fermi surface can be reduced to the classical $1/N$ behaviour by averaging over the twist of the wavefunction. We have implemented the twist averaged boundary conditions in the calculation of the energy differences needed to make the protonic moves in the CEIMC. We average over 1000 different twist angles in the three dimensional interval $(-\pi, \pi)$ found to be sufficient in the electron gas[42]. The additional issue of whether optimization of the variational parameters need to be done with or without twist averaging was investigated. We compare in Table 17.5 the results of single phase optimization and phase averaging optimizations for protons in the bcc lattice and at T=5000 K. The fourth row is the result of an optimization with twist averaging at T=0 K, while the fifth row is a run with the values of the variational parameters optimal for the Γ point, always at T=0 K. We observe an excellent agreement of the energies and we conclude that we can safely optimize the wave function using the Γ point and use the obtained variational parameters for all twist angles.

The sixth row in the table is the result of a simulation at T=5000 K using the values of the variational parameters optimal for T=0 K (bcc lattice). The energy should be compared with the result of the entry SJE3B in Table 17.4. The difference in energy is within error bars and indicates that we can safely optimize the wave function on lattice configurations for use at finite temperature to avoid repeating the optimization at each temperature. In the last two rows of Table 17.5 we report results of two runs at T=5000 K with twist averaging. In the first run the variational parameters optimized in the bcc configuration and with the Γ point have been used. In the second

Table 17.5. $r_s = 1, N_p = N_e = 16$ spin unpolarized. Optimized values of the variational parameters for the VMC trial function. The values are obtained minimizing local energy and variance for 1000 different equilibrium configurations.

T(K)	#phases	λ_b	r_b	w_b	λ_3	r_3	w_3	λ_e	w_e	E(a.u.)
0	1	–	–	–	—	–	–	–	–	-0.1306(2)
0	1	-0.2574	-0.2172	0.7623	-2.3742	-1.8150	1.9694	-0.0496	1.7937	-0.1353(1)
0	1000	–	–	–	–	–	–	–	–	-0.1779(1)
0	1000	-0.2386	-0.1757	0.6613	-2.2609	-1.8326	3.3130	-0.0475	2.0337	-0.18254(3)
0	1000	-0.2574	-0.2172	0.7623	-2.3742	-1.8150	1.9694	-0.0496	1.7937	-0.18253(3)
5000	1	-0.2574	-0.2172	0.7623	-2.3742	-1.8150	1.9694	-0.0496	1.7937	-0.1237(2)
5000	1000	-0.2574	-0.2172	0.7623	-2.3742	-1.8150	1.9694	-0.0496	1.7937	-0.1708(3)
5000	1000	-0.4611	-0.7339	1.1287	-2.0402	-2.2098	3.0213	-0.0949	1.2389	-0.1709(3)

one new values of the parameters, obtained by optimization over a set of configurations stored in the previous twist-averaged run, have been used. The excellent agreement on the energy (and on the variance of the local energy, not shown in the table) confirms that optimization of the variational parameters can be safely performed in the lattice configuration and with a single phase.

17.8.2 Comparison with PIMC

In order to establish the accuracy of the CEIMC method, we compare CEIMC and PIMC results at high temperatures and pressures. To eliminate the "fermion sign problem", the R-PIMC technique for fermions assumes the nodal surfaces of a trial density matrix. In most of the applications, free particle nodal surfaces, either temperature dependent or in the ground state, have been used[43–45]. More recently, variational nodes which account for bound states have been implemented in the study of the plasma phase transition [16,33]. However, the use of temperature dependent nodes, which break the imaginary time translational symmetry, is limited to quite high temperature, $T \geq 0.1 T_F$ where $T_F(a.u.) = 1.84158/r_s^2$ is the electronic Fermi temperature. Below this threshold, the Monte Carlo sampling becomes extremely inefficient and the method impractical. This pathology is not encountered when using ground state nodes, which preserve the original imaginary-time symmetry and are expected to become as accurate as the temperature dependent nodes at low enough temperature. At $r_s = 1$ ($0.1 T_F = 0.18158 a.u. \approx 57300\,\mathrm{K}$), we perform calculation at T=10000 K and T=5000 K and we exploit the PIMC with free particle ground state nodes. In Table 17.6 we compare energies and pressure from PIMC and CEIMC simulations. At T=10000 K, two different PIMC studies are reported, with $M = 500$ and $M = 1000$ time slices respectively, which correspond to $\tau = 0.063 (a.u.)^{-1}$ and $\tau = 0.0315 (a.u.)^{-1}$. The smaller value satisfy the empirical criteria for good convergence $\tau \leq 0.05/r_s^2 (a.u.)^{-1}$ we have established in the plasma phase at higher temperature [44]. At

Table 17.6. $r_s = 1, N_p = N_e = 18$. Comparison between PIMC and CEIMC methods at T=10000 K and T=5000 K.

method	$T/10^3(K)$	M	K_e/N_e	K_t/N	V_N/N	E/N	P
PIMC	10	500	1.477(9)	0.763(5)	-0.7771(6)	-0.0141(6)	0.119(2)
PIMC	10	1000	1.48(1)	0.764(6)	-0.7820(8)	-0.0180(7)	0.119(2)
CEIMC	10	–	1.3767(4)	0.7121(2)	-0.7995(2)	-0.0874(4)	0.0994(1)
PIMC	5	1000	1.39(1)	0.707(5)	-0.791(1)	-0.084(6)	0.099(2)
CEIMC	5	–	1.317(1)	0.6703(3)	-0.7939(2)	-0.1236(2)	0.0870(8)

T=5000 K only $M = 1000$ has been used and therefore the convergence with the number of time slices is limited.

We see small differences between PIMC and CEIMC. In particular, the electronic kinetic energy in PIMC is always slightly higher than in CEIMC. At the same time, CEIMC determined potential energy is lower than the PIMC value and this results in a significantly lower total energy of CEIMC compared to PIMC. The difference between PIMC and CEIMC seems to decrease with temperature. To judge the quality of these results, we should keep in mind advantages and limitations of each method. PIMC uses the "exact" bosonic action, the electrons are at finite temperature and excited states are taken into account, although in a approximate way because of the simplified nodal restriction: its approximation for the nodal surface is a Slater determinant of plane waves. CEIMC instead assumes a trial functions which, at the correlation (bosonic) level is certainly an approximation to the true bosonic action used in PIMC. Moreover, the electrons are in their ground state by construction. However, the trial wavefunction in CEIMC is better (for the ground state) than the one used in PIMC. Because of these differences we think that the comparison between the two methods shows agreement although a more detailed investigation is in order.

Comparison between PIMC and CEIMC for the pair correlation functions at T=10000 K and T=5000 K is given in Figs. 17.7 and 17.8 respectively. In Fig. 17.7 we first note a good agreement between the two PIMC calculations which show that pair correlation functions are much less sensible to finite imaginary time step errors. In general for all correlation functions except the electron-proton ones, the agreement between PIMC and CEIMC is excellent.

At T=10000 K the electron–proton pair correlation function from PIMC is in very good agreement with the result of the CEIMC method where a simple Slater–Jastrow trial wave function is used. Improving the trial wave function as discussed in the previous subsection worsen the agreement. The opposite behavior is observed at T=5000 K where the better agreement between PIMC and CEIMC is observed with the improved wave function (SJE3B in the figure). We interpret this behaviour as follows: at lower temperature the improved trial wave function (17.18) provides the "correct" electron–proton correlation (through the combined effect of the optimized Jastrow and the electron–proton backflow, see the discussion relative to Fig. 17.6 in the previous subsection). At this temperature the electronic thermal effects on the electron–proton correlation are quite small and the electronic ground state as provided by CEIMC is quite accurate. Instead at higher temperatures electron–proton scattering is influenced by excited electronic states which are not considered in the CEIMC method. As a result the electron–proton pair correlation function shows a weaker correlation near the origin and it is in better agreement with the CEIMC result with the Slater–Jastrow trial function rather than with the CEIMC result for the improved trial function.

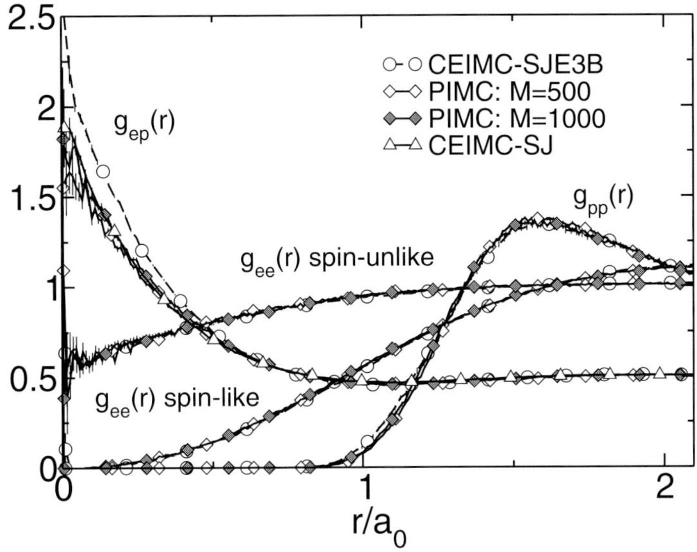

Fig. 17.7. Pair correlation functions at T=10000 K. Comparison between PIMC and CEIMC. $g_{ep}(r)$ have been shifted by -0.5 for sake of clarity.

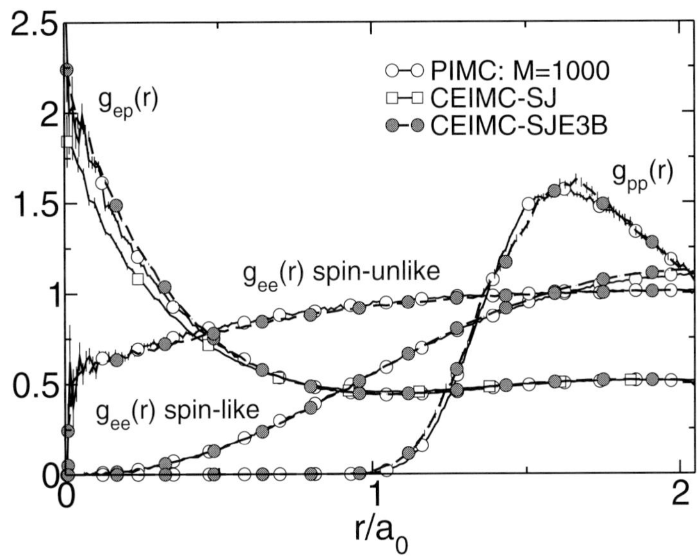

Fig. 17.8. Pair correlation functions at T=5000 K. Comparison between PIMC and CEIMC. $g_{ep}(r)$ have been shifted by -0.5 for sake of clarity.

17.8.3 Hydrogen Equation of State and Solid–Liquid Phase Transition of the Protons

We present in this subsection, results for the equation of state (EOS) of hydrogen in the metallic phase including the solid–liquid transition of protons. These results are preliminary in various respects. Firstly, the electrons are treated at the variational level and no use of Projection/Diffusion Monte Carlo was attempted. Secondly, the protons are considered as classical point particles although it is well known that zero point motion at such high pressure can be significant (at least around the phase transition). Finally, we have data at a single density ($r_s = 1$) and for a single system size, namely $N_e = N_p = 32$ (compatible with the fcc lattice) and we cannot address, at this stage, the issue of the relative stability of different crystal structures. Nonetheless, we believe these results are interesting because they show the applicability and provide a benchmark of the method.

Table 17.7. Simulation details: $r_s = 1, N_p = N_e = 32$. Δ_p is the maximum amplitude of the protonic step in units of the Bohr radius, M_{el} is the total number of electronic steps per protonic step, $(\beta\sigma)^2$ is the relative noise level entering in the penalty method, P_{acc} is the average acceptance of the protonic moves, η is the noise rejection ratio defined earlier, D_p is the diffusion constant in protonic configurational space with respect to CPU time.

$T(K)$	Δ_p	M_{el}	$(\beta\sigma)^2$	P_{acc}	η	$D_p \times 10^4$	time/step(sec)	#proc	machine
5000	0.03	15000	0.037(4)	0.80	0.0084	1.9(2)	5.96	32	beowulf
4000	0.03	15000	0.092(8)	0.77	0.013	3.8(3)	5.93	32	beowulf
3000	0.025	15000	0.10(2)	0.76	0.012	3.2(3)	10.3	16	origin3800
2000	0.03	15000	0.29(5)	0.68	0.033	2.6(3)	10.3	16	origin3800
1000	0.02	15000	0.30(3)	0.64	0.06	3.6(3)	10.3	16	origin3800
700	0.02	15000	0.418(4)	0.55	0.10	3.6(3)	9.90	16	origin3800
500	0.02	15000	0.747(5)	0.43	0.16	0.47(8)	8.93	32	platinum
300	0.015	18000	0.855(9)	0.39	0.21	0.18(1)	7.02	32	platinum

In Table 17.7, we report various details of the simulations such as the maximum amplitude of the protonic step in units of the Bohr radius Δ_p, the total number of electronic steps per protonic step M_{el}, the relative noise level for the penalty method $(\beta\sigma)^2$, the acceptance for the protonic move and the noise rejection ratio η [20]. A measure of the computational efficiency can be defined as the proton diffusion in configurational space with respect to the CPU time $D_p = (\sum_p [\Delta R_p]^2)/T_{CPU}$. In the table we report the values obtained in our simulations in units of a_0^2/sec. In addition, the CPU time per protonic step, the number of processors and the machine used are reported[1].

Note that $M_{el} = 15000$ is the minimum number of electronic steps needed to average over 1000 different twist angles. Except at the lowest temperatures ($T \leq 500$ K) such large number of electronic steps would not be necessary in order to reduce the noise level. Further improvements in efficiency could be gained by reducing the number of electronic steps or the number of angles averaged over for $T > 500\,K$.

Table 17.8. $r_s = 1, N_p = N_e = 32$, spin unpolarized

$T(K)$	K_{tot}	V_c	E_{tot}	σ_E^2	$P(Mbars)$	γ
5000	0.6241(2)	-0.7820(1)	-0.1579(2)	0.056(2)	21.72(2)	liquid
4000	0.0620(2)	-0.7821(2)	-0.1619(1)	0.055(3)	21.35(1)	liquid
3000	0.0616(1)	-0.7817(1)	-0.1662(2)	0.051(7)	20.93(1)	liquid
2000	0.06122(6)	-0.7842(1)	-0.1702(1)	0.050(2)	20.588(6)	liquid
1500	0.61113(7)	-0.7848(1)	-0.1737(1)	0.046(1)	20.374(6)	melted
1000	0.60847(6)	-0.78372(8)	-0.17525(4)	0.0446(5)	20.181(3)	liquid
1000	0.60894(5)	-0.78549(9)	-0.17655(7)	0.0416(5)	20.143(3)	0.137(4)
700	0.60787(3)	-0.78614(5)	-0.17817(8)	0.0402(6)	20.017(6)	0.109(2)
500	0.60811(3)	-0.78718(3)	-0.17913(5)	0.048(4)	19.985(3)	0.085(3)
300	0.60680(4)	-0.78686(2)	-0.18017(2)	0.042(3)	19.874(3)	0.083(1)

In Table 17.8 we report thermodynamic quantities for the system at various temperatures in the range $T \in$[300 K,5000 K]. The corresponding electron–protons and protons–protons pair correlation functions are given in Figs. 17.9 and 17.10 respectively. At each temperature, equilibrium runs of at least 20000 protonic steps have been performed. Statistics are collected every 20–50 steps. Besides energies and pressure we compute the Lindemann ratio γ for the fcc structure. In the upper part of the table, i.e. at higher temperature, we report in the last column the status of the system. At T=1500 K the system initially in the fcc configuration is found to melt after few thousands steps. Conversely at T=500 K we started the simulation in a disordered state

[1] Beowulf is a IBMx330 cluster with PentiumIII/1.13GHz in CINECA–ITALY (www.cineca.it/HPSystems/Resources/LinuxCluster), origin3800 is a SGI–origin3800 with R14000/500MHz in CINES–FRANCE (www.cines.fr) and platinum is a IBMx330 cluster with PentiumIII/1GHz in NCSA–USA (www.ncsa.uiuc.edu/UserInfo/Resources/Hardware/IA32LinuxCluster-/TechSummary).

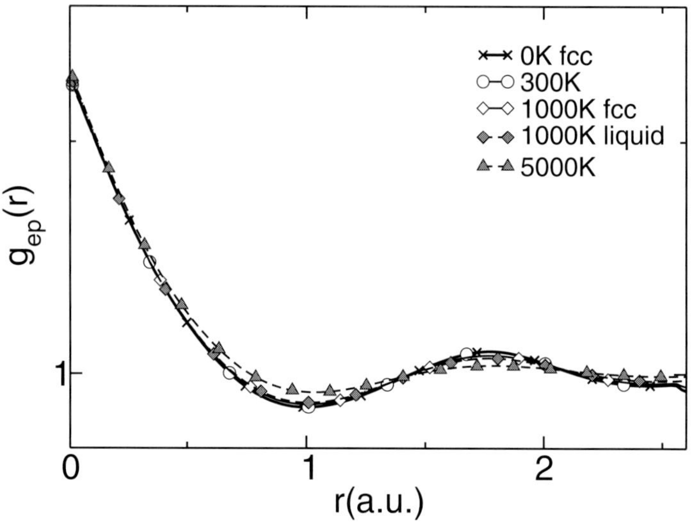

Fig. 17.9. $r_s = 1, N_e = N_p = 32$ spin unpolarized. Temperature dependence of the electron–proton pair correlation functions. The y scale is logarithmic.

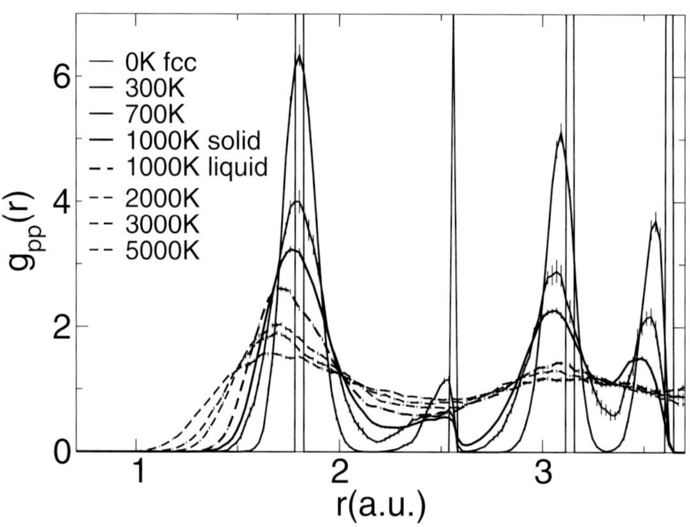

Fig. 17.10. $r_s = 1, N_e = N_p = 32$ spin unpolarized. Temperature dependence of the proton–proton pair correlation functions. Solid lines indicate solid state while dashed lines represent liquid situations. The delta like behaviour is the $T = 0$ correlation function. Reported solid and liquid temperature are given in the figure

and the system spontaneously ordered into the fcc structure. At T=1000 K instead, a system starting in a lattice configuration remains solid and a system starting from a liquid configuration remains liquid within the length of the runs. The Lindemann criterion for classical melting locates the transition at the temperature at which $\gamma \simeq 0.15$. From the result in the table the fcc–liquid transition temperature should be located between 1000 K and 1500 K. Previous investigation of such a transition has been performed by Car–Parrinello Molecular Dynamics [46]. This study suggests that at $r_s = 1$ the structure of the system at $T = 0K$ is hcp but for $T >$100 K the bcc structure is more favorable (as in this work, protons were considered as classical particles). The melting temperature of the bcc lattice has been estimated by the Lindemann criterion around 350 K, significantly lower than the present estimate. We are presently investigating the system of 54 protons in order to study the stability of the bcc structure and the melting transition temperature.

17.9 Conclusions and Outlook

In this article we have discussed the CEIMC method, along with a number of supporting developments to make it computationally efficient. Using the penalty method, we have shown how it is possible to formulate a classical Monte Carlo, with the energy difference having statistical noise, without affecting the asymptotic distribution of the protons. We have made significant progress on several related issues: the computation of energy differences, the development of wavefunctions that do not require optimization of variational parameters and use of twist averaged boundary conditions. We have applied the method to an important many-body system, molecular and metallic hydrogen at high pressure. We have shown that the method is feasible on current multi-processor computers.

One of the advantages of QMC over DFT, in addition to higher accuracy, is the different way basis sets enter. Single particle methods usually work in a "wave basis", where the wave function is expanded in plane waves or Gaussian orbitals. In contrast QMC uses a particle basis. A smooth basis (the trial wave function) is indeed used within VMC, however, cusps and other features are easily added without slowing down the computation. For this reason, the bare Coulomb interaction can be easily treated, while in LDA, typically a smooth pseudopotential is needed, even for hydrogen, to avoid an excessive number of basis functions.

As shown in the example of twist-averaged boundary conditions, the CEIMC method has further advantages when additional averages are to be performed. In the present calculation, we assumed classical protons for simplicity. Of course, quantum effects of the protons are important and have been included in previous QMC and LDA calculations. But it is not hard to see that it is possible to do path integrals of the nuclei within the penalty method for very little increased cost over classical nuclei. A path integral simulation creates a path of M slices, with each slice at an effective temperature of MT. We then need to perform M separate electronic simulations, one for each slice. However, the penalty method requires the error level to be approximately k_BT. Then the required error level at each slice is Mk_BT, so each of the M separate QMC simulations need not be as accurate. In contrast, for PI-LDA calculations, M time slices will take M times as long.

Our impression is that the CEIMC method on this application of high pressure hydrogen has the same order of computational demands as Car-Parrinello plane-wave methods: our results suggest that the CEIMC method may turn out to be both more accurate and faster. The processing power of current multi-processors is enough that significant applications are being pursued, giving much more accurate results for systems of hydrogen and helium including all effects of electron correlation and quantum zero point motion. In general, we expect the CEIMC method to be most useful when there are additional averages to be performed perhaps due to disorder or quantum effects. On the other hand DFT methods are more efficient for optimizing molecular geometries where the existing functional are known to be locally accurate or for dynamical studies outside the scope of CEIMC.

Tests for non-hydrogenic systems are needed to find the performance of the algorithms on a broader spectrum of applications. The use of pseudopotentials within QMC to treat atoms with inner core is well tested. What is not clear is how much time will be needed to generate trial functions, and to reduce the noise level to acceptable limits. Also interesting is to develop a dynamical version of CEIMC, i.e. CEIMD. Many of the techniques discussed here such as twist averaging, advanced trial functions and energy difference methods are immediately applicable. However, while it is possible within MC to allow quantum noise, it is clear that the noise level on the forces must be much smaller, since otherwise the generated trajectories will be quite different. The effect of the quantum noise, in adding a fictitious heat bath to the classical system, may negate important aspects of the rigorous approach we have followed. One possible approach is to locally fit the potential surface generated within QMC to a smooth function, thereby reducing the noise level. Clearly, further work is needed to allow this next step in the development of microscopic simulation algorithms.

Acknowledgments

This work has been supported by NSF DMR01-04399 and the computational facilities at NCSA Urbana (Illinois-USA) and at CINES, Montpellier (France). Thanks to M. Marechal for extending the facilities of CECAM. C.P. acknowledges financial support from CNRS and from ESF through the SIMU program. This work has been supported by the INFM Parallel Computing Initiative.

References

1. N. Metropolis, A. W. Rosenbluth, M. N. Rosenbluth, A. H. Teller, and E. Teller: J. Chem. Phys. **21**, 1087 (1953)
2. R. G. Parr and W. Yang. *Density Functional Theory of Atoms and Molecules*, Oxford, 1989.
3. R. Car and M. Parrinello: Phys. Rev. Lett. **55**, 2471 (1985)
4. M. C. Payne, M. P. Teter, D. C. Allan, T. A. Arias, and J. D. Joannopoulos: Rev. Mod. Phys **64**, 1045 (1992)
5. D. Marx and M. Parrinello: J. Chem. Phys. **104**, 4077 (1996)
6. M. Sprik: J. Phys.: Condens. Matter **12**, A161 (2000)
7. M. E. Tuckerman and G. J. Martyna: J. Phys. Chem. B **104**, 159 (2000)
8. W. M. C. Foulkes et al.: Rev. Mod. Phys. **73**, 33 (2001)
9. B. L. Hammond, Jr. W. A. Lester, and P. J. Reynolds. *Monte Carlo Methods in Ab Initio Quantum Chemistry*, World scientific lecture and course notes in chemistry, (World Scientific, Singapore, 1994)
10. J. B. Anderson:'Exact quantum chemistry by Monte Carlo methods' In *Quantum Mechanical Electronic Structure Calculations with Chemical Accuracy*, ed. S. R. Langhoff, (Kluwer Academic, 1995)
11. D. M. Ceperley and L. Mitas:'Quantum Monte Carlo methods in chemistry'. In *Advances in Chemical Physics* ed. by I. Prigogine and S. A. Rice, (Wiley and Sons, 1996)
12. J. C. Grossman, L. Mitas, and K. Raghavachari: Phys. Rev. Lett. **75**, 3870 (1995)
13. M. Dewing: Monte Carlo Methods: Application to hydrogen gas and hard spheres. PhD thesis, University of Illinois at Urbana-Champaign (2000). Available as arXiv:physics/0012030.
14. W. B. Hubbard and D. J. Stevenson: 'Interior structure'. In *Saturn*, ed. by T. Gehrels and M. S. Matthews (University of Arizona Press, 1984)
15. D. J. Stevenson:'The role of high pressure experiment and theory in our understanding of gaseous and icy planets'. In *Shock waves in condensed matter*, ed. by S. C. Schmidt and N. C. Holmes (Elsevier, 1988)
16. B. Militzer and E. L. Pollock: Phys. Rev. E **61**, 3470 (2000)
17. D. M. Ceperley and B. J. Alder: Phys. Rev. B **36**, 2092 (1987)
18. V. Natoli, R. M. Martin, and D. M. Ceperley: Phys. Rev. Lett. **70**, 1952 (1993)
19. V. Natoli, R. M. Martin, and D. M. Ceperley: Phys. Rev. Lett. **74**, 1601 (1995)
20. M. Dewing and D. M. Ceperley: 'Methods for Coupled Electronic-Ionic Monte Carlo'. In: *Recent Advances in Quantum Monte Carlo Methods, II*, ed. by S. Rothstein (World Scientific, Singapore), submitted Jan 2001.
21. P. J. Reynolds, D. M. Ceperley, B. J. Alder, and W. A. Lester: J. Chem. Phys. **77**, 5593 (1982)
22. D. M. Ceperley and M. Dewing: J. Chem. Phys. **110**, 9812 (1999)
23. D. M. Ceperley: Rev. Mod. Phys. **67**, 279 (1995)
24. D. M. Ceperley, M. Holzmann, K. Esler and C. Pierleoni: Backflow Correlations for Liquid Metallic Hydrogen, to appear.
25. Y. Kwon, D.M. Ceperley and R. M. Martin: Phys. Rev. **B 48**, 12037 (1993)
26. Y. Kwon, D.M. Ceperley and R. M. Martin: Phys. Rev. **B 58**, 6800 (1998)
27. D. M. Ceperley and B. J. Alder: J. Chem. Phys. **81**, 5833 (1984)

28. T. Ogitsu: 'MP-DFT (multiple parallel density funtional theory) code, (2000) http://www.ncsa.uiuc.edu/Apps/CMP/togitsu/MPdft.html.
29. I. F. Silvera and V. V. Goldman: J. Chem. Phys. **69**, 4209 (1978)
30. P. Diep and J. K. Johnson: J. Chem. Phys. **112**, 4465 (2000)
31. P. Diep and J. K. Johnson: J. Chem. Phys. **113**, 3480 (2000)
32. W. Kolos and L. Wolniewicz: J. Chem. Phys. **41**, 1964.
33. B. Militzer and D. M. Ceperley: Phys. Rev. Lett. **85**, 1890 (2000)
34. B. Militzer. *Path Integral Monte Carlo Simulations of Hot Dense Hydrogen.* PhD thesis, University of Illinois at Urbana-Champaign, 2000.
35. N. C. Holmes, M. Ross, and W. J. Nellis: Phys. Rev. B **52**, 15835 (1995)
36. D. Saumon and G. Chabrier: Phys. Rev. A **44**, 5122 (1991)
37. D. Saumon and G. Chabrier: Phys. Rev. A **46**, 2084 (1992)
38. D. Saumon, G. Chabrier, and H. M. Van Horn: Astrophys. J. Sup. **99**, 713 (1995)
39. W. J. Nellis, A. C. Mitchell, M. van Theil, G. J. Devine, R. J Trainor, and N. Brown: J. Chem. Phys. **79**, 1480 (1983)
40. D. Hohl, V. Natoli, D. M. Ceperley, and R. M. Martin: Phys. Rev. Lett. **71**, 541 (1993)
41. J. Kohanoff, S. Scandolo, G. L. Chiarotti, and E. Tosatti: Phys. Rev. Lett. **78**, 2783 (1997)
42. C. Lin, F. H. Zong and D. M. Ceperley: Phys. Rev. E **64**, 016702 (2001)
43. D. M. Ceperley; Phys. Rev. Lett. **69**, 331 (1992)
44. C. Pierleoni, B. Bernu, D. M. Ceperley and W. R. Magro: Phys. Rev. Lett. **73**, 2145 (1994); W. R. Magro, D. M. Ceperley, C. Pierleoni, and B. Bernu: Phys. Rev. Lett. **76**, 1240 (1996)
45. D. M. Ceperley: 'Path integral Monte Carlo methods for fermions'. In *Monte Carlo and Molecular Dynamics of Condensed Matter Systems*, ed. by K. Binder and G. Ciccotti (Editrice Compositori, Bologna, Italy, 1996)
46. J. Kohanoff and J. P. Hansen: Phys. Rev. E **54**, 768 (1996)

Lecture Notes in Physics

For information about Vols. 1–563
please contact your bookseller or Springer-Verlag

Vol. 564: T. Pöschel, S. Luding (Eds.), Granular Gases. VIII, 457 pages. 2001.

Vol. 565: E. Beaurepaire, F. Scheurer, G. Krill, J.-P. Kappler (Eds.), Magnetism and Synchrotron Radiation. XIV, 388 pages. 2001.

Vol. 566: J. L. Lumley (Ed.), Fluid Mechanics and the Environment: Dynamical Approaches. VIII, 412 pages. 2001.

Vol. 567: D. Reguera, L. L. Bonilla, J. M. Rubí (Eds.), Coherent Structures in Complex Systems. IX, 465 pages. 2001.

Vol. 568: P. A. Vermeer, S. Diebels, W. Ehlers, H. J. Herrmann, S. Luding, E. Ramm (Eds.), Continuous and Discontinuous Modelling of Cohesive-Frictional Materials. XIV, 307 pages. 2001.

Vol. 569: M. Ziese, M. J. Thornton (Eds.), Spin Electronics. XVII, 493 pages. 2001.

Vol. 570: S. G. Karshenboim, F. S. Pavone, F. Bassani, M. Inguscio, T. W. Hänsch (Eds.), The Hydrogen Atom: Precision Physics of Simple Atomic Systems. XXIII, 293 pages. 2001.

Vol. 571: C. F. Barenghi, R. J. Donnelly, W. F. Vinen (Eds.), Quantized Vortex Dynamics and Superfluid Turbulence. XXII, 455 pages. 2001.

Vol. 572: H. Latal, W. Schweiger (Eds.), Methods of Quantization. XI, 224 pages. 2001.

Vol. 573: H. M. J. Boffin, D. Steeghs, J. Cuypers (Eds.), Astrotomography. XX, 434 pages. 2001.

Vol. 574: J. Bricmont, D. Dürr, M. C. Galavotti, G. Ghirardi, F. Petruccione, N. Zanghi (Eds.), Chance in Physics. XI, 288 pages. 2001.

Vol. 575: M. Orszag, J. C. Retamal (Eds.), Modern Challenges in Quantum Optics. XXIII, 405 pages. 2001.

Vol. 576: M. Lemoine, G. Sigl (Eds.), Physics and Astrophysics of Ultra-High-Energy Cosmic Rays. X, 327 pages. 2001.

Vol. 577: I. P. Williams, N. Thomas (Eds.), Solar and Extra-Solar Planetary Systems. XVIII, 255 pages. 2001.

Vol. 578: D. Blaschke, N. K. Glendenning, A. Sedrakian (Eds.), Physics of Neutron Star Interiors. XI, 509 pages. 2001.

Vol. 579: R. Haug, H. Schoeller (Eds.), Interacting Electrons in Nanostructures. X, 227 pages. 2001.

Vol. 580: K. Baberschke, M. Donath, W. Nolting (Eds.), Band-Ferromagnetism: Ground-State and Finite-Temperature Phenomena. IX, 394 pages. 2001.

Vol.581: J. M. Arias, M. Lozano (Eds.), An Advanced Course in Modern Nuclear Physics. XI, 346 pages. 2001.

Vol.582: N. J. Balmforth, A. Provenzale (Eds.), Geomorphological Fluid Mechanics. X, 579 pages. 2001.

Vol.583: W. Plessas, L. Mathelitsch (Eds.), Lectures on Quark Matter, XIII, 334 pages. 2002.

Vol.584: W. Köhler, S. Wiegand (Eds.), Thermal Nonequilibrium Phenomena in Fluid Mixtures. XVII, 470 pages. 2002.

Vol.585: M. Lässig, A. Valleriani (Eds.), Biological Evolution and Statistical Physics. XI, 337 pages. 2002.

Vol.586: Y. Auregan, A. Maurel, V. Pagneux, J.-F. Pinton (Eds.), Sound–Flow Interactions. XVI, 286 pages. 2002

Vol.587: D. Heiss (Ed.), Fundamentals of Quantum Information. Quantum Computation, Communication, Decoherence and All That. XIII, 265 pages. 2002.

Vol.588: Y. Watanabe, S. Heun, G. Salviati, N. Yamamoto (Eds.), Nanoscale Spectroscopy and Its Applications to Semiconductor Research. XV, 306 pages. 2002.

Vol.589: A. W. Guthmann, M. Georganopoulos, A. Marcowith, K. Manolakou (Eds.), Relativistic Flows in Astrophysics. XII, 241 pages. 2002

Vol.590: D. Benest, C. Froeschlé (Eds.), Singularities in Gravitational Systems. Applications to Chaotic Transport in the Solar System. XI, 215 pages. 2002

Vol.591: M. Beyer (Ed.), CP Violation in Particle, Nuclear and Astrophysics. XI, 334 pages. 2002

Vol.592: S. Cotsakis, L. Papantonopoulos (Eds.), Cosmological Crossroads. An Advanced Course in Mathematical, Physical and String Cosmology. XVI, 477 pages. 2002

Vol.593: D. Shi, B. Aktaş, L. Pust, F. Mikhailov (Eds.), Nanostructured Magnetic Materials and Their Applications. XII, 289 pages. 2002

Vol.594: S. Odenbach (Ed.),Ferrofluids. Magnetical Controllable Fluids and Their Applications. XI, 253 pages. 2002

Vol.595: C. Berthier, L. P. Lévy, G. Martinez (Eds.), High Magnetic Fields. Applications in Condensed Matter Physics and Spectroscopy. X, 493 pages. 2002

Vol.596: F. Scheck, H. Upmeier, W. Werner (Eds.), Noncommutative Geometry and the Standard Model of Elememtary Particle Physics. XII, 346 pages. 2002

Vol.597: P. Garbaczewski, R. Olkiewicz (Eds.), Dynamics of Dissipation. X, 512 pages. 2002

Vol.598: K. Weiler (Ed.), Supernovae and Gamma-Ray Bursters. X, 350 pages. 2002

Vol.600: K. Mecke, D. Stoyan (Eds.), Morphology of Condensed Matter. Physcis and Geometry of Spatial Complex Systems. XVIII, 439 pages. 2002

Vol.601: F. Mezei, C. Pappas, T. Gutberlet (Eds.), Neutron Spin Echo Spectroscopy. Basics, Trends and Applications. XVII, 351 pages. 2003

Vol.604: J. Frauendiener, H. Friedrich (Eds.), The Conformal Structure of Space-Time: Geometry, Analysis, Numerics. XIV 373pages. 2002

Vol.605: G. Ciccotti, M. Mareschal, P. Nielaba (Eds.), Bridging Time Scales: Molecular Simulations for the Next Decade. XXVI, 500 pages. 2002

Vol.608: F. Courbin, D. Minniti (Eds.), Gravitational Lensing:An Astrophysical Tool. XI, 191 pages. 2002

Vol.611: A. Buchleitner, K. Hornberger (Eds.), Coherent Evolution in Noisy Environments. XV, 293 pages. 2002

Monographs

For information about Vols. 1–29 please contact your bookseller or Springer-Verlag

Vol. m 30: A. J. Greer, W. J. Kossler, Low Magnetic Fields in Anisotropic Superconductors. VII, 161 pages. 1995.

Vol. m 31 (Corr. Second Printing): P. Busch, M. Grabowski, P.J. Lahti, Operational Quantum Physics. XII, 230 pages. 1997.

Vol. m 32: L. de Broglie, Diverses questions de mécanique et de thermodynamique classiques et relativistes. XII, 198 pages. 1995.

Vol. m 33: R. Alkofer, H. Reinhardt, Chiral Quark Dynamics. VIII, 115 pages. 1995.

Vol. m 34: R. Jost, Das Märchen vom Elfenbeinernen Turm. VIII, 286 pages. 1995.

Vol. m 35: E. Elizalde, Ten Physical Applications of Spectral Zeta Functions. XIV, 224 pages. 1995.

Vol. m 36: G. Dunne, Self-Dual Chern-Simons Theories. X, 217 pages. 1995.

Vol. m 37: S. Childress, A.D. Gilbert, Stretch, Twist, Fold: The Fast Dynamo. XI, 406 pages. 1995.

Vol. m 38: J. González, M. A. Martín-Delgado, G. Sierra, A. H. Vozmediano, Quantum Electron Liquids and High-Tc Superconductivity. X, 299 pages. 1995.

Vol. m 39: L. Pittner, Algebraic Foundations of Non-Com-mutative Differential Geometry and Quantum Groups. XII, 469 pages. 1996.

Vol. m 40: H.-J. Borchers, Translation Group and Particle Representations in Quantum Field Theory. VII, 131 pages. 1996.

Vol. m 41: B. K. Chakrabarti, A. Dutta, P. Sen, Quantum Ising Phases and Transitions in Transverse Ising Models. X, 204 pages. 1996.

Vol. m 42: P. Bouwknegt, J. McCarthy, K. Pilch, The W3 Algebra. Modules, Semi-infinite Cohomology and BV Algebras. XI, 204 pages. 1996.

Vol. m 43: M. Schottenloher, A Mathematical Introduction to Conformal Field Theory. VIII, 142 pages. 1997.

Vol. m 44: A. Bach, Indistinguishable Classical Particles. VIII, 157 pages. 1997.

Vol. m 45: M. Ferrari, V. T. Granik, A. Imam, J. C. Nadeau (Eds.), Advances in Doublet Mechanics. XVI, 214 pages. 1997.

Vol. m 46: M. Camenzind, Les noyaux actifs de galaxies. XVIII, 218 pages. 1997.

Vol. m 47: L. M. Zubov, Nonlinear Theory of Dislocations and Disclinations in Elastic Body. VI, 205 pages. 1997.

Vol. m 48: P. Kopietz, Bosonization of Interacting Fermions in Arbitrary Dimensions. XII, 259 pages. 1997.

Vol. m 49: M. Zak, J. B. Zbilut, R. E. Meyers, From Instability to Intelligence. Complexity and Predictability in Nonlinear Dynamics. XIV, 552 pages. 1997.

Vol. m 50: J. Ambjørn, M. Carfora, A. Marzuoli, The Geometry of Dynamical Triangulations. VI, 197 pages. 1997.

Vol. m 51: G. Landi, An Introduction to Noncommutative Spaces and Their Geometries. XI, 200 pages. 1997.

Vol. m 52: M. Hénon, Generating Families in the Restricted Three-Body Problem. XI, 278 pages. 1997.

Vol. m 53: M. Gad-el-Hak, A. Pollard, J.-P. Bonnet (Eds.), Flow Control. Fundamentals and Practices. XII, 527 pages. 1998.

Vol. m 54: Y. Suzuki, K. Varga, Stochastic Variational Approach to Quantum-Mechanical Few-Body Problems. XIV, 324 pages. 1998.

Vol. m 55: F. Busse, S. C. Müller, Evolution of Spontaneous Structures in Dissipative Continuous Systems. X, 559 pages. 1998.

Vol. m 56: R. Haussmann, Self-consistent Quantum Field Theory and Bosonization for Strongly Correlated Electron Systems. VIII, 173 pages. 1999.

Vol. m 57: G. Cicogna, G. Gaeta, Symmetry and Perturbation Theory in Nonlinear Dynamics. XI, 208 pages. 1999.

Vol. m 58: J. Daillant, A. Gibaud (Eds.), X-Ray and Neutron Reflectivity: Principles and Applications. XVIII, 331 pages. 1999.

Vol. m 59: M. Kriele, Spacetime. Foundations of General Relativity and Differential Geometry. XV, 432 pages. 1999.

Vol. m 60: J. T. Londergan, J. P. Carini, D. P. Murdock, Binding and Scattering in Two-Dimensional Systems. Applications to Quantum Wires, Waveguides and Photonic Crystals. X, 222 pages. 1999.

Vol. m 61: V. Perlick, Ray Optics, Fermat's Principle, and Applications to General Relativity. X, 220 pages. 2000.

Vol. m 62: J. Berger, J. Rubinstein, Connectivity and Superconductivity. XI, 246 pages. 2000.

Vol. m 63: R. J. Szabo, Ray Optics, Equivariant Cohomology and Localization of Path Integrals. XII, 315 pages. 2000.

Vol. m 64: I. G. Avramidi, Heat Kernel and Quantum Gravity. X, 143 pages. 2000.

Vol. m 65: M. Hénon, Generating Families in the Restricted Three-Body Problem. Quantitative Study of Bifurcations. XII, 301 pages. 2001.

Vol. m 66: F. Calogero, Classical Many-Body Problems Amenable to Exact Treatments. XIX, 749 pages. 2001.

Vol. m 67: A. S. Holevo, Statistical Structure of Quantum Theory. IX, 159 pages. 2001.

Vol. m 68: N. Polonsky, Supersymmetry: Structure and Phenomena. Extensions of the Standard Model. XV, 169 pages. 2001.

Vol. m 69: W. Staude, Laser-Strophometry. High-Resolution Techniques for Velocity Gradient Measurements in Fluid Flows. XV, 178 pages. 2001.

Vol. m 70: P. T. Chruściel, J. Jezierski, J. Kijowski, Hamiltonian Field Theory in the Radiating Regime. VI, 172 pages. 2002.

Vol. m 71: S. Odenbach, Magnetoviscous Effects in Ferrofluids. X, 151 pages. 2002.

Vol. m 72: J. G. Muga, R. Sala Mayato, I. L. Egusquiza (Eds.), Time in Quantum Mechanics. XII, 419 pages. 2002.